Topics in Mining, Metallurgy and Materials Engineering

Series editor

Carlos P. Bergmann, Porto Alegre, Brazil

More information about this series at http://www.springer.com/series/11054

Shrikrishna N. Joshi · Uday Shanker Dixit
Editors

Lasers Based Manufacturing

5th International and 26th All India Manufacturing Technology, Design and Research Conference, AIMTDR 2014

Springer

Editors
Shrikrishna N. Joshi
Department of Mechanical Engineering
Indian Institute of Technology Guwahati
Guwahati, Assam
India

Uday Shanker Dixit
Department of Mechanical Engineering
Indian Institute of Technology Guwahati
Guwahati, Assam
India

ISSN 2364-3293 ISSN 2364-3307 (electronic)
Topics in Mining, Metallurgy and Materials Engineering
ISBN 978-81-322-2351-1 ISBN 978-81-322-2352-8 (eBook)
DOI 10.1007/978-81-322-2352-8

Library of Congress Control Number: 2015934047

Springer New Delhi Heidelberg New York Dordrecht London

Printed on acid-free paper

Springer (India) Pvt. Ltd. is part of Springer Science+Business Media (www.springer.com)

Preface

The manufacturing industry contributes a major share in the development of a nation. However, it is now facing several challenges such as rapid product development, flexibility, low to medium volume, and low-cost production. Many advanced/unconventional technologies/tools are being developed worldwide to face these challenges. Among these technologies laser has become quite popular due to its ability for precise focusing and ease in controlling the heat source. Researchers worldwide are now focusing their attention on improving productivity and product quality in the application of laser technology for the manufacture of components required for consumer electronics, aerospace systems and vehicles, bio-medical instrumentation, automobiles, and shipbuilding applications. Scientists, researchers, and engineers across the globe are striving to achieve excellence in the area of laser-based manufacturing by carrying out systematic experimental as well as numerical studies. The present book represents some of these efforts.

The chapters in this book present the basic and advanced topics in laser-based manufacturing. These chapters focus on theoretical as well as practical aspects of laser technology for various applications such as sheet metal bending, welding, sintering and micro-manufacturing, viz., micro-channelling, micro-turning, micro-drilling, etc. These chapters are the extended versions of peer-reviewed manuscripts presented at the 5th International and the 26th All India Manufacturing Technology, Design and Research (AIMTDR) conference held during December 12–14, 2014 at the Indian Institute of Technology Guwahati, India. The book will be useful to researchers and practicing engineers working in the area of manufacturing in general and laser-based manufacturing in particular.

The book comprises 22 chapters. First six chapters present the application of lasers for bending of metal sheets. The fundamentals of laser bending are discussed at length in these chapters. The research work carried out on numerical as well as experimental aspects is presented. In Chapters "Surface Alloying of Aluminium with Copper Using CO_2 Laser" and "Effect of Pulsed Nd:YAG Laser Parameters in Preplaced TiC Coating on Aluminium Substrate," the surface modification by laser processing is presented. Chapter "Finite Element Simulation of Laser Cladding for Tool Steel Repair" deals with finite element simulation of laser cladding for repair of cutting tools. Chapters "Excimer Laser Micromachining and Its Applications" and next six chapters present laser-based advanced technologies developed for manufacturing of micro-sized features such as channels. The state of the art on the employment of various types of lasers, viz., Nd-YAG, Excimer for processing of advanced materials such as Ti-6Al-4V, zirconia, alumina, and PMMA are presented in the respective chapters. The application of laser for micro-drilling of SiC-30BN nanocomposite is presented in the Chapter "Nd:YAG Laser Microdrilling of SiC-30BN Nanocomposite: Experimental Study and Process Optimization." Chapter "Pulsed Nd:Yag Laser Micro-Turning Process of Alumina Ceramics" depicts experimental studies on Pulsed Nd:YAG laser micro-turning alumina ceramic. Chapter "A Literature Review on CO_2 Laser Welding" presents an overview of the recent trends in laser welding, while Chap. "Fiber Laser Welding in a Controlled Inert Gas Atmosphere: An Experimental and Numerical Investigation" presents numerical as well as experimental investigations on fiber laser welding. Chapter "A 3-D Finite Element Analysis of Transient Temperature Profile of Laser Welded Ti-6Al-4V Alloy" discusses the finite element method-based methodology of 3-D simulation of laser welding of Ti-6Al-4V alloy. Finally, Chap. "Selective Laser Sintering: A Case Study of Tungsten Carbide and Cobalt Powder Sintering by Pulsed Nd:YAG Laser" presents an important application of laser, i.e., sintering of tungsten carbide and cobalt powder using pulsed Nd:YAG Laser. Overall, the coverage is very wide, encompassing different manufacturing processes.

The authors and editors have taken utmost care in presenting the information and acknowledging the original sources wherever necessary. The editors express their gratitude toward the authors, organizers of AIMTDR, and staff of Springer (India) for making possible the publication of this research book. Readers are requested to provide their valuable feedback on the quality of the presentation and inadvertent errors or omission of information if any. We expect that the book will be welcomed by students as well as practising engineers/researchers.

Shrikrishna N. Joshi
Uday Shanker Dixit

About the Conference

AIMTDR series of conferences is a highly prestigious biennial event in the field of mechanical engineering. It has a glorious history. The first conference entitled All India Machine Tools Design and Research Conference was held at Jadavpur University, Kolkata. Later it was renamed All India Manufacturing Technology, Design and Research Conference in order to widen its scope. It became an international event in 2006 with the first international conference held at IIT Roorkee. The conference aims to bring together academicians and industry professionals working in the field of manufacturing for the exchange of ideas.

The Indian Institute of Technology, Guwahati organized the 5th International and the 26th All India Manufacturing, Technology, Design and Research (AIMTDR) conference during 12–14 December 2014. The conference was attended by more than 400 delegates from academia and industry. Apart from the addresses by Chief Guest Prof. Amitabaha Ghosh, former Director of IIT Kharagpur and Mr. P.K. Borthakur, Former Director (offshore), ONGC, a total of 22 invited lectures were presented at the conference. Some of the notable speakers from academia were Prof. Kornel Ehmann, Prof. Shiv G. Kapoor, and Prof. Marc Madou from USA, Prof. Hocheng from Taiwan, and Prof. V. Radhakrishnan from India. Some of the invited speakers from industry/R&D organizations were Dr. V.K. Suri from BARC, Mr. Khongwir from IOCL Bongaigaon, Dr. Sumitesh Das from Tata Steel, Dr. Sanyal from CSIR, and Dr. K. Ramesh from DRDL Hyderabad. Some manufacturing organizations also gave presentations on their products. A total of 423 papers were published in the proceedings of the conference by young and senior engineers/scientists.

Professor M. Rahman of National University of Singapore delivered the Prof. Amitabha Bhattachraya Memorial Lecture. He advocated sustainable manufacturing that should not harm the environment. He emphasized the role of revolutionary and evolutionary innovation to achieve sustainable manufacturing. Overall, the conference was a huge success in terms of participation by delegates from premier educational and research institutions from India and abroad.

About the Conference

Editorial Acknowledgments

We extend our thanks to all the authors for contributing to this book by sharing their valuable research findings. We especially thank a number of reviewers for promptly reviewing the papers submitted to the conference. We are grateful to the volunteers, invited speakers, sponsors, subcommittee members, members of the national advisory committee, and members of the scientific advisory committee for successful conduct of the conference. The editors express their heartfelt gratitude toward Prof. Gautam Biswas, Director IIT Guwahati, Prof. Gautam Barua, former Director IIT Guwahati, Prof. Anoop K. Dass, Head Mechanical Engineering, and Prof. P. Mahanta, former Head Mechanical Engineering IIT Guwahati for their encouragement and motivation to organize such a prestigious event that paved the way for this book on laser-based manufacturing. Lastly, we express our sincere gratitude toward the staff members of Springer (India) who helped in publishing this book.

Contents

About the Editors

Dr. Shrikrishna N. Joshi has completed his doctoral studies in the area of "Intelligent modeling and optimization of electric discharge machining process" from IIT Bombay in the year 2009. Since then he is working as an Assistant Professor in the Department of Mechanical Engineering, IIT Guwahati. His research interests are Micro-machining and Micro-bending using Lasers; Computer aided design and manufacturing (CAD/CAM); Manufacturing process modeling and optimization; and Mechatronics. He is guiding five Ph.D. students those who are working on various research areas such as laser bending, laser induced plasma micro-machining, thin-wall milling and single point diamond turning. Dr. Joshi has about 25 papers published in international journals and conferences of national/ international reputes.

Dr. Uday Shanker Dixit obtained a bachelor's degree in Mechanical Engineering from the University of Roorkee (now Indian Institute of Technology Roorkee) in 1987, an M.Tech. in Mechanical Engineering from Indian Institute of Technology (IIT) Kanpur in 1993, and a Ph.D. in Mechanical Engineering from IIT Kanpur in 1998. A Professor in the Department of Mechanical Engineering, Indian Institute of Technology Guwahati, Dr. Dixit has published numerous papers and five books. He has also edited a book on Metal Forming, guest-edited a number of special journal issues, and is an associate editor for the Journal of Institution of Engineers Series C.

About the Editors

A Simple Analytical Model of Laser Bending Process

A. Eideh, Uday S. Dixit and Raghu Echempati

Abstract Laser bending is a process of bending a sheet by the irradiation of laser beam on the surface of the sheet. A number of analytical and numerical methods have been proposed for the estimation of bend angle. A brief review of these methods is presented. A finite element analysis of laser bending process is carried out with ABAQUS package for the purpose of understanding the physics of the laser bending. Afterwards, a simple analytical model is developed to evaluate the bending angle in laser bending of metal sheet. The model is based on the elastic-plastic bending of sheet. It is ascertained from the experimental results available in literature that the model provides reasonably good prediction of bend angle. It is also shown that the model can be used for the quick estimation of yield stress of the material during laser bending process.

Keywords Laser forming · Inverse problem · FEM · Elastic-plastic bending

1 Introduction

Laser forming is a thermo-mechanical process used to deform sheets by means of thermal heating by a laser beam. The process is suitable for rapid prototyping and deforming the low ductility materials. Since last two decades, several papers have been published on laser forming process, amongst them straight line bending by laser beam has been researched extensively. The researchers tried to understand the physics of the process and produced various models for the prediction of bend angle.

A. Eideh · U.S. Dixit (✉)
Department of Mechanical Engineering, Indian Institute of Technology Guwahati, Guwahati, India
e-mail: uday@iitg.ac.in

R. Echempati
Kettering University, Flint, MI 48504, USA

S.N. Joshi and U.S. Dixit (eds.), *Lasers Based Manufacturing*,
Topics in Mining, Metallurgy and Materials Engineering,
DOI 10.1007/978-81-322-2352-8_1

Some of these models are analytical and some are based on numerical methods like finite element method (FEM).

Shen and Vollertsen (2009) reviewed the modeling of laser forming. There are three prominent mechanism of laser bending. These are temperature gradient mechanism (TGM), buckling mechanism (BM) and upsetting mechanism (UM). The temperature gradient mechanism activates when laser beam diameter is of the order of sheet metal thickness and scan speed is high. In buckling mechanism, the beam diameter is relatively larger and scan speed is lower. In the upsetting mechanism, the beam diameter is much smaller than the thickness of the sheet.

Vollertsen (1994) derived an expression for the bending angle for TGM. The analytical expression is giving by

$$\alpha_b = \frac{3\alpha_{th} P\eta}{\rho c_p v h^2} \tag{1}$$

where α_b is the bending angle, α_{th} is the coefficient of thermal expansion of the work piece, P is the laser power, η is the absorption coefficient, ρ is the density, c_p is the specific heat capacity, v is the velocity and h is the sheet thickness. The Vollertsen's model does not include the effect of yield stress of material. Yau et al. (1997) included this effect. In their model, the bending angle is giving by

$$\alpha_b = \frac{21\alpha_{th} P\eta}{2\rho c_p v h^2} - \frac{36 l \sigma_y}{hE} \tag{2}$$

where l is the half length of heated zone, E is the Young's modulus and σ_y is the yield stress.

A mathematical model was developed by Kyrsanidi et al. (2000) who considered non-uniform temperature distribution throughout the thickness of the plate due to the developed plastic strains. This model, although computationally efficient, requires programming and includes iterative steps. Cheng et al. (2006) proposed analytical model for plate with varying thickness. The bending angle of the plate at the location with $h(x)$ thickness is

$$\alpha_b = b(1 - v^2)\varepsilon_{\max}\left(\frac{3f(x)\pi}{2h^2(x)} - \frac{4f^2(x)}{h^3(x)}\right), \tag{3}$$

where

$$b = c_1\sqrt{P/v}, \tag{4}$$

and

$$f = c_2 P/(vh) \tag{5}$$

Here v is scan speed, c_1 and c_2 are constants dependent on materials properties, $h(x)$ is the thickness of the sheet and ν is the Poisson's ratio. In Eq. (3), $\varepsilon_{\max}$ is the maximum plastic strain at the heated surface given by

$$\varepsilon_{\max} = \alpha_{th}\theta_{\max} - \sigma_y/E, \tag{6}$$

where θ_{max} is the maximum temperature increase.

Shen et al. (2006) proposed analytical model based on the assumption that the plastic deformation is generated only during heating, while during cooling the plate undergoes only elastic deformation. According to them, the bending angle is giving by

$$\alpha_b = \left[4\beta_v + \frac{12k\sigma_y}{E}\frac{2a_p r}{h(h-a_p)}\right]\frac{ha_p}{(h-a_p)^2}, \tag{7}$$

where β_v is the bending angle derived by Eq. (1), r is the laser beam radius, k is a reduction coefficient to account for the variation of yield strength and Young's modulus of elasticity with temperature and a_p is the characteristic length of plastic zone. The constants k and a_p need to be evaluated empirically. The model is valid for TGM as well as BM.

Lambiase (2012) proposed expression for the bending angle based on assumption of elastic-bending theory without taking into account plastic deformation during heating and cooling phases. The bending angle is given by

$$\alpha_b = \frac{3\eta P(h-h_1)\alpha_{th}}{\rho v c_p h(h^2 - 3hh_1 + 3h_1)}, \tag{8}$$

where h is sheet thickness and h_1 is the thickness of heated volume, which is invariably estimated empirically. Lambiase and Ilio (2013) developed a more rigorous analytical model to predict the deformation of thin sheets.

Several other analytical models have been proposed. Shi et al. (2007) provided a model for estimating the bend angle in an in-plane axis perpendicular to the scan direction. Vollertsen (1994) provided an expression for estimating the bend angle during BM mechanism. Kraus (1997) provided a closed-form expression for estimating the bend angle during upsetting mechanism.

Several FEM methods have also been proposed for laser forming (Chen et al. 1999; Kyrsanidi et al. 1999; Chen and Xu 2001; Hu et al. 2001). However, FEM simulation takes several hours and is not suitable for online optimization and control (Kyrsanidi et al. 2000). Some finite difference models have been proposed, which also require a large computational time.

In the present work a simple analytical model based on the elastoplastic bending of beams is proposed. This model takes into account the effect of the yield stress of the material. The temperature is estimated by a series solution. The method is very efficient and matches well with FEM simulations.

2 Modeling by ABAQUS

FEM simulations were carried out using ABAQUS package. 8-noded brick elements (DCC3D8) were used for the uniform mesh. The material properties of D36 shipbuilding steel were chosen based on the table provided in Dixit et al. (2012). However, the properties were considered as temperature independent. A plate of size 80 mm × 40 mm × 6 mm, laser power P = 1000 W, η = 0.3, scan velocity = 8 mm/s and laser beam radius r = 8 mm were chosen. The temperature distribution is assumed to be Gaussian and the heat flux is given by

$$q = \frac{2\eta P}{\pi r^2} \exp\left(-2\frac{x^2 + y^2}{r^2}\right). \tag{9}$$

The coefficient of thermal expansion is taken as constant and is equal to 12×10^{-6}/°C and the room temperature is taken as 20 °C. In this work, FEM model is developed for the prediction of bend angle of finite thickness plate, which is heated by a moving heat source having a Gaussian distribution. Laser beam is assumed to traverse along the width direction and it is sufficiently far away from the edges in the length direction. A schematic diagram of this arrangement is shown in Fig. 1. The sheet is clamped at the right edge parallel to scan direction, but the clamping arrangement is not shown in the figure.

No external loads are applied on the sheet. The right edge of the sheet is fixed to avoid the rigid body movements. The boundary condition of zero displacement at the fully constrained right edge of metal plate is imposed. The plate is divided into 80 elements along the length side (y-direction) of the plate, 40 elements along the width side (x-direction) of the plate and 6 elements through the thickness direction of the plate. The total number of elements is 19,200. The decision was based on the mesh sensitivity analysis.

Three FEM simulations were carried out with different values of yield stress to show the effect of yield stress on the bend angle. The calculation time required for

Fig. 1 Schematic diagram of straight line bending process (clamping of the sheet not shown)

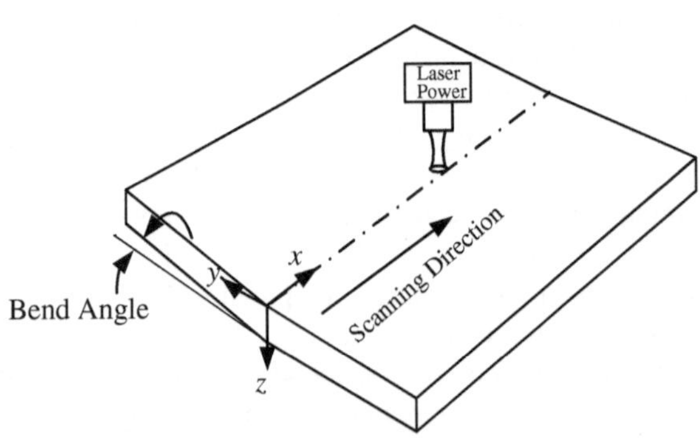

Table 1 Variation of bend angle with yield stress

Yield stress (MPa)	Bend angle (°)
50	0.79
335	0.51
495	0.0

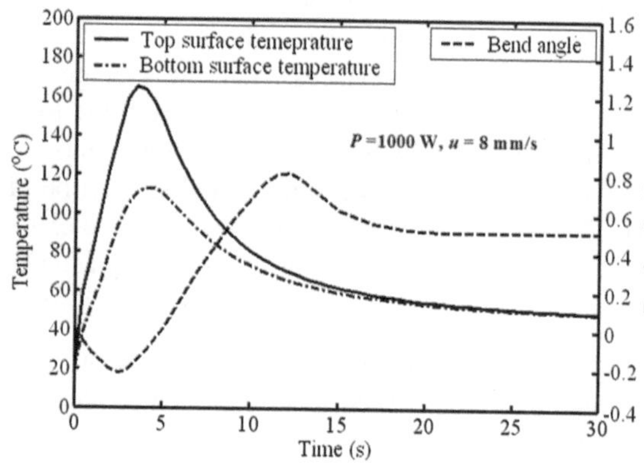

Fig. 2 Variation of temperature with time on the top and bottom surface of the sheet and variation of bend angle during heating and cooling

the finite element analyses presented in this work is 6–7 h on a Pentium IV PC. Table 1 shows the bend angle for three different values of yield stress.

It is observed that the yield stress has significant effect on the bend angle and neglecting the effect of yield stress in an analytical model may not be appropriate. The maximum temperature in the sheet was about 165 °C. Thus, the maximum increase in the temperature is about 145 °C. If the locally heated material is sufficiently constrained, the maximum thermal stress is $145\alpha E = 145 \times 12 \times 10^{-6} \times 200 \times 10^{9} = 358 \times 10^{6}$ Pa. If the yield stress is more than 358 MPa, no plastic deformation takes place and the bend angle should be zero. Table 1 also shows that at a yield stress of 495 MPa, the bend angle is zero.

Figure 2 shows the variation of top and bottom surface temperatures with time for a material with yield stress of 335 MPa. Initially, the temperature of both the surfaces increases, which causes counter bending due to the temperature difference in the temperatures of top and bottom surfaces. As the material underneath the laser beam is surrounded by the cold material and the yield stress drops due to temperature, there is some localized plastic deformation. After that cooling starts, and both top and bottom surfaces contract. The relative contraction on the top surface is more than that at bottom surface, causing the upward bending (towards laser beam side). After a certain time, the temperature at bottom and top surfaces becomes equal. There is a small decrease in the final bend angle, when the cooling beneath the surface is going on, whilst the top surface has already cooled.

3 Model for the Prediction of Bending Angle

The model for the prediction of bending angle is based on the following assumptions:

(1) During heating, the material beneath the laser beam undergoes plastic deformation. No counter bending is assumed.
(2) During the cooling phase, the top surface contracts and the bottom surface expands as both surfaces attain the average temperature.
(3) The strains during cooling are caused due to equivalent mechanical stresses. If a surface cools by ΔT amount, it amounts to applying $\alpha_{th} E \Delta T$ compressive stress to surface. The similar analogy can be given when the surface heats.
(4) The equivalent mechanical stresses cause elastoplastic bending and the bend angle can be obtained by usual elastoplastic theory of bending described in standard textbooks, e.g., Chakrabarty (2006).

The temperature is estimated by a model provided in Mishra and Dixit (2013). The heat flux due to laser beam is assumed as Gaussian of the following form in Mishra and Dixit (2013):

$$q(x,y) = \frac{\eta P}{\pi r^2} \exp\left\{-\frac{(x^2+y^2)}{r^2}\right\} \tag{10}$$

However, in the present work the heat flux is assumed to follow Eq. (9). Accordingly the expressions are modified. For a heat source moving with velocity u parallel to x-axis, the time to scan the sheet is denoted by t_1. Temperature at a point (x, y, z) at time $t > t_1$ is given as

$$\begin{aligned} T - T_\infty = {} & \frac{\eta P}{\pi c_p h} \int_0^{t_1} \frac{2}{8\alpha(t-t') + r^2} \exp\left(-\frac{2\left(\{x - u(t-t')\}^2 + y^2\right)}{8\alpha(t-t') + r^2}\right) \\ & \times \left\{1 + 2\sum_{n=1}^{n=\infty} \exp\left(-\frac{\alpha n^2 \pi^2 (t-t')}{h^2}\right) \cos\frac{n\pi z}{h}\right\} dt', \end{aligned} \tag{11}$$

where T is the temperature at time t at point (x, y, z), P is the laser power, η is the absorptivity of the material, r is the laser beam radius at the surface of the material, α is the thermal diffusivity and h is the thickness of the sheet. Temperature at time $t < t_1$ at a point (x, y, z) is given as

$$T - T_\infty = \frac{\eta P}{\pi c_p h} \int_0^t \frac{2}{8\alpha(t-t') + r^2} \exp\left(- \frac{2\left(\{x - u(t-t')\}^2 + y^2\right)}{8\alpha(t-t') + r^2} \right) \times \left\{ 1 + 2\sum_{n=1}^{n=\infty} \exp\left(- \frac{\alpha n^2 \pi^2 (t-t')}{h^2} \right) \cos\frac{n\pi z}{h} \right\} dt'. \quad (12)$$

Figure 3 shows the typical variation of temperature along thickness direction of the sheet. The average temperature can be calculated as

$$T_{av} = \frac{\int_0^h T dz}{h}. \quad (13)$$

Then, the equivalent longitudinal thermal stresses at any location along the thickness is given as

$$\sigma_x = \alpha_{th} E (T_{av} - T), \quad (14)$$

where α_{th} is coefficient of thermal expansion and E is Young's modulus of elasticity. The moment resulting from thermal stresses is given as

$$M = \int_0^h \sigma_x z dz. \quad (15)$$

Figure 4 shows schematic diagram of plain strain bending of the sheet of thickness h, where line corresponding to T_{av} is a neutral axis, c is the distance from

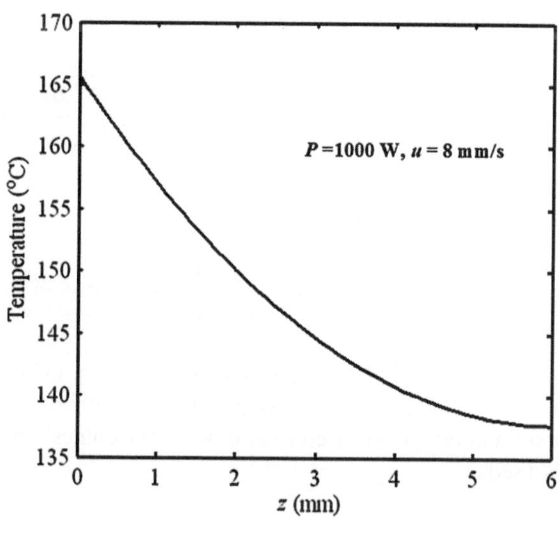

Fig. 3 Variation of temperature along thickness direction

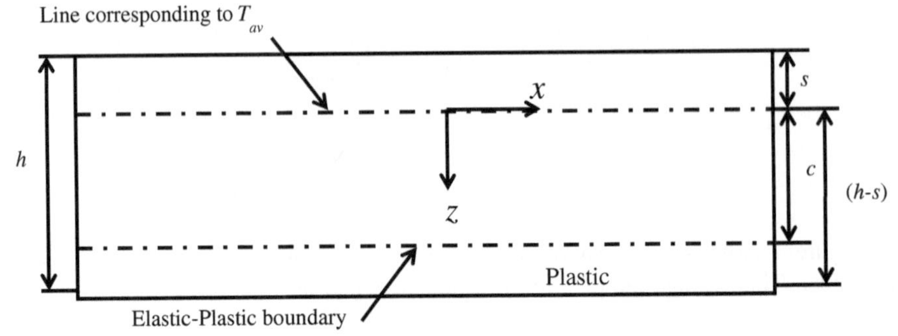

Fig. 4 Schematic diagram of plain strain bending

neutral axis to elastic plastic boundary, s is the distance from neutral axis to the top surface of the sheet, $(h - s)$ is the distance between neutral axis and elastic plastic boundary. When $c > s$, the moment during elastic plastic bending is given in Eq. (16).

$$M = \int_0^s \frac{\sigma_y z_1^2}{c\sqrt{1 - \nu + \nu^2}} dz_1 + \int_0^c \frac{\sigma_y z_1^2}{c\sqrt{1 - \nu + \nu^2}} dz_1 + \int_c^{(h-s)} \frac{2\sigma_y z_1}{\sqrt{3}} dz_1, \tag{16}$$

where ν is the Poisson's ratio. After simplification,

$$M = \frac{2\sigma_y}{\sqrt{3}} \left\{ \frac{1}{3c}(s^3 + c^3) + \frac{1}{2}\left[(h - s)^2 - c^2\right] \right\}, \tag{17}$$

when $c < s$, then in both sides of the neutral axis, there are elastic as well as plastic zones and the moment is given as

$$M = 2\int_0^c \frac{\sigma_y z_1^2}{c\sqrt{1 - \nu + \nu^2}} dz_1 + \int_c^{(h-s)} \frac{2\sigma_y z_1}{\sqrt{3}} dz_1 + \int_c^s \frac{2\sigma_y z_1}{\sqrt{3}} dz_1. \tag{18}$$

After simplification,

$$M = \frac{2\sigma_y}{\sqrt{3}} \left[\frac{1}{2}\left\{(h - s)^2 + s^2\right\} - \frac{c^2}{3} \right]. \tag{19}$$

The value of M can be calculated from Eq. (15). Thereafter, c can be obtained from Eq. (17) or Eq. (19) using bisection method. To begin with, it is assumed that $c > s$ and Eq. (17) is used to find out c. If c comes out to be less than s, then Eq. (19) is used.

For the plane strain elastic bending, the longitudinal stress is given as (Chakrabarty 2006),

$$\sigma_{x_1} = \frac{Ez_1}{(1 - \nu^2)R}, \tag{20}$$

where R is the radius of curvature. Applying this equation at the elastoplastic boundary and using von-Mises yield criterion

$$\frac{2\sigma_y}{\sqrt{3}} = \frac{Ec}{(1 - \nu^2)R}. \tag{21}$$

When the yielding just begins on the bottom surface

$$\frac{2\sigma_y}{\sqrt{3}} = \frac{E(h - s)}{(1 - \nu^2)R_e}, \tag{22}$$

where R_e is the radius of curvature when the yielding just commences. From Eqs. (21) and (22), the radius of curvature during elastic plastic bending is given as

$$R = \frac{c}{(h - s)} R_e, \tag{23}$$

where R_e is given as

$$R_e = \frac{\sqrt{3}E(h - s)}{2\sigma_y(1 - \nu^2)} \tag{24}$$

Hence, the final radius of the curvature during the elastic plastic bending is given as

$$R = \frac{\sqrt{3}Ec}{2\sigma_y(1 - \nu^2)}. \tag{25}$$

Referring to Fig. 5, the bend angle is calculated as

$$\theta = \frac{d}{R} \tag{26}$$

where d is the diameter of laser beam. Here, it is assumed that R is large compared to the thickness of the sheet.

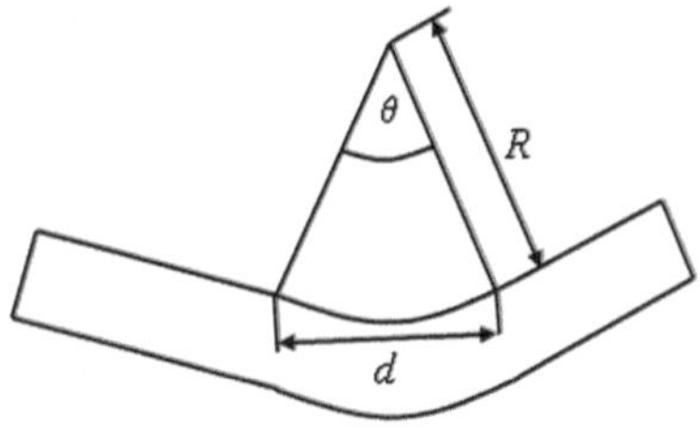

Fig. 5 Schematic representation of bend angle

4 Validation of Model

For validating the model, the analytical results have been compared with experimental results for four different laser powers, reported by Kyrsanidi et al. (2000). In all cases, the dimension of the sheet is equal to 300 mm × 150 mm × 6 mm, and it is made of 1.0584 (D36) shipbuilding steel. The scanning velocity ranges between 0.03 and 1.4 m/min and the power between 1 and 3 kW. The laser beam diameter is equal to 16 mm. The material properties are as follows: specific heat = 427 J/(kg °C), thermal conductivity = 35.1 W/(m °C), density = 7860 kg/m^3, absorption coefficient of the material = 0.3, coefficient of thermal expansion 12 × 10^{-6} 1/°C, Young's modulus = 200 GPa, Poisson ratio = 0.3. The values of yield stress are taken from Dixit et al. (2012) based on the average temperature. Average temperature is taken as the mean of maximum temperature and room temperature. Tables 2, 3, 4 and 5 compare the experimental and analytical results. Table 2 shows that at 1 kW laser power, the bend angle prediction matches well with the experimental results except at the highest and the lowest bend angle cases. Table 3 shows that at 1.5 kW percentage deviation between experimental and analytical results is less than 25 %. Table 4 shows that for many scan speed, the predicted bend angles match very closely with experimental bend angles. Table 5 also shows the similar results. The large percentage deviation in many cases may be due to uncertainty in the coefficient of absorption.

Table 2 Comparisons between present model and experimental data of Kyrsanidi et al. (2000) for 1 kW laser power

Scanning velocity (m/min)	Experimental bend angle (°)	Predicted bend angle (°)	% deviation
0.05	1.067	0.93	12.84
0.1	1.153	0.81	29.75
0.15	0.7	0.63	10
0.25	0.367	0.485	11.8
0.3	0.268	0.3	11.94
0.45	0.048	0.11	129.17

Table 3 Comparisons between present model and experimental data of Kyrsanidi et al. (2000) for 1.5 kW laser power

Scanning velocity (m/min)	Experimental bend angle (°)	Predicted bend angle (°)	% deviation
0.05	0.718	1.32	83.84
0.15	1.07	1.19	11.21
0.2	0.81	0.95	17.28
0.25	0.7	0.845	20.76
0.3	0.48	0.59	22.92
0.45	0.258	0.319	23.64
0.6	0.124	0.23	85.48
0.75	0.105	0.16	52.38

Table 4 Comparisons between present model and experimental data of Kyrsanidi et al. (2000) for 2 kW laser power

Scanning velocity (m/min)	Experimental bend angle (°)	Predicted bend angle (°)	% deviation
0.1	1.1	1.17	6.3
0.15	1.15	1.08	6.1
0.3	0.96	0.9	6.25
0.35	0.75	0.84	12
0.5	0.34	0.47	38.23
0.65	0.21	0.33	57.14
0.8	0.2	0.25	25

Table 5 Comparisons between present model and experimental data of Kyrsanidi et al. (2000) for 3 kW laser power

Scanning velocity (m/min)	Experimental bend angle (°)	Predicted bend angle (°)	% deviation
0.1	0.81	1.62	98.76
0.15	1.48	1.54	4.05
0.3	1.24	1.31	5.65
0.45	0.88	0.95	7.95
0.6	0.66	0.74	12.12
0.9	0.11	0.31	181.8
1.2	0.28	0.22	21.42

To further validate the proposed model, the present analytical solution is compared with the experimental results from Shen et al. (2006). Here, the line energy (*LE*) which is defined as $LE = P/v$ applied to heated surface is constant. The material properties are as follows: Young's modulus = 210 GPa, yield stress = 275 MPa, mass density = 7860 kg/m^{3}' specific heat = 427 J/(kg °C),

Table 6 Comparisons between present model and Shen et al. (2006) model

Power (W)	Scan velocity (mm/s)	Experimental bend angle (°)	Predicted bend angle (°)	% deviation
800	40	0.41	0.463	12.93
1000	50	0.47	0.542	15.32
1300	65	0.49	0.58	18.37

thermal conductivity = 35.1 W/(m °C). The absorption coefficient = 0.8. The sheet size is 80 mm × 80 mm × 2.3 mm. The diameter of laser beam is 4 mm and *LE* = 20 J/mm. The scan velocity is varied between 40 and 65 mm/s. Table 6 shows comparisons between proposed model and experimental results for three different values of laser power and scan velocity. It shows that the results predicted by the proposed model are in good agreement with the experimental values reported by Shen et al. (2006).

Further, the proposed model is compared with experimental data in Lambiase (2012). The dimension of the sheet is 80 mm × 40 mm × 2 mm, which is made of AISI 304 stainless steel. A diode laser with a maximum power of 1.5 kW, over a minimum spot size of 3.6 mm × 0.8 mm is used in the experiment. In the proposed model it is considered that the laser beam is circular and equal in area to the spot area. Consequently the equivalent diameter of laser beam is 1.91 mm. The scanning speed is varied between 15 and 50 mm/s. The material properties required for the model are as follows: specific heat = 500 J/(kg °C), thermal conductivity = 19 W/(m °C), density = 7850 kg/m^3, absorption coefficient of the material = 0.42, coefficient of thermal expansion 17×10^{-6} 1/°C, Young's modulus 210 GPa, Poisson ratio 0.3 and yield stress = 215 MPa. Four cases have been compared at different values of laser power and scan velocity. Laser power is 370, 420, 475 and 550 W respectively with variation of scan speed in range 15–50 mm/s for four cases. Tables 7, 8, 9 and 10 show comparison between proposed model and experimental results in Lambiase (2012). In all cases, the agreement between predicted and experimental bend angle is good, although in some cases the error is more than 30 %.

Table 7 Comparisons between present model of first case (370 W) and Lambiase (2012)

Scanning velocity (mm/s)	Experimental bend angle (°)	Predicted bend angle (°)	% deviation
15	0.95	1.16	22.1
30	0.54	0.63	16.67
40	0.31	0.42	35.48
50	0.23	0.3	30.43

Table 8 Comparisons between present model of second case (420 W) and Lambiase (2012)

Scanning velocity (mm/s)	Experimental bend angle (°)	Predicted bend angle (°)	% deviation
15	1.18	1.32	11.86
30	0.63	0.75	19.05
40	0.44	0.51	15.9
50	0.38	0.44	15.79

Table 9 Comparisons between present model of third case (475 W) and Lambiase (2012)

Scanning velocity (mm/s)	Experimental bend angle (°)	Predicted bend angle (°)	% deviation
15	1.36	1.51	11.02
30	0.73	0.81	10.96
40	0.52	0.64	23.07
50	0.44	0.52	18.18

Table 10 Comparisons between present model of fourth case (550 W) and Lambiase (2012)

Scanning velocity (mm/s)	Experimental bend angle (°)	Predicted bend angle (°)	% deviation
15	1.54	1.72	11.68
30	0.94	1.02	8.51
40	0.83	0.77	7.23
50	0.5	0.61	22

5 Estimation of Yield Stress by Inverse Analysis and Model Updating

With the model presented in this work, the yield stress can be estimated in an inverse manner to match the experimental results. The bisection method can be used to find out the proper value of σ_y, that produces very small error between the model predicted and experimental bend angle. Table 11 uses the data of Kyrsanidi et al. (2000). It is observed that inverse estimated yield stress is slightly different from the

Table 11 Inverse estimation of yield stress

Power (kW)	Yield stress (MPa)	
	Used in direct model	Predicted by inverse analysis
3	287.35	296.25
3	319.9	326.25
3	344.7	336.25
2	359.54	374.75

yield stress values used in direct model. This difference is not more than ±5 %. In fact this much uncertainty is always present in yield stress of any typical material. Thus, the proposed model can be used for the quick estimation of yield stress of the sheet material.

6 Conclusion

In this work a simple analytical model has been developed for estimating bend angle during laser forming of the sheet based on elastic-plastic theory. This model has been compared with some experimental results. A good agreement is obtained between experimental and predicted results. Although in some cases, the error between predicted and experimental results seems to be high, but as per the review paper of Shen and Vollertsen (2009) the average model prediction error by the best available analytical model is about 40 % and generally the error is always more than 20 %. Considering this, the present model provides a very good estimate and is fairly simple. The model can also be used for the quick estimate of yield stress.

References

Chakrabarty, J. (2006). *Theory of plasticity* (3rd ed.). Burlington: Butterworth-Heinemann.

Chen, G., & Xu, X. (2001). Experimental and 3D finite element studies of CW laser forming of thin stainless steel sheets. *Journal of Manufacturing Science and Engineering, 123*, 66–73.

Chen, G., Xu, X., Poon, C. C., & Tam, A. C. (1999). Experimental and numerical studies on microscale bending of stainless steel with pulsed laser. *Transaction of the ASME Journal of Applied Mechanics, 66*, 772–779.

Cheng, P., Fan, Y., Zhang, J., Mika, D. P., Zhang, W., Graham, M., et al. (2006). Laser forming of varying thickness plate-part1: Process analysis. *ASME Journal of Manufacturing Science and Engineering, 128*, 634–641.

Dixit, U. S., Joshi, S. N. & Kumar, V. H. (2012). Microbending with laser. In V. K. Jain (Ed.), *Micromanufacturing processes*. Boca Raton, Florida: CRC Press.

Hu, Z., Labudovic, M., Wang, H., & Kovacevic, R. (2001). Computer simulation and experimental investigation of sheet metal bending using laser beam scanning. *International Journal of Machine Tools and Manufacture, 41*, 589–607.

Kraus, J. (1997). Basic processes in laser bending of extrusions using the upsetting mechanism. *Laser Assisted Net shape Engineering 2, Proceeding of the LANE, 2*, 431–438.

Kyrsanidi, A. K., Kermanidis, T. B., & Pantelakis, S. G. (1999). Numerical and experimental investigation of the laser forming process. *Journal of Materials Processing Technology, 87*, 281–290.

Kyrsanidi, A. K., Kermanidis, T. B., & Pantelakis, S. G. (2000). An analytical model for the predictions of distortions caused by laser forming process. *Journal of Materials Processing Technology, 104*, 94–102.

Lambiase, F. (2012). An analytical model for evaluation of bending angle in laser forming of metal sheets. *Journal of Materials Engineering and Performance, 21*, 2044–2052.

Lambiase, F., & Ilio, A. D. (2013). A closed-form solution for thermal and deformation fields in laser bending process of different materials. *International Journal of Advanced Manufacturing Technology,* doi:10.1007/s00170-013-5084-9.

Mishra, A., & Dixit, U. S. (2013). Determination of thermal diffusivity of the material, absorptivity of the material and laser beam radius during laser forming by inverse heat transfer. *Journal of Machining and Forming Technology, 5*, 208–226.

Shen, H., Shi, Y., Yao, Z., & Hu, J. (2006a). An analytical model for estimating deformation in laser forming. *Computational Materials Science, 37*, 593–598.

Shen, H., & Vollertsen, F. (2009). Modeling of laser forming—A review. *Computational Material Science, 46*, 834–840.

Shen, H., Yao, Z., Shi, Y., & Hu, J. (2006b). An analytical formula for estimating the bending angle by laser forming. *Proceedings of the Institution of Mechanical Engineering, Part C: Journal of Mechanical Engineering Science, 220*, 243–247.

Shi, Y., Shen, H., Yao, Z., & Hu, J. (2007). Temperature gradient mechanism in laser forming of thin plates. *Optics & Laser Technology, 39*, 858–863.

Vollertsen, F. (1994). An analytical model for laser bending. *Laser in Engineering, 2*, 261–276.

Yau, C. L., Chan, K. C. & Lee, W. B. (1997). A new analytical model for laser bending. In M. Geiger & F. Vollertsen (Eds.), Laser assisted net shape engineering (Vol. 2, pp. 357–366), Proceedings of the LANE'97. Meisenbach, Bamberg.

Laser Forming of Mild Steel Sheets Using Different Surface Coatings

Sachin S. Gautam, Sunil K. Singh and Uday S. Dixit

Abstract Laser forming is a comparatively recent technique which has been developed to shape metal components. It has also been applied to the forming of composite materials. In this process, thermal stresses are produced in the sheet metal by irradiating the surface of the sheet using a controlled defocused laser beam. This causes the sheet to bend, usually, towards the side of the sheet which is exposed to laser irradiation. The laser beam power, scan speed, beam size, and number of laser passes determine the shape of the final product. Laser forming is different when compared to the traditional forming techniques like bending, drawing, pressing, stamping, etc. in the sense that it is a non-contact technique. Hence, the advantage of the laser forming process for a small batch of components is its flexibility and reduction in cost and time. Research to-date on laser forming has been done theoretically, experimentally, and numerically. Effect of almost all the parameters—geometric, material and energy—affecting the shape of final product have been studied at length. Some experimental studies have been carried out using surface coating like graphite on the face of the sheet where the laser is passed. However, there are limited studies on the effect of different surface coatings on laser forming of sheets. Hence, the objective of the present work is to investigate the effect of different surface coatings on the laser forming of metal sheets. In the present work, mild steel is selected as the work material. Two different coatings, viz., commercial lime and cement are selected. First, the effect of the coatings is studied on simple line bending operation. The experiments are carried out at different values of laser power and laser scan speed for each coating. The results are compared with the results obtained from line bending of uncoated specimen. It is found that the cement coating performs better than the lime coating. Next, the effect

S.S. Gautam (✉) · S.K. Singh · U.S. Dixit
Department of Mechanical Engineering, Indian Institute of Technology Guwahati, Guwahati 781039, India
e-mail: ssg@iitg.ernet.in

S.K. Singh
e-mail: sunil.singh@iitg.ernet.in

U.S. Dixit
e-mail: uday@iitg.ernet.in

S.N. Joshi and U.S. Dixit (eds.), *Lasers Based Manufacturing*,
Topics in Mining, Metallurgy and Materials Engineering,
DOI 10.1007/978-81-322-2352-8_2

of coatings is investigated on laser forming of complex shapes like dome and bowl shaped surfaces. For each coating the experiments are carried out at different values of laser power. It is found that the cement coated specimens can undergo more deformation than the lime coated specimens.

Keywords Laser forming · Surface coatings · Lime coating · Cement coating · Dome shaped surfaces

1 Introduction

1.1 What Is "Laser"?

The word "laser" is an acronym for "light amplification by stimulated emission of radiation". Radiation refers to "electromagnetic radiation" of which light is a special case. Laser light is different from the ordinary light because it has photons of same frequency, same wavelength and phase. While ordinary light is multidirectional a laser beam is highly unidirectional. Hence, laser has high power density and better focusing characteristics. The output of the laser can be continuous or pulsed. Continuous means that the power output is constant whereas it is not in the case of pulsed. An overview of the short pulsed laser is given in Table 1 (Meijer 2004).

Lasers are classified in different types depending on the type of lasing medium employed. The lasing medium can be solid, gas, liquid, semiconductor. Some of the examples are as follows:

(a) Solid-state lasers where the lasing material is distributed in a solid matrix such as the ruby and Nd: Yag (neodymium doped yttrium–aluminium garnet) lasers.
(b) Gas lasers (argon and helium-neon) have a primary output in the visible red light. CO_2 lasers emit energy in the far-infrared, and are used for cutting hard materials.
(c) Excimer lasers use reactive gases (e.g., chlorine and fluorine) mixed with inert gases (e.g., argon, krypton, or xenon).
(d) Dye lasers use organic dyes (e.g., rhodamine 6G) in liquid solution as lasing media, and
(e) Semiconductor lasers (also called diode lasers) are different from solid-state lasers. They use low power and are small in size. They find application in laser printers/CD players.

The CO_2 laser is one of the earliest gas lasers to be developed (invented by an Indian named Dr. C.K.N. Patel of Bell Labs in 1964). It is also one of the highest power continuous wave lasers and the most useful. It is also very efficient: the ratio of output power to pump power can be as large as 20 %. As shown in Table 1 the CO_2 laser produces a beam of infrared light with the wavelength bands centering

Table 1 Overview of short pulsed lasers

Laser type	Wavelength (nm)	Pulse length	Frequency (kHz)
CO_2	9400–10,600	200 μs	5
Nd: YAG	1060, 532, 355, 266	100, 10 ns	50
Excimer	193–351	20 ns	0.1–1
Copper vapor	611–578	30 ns	4–20
Ti sapphire	775	111 fs	1–250

around 9.4 and 10.6 μm. The active lasing medium (laser gain/amplification medium) is a gas discharge which is air-cooled (water-cooled in higher power applications). The filling gas within the discharge tube consists primarily of: Carbon dioxide (CO_2) (around 20–35 %), Nitrogen (N_2) (around 10–20 %), Hydrogen (H_2) and/or xenon (Xe) (a few percent; usually only used in a sealed tube.), and Helium (He) (which forms the remainder of the gas mixture). The specific proportions vary according to the particular laser. Because of the high power levels available combined with reasonable cost for the laser, CO_2 lasers are frequently used in material processing applications like cutting and welding, surgical applications (as water absorbs this frequency of light very well) like laser surgery, skin resurfacing, treatment of skin conditions, construction of microfluidic devices, military range finding using LIDAR techniques, enriching uranium using Silex process etc. In the present work, CO_2 laser (Model: Orion, Make: LVD) is used for all the studies.

1.2 Forming Process

Forming process is a process that changes the shape of the material by producing plastic deformation. This is carried out by applying compressive stress or tensile stress or shear stress or combination of stresses on the material. There is no removal of material in this process. The material is only deformed in this process. Forming processes may be of different types, for example: compressive forming, tensile forming, bending (combined tensile and compressive forming), shearing (combined tensile and compressive forming).

1.3 Manufacturing Through Lasers

Laser in manufacturing is further classified into two main categories namely

(a) laser forming and,
(b) laser assisted forming.

In laser forming, the final shape of the part is obtained by the exclusive use of the laser technology, eliminating other operations of the traditional forming. On the other hand laser assisted forming combines the features of both the laser forming and the traditional forming operations.

2 Laser Forming

Traditionally, machining and metal forming are two prominent methods of making a finished product from raw material. Machining is a process in which material is removed in the form of chips by means of a tool in machines like lathe, shaper etc. Some of the examples of machining process are cutting, shaping, milling, etc. On the other hand, in metal forming finished product is obtained by inducing plastic deformation in the material. During this process no material is removed. However, some material is removed during finishing operation to obtain the final product. Some of the examples of forming process are deep drawing, bending, spinning, etc. Metal forming can again be classified into two process namely bulk metal forming and sheet metal forming. In bulk metal forming process the work piece has a high volume to surface area ratio. Examples are rolling, extrusion, wire drawing, forging, etc. In sheet metal forming the work piece has a low volume to surface ratio. The deformation in the thickness direction is not desirable. Some of the examples of sheet metal forming process are deep drawing, stretch forming, bending, spinning, etc. Despite having low process time, accurate results are difficult to obtain through these conventional techniques (Dixit and Dixit 2008).

Laser forming is one of the newer forms of unconventional form of shaping a metallic or a non-metallic material. In this process, work piece, usually in the form of sheets, is deformed into a desired shape by irradiating the surface using high powered laser beam. This leads to development of thermal stresses especially near the heated area. The work piece expands and when it is restrained by the surrounding material localized plastic deformation takes place. On cooling, the bending takes place towards the laser beam. The desired shape of the work piece is achieved by controlled plastic and elastic deformation. Figure 1 shows the schematic diagram of the laser forming process. Laser forming is different when compared to the traditional forming techniques like drawing, stamping, pressing, etc. in the sense that it is a non-contact technique. Also, it is a highly flexible and low-volume manufacturing process.

Laser forming can also be used for cutting, welding, and drilling operation by changing the processing parameters. It is used to deform sheets of various metals e.g., stainless steels, aluminium alloys, magnesium alloys etc., which have high coefficient of thermal expansion. Also, many hard and brittle materials, such as titanium and nickel alloys, ceramics, glass, semiconductors, etc. also can be formed with little distortion or degradation.

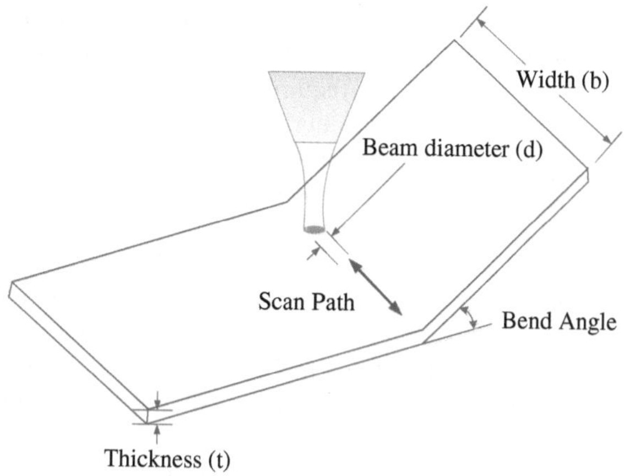

Fig. 1 Schematic view of a straight line bending of a flat metal sheet by a defocused laser beam

2.1 Parameters Affecting Laser Forming

In laser forming, the shape and the position of the zone, intended to be deformed into desired shape, is determined by a number of parameters which can broadly be divided into three groups- energy parameters, material parameters, and work piece related parameters. Next, each group is discussed in more detail.

(a) **Energy based parameters**: These parameters are directly related to the laser being used to form the product. The forming process is affected by the laser power, scanning speed (or the feed rate), geometry and size of laser beam, number of passes, absorption coefficient of the material, cooling conditions etc.
(b) **Material parameters**: The material properties of the material being formed have a direct influence on the forming process. Some of these are Young's modulus, Poisson's ratio, thermal expansion coefficient, thermal conductivity and heat capacity.
(c) **Work piece related parameters**: The geometric parameters of the sheet being formed also affect forming. This includes the sheet thickness, length and width. The sheet thickness is considered the most important geometric parameter (Singh 2013) in forming process.

2.2 Advantages and Disadvantages of Laser Forming

Laser forming process induces thermal stresses in the material using high powered laser beam without making any physical contact. Hence, the laser forming technique has the following advantages over the conventional metal forming processes.

(a) The lead time, i.e., the time required to design a product before it is actually manufactured is eliminated as conventional tools like die and punch, hammer, mallet etc. are not required.
(b) The cost of the forming process is greatly reduced because no tool or external forces are involved in the process.
(c) It is good for small batches and for a variety of sheet metal components.
(d) Quality of the product is improved by this process. Accurate deformations and patterns can be achieved because there is no spring-back effect.
(e) Even unreachable areas of a complex shaped work piece can be given the desired shape because of the non-contacting nature of the process.
(f) Laser bending is suitable for materials which are difficult to form by mechanical approaches. Processing of brittle, hard, and heavy materials such as titanium alloy, nickel alloy, ceramics, etc. can be carried out at hot condition.
(g) By applying specific irradiation patterns, a metal sheet can be formed into complex curved shapes like dish, screw, cone, saddle etc.
(h) It can be used for the forming of the laser parts at micro scale. Such a forming is called micro forming (μforming).
(i) Laser forming can be used for automated production as it is easy to control the process parameters.

Although laser forming has the aforementioned advantages, it does suffer from certain drawbacks which are discussed below.

(a) The laser forming process is a slow process compared to mechanical process.
(b) This process is energy consuming because of the low energy conversion factor of the laser source.
(c) Protective gears must be worn by the personnel and safety measures must be followed strictly because the reflection of the laser beam in all directions from the metal surface may cause damage.

3 Literature Review of Some Recent Work

Application of laser forming in the field of manufacturing specially forming is relatively new. One of the first references to the use of laser as the heat source for bending was in the work of Scully (1987). Further details and references on the laser forming process and process mechanisms can be found in Steen and Mazumder (2010) and references cited therein. A recent review on modelling of laser forming has been presented by Shen and Vollersten (2009) and some recent developments applicable to macro and micro laser forming have been discussed by Dearden and Edwardson (2003). In the present work, a brief overview of literature on laser forming specifically using some form of coating on the material surface during the laser bending process is presented.

Barletta et al. (2006) defined and investigated a new hybrid forming process consisting of a fluidized bed pre-treatment to coat with Al_2O_3 and pre-curve aluminium thin sheets and subsequently diode laser forming to perform the proper line bending of fluidized bed pre-treated sheets. A comparison with the diode laser forming of untreated aluminium thin sheets was carried out. Their experimental results showed that the bending angles up to 45° could be achieved on Al_2O_3 coated aluminium thin sheets employing output power in the range of 150–250 W. The uncoated specimens exhibited maximum bending angles of 25° employing output power of about 400 W. Recently, Roohi et al. (2012) studied the effect of external forces assisting the laser bending—known as "External Force Assisted Forming Process". They used anodised and graphite coated 5 mm thick Al-5005 alloy. It was found that the bending angle increased by about 30 %, in laser assisted bending in comparison to laser bending at the same values of the parameters. A numerical simulation using ABAQUS package was also performed which found that the equivalent plastic strain in laser forming process increases in a step wise manner with increasing the number of scanning passes. Griffiths et al. (2010) employed finite element modelling to ascertain the quantitative contribution of various process parameters such as thermal, geometrical and coefficient of absorption towards the variation in the bend angle per pass with multiple irradiations. They found that the thermal parameters have lesser influence than the geometrical parameters and coefficient of absorption. Burning-off of the graphite coating due to continued irradiation affects the achievement of bend angle significantly. The effect of different coatings, viz., lime, graphite and grease, was studied by Singh (2013). He studied the effect of coatings on the absorption of CO_2 laser beam in line bending of mild steel sheets. The bend quality was compared with mechanically bent sheets. It was observed that all the coatings increased the coefficient of absorption. More than 100 % increase in the coefficient of absorption for the case of graphite coated sheets was observed. The increase for lime coated sheets was found to be in the range of 400–1500 %.

4 The Objective of the Present Work

Based on the review of some recent work in Sect. 3, it can be seen that the use of coatings in laser forming process is very important. Use of coatings increases the coefficient of absorption of laser, which in turn causes greater deformation. Singh (2013) studied the effect of coatings on line bending. However, it is felt that there is need to study the effect of coatings further using some more coatings. Also, the effect of coatings needs to be studied for more complex shapes. The present work, thus, extends the work of Singh (2013) and explores these aspects further. The specific objectives of this work are

(a) to study the effect of different coatings on the line bending of mild steel sheets at various levels of laser power, and scan speeds,

(b) to identify a suitable coating, which leads to maximum bending angle, and
(c) to study the effect of the coating identified in (b) on the forming of complex shapes like bowl and dome.

5 Experimental Details

In the present work, commercially available mild steel is used for all the studies. All the specimens have the thickness of 2 mm. First, tensile tests are carried out. For this flat tensile test specimens are prepared along the three different directions of the sheet namely 0°, 45°, and 90° respectively. In each direction three specimens were prepared. Figure 2 shows the different tensile test specimens, and tensile test setup. Figures 3, 4 and 5 show the stress-strain curves for the tensile test specimens in three different directions. From the experimental results, the yield strength, ultimate strength, and fracture strength of the mild steel used in the current work was found to be 274 MPa (standard deviation: 9.21 MPa), 383 MPa (standard deviation: 5.74 MPa), 305 MPa (standard deviation: 30.54 MPa) respectively.

The specimens used for the experimental work were cut by a CNC 2.5 kW CO_2 laser cutting machine (Model: Orion 3015, Make: LVD). They were cleaned with acetone to remove any foreign particle from the surface. The same laser machine was used to impinge laser beam on the specimen surface. Figure 6 shows the photograph of the experimental setup used in the present work with different parts labelled. A circular laser beam of 3 mm diameter with a stand-off distance of 30 mm was used in Gaussian mode. Various levels of laser power and scanning speed were used in the experiments. The specific values are mentioned in respective sections.

As mentioned in the previous section, application of surface coating during laser forming leads to increase in the coefficient of absorption. This results in more deformation when compared to uncoated specimens. Singh (2013) found that lime

Fig. 2 Tensile test: **a** specimens, **b** setup, and **c** close up view of the setup

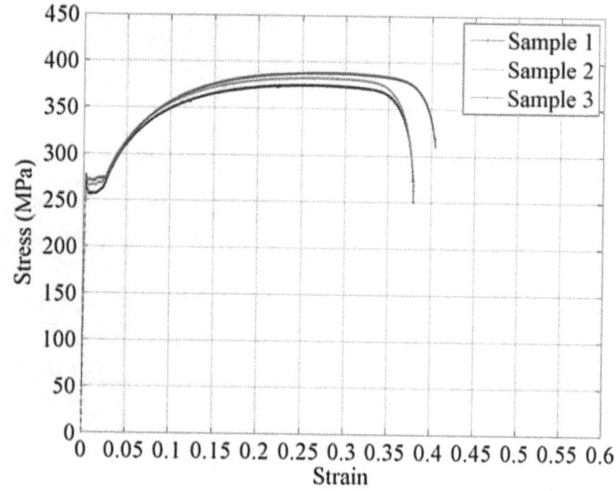

Fig. 3 Stress–strain curves for mild steel sheet along 0° direction for three different specimens

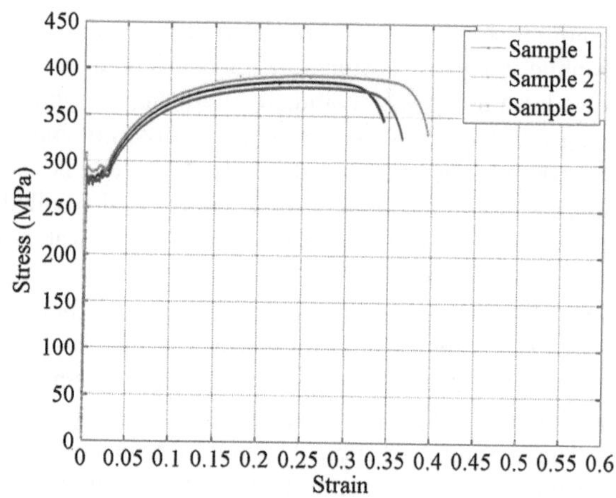

Fig. 4 Stress–strain curves for mild steel sheet along 45° direction for three different specimens

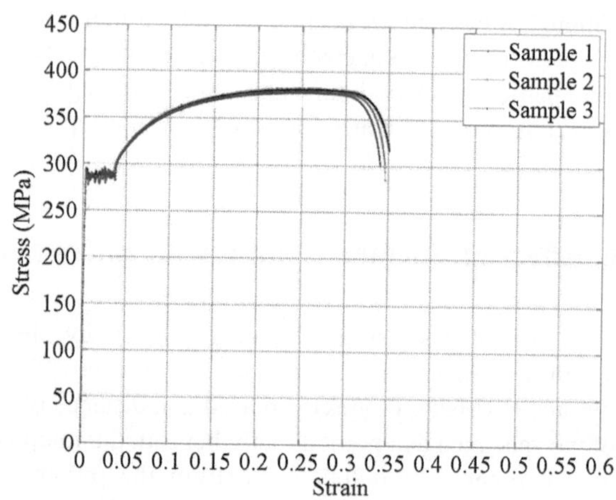

Fig. 5 Stress–strain curves for mild steel sheet along 90° direction for three different specimens

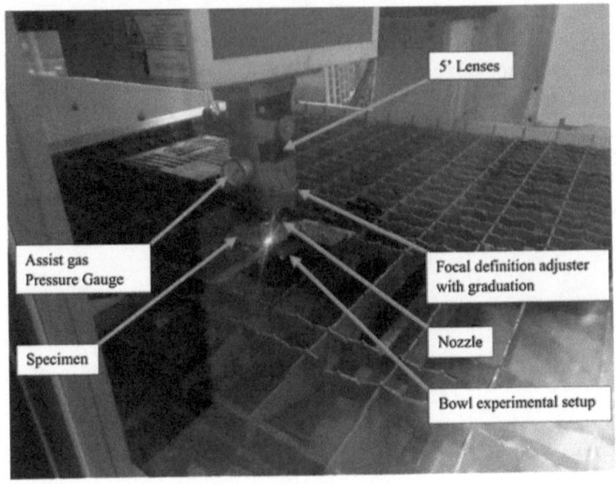

Fig. 6 A photograph of the experimental setup

coating performed better than the graphite and grease coatings. Hence, in this work, graphite and grease coatings are not considered. For the present study, lime coating is considered along with commercially available cement. Lime coating was prepared using following procedure. Commercially available lime (CaO) was put in a beaker and 100 ml water was added slowly. It is well known that when lime is mixed with water, it forms calcium hydroxide, which is also called slaked lime. The temperature of the mixture rises considerably on addition of water. In the meantime, the mixture is stirred well to ensure a homogeneous mix. Then, the mixture is kept for about one hour after which the lime precipitates at the bottom leaving some water at top. The water is then gently removed. The mixture which remains later is used as a coating. The cement coating is prepared in a similar manner. The amount of heat released is considerably lower when preparing cement coating.

The coatings are applied manually using a brush uniformly on the mild steel specimens. About 0.2 mm thick coating is applied in the middle of all the specimens. After applying the coating, the samples are dried to remove the moisture, although it has already been reported by Singh (2013) that moisture does not affect the performance of the coatings. Some experiments were also conducted by mixing sand particles with both lime and cement coatings. However, no substantial changes were observed and hence the results are not reported.

6 Effect of Surface Coatings on Energy Absorption

First, the effect of surface coatings on energy absorption is studied. Mild steel samples of size 50 mm × 50 mm and thickness 2 mm are prepared. An infrared pyrometer (Make: Raytek) is placed at a distance of 2 mm from the bottom surface of the specimen at a point directly beneath the midpoint to measure the temperature at that point. The output cable from the pyrometer is directly connected to a computer where DATATEMP software records the temperature as a function of time. For each data point the experiment is repeated three times. Number of passes is taken as 10.

Figures 7, 8 and 9 show the variation of maximum temperature of the bottom surface of the sheet for first three passes at three different values of laser power and laser scan speed. As expected, the temperature rises with increasing the laser power and decreases with increase in scan speed. The error bar shows significant deviation at higher value of laser power. One of the reasons may be due to the non-uniformity of the sheet thickness over different specimens. The repeatability is better at the lower values of laser power.

Figures 10, 11 and 12 show the effect of hydrated lime coating on the maximum temperature at the bottom surface of the specimen for first three passes at three different values of laser power and laser scan speed. As in the case of uncoated

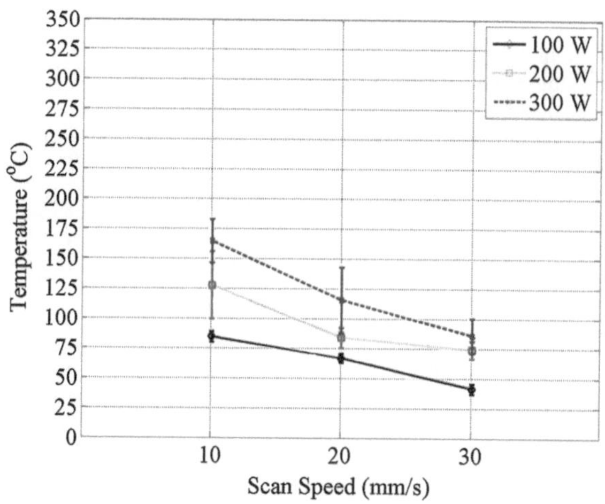

Fig. 7 Variation of maximum temperature with scan speed for uncoated sheet at different laser powers for first pass

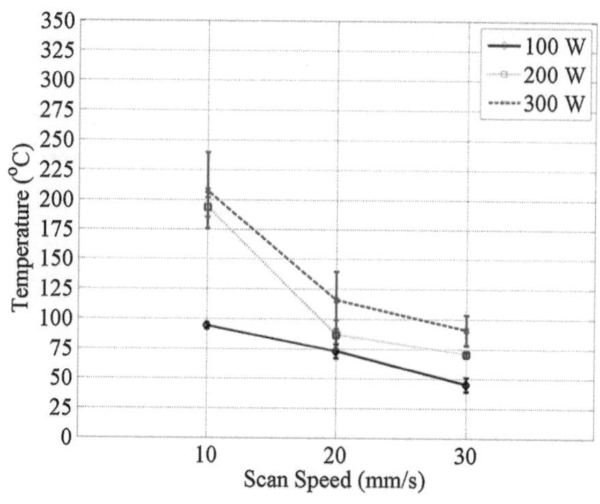

Fig. 8 Variation of maximum temperature with scan speed for uncoated sheet at different laser powers for second pass

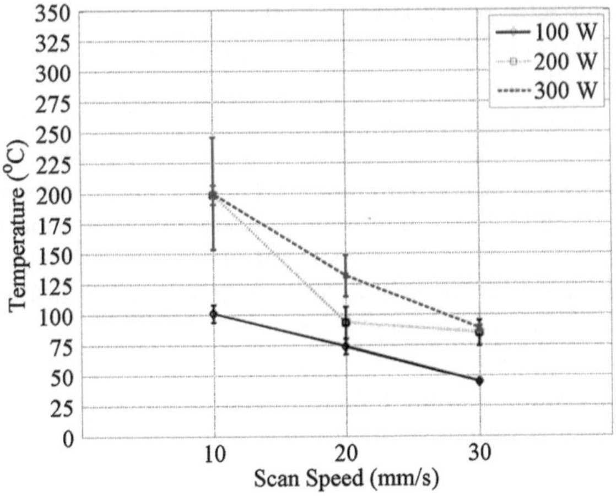

Fig. 9 Variation of maximum temperature with scan speed for uncoated sheet at different laser powers for third pass

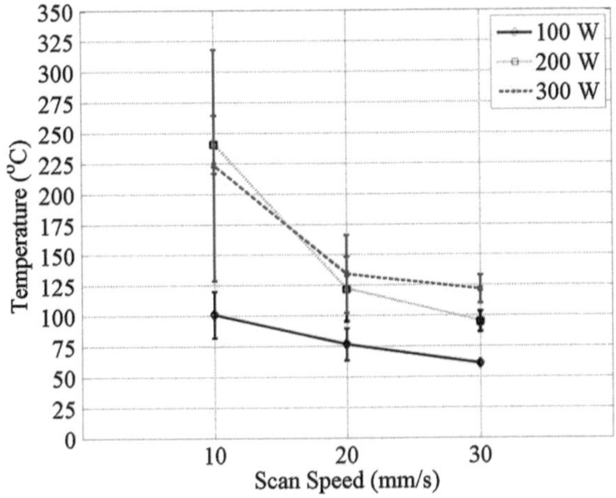

Fig. 10 Variation of maximum temperature with scan speed for lime coated sheet at different laser powers for first pass

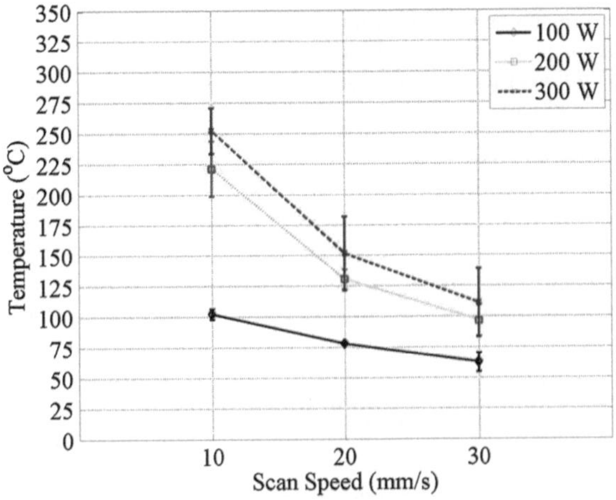

Fig. 11 Variation of maximum temperature with scan speed for lime coated sheet at different laser powers for second pass

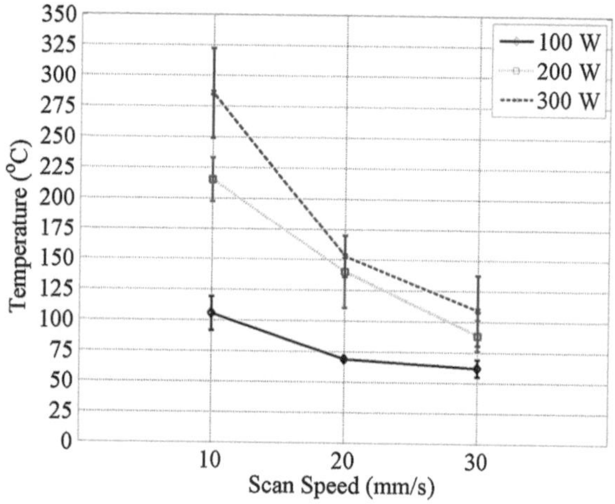

Fig. 12 Variation of maximum temperature with scan speed for lime coated sheet at different laser powers for third pass

sheet, the temperature increases with increasing laser power and decreases with increasing scan speed. As compared with the uncoated specimens it can be clearly seen that the maximum temperature for a given laser power and scan speed is higher especially for 200 and 300 W case.

Figures 13, 14 and 15 show the effect of cement coating on the maximum temperature at the bottom surface of the specimen for first three passes at three different values of laser power and laser scan speed. It can be clearly seen that for any given laser power and scan speed the maximum temperature is higher as compared to the uncoated and lime coated specimens. This is because unlike the lime coating which wears off after second pass the cement coating wears off slowly. Again, the maximum temperature increases with increase in laser power and decreases with increase in scan speed.

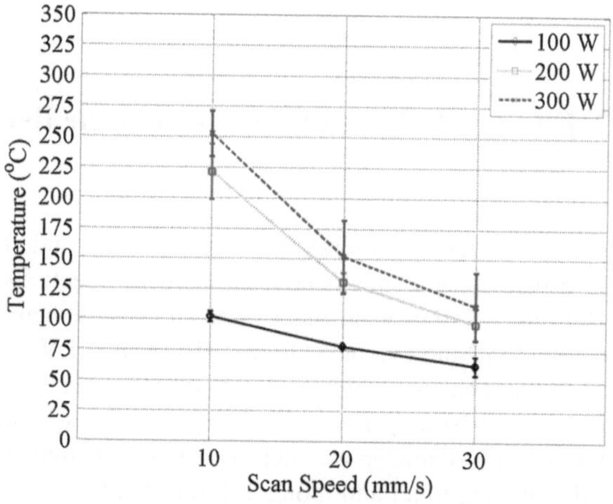

Fig. 13 Variation of maximum temperature with scan speed for lime coated sheet at different laser powers for first pass

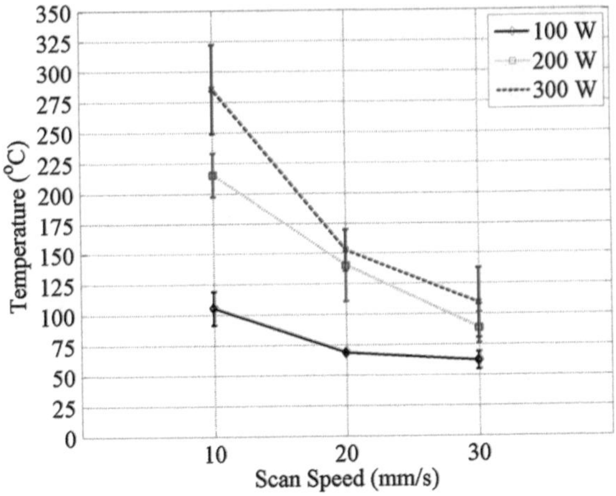

Fig. 14 Variation of maximum temperature with scan speed for lime coated sheet at different laser powers for second pass

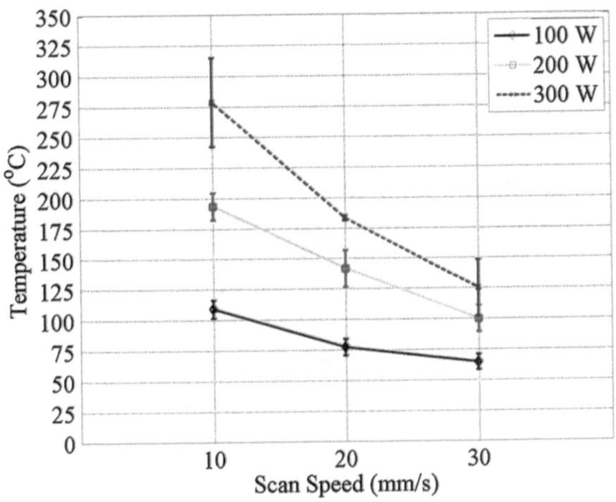

Fig. 15 Variation of maximum temperature with scan speed for lime coated sheet at different laser powers for third pass

7 Effect of Different Coatings on Line Bending of Mild Steel Sheets

Next, the effect of different coatings on simple line bending of mild steel sheets is investigated. For this study, mild steel sheet specimens of size 150 mm × 100 mm × 2 mm were made. The specimens were cut by the 2.5 kW CO_2 laser cutting machine (Model: Orion 3015, Make: LVD) and were cleaned in acetone. The specimens were held as a cantilever beam along the 100 mm edge and built-in length was 10 mm. For each data point the experiment was repeated three times. Laser beam diameter was kept constant at 6 mm. Laser beam path was 100 mm from the fixed end and along 100 mm width.

Fig. 16 Variation of bending angle with laser power for uncoated specimens for three different scan speeds after 10 passes

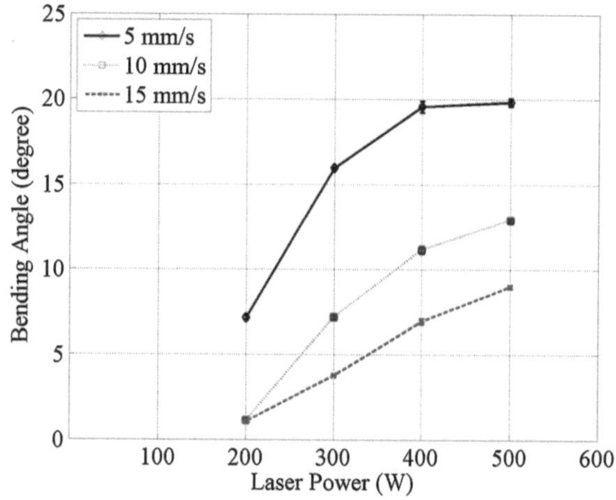

Figures 16, 17 and 18 show the variation of bending angle with laser power for uncoated, lime coated, and cement coated specimens respectively. It can be seen that for all the coatings the bending angle increases with increasing the laser power and decreasing the scan speed. It can be clearly seen from the figures that for a given laser power and scan speed the bending angle for the cement coated specimen is more as compared to uncoated and lime coated specimens.

Figures 19, 20 and 21 show the line bent specimens for uncoated, lime coated, and cement coated specimens respectively at 300 W laser power, 5 mm/s scan speed, and total of 10 passes.

To summarize, first, the effect of lime and cement coatings is studied for absorption. It is found that the use of coatings leads to increase in absorption which is evident by increase in the temperature measured at the bottom surface of a specimen. Also, it is found that the temperature measured at the bottom surface of a specimen is the highest for the cement coated specimen for all values of laser power

Fig. 17 Variation of bending angle with laser power for lime coated specimens for three different scan speeds after 10 passes

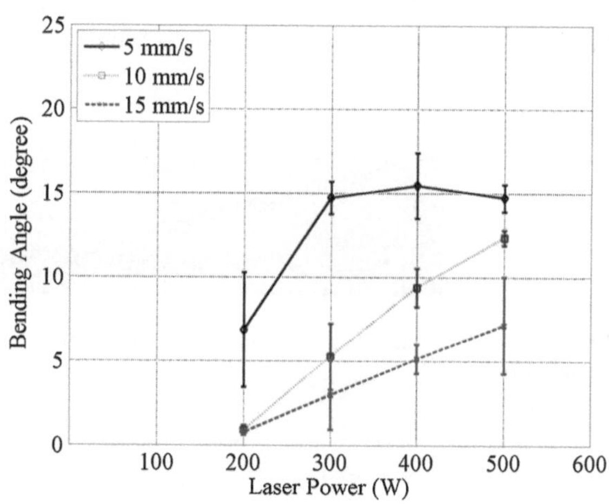

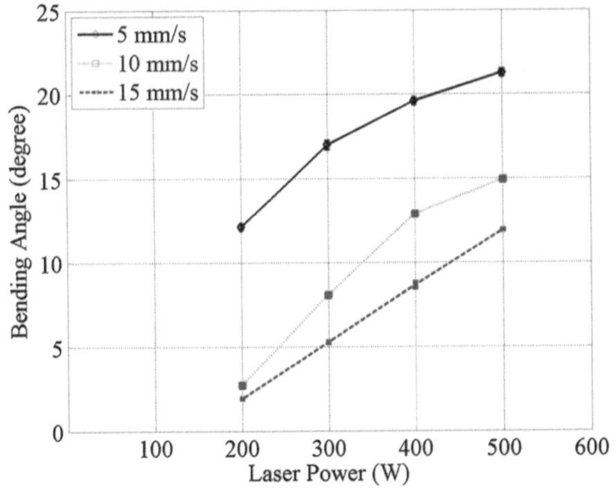

Fig. 18 Variation of bending angle with laser power for cement coated specimens for three different scan speeds after 10 passes

Fig. 19 Uncoated line bent specimen (power: 300 W, scan speed: 5 mm/s, number of passes: 10)

Fig. 20 Lime coated line bent specimen (power: 300 W, scan speed: 5 mm/s, number of passes: 10)

Fig. 21 Cement coated line bent specimen (power: 300 W, scan speed: 5 mm/s, number of passes: 10)

and scan speeds. Then, the effect of lime and cement coatings is studied on a simple line bending operation. Again, it is observed that the cement coated specimens show higher values of bending angles as compared to lime coated and uncoated specimens. Hence, the conclusions are: (a) use of surface coatings leads to increase in the absorption, and (b) cement coating performs better than the lime coating.

8 Effect of Surface Coatings on Generation of Complex Shaped Surfaces

In the previous section, the effect of different coatings was studied on simple line bending operation. It was found that the use of coatings, in general, leads to increase in the absorption. Also, cement coating was found to be better than the lime coating. Laser forming a three dimensional shape from metal sheets requires multiple irradiation strategy i.e., passing the laser along different paths. The present section presents the effect of cement coating on laser forming of three dimensional shaped surfaces viz., dome and bowl. First, the effect of cement coating is studied on forming of dome shaped surface and then on bowl shaped surface.

8.1 Dome Shaped Surfaces

Maji et al. (2014) investigated the forming of dome shaped surface under temperature gradient mechanism. In the present section, we study the effect of cement coating on forming of dome shaped surface from a square metal sheet. Again, a CNC controlled 2.5 kW CO_2 laser machine (Model: Orion, Make LVD) is used to impinge the laser beam on the surface. Circular laser beam of 3 mm diameter at a stand-off distance of 30 mm is used in Gaussian mode. Four levels of laser power—200, 300, 400, and 500 W—are used. Mild steel sheet specimens of dimension 120 mm × 120 mm × 2 mm were cut from a large sheet using laser cutting. The specimens were cleaned in acetone to remove any unwanted dirt and grease before they are used as work piece in the experiments. The work piece was kept freely on a flat surface on the workbench of the laser machine. The total number of laser pass is kept at one and scan speed chosen is 5 mm/s.

As stated earlier, forming of complex 3D shaped surfaces requires multiple irradiation strategies. Hennige (2000) has described various scanning strategies using both simple and curved laser scan paths for laser forming of dome shaped surface from flat sheet metals. He also investigated the forming behaviour of sheet metal in case of simple and curved scan paths. The differences were identified and explained. Also, some finite element analysis results for curved laser forming were presented for better understating of the 3D laser forming process. Kim and Na (2009) proposed two methods, distance-based and angle-based, to generate an irradiation strategy and to form angles at each irradiation path.

In the present work, the irradiation strategy mentioned in Yang et al. (2010) is adopted. Figure 22 shows the schematic of the laser scan strategy used in the present work. This scan strategy involves two types of laser scans i.e., diagonal and axial which are used alternatively. The scan starts at the point marked 'Start' and moves diagonally to point 2 as shown in the Fig. 2. Then, the laser is moved from point 2 to point 3 axially downwards and then moved diagonally to point 4 followed by axial scan to point 5 followed by diagonal scan to 6 and so on until the laser reaches the point 9 which also happens to be the starting point. From there the laser is moved to point 10 and the previous steps are repeated until the laser reaches the center of the sheet. In the present work, the scan strategy was first drawn in AUTOCAD software from which the CNC code was generated using CADMAN-PL software. A total of three diagonal and three axial scan per quadrant of the sheet were used.

Figure 23 shows the dome height measured at the center of the dome with different laser power for both the uncoated and cement coated specimens. All the heights were measured using coordinate measuring machine. It can be seen that the cement coated specimens have higher deformation level. However, the difference is not significant except at 400 W laser power. One of the possible reason is that only one pass has been used. It would be interesting to see the effect of multiple irradiations and the investigation is left for future work. The final shape of the specimens is shown in Fig. 24 for both uncoated and cement coated case. The scan lines can be clearly seen. The front view of the dome is shown in Fig. 25 where the red lines show the dome height.

Fig. 22 Laser irradiation strategy proposed by Yang et al. (2010) is considered in the present experimental work for forming dome shaped surface

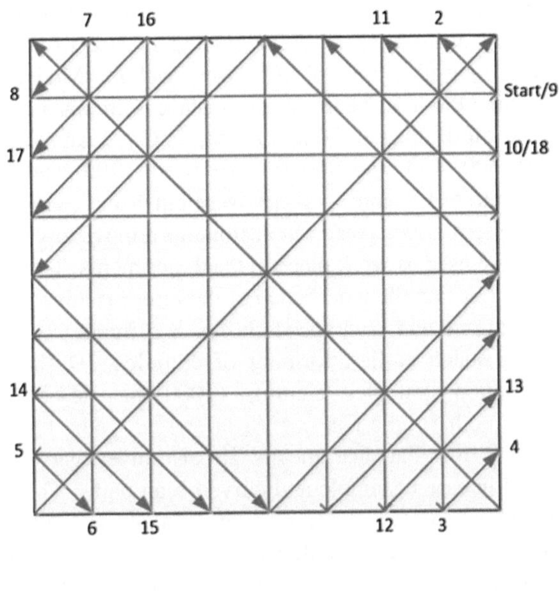

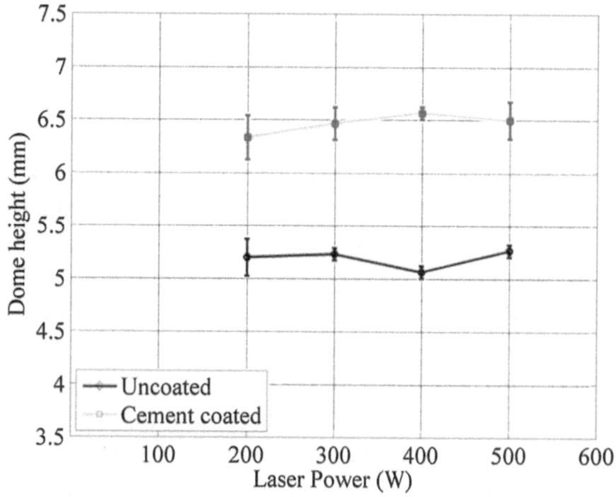

Fig. 23 Height of dome at the center for different values of laser power for uncoated and cement coated specimens. For each data point three specimens were tested

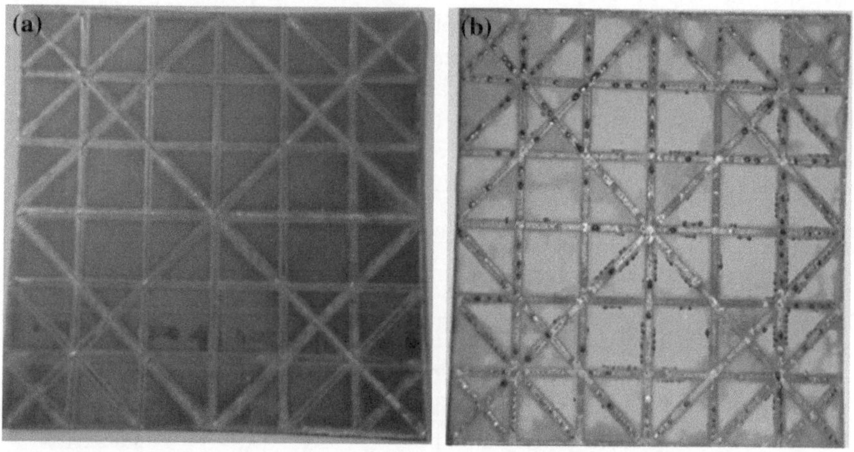

Fig. 24 Dome shaped specimens: **a** uncoated specimen, **b** cement coated specimen

Fig. 25 Dome shaped specimen after laser forming. The *horizontal red lines* show the dome height measured from the base to the top most point at the center of the specimen

8.2 Bowl Shaped Surfaces

Chakraborty et al. (2012) investigated the effect of process parameters, viz., laser spot diameter, laser power, and scan speed on the in-plane and out-of-plane forming of circular stainless steel specimens for various circular and radial scan schemes. Shaping of a circular specimen into a bowl requires both in plane thickening and out-of-plane bending. It offers considerable challenge for the laser forming. Hence, in the present work, this case is selected for studying the effect of cement coating. Again, a CNC controlled 2.5 kW CO_2 laser machine (Model: Orion, Make LVD) is used to impinge the laser beam on the coating surface. Circular laser beam of 3 mm diameter at a stand-off distance of 30 mm is used in Gaussian mode. Four levels of laser power (200, 300, 400, and 500 W) are used. Circular specimens of mild steel sheet of diameter 100 mm × 2 mm were cut from a large sheet using laser cutting. The specimens were cleaned in acetone to remove any unwanted dirt and grease before they are used as work piece in the experiments. To ensure that the specimen does not move during the forming operation a circular hole of diameter 6 mm was made at the center of specimen through which the specimen was held on a base plate with the help of screw and nut. As suggested in Chakraborty et al. (2012), a gap was maintained between the base plate and the specimen by use of washers.

In the present section, the irradiation strategy mentioned in Chakraborty et al. (2012) is adopted. Figure 26 shows the schematic of the laser scan strategy used in the present work. Three circular scans (scans 2, 4, and 6) and three radial scans (scans 1, 3, and 5) are used alternatively to achieve the final shape. Further details can be found in Chakraborty et al. (2012).

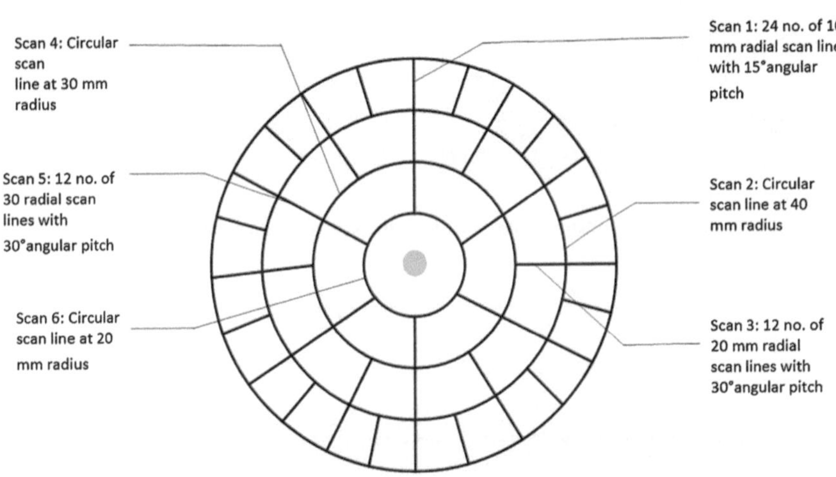

Fig. 26 Laser scan scheme proposed by Chakraborty et al. (2012) considered in the present experimental work for forming bowl shaped surface

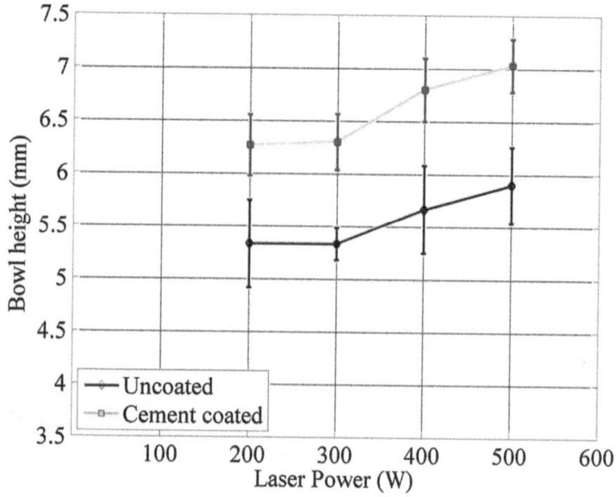

Fig. 27 Height of bowl at the center for different values of laser power for uncoated and cement coated specimens. For each data point three specimens were tested

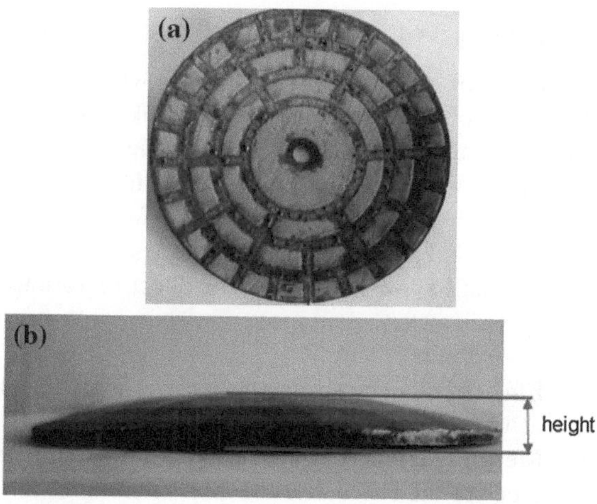

Fig. 28 Bowl shaped specimens: **a** the top view, and **b** shows the side view of cement coated specimen. In **b** the *horizontal red lines* show the dome height measured from the base to the top most point at the center of the specimen

Figure 27 shows the bowl height at the center of both the uncoated and cement coated specimen at different values of laser power. As in laser forming of dome shaped surface, the cement coated specimens have higher deformations. Figure 28a, b show the top view and front view of the bowl shaped specimen after the completion of the laser scan.

9 Conclusion

In the present work, the experimental investigations on the performance of lime and cement coating in laser forming are carried out. Mild steel is used as the working material for all the experiments. The sheet thickness and laser spot diameter is kept constant in all the experiments. First, the effect of coatings is studied on simple line bending operation. Then, the effect of cement coating is studied on laser forming of more complex shapes like bowl and dome. The conclusions, based on the above studies, are presented in the next section. On the basis of the results presented in Sects. 7 and 8, the following specific conclusions are made:

(a) The use of surface coatings leads to increase in the absorption of laser. This is evident by the increase in the temperature measure at the bottom of the sheet as compared to the uncoated specimen.
(b) The increase in temperature at the bottom of the sheet for cement coated specimen is found to be greater than that of the lime coated specimen.
(c) The bending angle increased the most when cement coating was applied.
(d) During the forming of complex shaped 3D surfaces like bowl and dome it is found that the use of cement coating leads to increase in the deformation as compared to uncoated specimen. However, the deformation does not increase with increasing laser power.

Following are the areas which offer scope of future work for other investigators:

(a) Effect of surface coating on sheets with varying thickness can be studied.
(b) Effect of different beam geometry like square, rectangle can be investigated.
(c) Effect of surface coating can be studied on deformation of some more complex shapes like saddle shape, tube bending etc.
(d) Effect of surface coating on the deformation can be investigated on different materials like metals and composites.
(e) A detailed investigation of the effect of various process parameters like laser spot diameter, scan speed, number of passes, and laser power need to be carried out.
(f) Effect of different scanning strategies other than those used in the current work can be carried out.
(g) Metallurgical and mechanical studies should be carried out to see the properties of the sheet after the laser forming.
(h) Numerical studies like finite element simulation need to be carried out to further understand the effect of coating on laser forming process.

References

Barletta, M., Casamichele, L., & Tagliaferri, V. (2006). Line bending of Al_2O_3 coated and uncoated aluminium thin sheets. *Surface & Coatings Technology, 201*, 660–673.

Chakraborty, S. S., Racherla, V., & Nath, A. K. (2012). Parametric study on bending and thickening in laser forming of a bowl shaped surface. *Optics and Lasers in Engineering, 50*, 1548–1558.

Dearden, G., & Edwardson, S. P. (2003). Some recent developments in two- and three-dimensional laser forming for 'macro' and 'micro' application. *Journal of Optics A: Pure ad Applied Optics, 5*, S8–S15.

Dixit, P. M. & Dixit, U. S. (2008). Modelling of metal forming and machining processes by finite element and soft computing methods, Springer-Verlag, London, UK.

Dixit, U. S., Joshi, S. N., & Hemanth, K. V. (2013). Microbending with laser. In V. K. Jain (Ed.), *Micro manufacturing processes*. Boca Raton, FL: CRC Press.

Griffiths, J., Edwardson, S. P., Dearden, G., & Watkins, K. G. (2010). Finite element modelling of laser forming at macro and microscales. *Physics Procedia, 5*, 371–380.

Hennige, T. (2000). Development of irradiation strategies for 3D-laser forming. *Journal of Materials Processing Technology, 103*, 102–108.

Kim, J., & Na, S. J. (2009). 3D laser forming strategies for sheet metal by geometrical information. *Optics & Laser Technology, 41*, 843–852.

Maji, K., Pratihar, D. K., & Nath, A. K. (2014). Laser forming of a dome shaped surface: Experimental investigations, statistical analysis and neural network modelling. *Optics and Lasers in Engineering, 53*, 31–42.

Meijer, J. (2004). Laser beam machining (LBM), state of the art and new opportunities. *Journal of Materials Processing Technology, 149*, 2–17.

Roohi, A. H., Gollo, M. H., & Naeini, H. M. (2012). External force-assisted laser forming process for gaining high bending angles. *Journal of Manufacturing Processes, 14*, 269–276.

Scully, K. (1987). Laser line heating. *Journal of Ship Production, 3*, 237–246.

Shen, H., & Vollertsen, F. (2009). Modelling of laser forming—a review. *Computational Materials Science, 46*, 834–840.

Singh, K. (2013). Effect of lime coating on laser bending process. M. Tech Thesis, IIT Guwahati, Guwahati, India.

Steen, W., & Mazumder, J. (2010). *Laser material processing*. London: Springer.

Yang, L., Wang, M., Wang, Y., & Chen, Y. (2010). Dynamic analysis on laser forming of square metal sheet to spherical dome. *International Journal of Advanced Manufacturing Technology, 51*, 519–539.

Finite Element Simulations of Laser Bending of Small Sized Sheets

Besufekad N. Fetene and Uday S. Dixit

Abstract Recently, laser bending has received the attention for a wide variety of applications in industries due to its excellent bend quality with high productivity and flexibility. In this work, finite element simulations of bending of small sized sheets are carried out using ABAQUS package. The temperature and strain-rate dependent material properties of D36 shipbuilding steel sheet are considered. Simulation results throw light on the bending behavior of small sized sheet components.

Keywords Laser bending · Stationary heat source · Moving heat source · FEM

1 Introduction

Laser bending is a process of bending the work piece by using controlled laser heat source, which induces thermal stress through a specified path to shape the work piece without any mechanical forces. Geiger and Vollertsen (1993) identified three most relevant mechanisms to explain the thermo-mechanical behavior in laser forming depending on specific combinations of component geometries and laser process parameters. These are temperature gradient mechanism (TGM), buckling mechanism (BM) and upsetting mechanism (UM). TGM is the most common mechanism of sheet metal bending and is activated when the beam diameter is much smaller than the work piece thickness. In the case of the BM, the laser beam diameter is much larger than the sheet thickness and laser scan speed is slower than that in TGM. Here, the part can be made to bend either away from laser beam or towards laser beam depending on a number of factors. UM is activated when the

B.N. Fetene (✉) · U.S. Dixit (✉)
Department of Mechanical Engineering, Indian Institute of Technology Guwahati, Guwahati 781039, India
e-mail: b.negash@iitg.ernet.in

U.S. Dixit
e-mail: uday@iitg.ernet.in

S.N. Joshi and U.S. Dixit (eds.), *Lasers Based Manufacturing*,
Topics in Mining, Metallurgy and Materials Engineering,
DOI 10.1007/978-81-322-2352-8_3

beam diameter is much smaller than the sheet thickness of the work piece. Additionally, a high thermal conductivity of the sample material facilitates the activation of UM (Pretorius 2009).

The laser bending process has been investigated for many materials, such as mild steel (Thomson and Pridham 2001), stainless steels (Liu et al. 2009; Vásquez-Ojeda and Ramos-Grez 2009), aluminum and its alloys (Merklei et al. 2001; Labeas 2008; Smith et al. 2012), titanium and its alloys (Walczyk et al. 2000) and silicon (Wang et al. 2011). These studies considered large sized sheets. The geometry parameters like sheet thickness, length and width influence the bend angle. For example, the bending angle is approximately inversely proportional to the square of the sheet thickness (Geiger and Vollertsen 1993). The length of the work piece has little influence on the bend angle (Shichun and Jinsong 2001). However, in the bending of small components, all geometric dimensions will have profound influence on the bend angle. The direction of bend is also influenced by a number of parameters. Li and Yao (2001) suggested that direction of bending can be assessed by evaluating the Fourier number given by

$$F_0 = \frac{\alpha_d D}{s^2 v}, \tag{1}$$

where α_d is the thermal diffusivity, D is the diameter of beam at the surface of the sheet, s is the sheet thickness and v is the scan speed.

Computational and experimental investigations of laser forming have been extensively carried out during the recent years. The majority of computational models for laser forming use the finite element method (FEM) and finite difference method (FDM) for achieving highly accurate laser forming predictions. Geiger and Vollertsen (1993) first modeled the laser bending by using both the FDM and FEM. Ji and Wu (1998) mainly dealt with FEM simulation of the temperature field during the laser forming of the sheet metal. Hu et al. (2001) carried out computer simulation and experimental investigation of sheet metal bending using laser beam scanning. Zhanga et al. (2004) investigated the finite element modeling discretization requirements for the laser forming process. These studies also considered large sized sheets.

The objective of this work is to evaluate the effect of laser power, work piece geometry and laser diameter on the bending angle of the small sized sheets for the stationery and the moving heat sources. Simulations of the 3D laser bending were performed by using ABAQUS package in order to investigate the influence of laser forming with variation of laser parameters on small sized work piece.

2 Finite Element Modeling Using ABAQUS

Figure 1 shows the schematic diagram of laser bending. Numerical simulations are carried out using ABAQUS package in order to analyze the variations in the bend angle and temperature by applying stationery and moving heat sources with different laser parameters. In this work, a coupled thermo-mechanical analysis is carried out using ABAQUAS/Explicit. In ABAQUAS/Explicit, the heat transfer equations are integrated using an explicit forward-difference time integration rule and the mechanical solution response is obtained using an explicit center-difference integration rule. The strategies for time control are fixed time incrementation. Nonlinear system is solved using full Newton solution technique.

A laser beam scans the top surface parallel to the fixed side with a constant laser power and a constant scanning speed as shown in Fig. 1. D36 shipbuilding steel specimens of size 25 mm × 20 mm × 2 mm and 15 mm × 10 mm × 2 mm were considered. Figure 2a and b shows the top views of work-pieces indicating the locations of moving and stationary heat sources. For each work-piece, 5 mm length is used for clamping at one side of the work piece parallel to scanning direction.

Heat convection is modeled by Newton's law given by

$$q = h(T_s - T), \tag{2}$$

where q is the heat flux, h is the heat transfer coefficient, T_s is the sheet metal temperature and T is the ambient temperature, taken as 20 °C. Radiation is neglected. As shown in Fig. 3, clamped surfaces are considered to be insulated and h = 10 W m^{-2} K^{-1} is used for the exposed surfaces.

The non-linear transient thermo-mechanical analysis is carried out. The temperature dependent thermal property and mechanical properties of the D36 shipbuilding steel are taken from Dixit et al. (2013). During the simulation, the thermal load is given in the form of heat flux that obeys a normal Gaussian distribution as follows:

$$q(x, y) = \frac{c\eta P}{\pi r^2} \exp\left(-c\frac{(x^2 + y^2)}{r^2}\right), \tag{3}$$

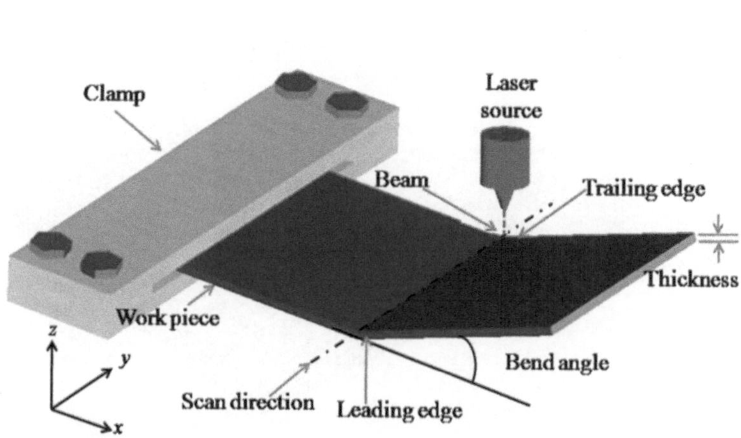

Fig. 1 Schematic of laser bending

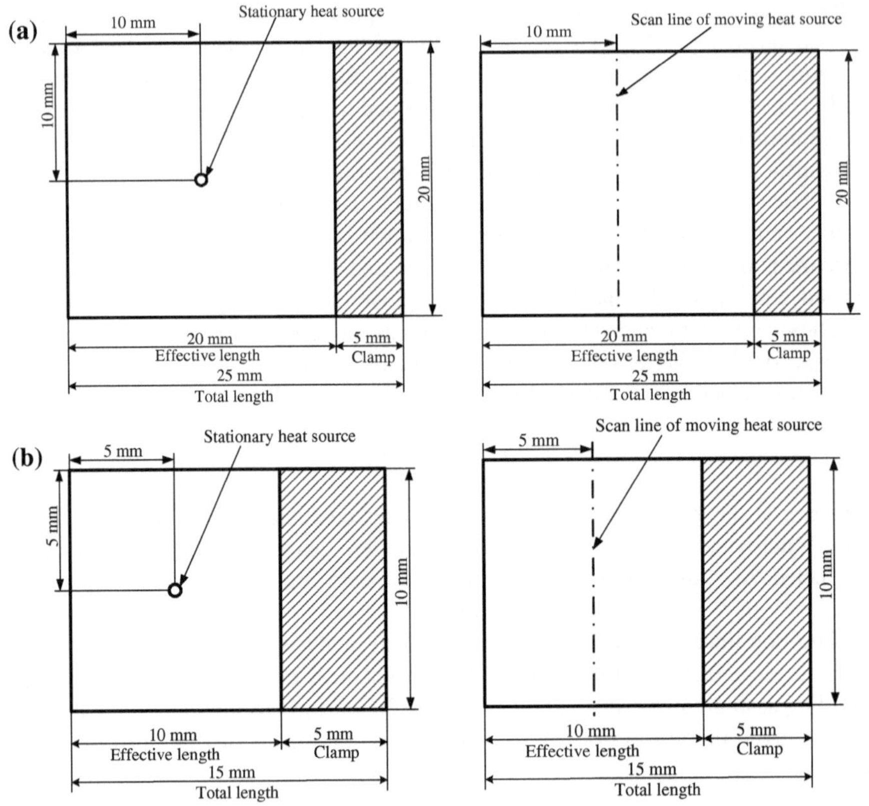

Fig. 2 The work-pieces for movable and stationary heat source **a** 25 mm × 20 mm × 2 mm sheet. **b** 15 mm × 10 mm × 2 mm sheet

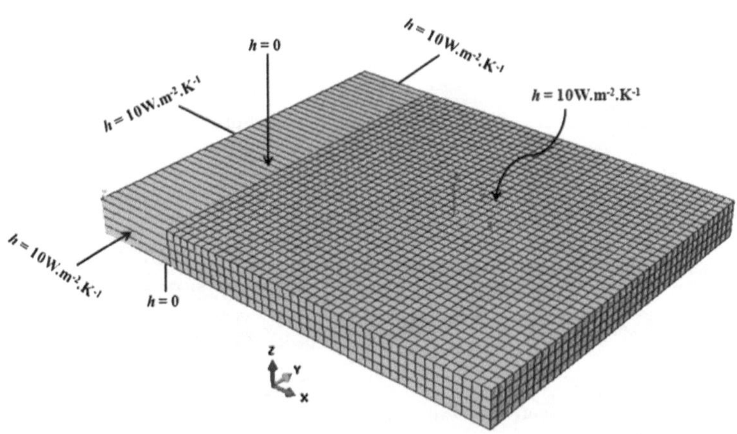

Fig. 3 Assignment of heat transfer coefficient

Table 1 Mesh sensitivity study

Type of heat source: stationary
Work piece size: 25 mm × 20 mm × 2 mm
Laser power: 225 W
Laser diameter: 8 mm

Element size	Number of elements	Simulation time	Bend angle
1 mm × 1 mm × 1 mm	1000	2 min	–0.016
0.5 mm × 0.5 mm × 1 mm	3600	10 min	–0.1
0.5 mm × 0.5 mm × 0.5 mm	6560	20 min	–0.2
0.5 mm × 0.5 mm × 0.25 mm	15,360	1 h	–0.218
0.25 mm × 0.25 mm × 1 mm	13,600	45 min	–0.108
0.15 mm × 0.15 mm × 1 mm	36,708	2 h	–0.115

Table 2 FEM models for work piece of different geometry size

Case	Geometry of work piece	Element size	Total no. elements
1	25 mm × 20 mm × 2 mm	0.5 mm × 0.5 mm × 0.5 mm	6560
2	15 mm × 10 mm × 2 mm	0.5 mm × 0.5 mm × 0.5 mm	1760

where q is the thermal heat flux density of laser beam, η is the absorptivity of the sheet material $\eta = 0.8$, P is the power of the laser (100–250 W), r is the radius of the laser beam and c is constant value between 1 and 3. For this work c has been taken equal to 2.

Moving heat source along a straight line is modeled as a moving heat flux with small steps. The heating time for every single load step is 0.05 s for both the work pieces. The jumping of consecutive heat flux node is kept to a minimum of 1 mm for both 4 and 8 mm laser diameters. The total heating time depends on the scanning speed and the width of the work piece. The same scanning speed 20 mm/s is used for both the work pieces. The stationery heat source is modeled as a stationery heat flux at the center of the effective length.

3D thermally coupled brick elements (C3D8RT) with 8 nodes are used. As illustrated in Table 1, mesh sensitivity study is carried out by varying the number of elements and comparing the simulation time. After mesh sensitivity study, the element of size 0.5 mm × 0.5 mm × 0.5 mm is chosen throughout for all the cases (Table 2).

3 Effect of Parameters on Bending Angle

The behavior of material components under this process is influenced by a specific combination of laser process parameters, geometry and material parameters. The bending angle is calculated at the edge after the steady state has reached. The steady state is arrived after cooling for 15 s.

3.1 *Effect of Laser Power*

The effects of laser power on the bending of 25 mm × 20 mm × 2 mm work piece size is observed for stationery and moving heat sources. The scanning speed is 20 mm/s and heating time is 1 s. For one case, the laser diameter is 8 mm and for the other case it is 4 mm. The simulation result reveals that the bend angle is critically dependent on the laser power. As seen in Fig. 4a for 8 mm beam diameter, the bending direction is negative (away from laser source) from 100–200 W powers, but for more than 200 W powers, the bending direction is positive during stationary heat source. In the case of moving heat source, the bending direction is negative (away from laser source) even though the bend angle is very small for 100–250 W laser powers.

As shown in Fig. 4b, at stationary heat source of 4 mm beam diameter, the bend direction is negative at 100 W, after which it tends to positive direction increasing up to 250 W laser powers. However in the case of scanning laser, the bending angle is negative up to 200 W, but starts increasing beyond 200 W towards positive direction.

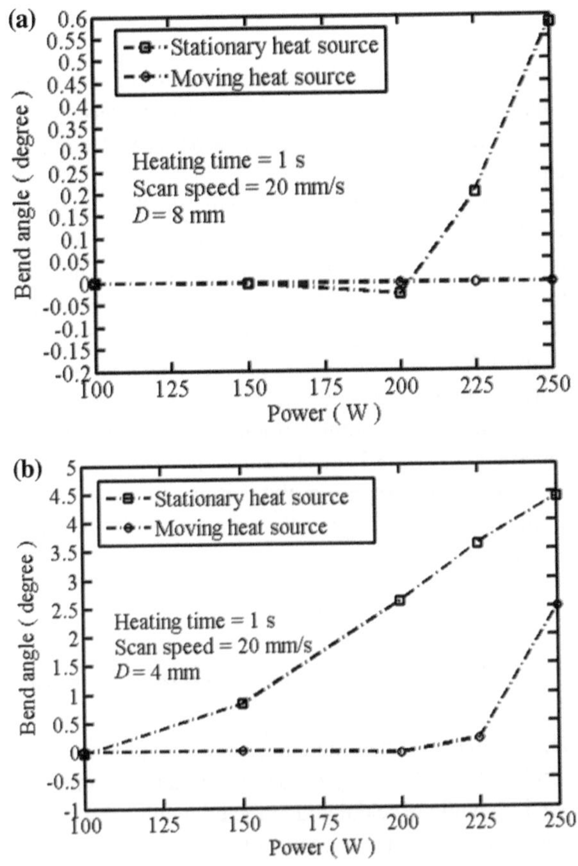

Fig. 4 Computational results for the effects of laser power on the bending angle by applying stationary and moving heat source on 25 mm × 20 mm × 2 mm work piece **a** 8 mm beam diameter. **b** 4 mm beam diameter

The similar study is carried out for 15 mm × 10 mm × 2 mm work piece. Here, for 8 mm laser beam diameter, the bend angle is very small and the bending is away from the heat source for both stationary and moving heat source as shown in Fig. 5a. The trend is different for 4 mm laser beam diameter as shown in Fig. 5b. The bending direction is negative from 100–150 W powers, but starts increasing beyond 150 W towards positive direction during stationary heat source. In the case of moving heat source, the bending direction is negative for 100–250 W laser powers.

In all the cases, the bend angle is greater in the case of stationary heat source. However, in the case of stationary heat source, the bend angle is not uniform across the width as shown in Fig. 6.

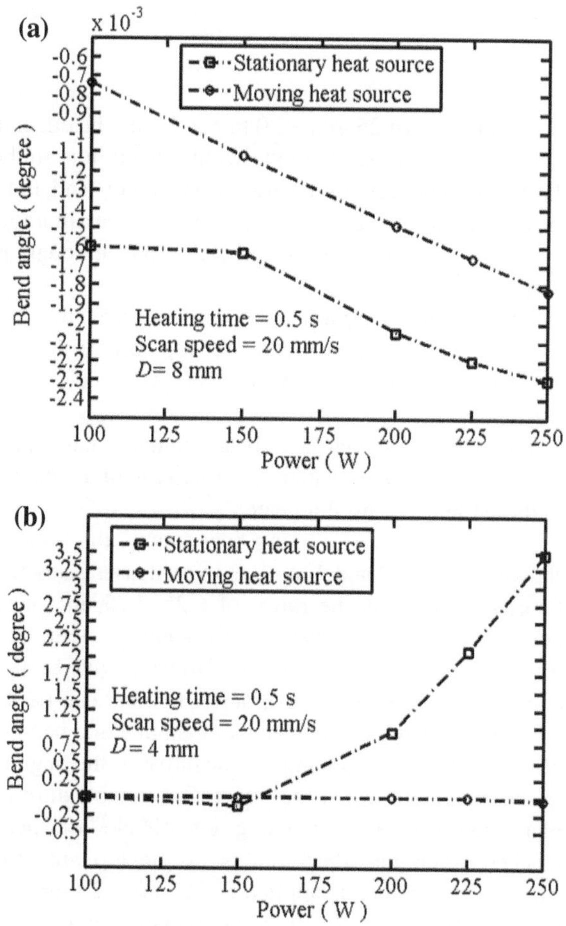

Fig. 5 Computational results for the effects of laser power on the bending angle by applying stationary and moving heat source on 15 mm × 10 mm × 2 mm work piece **a** 8 mm laser diameter. **b** 4 mm laser diameter

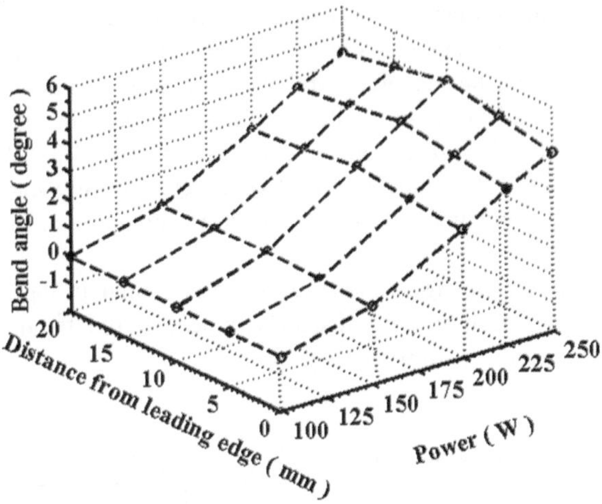

Fig. 6 The effects of stationary heat source on bending angle along width direction by using 4 mm laser diameter on 25 mm × 20 mm × 2 mm work piece

For the case of 25 mm × 20 mm × 2 mm, the effect of change in the power on the bend angle variation along the width direction is studied. For 250 W laser powers, the bend angle variation between stationary and moving heat source is approximately 2 degrees and the bend angle is the maximum in the middle (Fig. 7a). This is not the case for 100 W moving heat source, although the bend angles are very small (Fig. 7b).

Figure 8 shows the top and bottom surface temperatures versus laser power. Rate of increase of power is greater for top surface than for bottom surface. The maximum temperature at 250 W is 1200 °C, which is below the melting point of the work piece.

Figure 9 shows that laser beam diameter has a profound effect on temperature. At small laser beam diameter, the maximum temperature is much higher compared to that at large beam diameter.

In all the cases of moving heat source, the Fourier number is less than one as shown in Tables 3 and 4 for 4 and 8 mm laser diameter. Li and Yao (2001) obtained Fourier numbers in the range of 6.25–7.75, in which the smaller values provide positive (towards the laser beam) bending. Here, also the small Fourier numbers tend to provide positive bending. However, due to want of enough data, no definite conclusions with respect to Fourier number can be drawn at this stage.

Figure 10a and b shows the temperature and bend angle variation with time. It is observed that the bend angle is negative in the beginning and becomes the highest in the magnitude when the top surface temperature is the highest. During cooling phase, the sheet starts bending towards the laser beam source. The steady state is reached when the temperatures of the top and bottom surface become equal. Moving heat source is applied at 25 mm × 20 mm × 2 mm size work piece at the power 250 W and 8 mm laser diameter, it is observed that the bend angle is negative. For stationary heat source, the bend angle is positive.

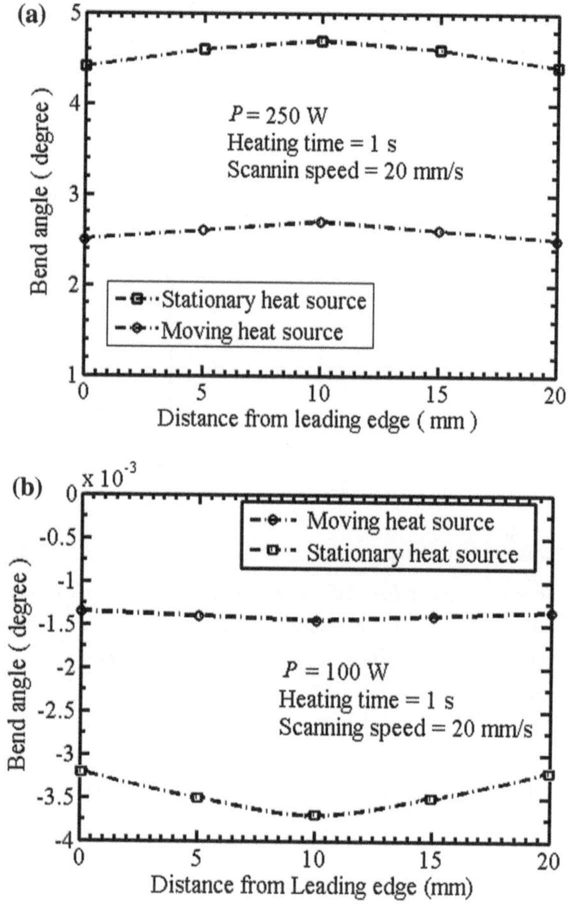

Fig. 7 Effect of laser power on bend angle variation for 4 mm beam diameter and 25 mm × 20 mm × 2 mm work piece **a** 250 W power. **b** 100 W power

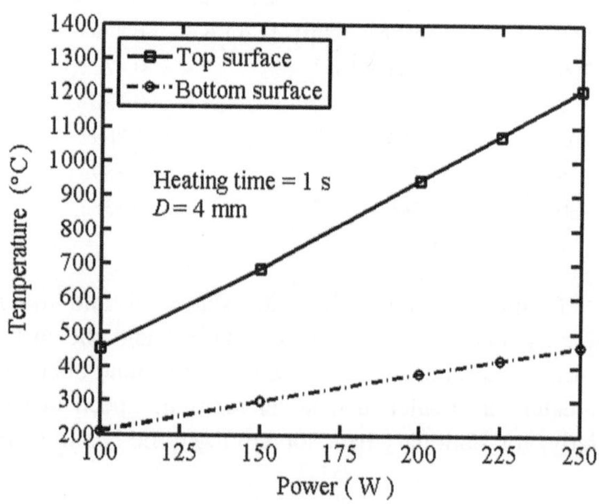

Fig. 8 Variation of *top* and *bottom* surface temperature with laser power on 25 mm × 20 mm × 2 mm work piece

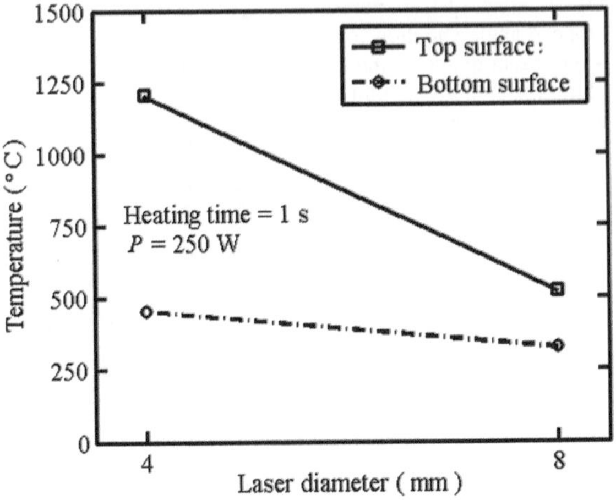

Fig. 9 Variation of *top* and *bottom* surface temperature with beam diameter on 25 mm × 20 mm × 2 mm work piece

Table 3 Calculated Fourier number 25 mm × 20 mm × 2 mm work piece at 4 mm laser diameter

Scanning speed: 20×10^{-3} m/s and thickness of work piece: 2×10^{-3} m			
Power (W)	Thermal diffusivity (m^2/s)	Fourier number	Bending direction
100	9.404×10^{-6}	0.9404	Away from laser direction
150	9.404×10^{-6}	0.9404	Away from laser direction
200	9.404×10^{-6}	0.9404	Away from laser direction
225	9.404×10^{-6}	0.9404	Toward the laser beam
250	8.655×10^{-6}	0.8654	Toward the laser beam

Table 4 Calculated Fourier number for 25 mm × 20 mm × 2 mm at 8 mm laser diameter

Scanning speed: 20×10^{-3} m/s and thickness of work piece: 2×10^{-3} m			
Power (W)	Thermal diffusivity (m^2/s)	Fourier number	Bending direction
100	9.404×10^{-6}	0.4702	Away from laser direction
150	8.655×10^{-6}	0.4327	Away from laser direction
200	7.784×10^{-6}	0.3892	Away from laser direction
225	7.784×10^{-6}	0.3892	Away from laser direction
250	6.766×10^{-6}	0.3383	Away from laser direction

Figure 11a and b show the stationary and moving heat source temperatures versus time at a point on top surface and the just opposite point at the bottom surface. Laser scanning speed has a profound effect on temperature. Temperature is greater for smaller than larger scanning speed in moving heat source case. Similarly, for stationary heat source, more the heating time, greater is the temperature.

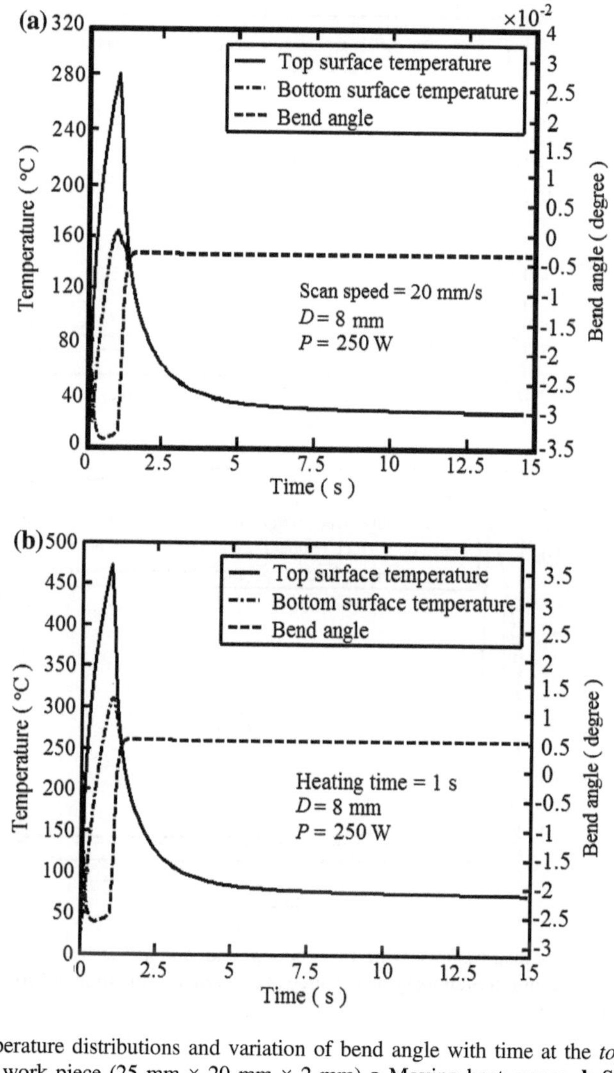

Fig. 10 Temperature distributions and variation of bend angle with time at the *top* and *bottom* surface of the work piece (25 mm × 20 mm × 2 mm) **a** Moving heat source. **b** Stationary heat source

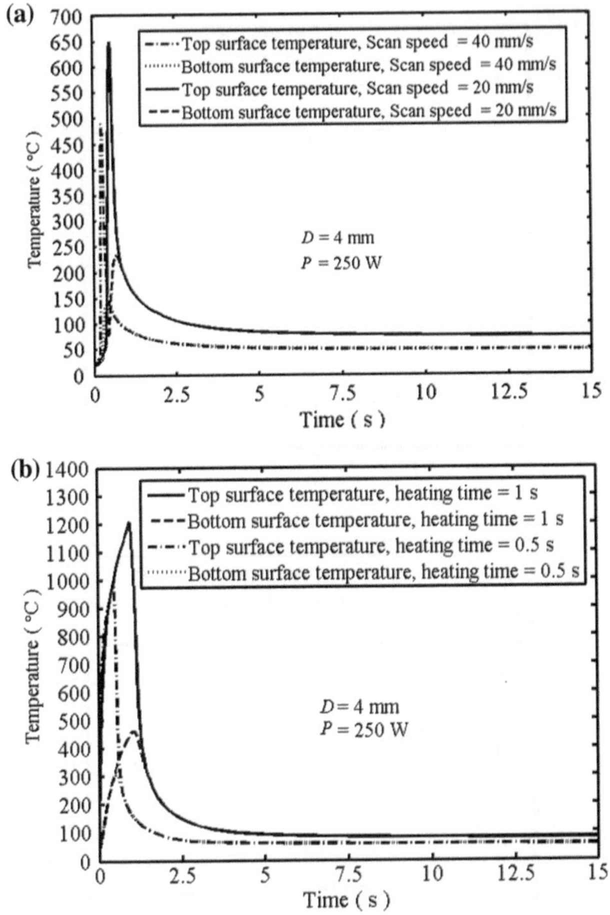

Fig. 11 Temperature distributions with time at *top* and *bottom* surface of the work piece (25 mm × 20 mm × 2 mm) **a** Stationary heat source at different heating time. **b** Moving heat source at different scan speed

4 Conclusions

Following conclusions can be drawn:

- The bending angle is highly dependent on laser power; as the power increases the bend angle also increases for both stationary and moving heat sources.
- The bending angle is highly dependent on type of heat sources, the stationary heat source provides more bend.
- Changing laser beam diameter from 4 to 8 mm changes the bend angle and it is greater for small diameter.

- The pattern of bend angle variation along width direction is different for laser power of 100 and 250 W.
- With stationary heat source, the bend angle is non-uniform across the width.
- The temperature difference in the thickness direction is increased with increase of the laser power (100–250 W).
- Laser scanning speed has a profound effect on temperature.

References

Dixit, U. S., Joshi, S. N., & Kumar, V. H. (2013). Microbending with lasers. In V. K. Jain (Ed.), *Micromanufacturing processes*. Boca Raton: CRC Press.

Geiger, M., & Vollertsen, F. (1993). The mechanisms of laser forming. *CIRP Annals Manufacturing Technology, 42*, 301–304.

Hu, Z., Labudovic, M., Wang, H., & Kovacevic, R. (2001). Computer simulation and experimental investigation of sheet metal bending using laser beam scanning. *International Journal of Machine Tools and Manufacture, 41*, 589–607.

Ji, Z., & Wu, S. (1998). FEM simulation of the temperature field during the laser forming of sheet metal. *Journal of Materials Processing Technology, 74*, 89–95.

Labeas, G. N. (2008). Development of a local three-dimensional numerical simulation model for the laser forming process of aluminium components. *Journal of Materials Processing Technology, 207*, 248–257.

Li, W., & Yao, Y. L. (2001). Numerical and experimental investigation of convex laser forming process. *Journal of Manufacturing Processes, 3*, 73–81.

Liu, J., Sun, S., & Guan, Y. (2009). Numerical investigation on the laser bending of stainless steel foil with pre-stresses. *Journal of Materials Processing Technology, 209*, 1580–1587.

Merklei, M., Hennige, T., & Geiger, M. (2001). Laser forming of aluminum and aluminum alloys microstructural investigation. *Journal of Materials Processing Technology, 115*, 159–165.

Pretorius, T. (2009). Laser forming. In J. Dowden (Ed.), *The theory of laser materials processing* (Vol. 119, pp. 281–314). UK: Springer.

Shichun, W., & Jinsong, Z. (2001). An experimental study of laser bending for sheet metals. *Journal of Materials Processing Technology, 110*, 160–163.

Smith, T.M., Michaleris, P., Reutzel, E.W., & Hall, B. (2012). Finite element model of pulsed laser forming. In *the 13th International Symposium on Laser Precision Microfabrication.*

Thomson, G., & Pridham, M. (2001). Material property changes associated with laser forming of mild steel components. *Journal of Materials Processing Technology, 118*, 40–44.

Vásquez-Ojeda, C., & Ramos-Grez, J. (2009). Bending of stainless steel thin sheets by a raster scanned low power CO_2 laser. *Journal of Materials Processing Technology, 209*, 2641–2647.

Walczyk, D. F., Vittal, S., & York, N. (2000). Bending of titanium sheet using laser forming. *Journal of Manufacturing Processes, 2*, 258–269.

Wang, X., Xu, W. X., Xu, W. J., Hu, Y. F., Liang, Y. D., & Wang, L. J. (2011). Simulation and prediction in laser bending of silicon sheet. *Transactions of Nonferrous Metals Society of China, 21*, s188–s193.

Zhanga, L., Reutzelb, E. W., & Michalerisc, P. (2004). Finite element modeling discretization requirements for the laser forming process. *International Journal of Mechanical Sciences, 46*, 623–637.

Numerical and Experimental Studies on Pulsed Laser Forming of Sheet Metal

Kuntal Maji, D.K. Pratihar and A.K. Nath

Abstract Pulsed laser forming is a non-contact thermal forming process, where sheet metal gets plastically deformed by thermal residual stresses induced by controlled discontinuous laser irradiations. The temperature and deformation fields have been determined using finite element analysis under different processing conditions. Two types of pulsed laser forming processes, i.e., overlapped and discrete spot forming have been identified depending on the combinations of process parameters. Bending angle is found to increase with the degree of overlap and decrease with the increase of gap in case of the two types of spot forming processes. A comparative study between pulsed and continuous laser forming has also been performed using both finite element simulations and experiments. Bending angle in case of discrete spot pulsed laser forming is found to be more compared to the continuous laser forming. The results of finite element simulations have been found to be in good agreement with the experimental results.

Keywords Pulsed laser forming · Finite element analysis · Experimental study

1 Introduction

Laser forming (LF) is a non-traditional forming process for flexible shaping of sheet metal components by laser induced thermal stress instead of applying any external mechanical force. In this process, thermal stresses can be induced by a controlled

K. Maji · D.K. Pratihar (✉) · A.K. Nath
Department of Mechanical Engineering, Indian Institute of Technology, Kharagpur, Kharagpur 721302, India
e-mail: dkpra@mech.iitkgp.ernet.in

K. Maji
e-mail: kuntalmajiiitkgp@gmail.com

A.K. Nath
e-mail: aknath@mech.iitkgp.ernet.in

S.N. Joshi and U.S. Dixit (eds.), *Lasers Based Manufacturing*, Topics in Mining, Metallurgy and Materials Engineering,
DOI 10.1007/978-81-322-2352-8_4

laser beam either in continuous or pulsed modes of irradiations. It has potential applications in macro and micro scale manufacturing for rapid prototyping and precision adjustment in ship building, aerospace, automobile, microelectronics industries (Steen and Mazumder 2010). It is a complex thermo-elasto-plastic process and depends on interaction of a large number of process parameters related to laser irradiation, thermo-physical properties of workpiece material, sheet geometry, and others. Depending on the combination of process parameters, there are mainly three types of mechanisms, namely temperature gradient mechanism (TGM), buckling mechanism (BM) and upsetting mechanism (UM) (Steen and Mazumder 2010). TGM is the most commonly used mechanism in laser metal forming, where due to rapid localized heating of sheet metal, a steep temperature gradient is produced along the thickness direction, and after the cooling, the metal finally gets bent towards the laser beam due to compressive plastic strain at the top surface. The LF process, in its present stage, is far from automation and it has limited applications in small series production, rapid prototyping, shape correction and precision adjustments in sheet metal industries. Automation of the process requires integrations of a lot of software and hardware components, development of predictive models and control scheme for both continuous and pulsed laser forming processes. Deformation behavior and process parameters of pulsed laser forming are different from that of the continuous laser forming. The studies on pulsed laser forming of sheet metal carried out by different researchers are discussed below.

2 Literature Review

Extensive studies had been carried out by various researchers (Shen and Vollertsen 2009) to investigate the continuous laser bending process. Sheet metal forming using pulsed mode of laser irradiation was also investigated by some researchers to study the effects of different process parameters on the deformation and properties of the laser formed samples. Numerical and experimental investigations were carried out by various investigators (Chen et al. 1999; Lee and Lin 2002; Zhang et al. 2002; Hseih and Lin 2004) to determine the temperature and deformation fields, and others, in pulsed laser forming considering different process parameters, i.e., laser parameters, laser beam shapes, single and multiple pulses etc. Good correlations were found between the numerical and experimental results of deformation in their studies. Transient nonlinear 3D finite element (FE) simulation of pulsed laser forming process is computationally expensive because of finer mesh sizes at the laser irradiated zone and the small time steps required for the convergence and good accuracy. Zhang et al. (2002) proposed an efficient method to reduce the computational time in determining bending angle using FE method in pulsed laser bending of thin stainless steel sheet. Experimental investigations and empirical modeling were carried out by Gollo et al. (2008) in laser bending of sheet metal with a Nd: YAG pulsed laser. Taguchi experimental design was used and regression analysis was performed by considering the various process parameters like laser power, beam diameter, scan speed and pulse duration to express the bending angle in terms of process parameters. Yang et al. (2010) investigated the metallurgical changes of surface properties of stainless steel due to pulsed laser forming. The effects of different process parameters on the surface properties of heat affected zone (HAZ) like microstructure, micro-hardness etc. were studied with the metallographic microscope, scanning electron microscope (SEM) and micro-hardness tester, respectively. Gollo et al. (2011) investigated further the effects of different process parameters, such as laser power, beam diameter, scan speed, sheet thickness, number of passes and pulse duration on bending angle using FE simulations and experiments. Significant process parameters were identified through Taguchi experimental design method and regression analysis was also performed to predict the bending angle in pulsed laser forming process. However, a few studies had been carried out on the different types of processing conditions and their effects on deformation in pulsed laser forming process. Moreover, much less attention had been given on a comparative study of continuous and pulsed laser forming processes for energy efficiency. Recently, Maji et al. (2012, 2013) have performed empirical modeling of pulsed laser bending process using statistical and soft computing-based methods to predict bending angle and studied the effects of different process parameters. However, transient temperature and deformation fields determined through FE analysis could give better insight of the process.

This chapter deals with the studies related to the effects of different types of pulsed laser forming, i.e., effects of gap and overlapping on deformation in discrete spots and overlapped spots pulsed laser forming. The efficiency of continuous and pulsed laser forming processes is also investigated in terms of energy requirement and produced deformation. The different process parameters and processing conditions for their various combinations are discussed in the following section.

3 Theory of Pulsed Laser Forming

In pulsed laser forming process, the workpiece is actually heated by a modulated power laser beam in a discontinuous manner, forming a series of discontinuous or discrete, continuous and overlapping spots as shown schematically in Fig. 1. During pulsed laser forming process, the laser spot becomes elliptical in the scan direction (Tzeng 2000; Yang 2010). Depending on the combinations of process parameters like scan speed (v), spot diameter (*d*), frequency (*f*) and duty cycle (C_d), different

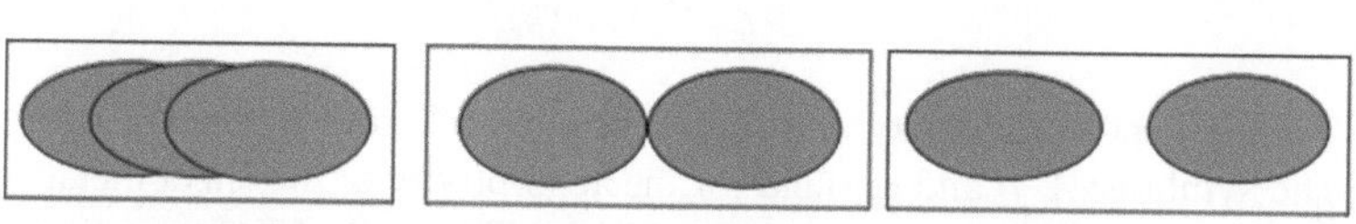

Fig. 1 Different types of spots (overlapped, continuous and discrete) formed in pulsed laser forming process

types of spots, i.e., overlapped, continuous and discontinuous or discrete may be formed as shown in Fig. 1. The major diameter of the elliptical spot can be determined as $d_m = d + v \times \tau$, where τ denotes pulse duration; and the distance moved by the spot during a cycle time (T_c) can be calculated as $d_c = v \times T_c$. For forming overlapping spots, d_m has to be greater than d_c ($d_m \geq d_c$). The degree or percent of overlapping can be calculated as $\left(\frac{d_m - d_c}{d_m}\right) \times 100\,\%$. Continuous spots are formed, when d_m is set equal to d_c.

Similarly, in case of discrete spots forming, the percent of gap can also be calculated as $\left(\frac{d_c - d_m}{d_m}\right) \times 100\,\%$. The frequency ($f$ = 50 Hz), duty cycle (C_d = 50 %), pulse duration (τ = 10 ms), laser power and beam diameter (d = 1.0 mm) have been kept constant during the study. Both finite element simulations and experiments have been performed to study the effects of gap between spots and their overlapping on bending angle. Finite element formulation for the pulsed laser forming and the details are stated as given in the following section.

4 Finite Element Formulation

Numerical simulation of pulsed laser forming process has been carried out to determine temperature distribution and deformations. Laser bending or forming is a weakly-coupled thermo-mechanical process. A nonlinear transient indirect coupled field analysis has been performed using the ANSYS APDL (2011). The simulation process consists of a few steps. The first step involves model generation and selection of input process parameters. The next step deals with the thermal and structural analyses. The thermal equilibrium equation for heat transfer analysis of an isotropic material can be written as given in Eq. (1).

$$k\left(\frac{\partial^2 T}{\partial x^2} + \frac{\partial^2 T}{\partial y^2} + \frac{\partial^2 T}{\partial z^2}\right) + \dot{q} = \rho c \dot{T}, \tag{1}$$

where ρ, c and k are the density, specific heat and thermal conductivity of the material, respectively, and $\dot{q}$ is the rate of heat generation per unit volume of the material. The heat flux density of moving laser beam (Fiber laser) is assumed to obey the normal distribution along radial direction as given in Eq. (2).

$$I = \frac{2AP}{\pi r_b^2} \exp\left(-\frac{2r^2}{r_b^2}\right), \tag{2}$$

where I is the heat flux density at a radial distance 'r' from the center of the laser beam. The symbols: A, P and r_b stand for the absorptivity of the sheet metal surface, laser power and laser beam radius, respectively. The basic FEM equation for the indirectly-coupled thermo-mechanical calculation can be written, as given in Eq. (3).

$$\begin{bmatrix} [0] & [0] \\ [0] & [C] \end{bmatrix} \begin{Bmatrix} \{\dot{u}\} \\ \{\dot{T}\} \end{Bmatrix} + \begin{bmatrix} [K] & [0] \\ [0] & [K_T] \end{bmatrix} \begin{Bmatrix} \{u\} \\ \{T\} \end{Bmatrix} = \begin{Bmatrix} \{F\} + \{F_{th}\} \\ \{Q\} \end{Bmatrix}, \quad (3)$$

where $[C] = \int_V \rho c[N][N]^T dV$ is the heat capacity matrix; $[K] = \int_V [B]^T [D^{ep}][B] dV$ is the tangential stiffness matrix; $[K_T] = \int_V k[B][B]^T dV$ represents the heat conduction matrix; $\{u\}$ and $\{T\}$ are the nodal displacement and temperature vector, respectively; $[N]$ and $[B]$ denote the shape function matrix and the general geometric matrix, respectively; $\lfloor D^{ep} \rfloor$ is the elasto-plastic stress-strain matrix; F and F_{th} are the applied nodal force and the force caused by the thermal strain, respectively; $\{Q\}$ is the heat flux vector. Solid 70 and solid 185 elements (ANSYS theory manual 2011) have been used for the thermal and structural analysis, respectively. The following assumptions have been made for the FE analysis of pulsed laser bending process.

(i) The workpiece material has been assumed to be isotropic, and considered to be flat and free of residual stresses.
(ii) The laser beam energy follows a Gaussian distribution.
(iii) Melting of the workpiece surface has been neglected and no external force has been applied to the sheet metal.
(iv) Cooling of the laser irradiated material occurred through free convection to air, and radiation has been neglected.
(v) Temperature dependent properties (i.e., thermal conductivity, specific heat, density, Young's Modulus, Poisson's ratio etc.) have been considered and their variations have been determined by means of linear interpolation of the data given in Cheng and Lin (2000).
(vi) Von Mises Criteria have been used for plastic yielding in the simulation process.
(vii) The hardening effect of material behaviour has been considered as bilinear isotropic hardening model.
(viii) The dissipation of energy due to plastic deformation has been neglected compared to energy input from the laser beam.

Computer simulation of the laser-forming process is complex and the main steps of the analysis are the thermal and structural analyses. In these analyses, the finite element code: ANSYS has been used to calculate the thermo-structural coupled field with a large plastic deformation. An indirect coupled-field analysis (i.e., two sequential analyses) has been performed. The flow-chart of the FEM simulation is shown in Fig. 2. In this method, the results from the transient thermal analysis for nodal temperatures are applied as loads for the structural analysis (Dhondt 2004; Nicholson 2008). Therefore, the calculations related to the temperature distributions are very much essential for the determination of the stress–strain distributions. The following assumptions have been made in this study:

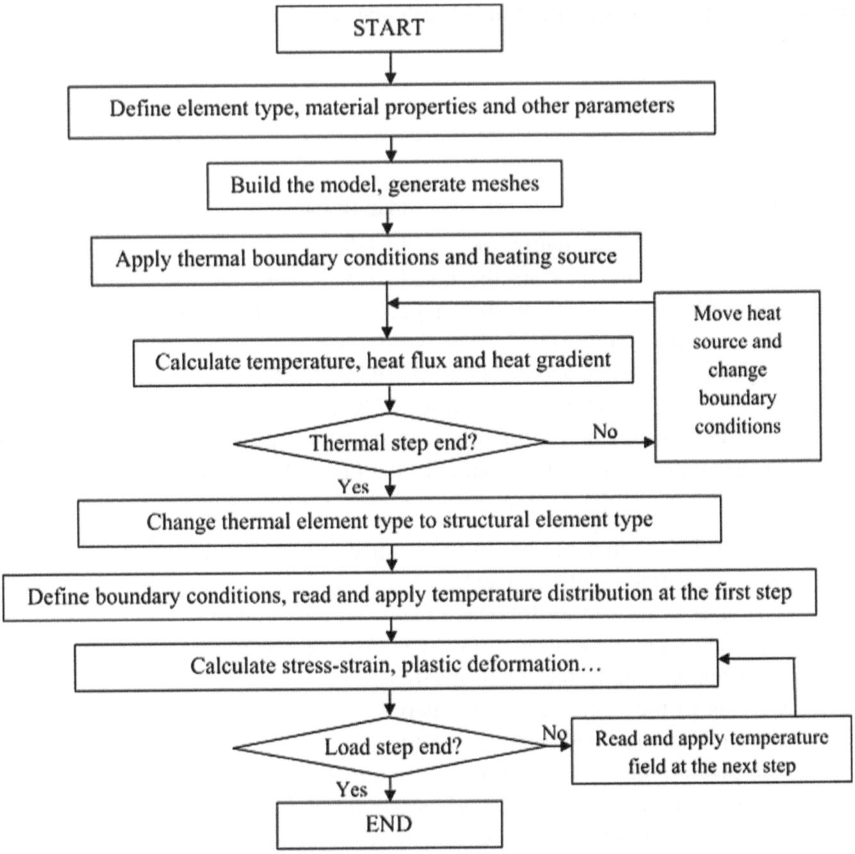

Fig. 2 Flow-chart of FEM simulation of laser forming process

(i) The workpiece material is isotropic.
(ii) Laser operates in a continuous wave mode.
(iii) There is no melting and no external forces are applied to the sheet metal.
(iv) Von Mises criterion is used as the yield criterion.
(v) Sheet metal is considered to be flat and free from any kind of residual stress.

For the thermal and structural analyses, the elements: SOLID 70 (eight-node 3D thermal solid with thermal conduction, convection, material nonlinearities) and SOLID 185 (eight-node 3D solid with extra shape functions, large elasto-plastic deformation, geometric and material nonlinearities) (ANSYS Release 12.0 Documentation 2011) have been used, respectively. An ANSYS parametric design language (APDL) has been used to model the moving laser beam. For the application of the proposed FE model, the required different parameters are the plate dimensions (length, width and thickness), laser power, laser beam diameter, scan speed, material absorption coefficient etc. The calculated outputs are mainly the transient temperature and deformation fields, and the final bending angle. The FE model results have been validated through experiments as discussed in the following sections.

Numerical simulations have been carried out for studying the effects of gap and overlap on bending angle in the two types of pulsed laser forming processes. Simulations have been conducted by taking the model size to be equal to that of workpiece excluding the clamped length, and laser irradiation has been done at the middle of the workpiece and parallel to the free edge of the sample, as shown in Fig. 3. Fine mesh size has been used in the laser irradiated zone and gradually coarse meshing has been utilized away from the irradiated zone to reduce the computational time as shown in Fig. 3.

The laser beam has been moved for a small distance at each time-step for simulating the moving pulsed laser beam. A small time-step has been taken during laser heating for convergence and gradually increasing time-steps have been considered during cooling to reduce the total computational time. For studying the effects of gap and overlap, other process parameters, i.e., laser power and spot diameter have been kept constant. AISI304 steel samples of the size of $100 \times 30 \times 0.5\ \text{mm}^3$ have been used for the study. One end of the sample has been clamped, therefore, this end has been kept fixed or the displacement of that end has been made zero as boundary condition. The bending angle has been calculated using the triangulation method from the displaced or deformed sample as obtained from the simulations. The results of the numerical simulations have been validated by performing real experiments as discussed in the following section.

5 Experimental Validations

Laser bending experiments have been conducted on a fiber laser (wavelength = 1.07 μm, Model YLR-2000) having a maximum laser power of 2.0 kW (refer to Fig. 4). A 10 mm diameter collimated laser beam is delivered through an optical fiber onto a focusing optical system consisting of a lens of 200 mm focal length. This

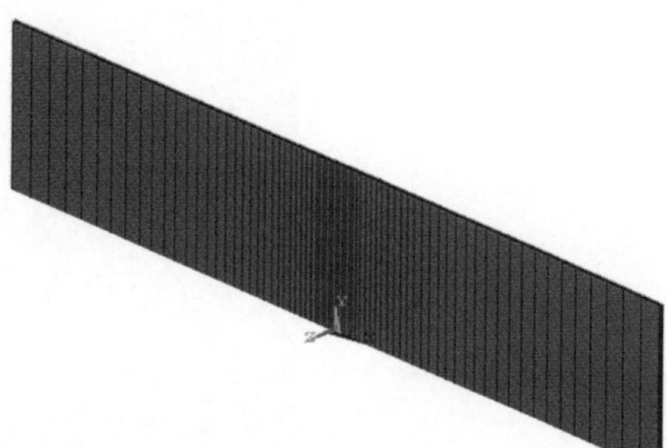

Fig. 3 Model of the sample with meshing

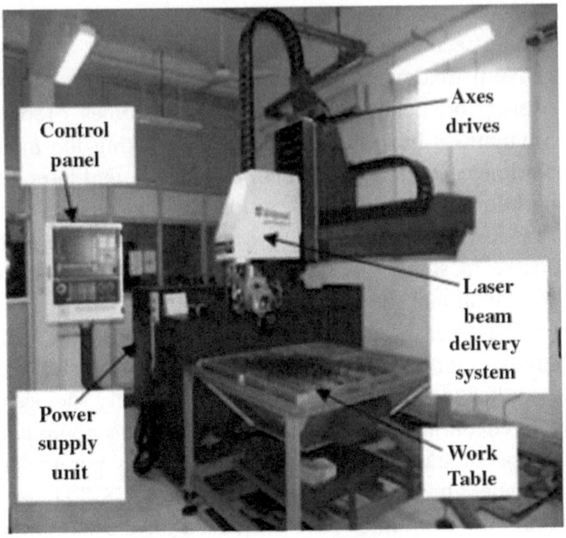

Fig. 4 2 kW fiber laser system

minimum spot diameter available is 250 µm at the focal plane, and nitrogen gas at 0.5 bar pressure has been used for shielding to protect the optical system.

The work-pieces are AISI 304 stainless steel sheets of 120 mm × 30 mm × 0.5 mm dimensions. The samples have been cleaned using acetone to remove any unwanted dirt and grease. A 10,000 W-Lp Ophir power meter has been used to measure the laser power. By measuring the incident and reflected laser power, the average absorption coefficient of the work-piece surface is estimated to be in the range of 0.35–0.45. During laser bending experiment, one end of the rectangular workpiece is held in a clamp and laser scans have been performed parallel to the free edge of the sample, as shown in Fig. 5.

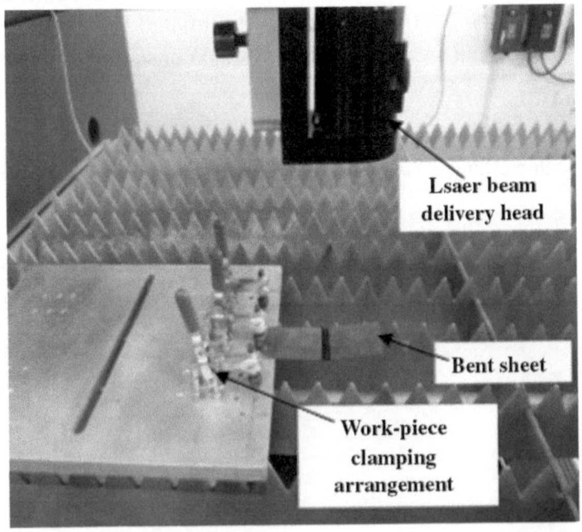

Fig. 5 Experimental setup

After laser irradiations, samples are allowed to cool in the atmosphere. After cooling, the samples have been measured using a laser displacement sensor (Make: Micro-Epsilon, Model: Opto-NCDT 1402). Measurements have been taken at 3–4 locations along the scanning direction and their average value has been calculated. Bending angle has been calculated by the triangulation method. Results have been compared with that obtained from the finite element simulations of the laser bending process as discussed in the following section.

6 Results and Discussion

Experiments and FE simulations have been performed to study the effects of gap between spots and their overlapping on bending angle in pulsed laser bending (PLB). The degrees of overlap and gap have been controlled by varying the scan speed after keeping the other parameters constant.

Figure 6 shows the temperature profile obtained during heating with a laser power of 600 W and 50 % gap. The corresponding deformation field is shown in Fig. 7.

Figure 8 shows the variations of bending angle, both experimentally obtained and FE calculated, with the degree of gap between spots, obtained at constant laser power (600 W) and laser spot diameter (1.0 mm). At constant laser power, as the

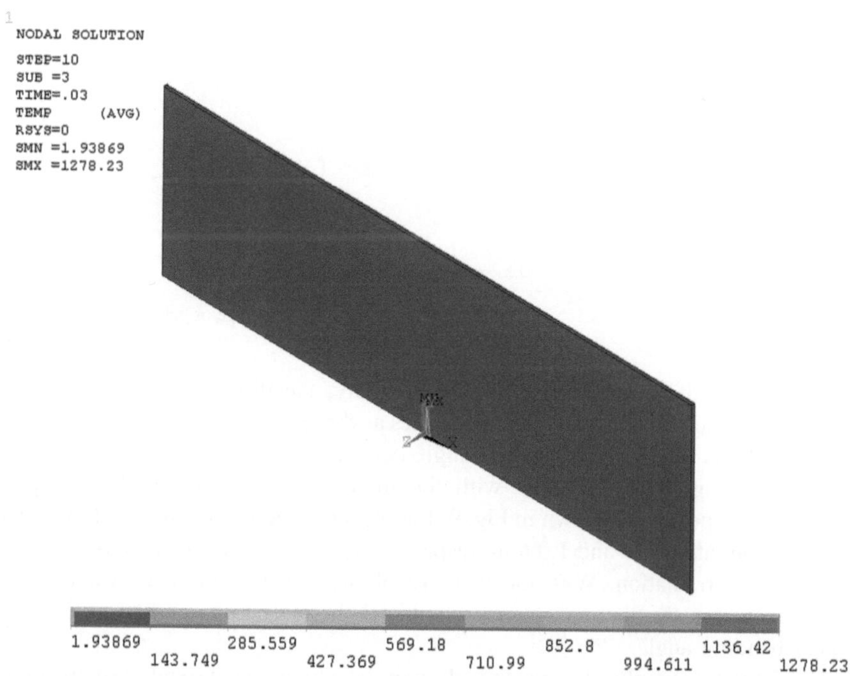

Fig. 6 Temperature profile during heating

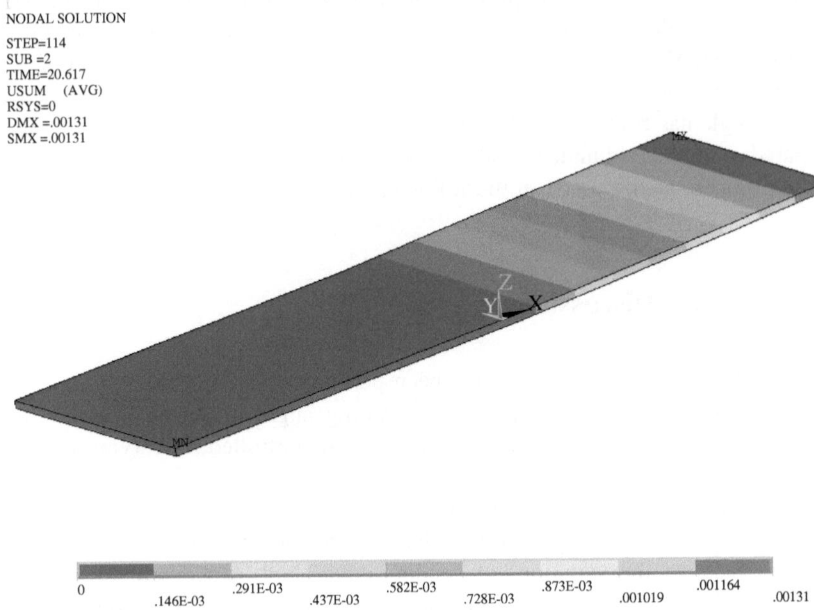

Fig. 7 Deformation field due to pulsed laser forming with 50 % gap and 600 W laser power

Fig. 8 Effect of gap on bending angle

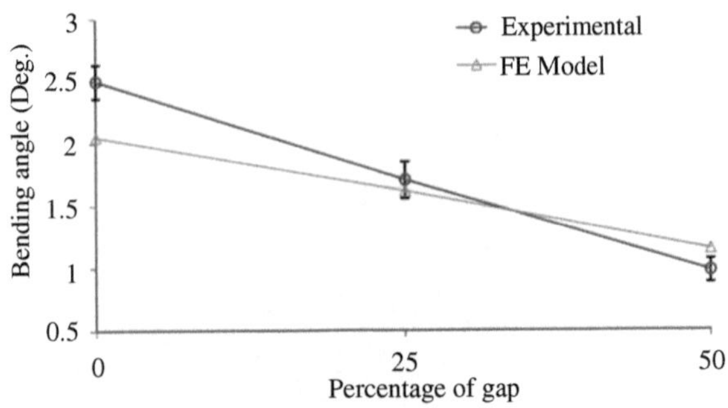

gap is increased by increasing the scan velocity, the discrete line energy input reduces, and the power density also decreases as the spot size increases at the higher scan speed. As a result, the bending angle decreases.

The bending angle increases with the increase of percentage of overlap at constant laser power, as shown in Fig. 9. Laser power and beam diameter have been kept constant at 300 W and 1.0 mm, respectively, to study the effects of overlapped pulsed laser irradiation. With the increase of overlap, the line energy increases, which increases the thermal energy input and thermal stress, and resulting in a higher bending angle.

FE simulation results also match well with the experimental trend of the bending angle with the gap and overlap (refer to Figs. 7 and 8). At the higher degree of

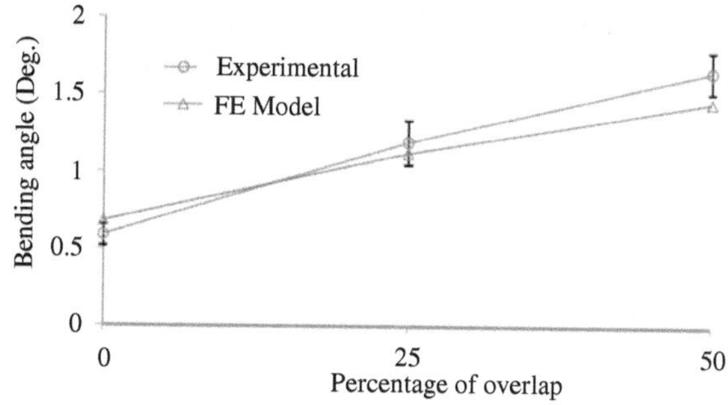

Fig. 9 Effect of overlap on bending angle

overlap, the workpiece surface melts and this effect has not been taken into account in the FE simulation to calculate the bending angle. Finite element analyses of pulsed laser irradiations have been carried out, and transient temperature and deformations have been calculated for different processing conditions. Bending angle is found to increase with the increase of overlap and decrease with the increase of gap.

FE simulation and experiments have been conducted to study the deformation and process efficiency in thermal forming of sheet metal under continuous and discrete laser heat inputs. The variation of bending angle with the line energy is plotted in Fig. 10 for both the cases. The line energy for continuous laser bending is given by $LE_c = p/v$, and that for pulsed laser bending is calculated as follows:

$$LE_p = \frac{p \times \tau}{v \times T} = \left(\frac{p}{v}\right) \times \left(\frac{\tau}{T}\right) = LE_c \times C_d, \tag{4}$$

where the symbols carry usual meaning as stated above. The bending angle obtained in case of pulsed laser forming is found to be more compared to that in continuous laser forming for the same line energy as shown in Fig. 10. The larger bending angle in case of discrete spots pulsed laser bending (gap between two spots has been kept 25 %) can be attributed to the fact that the discontinuous thermal energy input produces more thermal stress due to the higher resistance imposed by the cold material, which is not getting laser irradiation during laser pulse-off time. The rise in surface temperature at the irradiated spot is also expected to be high in

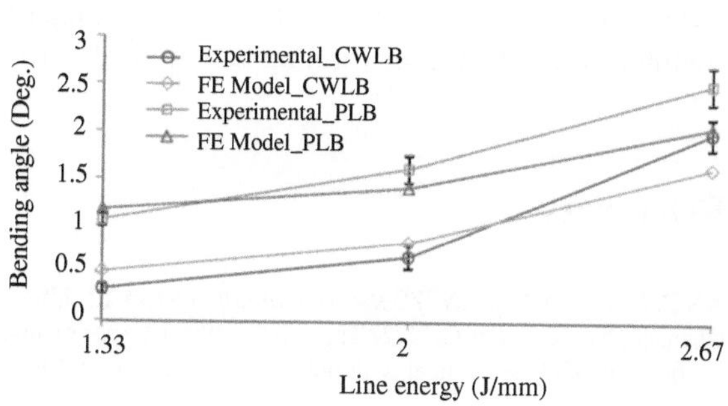

Fig. 10 Comparison of pulsed and continuous laser forming

case of pulsed laser forming compared to that of continuous laser forming at constant line energy (McBride et al. 2005). This could also be another reason for producing more bending in case of pulsed laser bending process.

Therefore, pulsed laser forming is found to be more energy efficient compared to continuous mode of laser forming. In future, a comparative study will be made for more number of processing conditions and for 3D forming also. The conclusions from the present study are summarized in the following section.

7 Summary

Finite element analyses of laser bending process with pulsed laser irradiations have been carried out. Transient temperature and deformations have been calculated for different processing conditions. The effects of different processing conditions on deformation have also been studied. Deformation in pulsed laser forming has been found to decrease and increase with the increase of gap and that of overlap between spots, respectively. A comparative study of continuous wave (CW) and pulsed laser bending processes has also been done through FE simulations and experiments. Discrete spot pulsed laser forming has been found to be more energy efficient compared to the continuous laser forming. Results of finite element simulations have been validated through experimental results and these are found to be satisfactory. The present study can be extended in different ways as mentioned in the scope for future study.

8 Scope for Future Study

Numerical simulations of laser bending process with both CW and pulsed laser irradiations have been carried out and compared. However, a detailed study on the effects of different other process parameters like frequency, duty cycle etc. can be investigated. A comparative study of CW and pulsed laser bending processes has been done through FE simulations and experiments. Discrete spot pulsed laser forming has been found to be more energy efficient compared to the continuous laser forming. A more detailed study on the comparisons of pulsed mode with the continuous mode of laser forming can be made in future for forming 3D shapes.

References

ANSYS Inc. (2011). *ANSYS theory manual release* 12, USA.

Chen, G., Xu, X., Poon, C. C., & Tam, A. C. (1999). Experimental and numerical studies on microscale bending of stainless steel with pulsed laser. *Journal of Applied Mechanics, 66*, 772–779.

Cheng, P. J., & Lin, S. C. (2000). An analytical model for the temperature field in laser forming of sheet metal. *Journal of Materials Processing Technology, 101*, 260–267.

Dhondt, G. (2004). *The finite element method for three-dimensional thermomechanical applications*. West Sussex, England: Wiley.

Gollo, H. M., Mahdavian, S. M., & Naeini, H. M. (2011). Statistical analysis of parameter effects on bending angle in laser forming process by pulsed Nd:YAG laser. *Optics & Laser Technology, 43*, 475–482.

Gollo, M. H., Naeini, H. M., Liaghat, G. H., Torkamany, M. J., Jelvani, S., & Panahizade, V. (2008). An experimental study of sheet metal bending by pulsed Nd:YAG laser with DOE method. *International Journal of Material Forming, 1*, 137–140.

Hsieh, H. S., & Lin, J. (2004). Thermal–mechanical analysis on the transient deformation during pulsed laser forming. *International Journal of Machine Tools and Manufacture, 44*, 191–199.

Lee, K. C., & Lin, J. (2002). Transient deformation of thin metal sheets during pulsed laser forming. *Optics & Laser Technology, 34*, 639–648.

Maji, K., Pratihar, D.K. & Nath, A.K. (2012). Modeling of pulsed laser bending of sheet metal using neuro-fuzzy system, In *4th International and 25th All India Manufacturing Technology, Design and Research (AIMTDR) Conference*, Jadavpur University, Kolkata, India (Vol. 1, pp. 46–51).

Maji, K., Pratihar, D. K., & Nath, A. K. (2013). Experimental investigations and statistical analysis on pulsed laser bending of AISI304 stainless steel sheet. *Optics & Laser Technology, 49*, 18–27.

McBride, R., Bardin, F., Gross, M., Hand, D. P., Jones, J. D. C., & Moore, A. J. (2005). Modeling and calibration of bending strains for iterative laser forming. *Journal of Physics D: Applied Physics, 38*, 4027–4036.

Nicholson, D. W. (2008). *Finite element analysis: Thermomechanics of solids* (2nd ed.). London, New York: CRC Press, Taylor & Francis Group.

Shen, H., & Vollertsen, F. (2009). Modeling of laser forming—An review. *Computational Materials Science, 46*, 834–840.

Steen, W. M., & Mazumder, J. (2010). *Laser material processing* (4th ed.). London: Springer.

Tzeng, Y. (2000). Parametric analysis of the pulsed Nd:YAG laser seam-welding process. *Journal of Materials Processing Technology, 102*, 40–47.

Yang, L. J., Tang, J., Wang, M. L., Wang, Y., & Chen, Y. B. (2010). Surface characteristics of stainless steel sheet after pulsed laser forming. *Applied Surface Science, 256*, 7018–7026.

Zhang, X. R., Chen, G., & Xu, X. (2002). Numerical simulation of pulsed laser bending. *Journal of Applied Mechanics, 69*, 254–260.

Experimental Studies on TGM and BM Dominated Curvilinear Laser Bending of Aluminum Alloy Sheets

Ravi Kant, Parag M. Bhuyan and S.N. Joshi

Abstract During Laser bending process, the worksheet bends by means of thermal stresses induced by the laser beam irradiation. It can be achieved by various mechanisms viz. temperature gradient mechanism (TGM), buckling mechanism (BM) and upsetting mechanism (UM). The interactive effect of process parameters viz. laser power, scanning speed, beam diameter and absorption coefficient decide the occurrence of bending mechanism during a laser bending operation. Literature reports experimental as well numerical studies on the effect of process parameters viz. laser power, scan speed, beam diameter on the process mechanism and process performance. However, a very few attempts have been made on the study of shape of laser irradiation path on the quality and productivity of laser bending operation. Curvilinear laser bending is generally used to produce complex shapes using lasers. In this chapter an experimental study on the curvilinear laser bending of aluminum sheets for TGM and BM mechanisms has been presented. Initially the basic principle of the laser bending process and TGM and BM are discussed. Then the experimental procedure, plans are presented. The results are discussed in terms of the effect of laser power and scan speed on the bend angle and edge effect during parabolic irradiation. The experiments are carried out for both thick as well as thin worksheets. It was found that, in thin sheets, the scanning path curvature does not have significant effect on the bend angle however, in thick sheets the bend angle increases with decrease in scanning path curvature. The deformation behavior of curvilinear laser bending was found to be different from that of straight line laser bending process. The presented results may be used as guidelines to generate complex shapes in aluminum and its alloys using lasers.

R. Kant · S.N. Joshi (✉)
Department of Mechanical Engineering, Indian Institute of Technology Guwahati, Guwahati 781039, India
e-mail: snj@iitg.ac.in

R. Kant
e-mail: r.kant@iitg.ac.in

P.M. Bhuyan
Engineering Research Center, Tata Motors Ltd., Pune 411018, India
e-mail: parag.bhuyan@tatamotors.com

S.N. Joshi and U.S. Dixit (eds.), *Lasers Based Manufacturing*,
Topics in Mining, Metallurgy and Materials Engineering,
DOI 10.1007/978-81-322-2352-8_5

Keywords Laser bending · Curvilinear irradiation · Edge effect · Temperature gradient mechanism · Buckling mechanism · Aluminum alloy

1 Laser Bending Process

In conventional sheet bending operation, bending is done with the help of mechanical punches and dies. The mechanical bending is economical and generally used for mass production of components. In general, the capital cost and time involved in alteration or adjustment of the tools is high in mechanical bending process. Therefore, it is not suitable for prototyping or low volume production. The spring-back effect, deformation of material having low ductility and the processing at the inaccessible region are some of the important issues which limit the application of mechanical bending operation in real practice. The lasers helped to overcome some of these limitations. Laser bending is a type of thermo-mechanical forming process, which uses a focused laser beam to deform the sheet metal worksheet by means of thermal stresses without the application of any external mechanical load. The laser beam irradiates over the worksheet surface as shown in Fig. 1. The temperature of the irradiated region increases which leads to the thermal expansion. The thermal expansion is restricted by the surrounding material which generates thermal stresses in the heated region. These thermal stresses result in permanent deformation during laser bending of the worksheet (Walczyk and Vittal 2000; Geiger et al. 2004; Kannatey-Asibu 2009).

In laser bending, the hard tooling such as punches, dies, and presses are not required. The laser beam works as a non-contact virtual tool to bend the worksheet (Li and Yao 2000). The non-contact nature of the process makes the process independent of tool inaccuracies that might result in worksheet form errors and undesired deflections. The lead time associated with replacement of tools and dies also gets reduced due to the application of lasers. The laser irradiation can easily be controlled by using the microprocessor based controllers. The laser beam can be

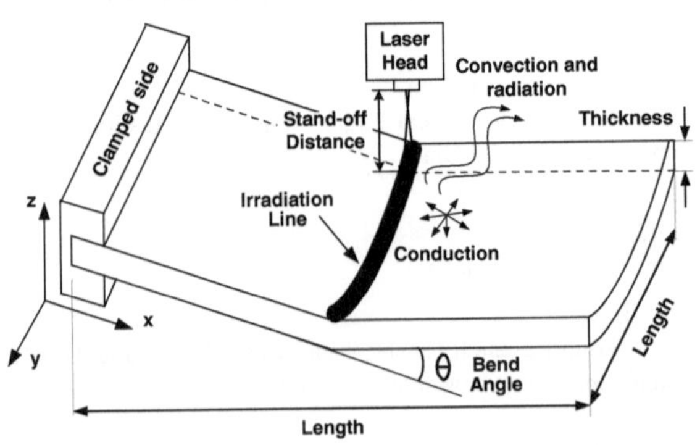

Fig. 1 Schematic of laser bending process

transferred easily by using fiber optics cables which makes the process suitable for the applications in which the mechanical tools are not accessible. The laser beam can be focused at a small spot (in the range of μm) which makes it suitable for the processing of small components (Jain 2012). The process can be used to bend sheets, plates, foils and pipes for a wide range of materials including metals, non-metals, composites and ceramics (Kant and Joshi 2013). The brittle, hard and thick materials can also be processed as the deformation occurs at elevated temperature. Precise small bend angles (of the order of 1 °C) can easily be produced which may not be possible in mechanical bending due to presence of spring-back effect (Lawrence et al. 2001; Chen and Xu 2001). The other advantages include the flexibility of the process, small heat affected zone due to narrow focus of the laser beam and ease of complex shapes generation with selected irradiation strategies etc. The laser bending uses localized heating to induce controlled deformation instead of entire work-piece heating. Therefore, it has the advantage of energy efficiency as compared with other thermal bending operations (Casamichele et al. 2007).

In spite of various advantages, the laser bending has some limitations too. For mass production, the laser bending process is slow in comparison with the traditional punch and die technique. The process is not suitable for the materials having high reflectivity. The application of suitable coating on the irradiation region can solve this problem (Singh 2013). The improper selection of laser parameters may lead to the surface melting which degrades the material properties. The mechanical and micro-structural properties of irradiated region may deteriorate due to presence of high temperature and thermal stresses in the heated region (Cheng and Yao 2001). The bend angle per laser scan is small and therefore, to get higher bend angle, multiple irradiations are required. The high capital cost of the laser machine makes the process more expensive than other forms of thermo-mechanical bending such as flame bending. The interaction of laser beam with human body may be dangerous and hence special safety precautions are essential (Yanjin et al. 2003; Kant and Joshi 2013).

Laser bending has many applications in several fields of industrial manufacturing which include—automotive, aerospace, shipbuilding, medical, micro-electronics and material processing. Laser bending is also useful in car body part straightening. Laser bending is used as an accurate and cost effective process to adjust or align the mating parts in welded constructions and ship building industry (Kant and Joshi 2014). Due to small size of the laser beam, and the ease of control, the laser bending is suitable for deformation of small meso-scaled components. The laser bending is now being used to manufacture small and precise bend angles in very small parts of micro-electro-mechanical systems (MEMS), chemical and sensor industries (Ocana et al. 2007).

Laser bending is the best option for low volume production requirements viz. forming of ship planks and production of aerospace fuselage. Due to high flexibility of the laser beam, it is well suited for the production of sheet metal components in space (Shen and Vollertsen 2009). Since it does not require external tools, dies or presses; the laser bending is suitable for rapid prototyping and rapid product development. It is also used for bending of brittle materials which is not possible with conventional bending operations (Li and Yao 2000; Wu et al. 2010).

2 Laser Bending Mechanisms

The laser bending occurs due to the plastic deformation resulted by the induced thermal stress generated in the heated region. It involves complex interaction of laser process parameters, workpiece material properties and the workpiece geometries. Variation in material properties, process parameters and workpiece geometries results in bending of work sheets through three different mechanisms. These mechanisms are: temperature gradient mechanism (TGM), buckling mechanism (BM) and shortening or upsetting mechanism (UM). TGM and BM are mainly responsible for worksheet bending and UM is responsible for shortening and thickening of the worksheet (Shi et al. 2006). These mechanisms are discussed in details in the following sections.

2.1 Temperature Gradient Mechanism (TGM)

Temperature gradient mechanism (TGM) is the most widely reported mechanism in the literature. It is used to bend the worksheet in the direction of the laser source. TGM occurs when energy parameters generate steep temperature gradient along the thickness direction. In TGM, the beam diameter is approximately equal to the worksheet thickness and the scanning speed is high enough to generate a steep temperature gradient along the worksheet thickness. To bend the material with high thermal conductivity using TGM, it is recommended that higher scanning speeds must be employed (Li and Yao 2001). This will help in generating higher temperature gradients due to lesser time for heat conduction into the work sheet.

Figure 2 shows the various steps involved in TGM. Initially, a steep temperature gradient generates along the thickness direction due to laser beam irradiation as shown in Fig. 2a. It results in non-uniform thermal expansion in the heated region. The temperature of the top surface is higher therefore the thermal expansion is more at the top surface which leads to the bending of the worksheet away from the laser source as shown in Fig. 2b. The thermal expansion of the heated region is restricted by the surrounding cooler material which generates compressive thermal stresses in the heated region and tensile stresses in the surrounding cooler region. When these thermal stresses exceed temperature dependent flow stress, the plastic deformation occurs. The flow stress decreases with temperature and hence, the compressive plastic deformation occurs in the heated region. The cooler region does not undergo any deformation as the flow stress is high at low temperature. The top irradiated surface has the highest temperature and therefore, the compressive deformation is more at the top surface. The plastic deformation is negligible at the bottom surface due to lower temperature. The material contracts during cooling. The plastic compressive deformation causes local shortening at top surface. Thus the worksheet finally bends towards the laser source as shown in Fig. 2c. Thus the bending occurs due to difference between plastic deformation at top and bottom surfaces

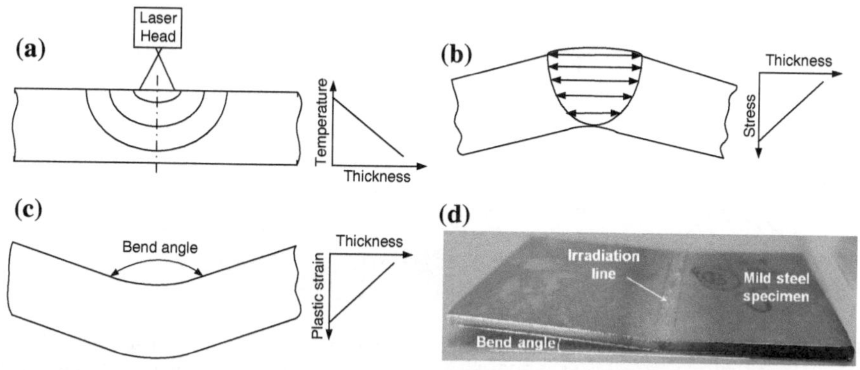

Fig. 2 Process steps of temperature gradient mechanism (process condition for figure (**d**): laser power = 400 W, scanning speed = 400 mm/min, beam diameter = 6.77 mm and total number of laser beam irradiations = 3). **a** Temperature distribution due to laser irradiation. **b** Thermal expansion and stress distribution due to laser heating. **c** Final bending towards laser source and plastic strain distribution. **d** Laser bent mild steel specimen with TGM

(Shi et al. 2006). A small plastic re-strain occurs during worksheet cooling as the yield stress and Young's modulus return to the higher level (Lawrence et al. 2001). Figure 2d shows a typical TGM dominated laser bent mild steel sheet with a small bend angle. In general, bend angle in the range of 0.1 to 3° can be achieved in a single laser beam irradiation.

2.2 Buckling Mechanism (BM)

The buckling mechanism (BM) occurs during laser bending of thin worksheets. In BM, temperature gradient between the top and bottom surfaces is negligible. BM generally occurs when a thin worksheet of high thermal conductivity material is irradiated with a laser beam of large beam diameter and low scanning speed. The beam diameter is about 10 times of the worksheet thickness (Hu et al. 2002).

Figure 3 shows the various steps involved in BM dominated laser bending process. The laser irradiation generates high temperature isotherms along the thickness direction as shown in Fig. 3a. It results in large amount of thermal expansion in the heated region. The thermal expansion is uniform along the thickness direction which is restricted by the surrounding material. This generates compressive stresses in the heated region. The buckling stiffness of the worksheet is less as the worksheet is thin. The buckling stiffness further reduces due to high temperature field caused by laser beam irradiation. Due to large beam diameter, the lateral expansion is more in the heated region. The combined effect of large thermal expansion and less buckling stiffness generates a buckle in the heated region as shown in Fig. 3b. The buckle tendency is more when the sheet is thin and coefficient of thermal expansion and temperature dependent flow stress are high.

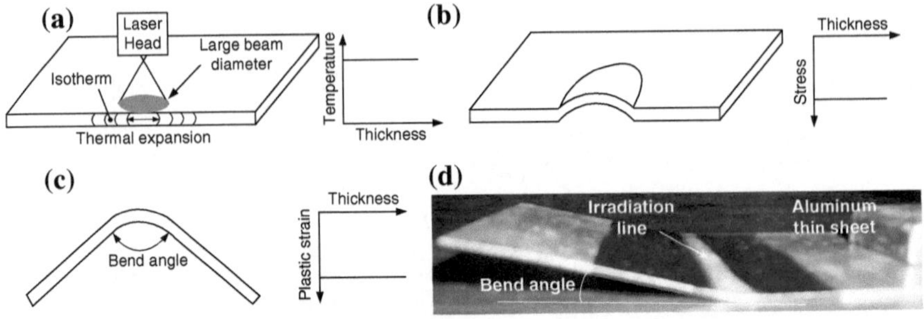

Fig. 3 Process steps of buckling mechanism (process condition for Figure (**d**): Laser power = 150 W, scanning speed = 200 mm/min, beam diameter = 6.77 mm). **a** Temperature isotherm due to laser irradiation. **b** Growth and development of buckling due to uniform thermal stress along the thickness. **c** Final bending occurs due to buckling. **d** Laser bent aluminum thin sheet with BM

Once buckling is initiated, it extends along with laser beam irradiation. When thermal stresses exceed temperature dependent flow stress, plastic deformation occurs in the buckle. Finally, based on the direction of the buckle, the worksheet bends towards or away from the laser source as shown in Fig. 3c.

Figure 3d shows a typical laser bent thin aluminum sheet with BM dominated process conditions. It can be seen that the bend angle is significantly higher as compared with that obtained with TGM process conditions. In general, the bend angle in the range of 1°–15° can be achieved in a single laser scan using BM dominated process conditions. Unlike TGM, the counter bending does not occur in BM. However it has been noted that the bending direction is not certain in BM. Bending in BM is also governs by pre-curvature of the sheet, internal stresses and external or gravitational forces acted on the worksheet (Shi et al. 2006).

Li and Yao (2001) proposed an irradiation scheme by which a certain convex bending (away from laser source) can be achieved in BM without application of pre-bending or external mechanical constraints. In this scheme, the laser irradiation was started near to the middle of the scanning path instead from an edge of the worksheet. Jamil et al. (2011) studied the effect of rectangular beam geometries with different transverse width to length aspect ratio on BM dominated laser bending of thin sheets. The beam geometry played an important role in temperature distribution and deformation behavior. Longer beam dimensions in the scanning direction (in relation to its lateral dimension) produced higher temperatures and also had a tendency to form a concave shape.

Shi et al. (2006) gave a critical condition to know whether the process is dominated by the TGM or by the BM. The condition was given as:

$$\frac{Pd^{1/2}}{h^2V^{1/2}} > \frac{\eta\pi^{7/2}k^{1/2}\rho^{1/2}c^{1/2}}{41.52(1+\mu)A\alpha_{th}} \tag{1}$$

where P, d, h, V, η, k, ρ, c, μ, A and α_{th} are laser power, beam diameter, sheet thickness, scanning speed, correction factor, thermal conductivity, density, specific heat, Poisson's ratio, absorption coefficient and coefficient of thermal expansion respectively. Also, the process conditions for the dominating mechanism can be obtained by using a Fourier number derived by Shi et al. (2008). It is given by,

$$F_0 = \alpha_d d / h^2 V \tag{2}$$

where α_d, h, d and V are thermal diffusivity, sheet thickness, beam diameter and scanning speed respectively. The smaller value of the Fourier number corresponds to a TGM dominated laser bending while a high Fourier number indicates the dominance of BM.

2.3 Upsetting Mechanism (UM)

In upsetting mechanism (UM), the worksheet shortens and therefore, it is also called as shortening mechanism. This mechanism is used for shortening of small frames, pipe bending of various kinds of cross-sections and alignment of micro-parts.

In UM, due to laser irradiation high temperature isotherms occur along the worksheet thickness as shown in Fig. 4a. It results in uniform thermal expansion along the thickness. Due to uniform thermal expansion, the counter-bending does not occur. The thermal expansion is restricted by the surrounding cooler bulk material which generates compressive thermal stresses in the heated region. The worksheet is thick and the beam diameter is small; therefore, the buckling is prevented by the worksheet. When the thermal stresses in the heated region exceed temperature dependent flow stress, the plastic deformation occurs. The plastic

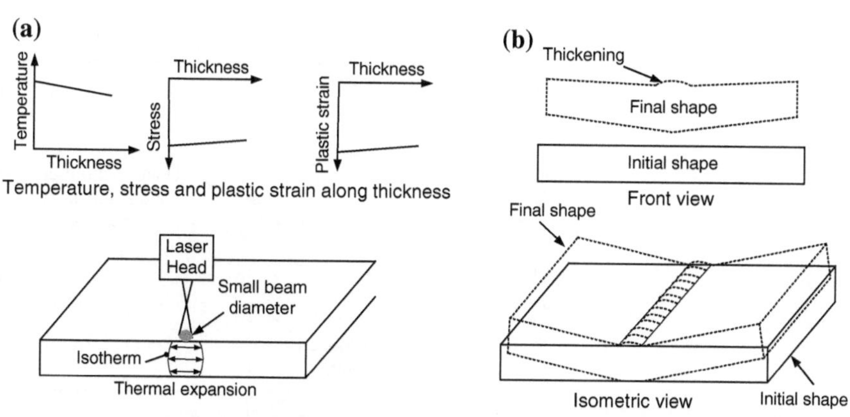

Fig. 4 Process steps of upsetting mechanism. **a** Temperature isotherm in thickness due to laser irradiation. **b** Final bending and thickening in UM

deformation is uniform and compressive along the worksheet thickness. The plastic deformation is almost uniform along the worksheet thickness. During cooling it results in local shortening and thickening of the worksheet (Shi et al. 2012). As the worksheet is thick, a small temperature gradient occurs in the thickness direction. This results in little bending of worksheet with thickening of the irradiated region as shown in Fig. 4b.

3 Edge Effect in Laser Bending Process

Heat conduction is the most significant parameter in laser bending. A schematic of variation in heat conduction during laser irradiation is shown in Fig. 5. It can be seen that the thermal conduction is low at the start and end of the irradiation line due to availability of less material. However it is uniform when laser beam is at the middle. This variation results in the higher surface temperatures near the edges and lower surface temperature at middle of the scanning path. The mechanical constraint provided by the surrounding cooler material also varies from one end to the other end of irradiation path. At start of laser irradiation path, the heated region is less. The surrounding cooler material provides higher restriction to the thermal expansion of the heated region. As the beam moves further, heat starts conducting in the surrounding cooler region and increases its temperature. This results in the reduction of flow stress of the material. Thus the plastic deformation in the heated region increases along the scan line. This can be seen in Fig. 6. When laser beam is just about to leave the worksheet surface, the worksheet material is heated all along the irradiation line. Therefore, the surrounding material provides less restriction to the thermal expansion caused by laser heating. In this way, the changes in thermal

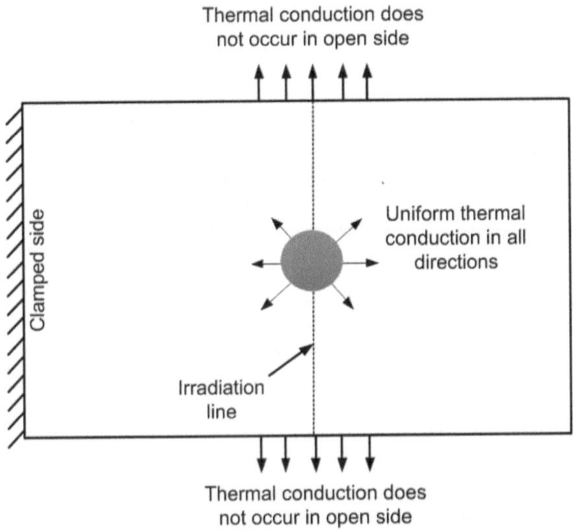

Fig. 5 Non-uniform thermal conduction during laser irradiation

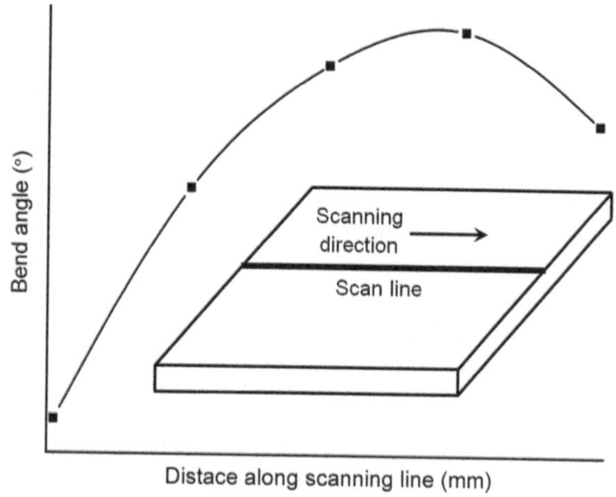

Fig. 6 Schematic to show variation in bend angle along the irradiation line

and mechanical constraints from start to end of the irradiation line affect the plastic deformation. Other thermal properties like convection coefficient and absorptivity also affect the distribution temperature into the worksheet (Bao and Yao 2001; Shen et al. 2010). The variation in thermal and mechanical constraints results in generation of non-uniform bend angle from one end to the other of the irradiation line as shown in Fig. 6. This variation in bend angle along the irradiation line is called as edge effect. In general, the edge effect is not desirable, but in some cases it can be utilized to generate complex shapes.

Literature reports a number of methods to reduce the edge effect. Shen et al. (2010) showed that scanning speed is important parameter to control the edge effect and the combination of acceleration and deceleration scanning scheme can minimize the edge effects. Hu et al. (2013) proposed two methodologies to reduce the edge effect; first, to maintain a constant peak temperature along the irradiation path and the second, to put external mechanical constraint in the form of clamping at both ends of the irradiation path. These methodologies significantly reduced the edge effect. Zahrani and Marasi (2013) showed that number of irradiations, worksheet thickness, scanning speed and laser power in order of their significance directly affect the edge effect. They found that edge effect decreases with increase in number of irradiations, sheet thickness, scanning speed and decrease in beam diameter.

4 Curvilinear Laser Bending

Straight line irradiation is used to produce uniform bending of simple parts about the irradiation path. However in many cases, work parts with complex or spatially curved geometries like spherical dome or ship hulls are required. Manufacturing of such complex geometries by using straight line irradiations is quite difficult. In such cases, instead of straight line, curvilinear irradiations are found to be more convenient and efficient (Zhang et al. 2007). The laser bending with curvilinear irradiation paths is called as curvilinear laser bending process. Hennige (2000) investigated the differences in the forming behavior of sheet metal parts using straight and curved irradiations. Various irradiation strategies were studied to generate the spherical dome shapes. It was suggested that the combination of radial and concentric irradiation lines can be used to produce spherical structures. The concentric lines were used for stabilizing the initial flat plate against wrinkling. Chen et al. (2004) studied the deformation behavior of laser curve bending of sheet and found that the deformation occurs only on one side of the scanning path along which the rigid constraint is relatively lower. Zhang et al. (2007) presented a finite element model of sheet metal forming using B-spline curve scanning of laser beam. The results showed that the peak temperature increases with increase in scanning path curvature and the laser curve bending produces a significant change of distortion under the same set of process conditions. The warping was found to be increased with scanning path curvature. Venkadeshwaran et al. (2010) studied the deformation of a circular plate subjected to a circular irradiation path with shifting of starting point. It was observed that discrete section heating in symmetry with shifting in starting point of irradiation in subsequent passes reduced the undesired waviness. Kant and Joshi (2014) found that the bending occurs outside of the scanning path curvature during circular irradiation of laser beam as shown in Fig. 7. They carried out numerical investigation on curvilinear laser bending of magnesium alloy using finite element method (FEM). The bending was offset near the edges while it was on the irradiation path at the middle of the scanning path. This behavior was found to be different from the straight line laser bending process where bending occurs on the irradiation line. In curvilinear irradiation, the bending offset was found to be increased with increase in beam diameter and laser power. It was also noted that the bend angle increases with increase in scanning path curvature. It may be due to absorption of more energy along the longer scanning path with higher scanning path curvature. Based on these initial numerical investigations, it was

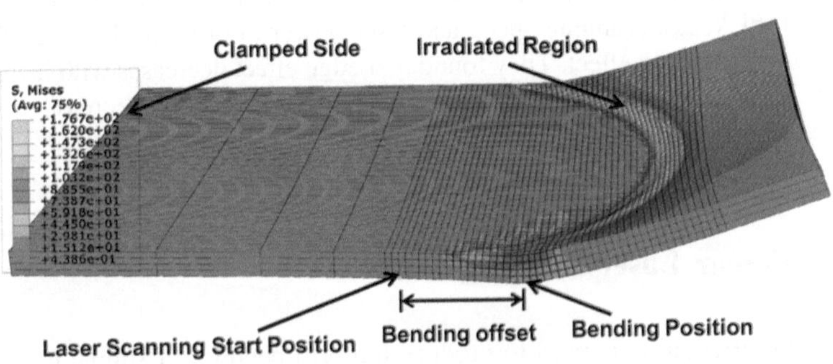

Fig. 7 Bending behavior in curvilinear laser bending process (Kant and Joshi 2014, Copyright with authors)

noted that a need exists for a detail experimental investigation on the effect of variation in the shape of laser path during TGM as well as BM based laser bending process. The next sections present the work carried out in this direction.

5 Experimental Studies on TGM and BM Curvilinear Laser Bending of Aluminum Alloy

5.1 Experimental Details

In this section, the details of experiments carried out on curvilinear laser bending of aluminum alloy thin and thick worksheets are presented. The experimental studies were performed on commercially available aluminum alloy worksheet. The composition of material is shown in Table 1. The experiments were performed for TGM and BM dominated process conditions. In general, the TGM occurs in thick sheet and BM occurs in the thin sheets. Therefore, the specimen of two thicknesses i.e. 1.45 and 0.5 mm were used during the experiments. The specimen of thickness 1.45 mm was used for TGM laser bending while the specimen of 0.5 mm thickness was used for BM dominated laser bending process. The specimens of size 80 mm length and 50 mm width were cut by using CO_2 laser cutting machine. The laser cutting was preferred over shear cutting as the former process provides dimensionally accurate specimen without pre-bending. Most of the metals have high reflectivity (low absorptivity). The absorptivity varies non-uniformly along the laser scanning path. This may be due to the presence of rust and other foreign particles on the surface of metal work-sheets. To increase the absorptivity, the specimens were coated with graphite spray prior to laser beam irradiation. The graphite spray coating increases the absorptivity and also helps to achieve uniform absorptivity along the laser scanning path. The coated specimens were allowed to dry for 1 h under normal room conditions.

The parabolic shaped curvilinear path was chosen for the experiments. The parametric equations of the parabola are given as:

$$x = 2at \tag{3}$$

$$y = 2at \times t \tag{4}$$

where t is the controlling parameter. Figure 8 shows a parabolic curve with point A as the vertex and S as the Focus. Distance AS is denoted as a. It can be seen that the radius of curvature increases with increase in the value of a. For an infinite

Table 1 Composition of commercially available aluminum

Elements	Mg	Al	Si	Ti	Mn	Cu	Zn
Weight (%)	0.03	97.22	1.65	0.33	0.38	0.35	0.6

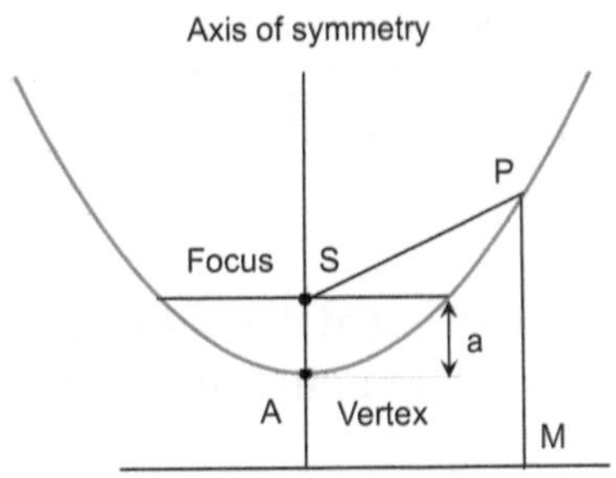

Fig. 8 Parabolic irradiation path

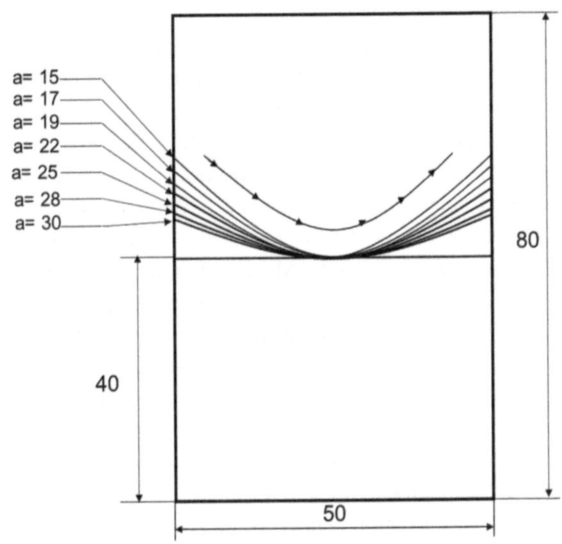

Fig. 9 Scheme of parabolic irradiations

value of *a,* the parabolic curve tends to be a straight line. As shown in Fig. 9, seven different values of *a* were considered to study the effect of scanning path curvature.

The laser heating was performed by using LVD Orion 3015 2.5 kW continuous wave CO_2 laser machine. Figure 10 shows the Laser unit and the experimental set-up. The specimen was clamped over the laser machine bed using a fixture as shown in Fig. 10b. The laser beam was irradiated along the predefined path. Process (laser) parameters viz. laser power (*P*), scanning speed (*V*) and beam diameter (*D*) were varied to control the operation. The heated specimens were allowed to cool naturally after the laser beam irradiation. Each experiment was repeated thrice to study the repeatability. The bent specimen and the damaged coating are shown in Fig. 11.

The bend angle produced due to laser beam irradiation was measured by using Zeiss make coordinate measuring machine (CMM). The bent specimen was kept on the anvil as shown in Fig. 12 and touch probe was moved in *x*, *y* and *z* direction to collect the data points on either side of the laser scanning path. The bend angle was computed at seven positions by recording coordinates of total twenty eight data points as shown in Fig. 13. Two points form a line. The bend angle was calculated

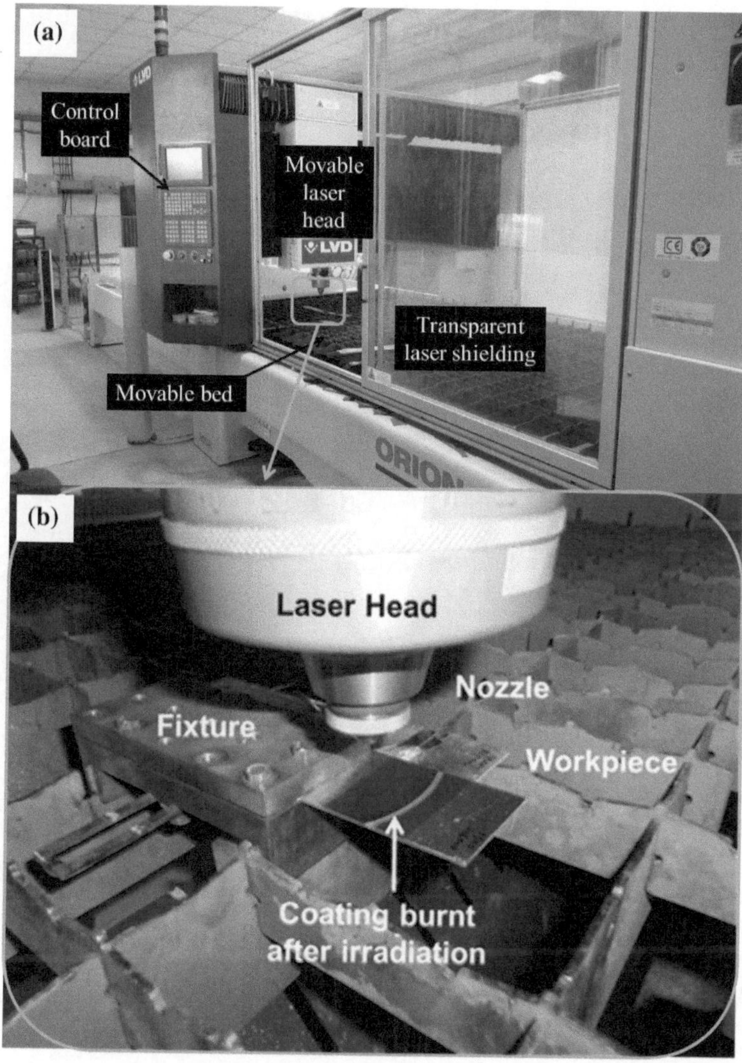

Fig. 10 Details of experimental setup **a** laser machine, **b** worksheet holding

between the two respective lines recorded on either sides of the laser beam irradiation. For a set of process conditions, the average of three trials was considered as the experimental value. The edge effect is computed as the relative variation in bend angle (*RVBA*) along the laser scanning path as (Zahrani and Marasi 2013):

$$RVBA = \frac{\theta_{\max} - \theta_{\min}}{\theta_{average}} \times 100 \tag{5}$$

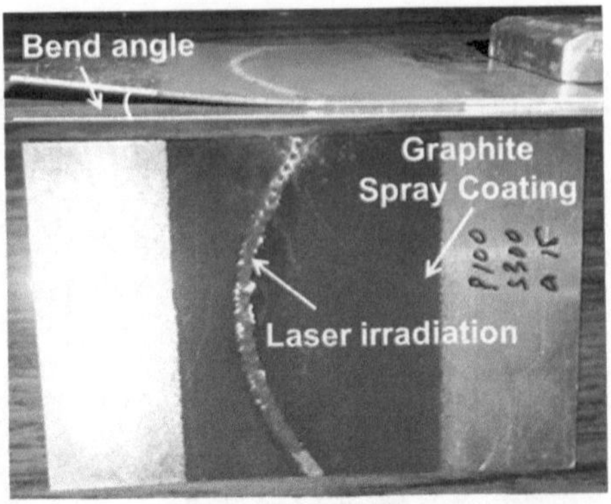

Fig. 11 Laser irradiated specimens

Fig. 12 Touch probe of CMM used to measure bend angle of laser bent specimen

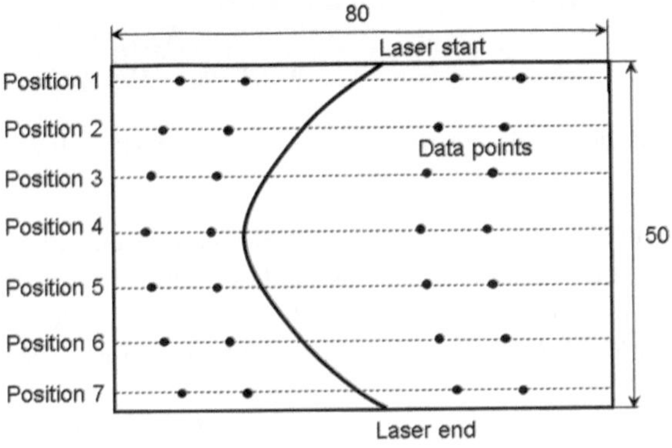

Fig. 13 Bend angle measurements at various positions

where $\theta_{\max}$, $\theta_{\min}$ and $\theta_{average}$ are the maximum, minimum and average bend angles along the scan line respectively. The edge effect is more when the value of *RVBA* is higher.

5.2 Results and Discussion

This section deals with the discussion on the results obtained during parabolic irradiations of laser over the aluminum work sheets. Effects of process parameters on bend angle and deformation behavior are presented for both TGM and BM process conditions in the following sections.

5.2.1 Laser Bending Using TGM

The laser bending of aluminum sheet of 1.45 mm thickness is carried out with TGM process conditions. The effects of scanning speed, laser power and scanning path curvature on bend angle are discussed as below.

Figure 14 shows the effect of scanning speed on bend angle for the sheet thickness of 1.45 mm. It can be observed that for low laser power (300 W), the bend angle decreases with increase in scanning speed. The scanning speed controls the energy input into the specimen and the temperature gradient along the specimen thickness. For low power, at lower scan speed the effects of temperature gradient and energy input are almost in balanced condition. Therefore there is not much significant effect of scanning speed on the bend angle. For medium to higher scan

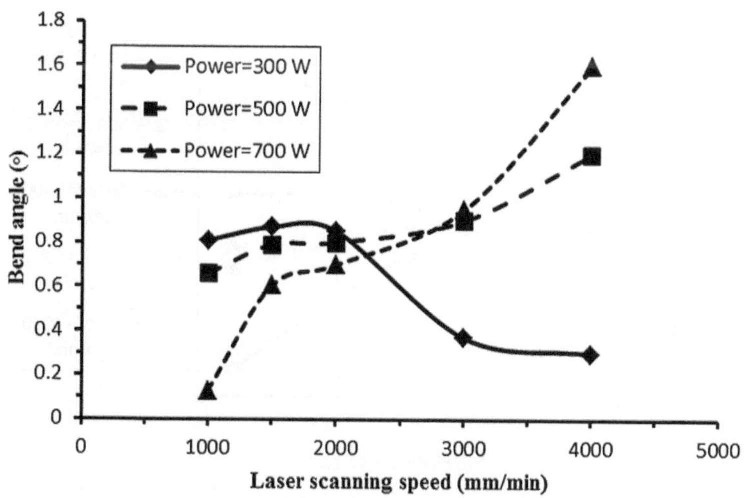

Fig. 14 Effect of laser scanning speed on bend angle for 1.45 mm thick sheet

speed, the bend angle decreases. It is due to less absorption of energy at higher scanning speed which reduces the peak temperature. This results in less plastic deformation in the heated region.

For higher laser power of 500 and 700 W, the bend angle increases with increase in scanning speed. It is because at higher laser power at higher scan speeds, the peak temperature at top surface is high enough to generate the required temperature gradient for the plastic deformation. At higher speeds, the time for energy absorption is less; therefore the temperature at the bottom surface is also less. This results in higher bend angles.

The laser power directly controls the energy input into the worksheet surface. The effect of laser power on bend angle for 1.45 mm thick sheet is shown in Fig. 15. It can be observed that the bend angle increases with increase in laser power when scanning speed is 4000 mm/min. It is due to more energy input at higher laser power and higher temperature gradient in the thickness direction at high scanning speed.

At lower scanning speeds (1000–2000 mm/min), bend angle decreases with increase in laser power. It is due to high thermal conductivity of aluminum, significant temperature gradient cannot be maintained at lower scanning speed. This results in high temperature and more plastic deformation at bottom surface. As the laser power increases, the plastic deformation at bottom surface also increases which results in the decrease of bend angle with increase in laser power. The scanning speed 3000 mm/min has the intermediate effect. The bend angle first increases and then becomes constant with the increase in power.

In curvilinear laser bending process, the scanning path curvature is an important parameter which affects the laser bending process. In the present work, the laser beam has been irradiated in parabolic curved path. The details are presented in Sect. 5.1. The scanning path curvature was controlled by parameter *a*. Path

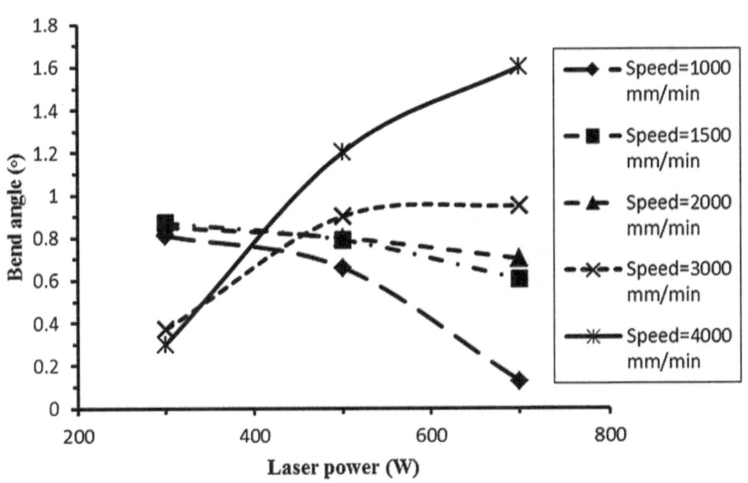

Fig. 15 Effect of laser power on bend angle for parabolic irradiation for 1.45 mm thick aluminum sheet

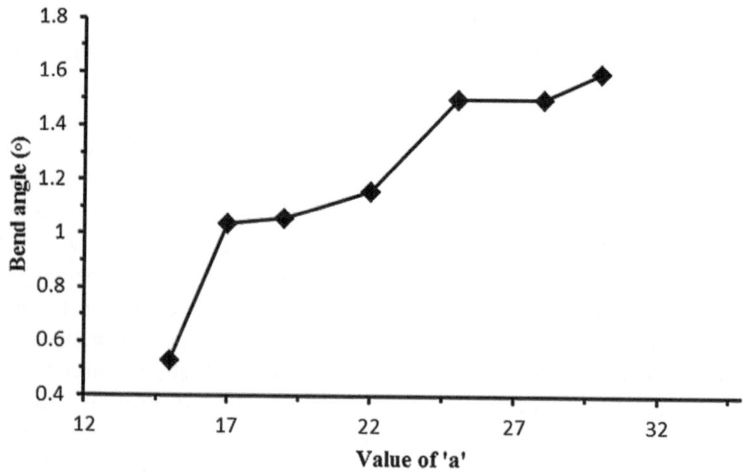

Fig. 16 Effect of scanning path curvature on bend angle for 1.45 mm thick aluminum sheet with parabolic irradiation

curvature increases with increase in value of *a* (Fig. 9). For sheet thickness of 1.45 mm, the effect of increase in scanning path curvature on bend angle is shown in Fig. 16. It can be observed that the bend angle increases with increase in scanning path curvature. This may be due to the fact that for lower path curvature, the scanning length is more. Thus the proceeding material gets more preheated when the curvature of path is less. This result in lower temperature gradients and less bend angles. Also, for higher scanning lengths, more energy will be absorbed into the workpiece which may lead to more plastic deformation at bottom surface.

For 1.45 mm sheet thickness, variation in bend angle along the laser scanning path for various scanning path curvatures is shown in Fig. 17. The bend angle was

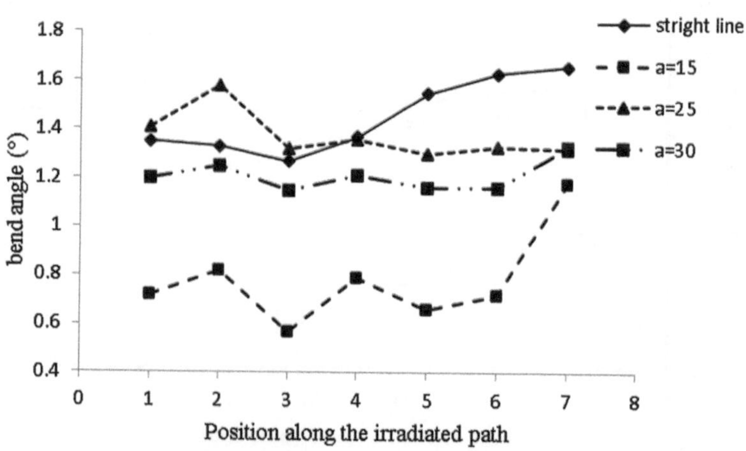

Fig. 17 Variation in bend angle along the laser scanning path for 1.45 mm thick aluminum sheet

measured at seven different locations along the scanning path as shown in Fig. 13. From Fig. 17, no uniform pattern or trend can be noticed. For $a = 15$, the profile has been noted as wavy, however for $a = 25$ and 30, the profiles are almost flat. In most of the cases, higher bend angle was noted near the edges. The edge effect is calculated along the laser scanning path by using Eq. (5). It is observed that the edge effect (*RVBA*) is highest (32.05 %) for a = 15 and lowest (8.27 %) for the case of a = 30. The edge effect for straight line laser bending is found to be 24.8 %. This can be attributed to the probable change in bending mechanism from the start to the end of the laser scanning path. Initially when laser irradiation starts, the temperature gradient is high and when laser moves in the forward direction, the temperature gradient decreases due to the preheating of the material. Thus the process leads to the BM dominated process condition. After reaching the highest (peak) position of the scanning path curvature, laser starts moving to its end position. During this period of movement from the peak of curvature to the end-point of scanning path, the temperature gradient increases due to availability of less material for heat conduction. Thus the bend angle increases near the edges.

5.2.2 Laser Bending of Thin Sheets Using BM

The laser bending of aluminum sheet of 0.5 mm thickness is carried out with BM dominated process conditions. The effects of scanning speed, laser power, and scanning path curvature are discussed as below.

Figure 18 shows the effect of scanning speed on bend angle for the sheet thickness of 0.5 mm. It can be observed that the bend angle decreases with increase

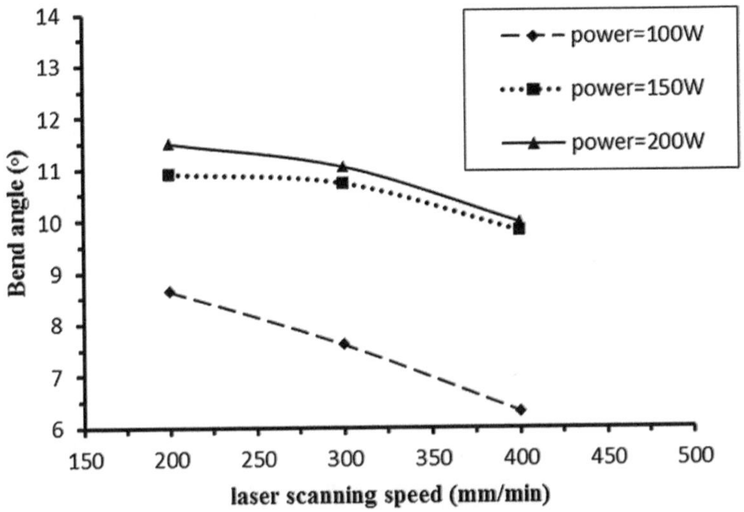

Fig. 18 Effect of laser scanning speed on 0.5 mm thin sheet

in scanning speed. It may be due to the combined effect of lesser energy input and high temperature gradient at higher scan speeds. In BM, bending occurs when the temperature gradient is negligible. The bend angle increases with increase in temperature isotherm along the worksheet thickness. It means the bend angle is more when the temperature gradient along the worksheet thickness is less. The increase in scanning speed increases temperature gradient along the thickness direction which further leads to the decrease in bend angle at higher scanning speed.

In BM, the laser power is an important parameter which controls the bending of the worksheet. Figure 19 shows the effect of laser power on bend angle for sheet thickness of 0.5 mm. The bend angle increases with increase in laser power. For higher values of laser power, the temperature gradient along the thickness direction decreases due to quick flow of heat into the worksheets. The reduction in temperature gradient leads to increase in bend angle.

Figure 20 shows the variation of bend angle with respect to scanning path curvature for sheet thickness of 0.5 mm. Figure 20 shows that for sheet thickness of 0.5 mm, the bend angle is almost constant with respect to the change in scanning path curvature. However, for a = 15, the bend angle it was found to be slightly less.

Figure 21 shows the variation in bend angle along the laser scanning path for various values of path curvature. It can be seen that for straight line heating the bend angle increases from the start to the end in laser pass. It may be due to the fact that the temperature gradient is highest when laser irradiation starts and as laser beam proceeds, the temperature gradient reduces due to the intense preheating of work sheet. This leads to increase in bend angle along the scanning path. In curvilinear laser bending, the bend angle increases from laser start position, attains a peak and then decreases towards the laser end position. The similar trend is

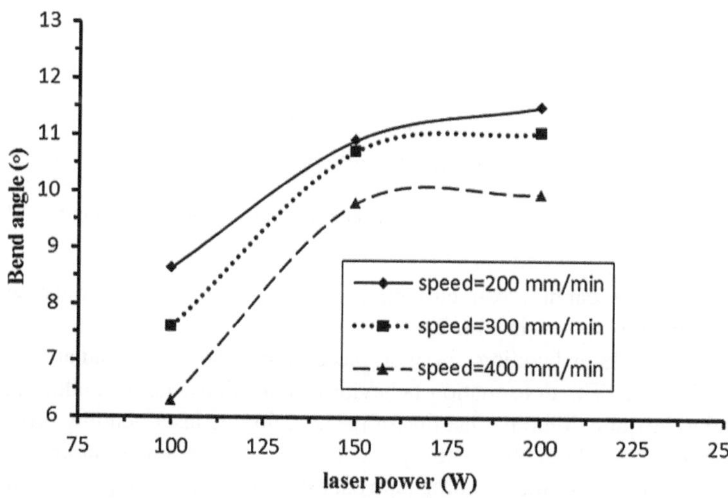

Fig. 19 Effect of laser power on bend angle on 0.5 mm thick aluminum sheet with parabolic irradiation

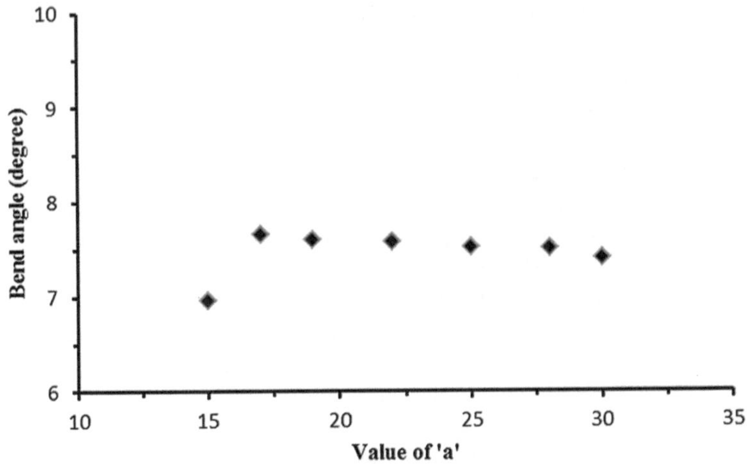

Fig. 20 Effect of curvature on bend angle for 0.5 mm thick aluminum sheet with parabolic irradiation

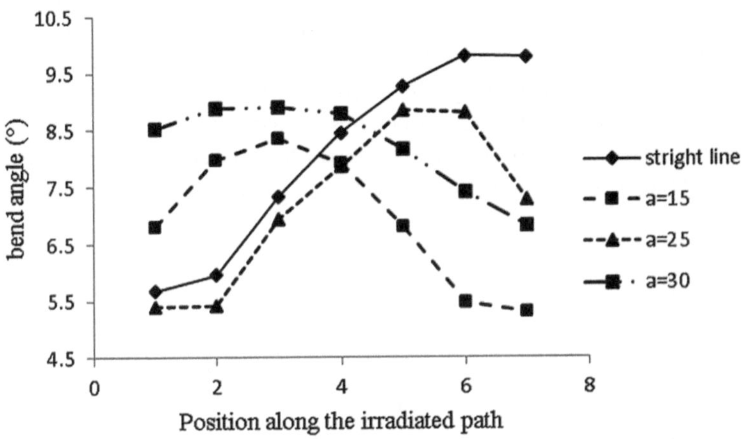

Fig. 21 Variation of bend angle along laser scanning path for 0.5 mm thick aluminum sheet

observed for all the scanning path curvatures (Fig. 21). It may be due to the bending offset phenomenon shown in Fig. 22.

In straight line laser bending, the workpiece bends about the irradiation line as shown in Fig. 22a. The deformation behavior of curvilinear laser bending process was found to be quite different than that of the straight line laser bending process. In curvilinear laser bending, the bending did not occur over the scanning path. It occurred at the outside of the scanning path curvature as shown in Fig. 22b. It may be due to peak temperature offset and tendency of the sheet to bend outside of the scanning path curvature. Similar observations were noted by Kant and Joshi (2013)

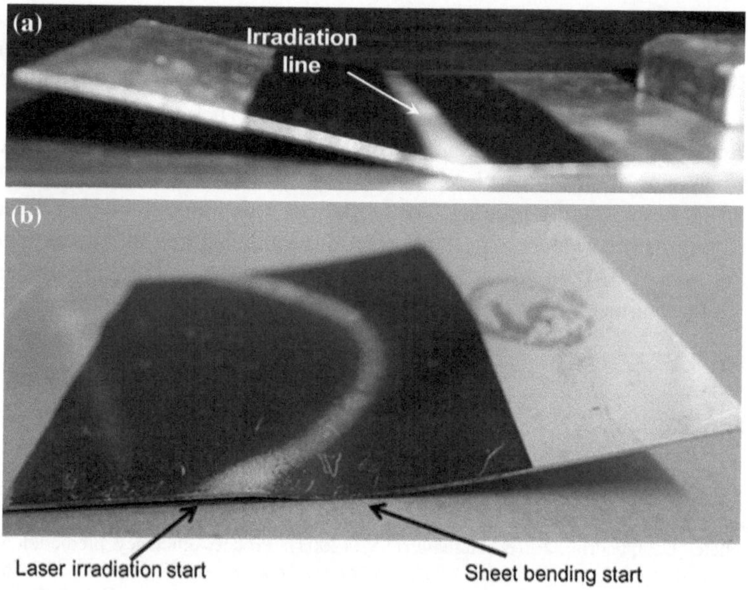

Fig. 22 **a** Bending behavior in straight line laser bending process with laser power = 150 W, scanning speed = 200 mm/min, beam diameter = 6.77 mm; **b** Bending behavior in curvilinear laser bending process with laser power = 250 W, scanning speed = 300 mm/min, beam diameter = 6.77 mm

during the numerical investigations on curvilinear laser bending (Sect. 4). However, at the middle of the scanning path, the bending was occurred over the irradiation path. For this study, the edge effect was calculated along the laser scanning path by using Eq. (5). It was observed that the edge effect (*RVBA*) is highest (51.24 %) for the straight line laser bending and lowest (25.43 %) for the case of $a = 30$. It was also observed that the edge effect in TGM is comparatively less with that of with BM.

6 Conclusions

Laser bending is an important manufacturing process which is gaining importance as it is evident from a large number of papers appearing in journals. In this chapter, basics of laser bending process, various bending mechanisms and the edge effect during laser bending process have been presented. Advantages and limitations of the laser bending process are discussed. In this work, experimental studies on curvilinear laser bending of aluminum alloy sheets have been carried out. For parabolic irradiation, the effects of various process parameters viz. laser power, scanning speed and scanning path curvature on bend angle have been studied. Edge effects for two different mechanisms i.e. temperature gradient mechanism (TGM) and buckling mechanism (BM) are studied.

The experimental study revealed that a combination of high scanning speed and high laser power produces higher bend angle in TGM dominated laser bending process. However in BM, a combination of low scanning speed and high laser power was found to be generating larger bend angle. The scanning path curvature was found to an influencing parameter in the laser bending process. The deformation behavior of curvilinear laser bending was found to be different from that of straight line laser bending process. The workpiece was found to be bent outside of the scanning path curvature. These observations validated the findings reported by Kant and Joshi (2014) based on their numerical work. The presented results may be useful to the engineers and scientists working in the area of Laser forming.

References

Bao, J., & Yao, Y. (2001). Analysis and prediction of edge effects in laser bending. *Journal of Manufacturing Science and Engineering, 123*, 53–61.

Casamichele, L., Quadrini, F., & Tagliaferri, V. (2007). Process-efficiency prediction in high power diode laser forming. *Journal of Manufacturing Science and Engineering, 129*, 868–873.

Chen, D., Wu, S., & Li, M. (2004). Deformation behaviours of laser curve bending of sheet metals. *Journal of Materials Processing Technology, 148*, 30–34.

Chen, G., & Xu, X. (2001). Experimental and 3D finite element studies of CW laser forming of thin stainless steel sheets. *Journal of Manufacturing Science and Engineering, 123*, 66–73.

Cheng, J., & Yao, Y. (2001). Cooling effects in multiscan laser forming. *Journal of Manufacturing Processes, 3*, 60–72.

Geiger, M., Merklein, M., & Pitz, M. (2004). Laser and forming technology—an idea and the way of implementation. *Journal of Materials Processing Technology, 151*, 3–11.

Hennige, T. (2000). Development of irradiation strategies for 3D-laser forming. *Journal of Materials Processing Technology, 103*, 102–108.

Hu, Z., Kovacevic, R., & Labudovic, M. (2002). Experimental and numerical modeling of buckling instability of laser sheet forming. *International Journal of Machine Tools and Manufacture, 42*, 1427–1439.

Hu, J., Xu, H., & Dang, D. (2013). Modeling and reducing edge effects in laser bending. *Journal of Materials Processing Technology, 213*, 1989–1996.

Jain, V. K. (Ed.). (2012). *Micromanufacturing processes* (pp. 283–303). CRC Press, Boca Raton.

Jamil, M., Sheikh, M., & Li, L. (2011). A study of the effect of laser beam geometries on laser bending of sheet metal by buckling mechanism. *Optics & Laser Technology, 43*, 183–193.

Kannatey-Asibu, E. (2009). *Principles of laser materials processing*. Hoboken, New Jersey: Wiley.

Kant, R., & Joshi, S. N. (2013). Finite element simulation of laser assisted bending with moving mechanical load. *International Journal of Mechatronics and Manufacturing Systems, 6*, 351–366.

Kant, R., & Joshi, S. N. (2014). Numerical modeling and experimental validation of curvilinear laser bending of magnesium alloy sheets. *Proceedings of The Institute of Mechanical Engineering Part B: Journal of Engineering Manufacture, 228*, 1036–1047.

Lawrence, J., Schmidt, M. J. J., & Li, L. (2001). The forming of mild steel plates with a 2.5 kW high power diode laser. *International Journal of Machine Tools and Manufacture, 41*, 967–977.

Li, W., & Yao, Y. (2000). Numerical and experimental study of strain rate effects in laser forming. *Journal of Manufacturing Science and Engineering, 122*, 445–451.

Li, W., & Yao, Y. (2001). Numerical and experimental investigation of convex laser forming process. *Journal of Manufacturing Processes, 3*, 73–81.

Ocana, J., Morales, M., Molpeceres, C., Garcia, O., Porro, J., & Garcia-Ballesteros, J. (2007). Short pulse laser microforming of thin metal sheets for MEMS manufacturing. *Applied Surface Science, 254*, 997–1001.

Shen, H., Hu, J., & Yao, Z. (2010). Analysis and control of edge effects in laser bending. *Optics and Lasers in Engineering, 48*, 305–315.

Shen, H., & Vollertsen, F. (2009). Modelling of laser forming—an review. *Computational Materials Science, 46*, 834–840.

Shi, Y., Liu, Y., Yao, Z., & Shen, H. (2008). A study on bending direction of sheet metal in laser forming. *Journal of Applied Physics, 103*, 053101.

Shi, Y., Liu, Y., Yi, P., & Hu, J. (2012). Effect of different heating methods on deformation of metal plate under upsetting mechanism in laser forming. *Optics & Laser Technology, 44*, 486–491.

Shi, Y., Yao, Z., Shen, H., & Hu, J. (2006). Research on the mechanisms of laser forming for the metal plate. *International Journal of Machine Tools and Manufacture, 46*, 1689–1697.

Singh, K. (2013). Effect of lime coating on laser bending process. M.Tech. thesis, IIT Guwahati.

Venkadeshwaran, K., Das, S., & Misra, D. (2010). Finite element simulation of 3-D laser forming by discrete section circle line heating. *International Journal of Engineering Science and Technology, 2*, 163–175.

Walczyk, D., & Vittal, S. (2000). Bending of titanium sheet using laser forming. *Journal of Manufacturing Processes, 2*, 258–269.

Wu, D., Zhang, Q., Ma, G., Guo, Y., & Guo, D. (2010). Laser bending of brittle materials. *Optics and Lasers in Engineering, 48*, 405–410.

Yanjin, G., Sheng, S., Guoqun, Z., & Yiguo, L. (2003). Finite element modeling of laser bending of pre-loaded sheet metals. *Journal of Materials Processing Technology, 142*, 400–407.

Zahrani, E. G., & Marasi, A. (2013). Experimental investigation of edge effect and longitudinal distortion in laser bending process. *Optics & Laser Technology, 45*, 301–307.

Zhang, P., Guo, B., Shan, D., & Ji, Z. (2007). FE simulation of laser curve bending of sheet metals. *Journal of Materials Processing Technology, 184*, 157–162.

Mathematical Formulation for Development of Compound Curve Surface by Laser Line Heating

Biplab Das and Pankaj Biswas

Abstract Line heating assisted with laser as a heat source is a flexible forming process that forms sheet metal by means of stresses induced by external heat instead of by means of external force. The process has the potential to be applied as a primary forming method for forming accurate shapes. However successful application of this process in industry is limited due to high equipment costs and safety requirements. The production of complex shapes requires the understanding of laser-material interaction. The chapter presents the mathematical formulation of development of smooth continuous curved surface. It is developed by deformation of sheet under plane stress condition by taking into account the strain distribution and the coefficient of first fundamental form of curve surface. Surface development is carried out along principal curvature direction along with the procedure for suitable determination of heating line pattern for the desired engineering surfaces.

Keywords Line heating · Strain field · Doubly curve surface · Scanning path

1 Introduction

Line heating is a suitable technique for obtaining desired shape under the influence of a suitable thermal load on a given metallic plate. The thermal load can be induced with the use of suitable heat source which includes: oxy-acetylene gas flame, high frequency induction heating or laser beam either one depending on the suitability. The process is controlled by the intensity of heat source, its speed of movement over the plate and the pattern of lines where the heat is to be applied.

B. Das · P. Biswas (✉)
Department of Mechanical Engineering, Indian Institute of Technology Guwahati, Guwahati 781039, India
e-mail: pankaj.biswas@iitg.ernet.in

B. Das
e-mail: das.biplab@iitg.ernet.in

S.N. Joshi and U.S. Dixit (eds.), *Lasers Based Manufacturing*,
Topics in Mining, Metallurgy and Materials Engineering,
DOI 10.1007/978-81-322-2352-8_6

Keeping all these conditions in account laser beam has been considered to be most suitable heat source as it provides ease for precision and effective control over the process. As the process is a non-contact type forming and does not require any external force of contact it reduces the hard tooling cost. As the process is recommendable for the development of complex shapes, using laser as a heat source in this technique is quiet justifiable for obtaining large volume production with in a quick period of time. It has a significant value to the industries that previously relied on expensive stamping processes. For past few years many approaches were taken into account in process design. Design based on Genetic algorithm (Shimizu 1997; Cheng and Yao 2001) and based on response surface methodology (Liu and Yao 2002) were reported. Ueda et al. (1994) addresses the issue of determination of the scanning path was done on the basis of FEM by determining the in-plane strain. A suitable algorithm was developed by Jang and Moon (1998) for the determination of heating lines by calculating the line of curvature and principal curvature of the desired surface. But this method is limited to simple surfaces only. Edwardson et al. (2001) have developed certain rules for the positioning and sequencing of scanning path which are required for the development of 3-D curved surface, but the work has been found to be solely dependent on the prior experience for the development of the pattern and it is found to be minimum effective when the shapes are to be formed of more complex shapes. For implementing the technique of laser line heating effectively a suitable mathematical approach for the development of compound curve surface by evaluating strains from the surface to its planar development corresponds to forming from planar shape to curved surface by the process of line heating has been dealt with detail.

2 Basics of Differential Geometry of Surfaces

2.1 Surfaces

We need to formulate mathematically the notion of a surface; it is defined to be a subset S of R^3 is a surface if, for every point $P \in S$, there is an open set U in R^2 and an open set W in R^3 containing P such that $S \cap W$ is homeomorphic (Pressley 2001).

2.2 Regular Surfaces

A regular surface can be analysed in R^3, such that A subset $S \subset R^3$ is a regular surface if, for each $P \in S$, there exists a neighbourhood V in R^3 and a map $x{:}U \to V \cap S$ of an open set $U \subset R^2$ onto $V \cap S \subset R^3$ such that, x is differentiable, homeomorphism and obeys the regularity condition (do Carmo 1976).

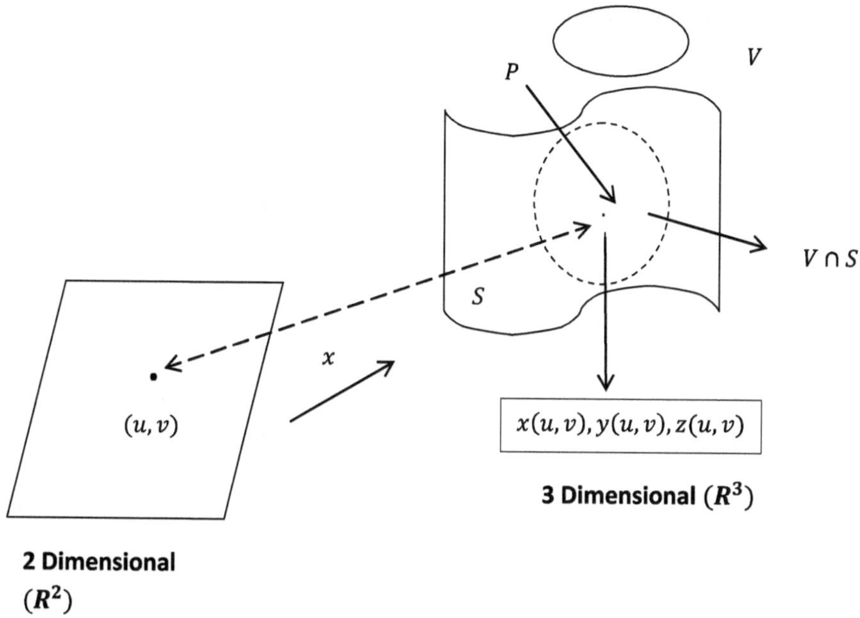

Fig. 1 Definition of regular surface

Fig. 2 Flow chart for obtaining curved surface by the process of line heating

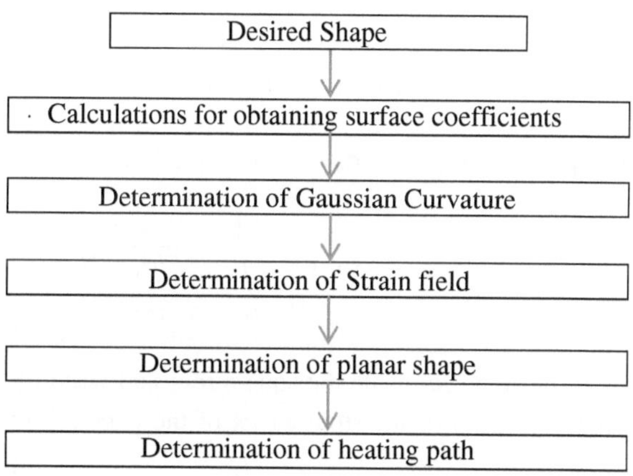

The mapping x is called a parameterization or a system of (local) coordinates in (a neighbourhood of) P as shown in Fig. 1. The neighbourhood $V \cap S$ of P in S is called a coordinate neighbourhood. For formulating the line heating process the following algorithm has been developed as shown in Fig. 2.

2.2.1 The First Fundamental Form

The first thing while working with a surface is to find the distance between the two points on a surface. The action to be conducted is usually be different from the distance between the points to be measured in three dimensional space, as the straight line segment which is covering the shortest path between the points in R^3 will generally be not present on the surface, the specified action which allows for computation of length, area is the first fundamental form of a surface.

Curve Length on a Surface

Considering $\gamma(t)$ is a curve in a surface patch (Pressley 2001) σ such that $\gamma(t) = \sigma(u(t), v(t))$, its arc length starts at a point $\gamma(t_0)$ is given by

$$s = \int_{t_0}^{t} \dot{\gamma}(u)du. \tag{1}$$

By chain rule, $\dot{\gamma} = \sigma_u \dot{u} + \sigma_v \dot{v}$ so,

$$\begin{aligned} \left\| \dot{\gamma}^2 \right\| &= (\sigma_u \dot{u} + \sigma_v \dot{v}) \cdot (\sigma_u \dot{u} + \sigma_v \dot{v}) \\ &= (\sigma_u \cdot \sigma_u)\dot{u}^2 + (\sigma_u \cdot \sigma_v)\dot{u}\dot{v} + (\sigma_v \cdot \sigma_u)\dot{v}\dot{u} + (\sigma_v \cdot \sigma_v)\dot{v}^2 \\ &= (\sigma_u \cdot \sigma_u)\dot{u}^2 + 2(\sigma_u \cdot \sigma_v)\dot{u}\dot{v} + (\sigma_v \cdot \sigma_v)\dot{v}^2 \\ &= E\dot{u}^2 + 2F\dot{u}\dot{v} + G\dot{v}^2 \end{aligned} \tag{2}$$

where, $E = \left\| \sigma_u^2 \right\|, F = \sigma_u \cdot \sigma_v, G = \left\| \sigma_v^2 \right\|$

$$\text{So, } s = \int_{t_0}^{t} \sqrt{(E\dot{u}^2 + 2F\dot{u}\dot{v} + G\dot{v}^2}dt. \tag{3}$$

So by bringing dt inside the square root and write $(\frac{du}{dt})^2(dt)^2 = du^2$, etc. we see that s is the integral of the square root of the expression:

$$Edu^2 + 2Fdudv + Gdv^2 \tag{4}$$

This is called the first fundamental form of $\sigma \cdot s = \int \sqrt{ds^2}$. From Eqs. (3) and (4) we can also write as:

$$ds^2 = Edu^2 + 2Fdudv + Gdv^2.$$

Surface Area

The area of the part σ(*R*) of surface patch $\sigma : U \rightarrow R^3$ corresponding to a region $R \subset U$ is

$$\iint \|\sigma_u \times \sigma_v\| du dv.$$

For finding the value of $\|\sigma_u \times \sigma_v\|$ we have used vector algebra. Considering *a*, *b*, *c* and *d* are vectors in R^3, then $(a \times b) \cdot (c \times d) = (a \cdot c)(b \cdot d) - (a \cdot d)(b \cdot c)$.

Applying it to

$$\begin{aligned} \|\sigma_u \times \sigma_v\|^2 &= (\sigma_u \times \sigma_v) \cdot (\sigma_u \times \sigma_v) \\ &= (\sigma_u \cdot \sigma_u)(\sigma_v \cdot \sigma_v) - (\sigma_u \cdot \sigma_v)^2 \\ &= EG - F^2 \end{aligned} \quad (5)$$

For regular surface, $EG - F^2 > 0$ everywhere, as for regular surface $\sigma_u \times \sigma_v$ is not zero. From the definition of area it can be written as: $\iint \sqrt{EG - F^2} du dv$.

2.2.2 Second Fundamental Forms

For introduction of the measurement of "Curvedness" of a surface "Second fundamental form" of surfaces need to be taken into consideration. For defining the curvatures of a surface S, considering a curve C on S passing through a point P as shown in Fig. 3, where t is the unit tangent vector, n. n is the unit normal vector and K. *K* is the curvature vector of the curve C at point P (Yu et al. 2000).

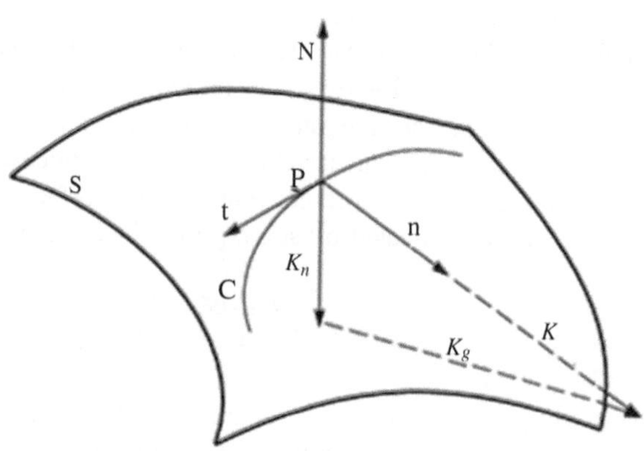

Fig. 3 Normal curvature

$$K = \frac{dt}{ds} = \text{Kn} = K_n + K_g \tag{6}$$

The unit normal vector at each point is given by

$$N = \frac{\sigma_u \times \sigma_v}{\|\sigma_u \times \sigma_v\|} \tag{7}$$

Normal Curvature and Principal Curvature

The normal curvature vector K_n can be expressed as

$$K_n = k_n N \tag{8}$$

The scalars K_n and K_g are called the normal curvature and the geodesic curvature respectively.

For an unit speed curve on a surface patch, its normal curvature is given by

$$K_n = L(u')^2 + 2Mu'v' + N(v')^2 \tag{9}$$

where L $= -\langle \sigma_u, N_u \rangle$, $M = -\langle \sigma_u, N_v \rangle = -\langle \sigma_v, N_u \rangle$, $N = -\langle \sigma_v, N_v \rangle$.

These are the coefficients of the second fundamental form.

The normal curvature can also be expressed as

$$K_n = \frac{L + 2M\lambda + N\lambda^2}{E + 2F\lambda + G\lambda^2} \tag{10}$$

where $\lambda = \frac{dv}{du}$.

The principle curvatures at a point of a surface are basically the maximum and minimum values of the normal curvature of all curves on the surface that passes through the point. Moreover, the principal vectors are the tangent vectors of the curves providing the maximum and minimum values. The basic geometric interpretation of the principal curvature is that it provides us information about the shape of the surface near a particular point on the surface. The principal curvatures of a surface can be evaluated by solving the quadratic equation:

$$\begin{vmatrix} L - kE & M - kF \\ M - kF & N - kG \end{vmatrix} = 0 \tag{11}$$

This will produce two roots k_1 and k_2 which are the principal curvatures at a point on a surface and are found to be real numbers.

2.2.3 Gaussian and Mean Curvatures

Gaussian and mean curvature together gives the same information as the two principal curvature gives. They have a greater significance towards the geometrical aspects of the surface. The Gaussian curvature has a remarkable property that it does not change when the surface is bent without stretching, which is not shared by principal curvatures. The knowledge of Gaussian curvature and mean curvature implies about the geometry of the surface.

From Eq. (10) the extreme values of K_n can be obtained by evaluating $\frac{dk_n}{d\lambda} = 0$, which gives after several algebraic manipulations:

$$k_n^2 - 2Hk_n + K = 0. \tag{12}$$

Here K and H are called Gauss (Gaussian) and mean curvature respectively. The values can be evaluated from the Eq. (11) of the principal curvature. Equation (11) can be solved as follows:

$$\begin{aligned} &\begin{vmatrix} L - kE & M - kF \\ M - kF & N - kG \end{vmatrix} = 0 \\ &\Rightarrow (L - kE)(N - kG) - (M - kF)^2 = 0 \\ &\Rightarrow \left(EG - F^2\right)k^2 - (LG - 2MF + NE)k + LN - M^2 = 0 \end{aligned} \tag{13}$$

Equation (13) is in the form of a quadratic equation which is of the form $ak^2 + bk + c = 0$, where sum of the roots is $-b/a$ and the product of the root is c/a. So $K = k_1k_2$ and $H = \frac{1}{2}(k_1 + k_2)$.

The values of K and H are found to be as follows:

$$K = \frac{LN - M^2}{EG - F^2} \tag{14}$$

$$H = \frac{EN + GL - 2FM}{2(EG - F^2)} \tag{15}$$

Alternatively, the Gaussian curvature K can be expressed as a function of E, F, G and their derivatives.

$$\begin{aligned} 4(EG - F^2)^2K = {} & E\left(E_vG_v - 2F_uG_v + G_u^2\right) \\ & + F(E_uG_v - E_vG_u - 2E_vF_v + 4E_uF_v - 2F_uG_u) \\ & + G\left(E_uG_u - 2E_uF_v + E_v^2\right) - 2(EG - F^2)(E_{vv} - 2F_{uv} + G_{uu}) \end{aligned} \tag{16}$$

3 Surface Development

Basically surface development can be done in two ways:

(i) It can be developed along iso parametric direction.
(ii) It can be developed along principal curvature direction.

Here surface developments based on strains along principal curvature directions are presented. As the principal curvature directions are independent of parameterization of surfaces and are unique except at umbilic points.

3.1 Determination of Strain Field

3.1.1 Formulation

Considering a surface r, which can be represented I parametric form i.e. $r(u, v)$, where u, v basically indicates the unit vector along u and v direction. The coefficients of the first fundamental form of the curved surface are given by:

$$E = r_u \cdot r_u,\ F = r_u \cdot r_v,\ G = r_v \cdot r_v \tag{17}$$

The strains due to development from curved surface to its plane development are $\varepsilon^s(u, v) \geq 0$ and $\varepsilon^t \geq 0$ along the maximum and minimum principal curvature directions respectively.

Therefore the small change in length $|r_s ds|$ changes to $(1 + \varepsilon^s)|r_s ds|$ and small length $|r_t dt|$ changes to $(1 + \varepsilon^t)|r_t dt|$, according to the definition of strain. So,

$$|R_s| = (1 + \varepsilon^s)|r_s|,\ |R_t| = (1 + \varepsilon^t)|r_t| \tag{18}$$

where R (u, v) is the planar development. R (u, v) can also be considered as a parametric surface with its first fundamental form coefficients defined by:

$$e = R_u \cdot R_u, f = R_u \cdot R_v,\ g = R_v \cdot R_v \tag{19}$$

Since,

$$R_s \cdot R_s = (R_u u_s + R_v v_s).(R_u u_s + R_v v_s) = eu_s^2 + 2fu_s v_s + gv_s^2 \tag{20}$$

and

$$\begin{aligned} r_s \cdot r_s &= (r_u u_s + r_v v_s).(r_u u_s + r_v v_s) \\ &= Eu_s^2 + 2Fu_s v_s + Gv_s^2 \end{aligned} \tag{21}$$

Using the relations (18)–(21), we obtain

$$eu_s^2 + 2fu_s v_s + gv_s^2 = (1 + \varepsilon^s)^2 (Eu_s^2 + 2Fu_s v_s + Gv_s^2) \tag{22}$$

Similarly, along minimum principal curvature direction, we have

$$eu_t^2 + 2fu_t v_t + gv_t^2 = (1 + \varepsilon^t)^2 (Eu_t^2 + 2Fu_t v_t + Gv_t^2) \tag{23}$$

Assuming that after development the principal curvature directions remain orthogonal, this gives,

$$R_s \cdot R_t = (R_u u_s + R_v v_s) \cdot (R_u u_t + R_v v_t) = 0 \tag{24}$$

(As the two vectors make an angle of 90° to each other, their dot product is zero) e, f and g are already defined in expression (19), so by simplifying Eq. (24) we get

$$\mathrm{e}u_s u_t + \mathrm{f}(u_s v_t + u_t v_s) + \mathrm{g}v_s v_t = 0 \tag{25}$$

The three linear equations (22), (23) and (25) and the solution is given by

$$\mathrm{e} = \frac{v_t^2 [Eu_s^2 + 2Fu_s v_s + Gv_s^2](1 + \varepsilon^s)^2}{(v_s u_t - u_s v_t)^2} + \frac{v_s^2 [Eu_t^2 + 2Fu_t v_t + Gv_t^2](1 + \varepsilon^t)^2}{(v_s u_t - u_s v_t)^2} \tag{26}$$

$$\mathrm{f} = \frac{u_t v_t [Eu_s^2 + 2Fu_s v_s + Gv_s^2](1 + \varepsilon^s)^2}{(v_s u_t - u_s v_t)^2} - \frac{u_s v_s [Eu_t^2 + 2Fu_t v_t + Gv_t^2](1 + \varepsilon^t)^2}{(v_s u_t - u_s v_t)^2} \tag{27}$$

$$\mathrm{g} = \frac{u_t^2 [Eu_s^2 + 2Fu_s v_s + Gv_s^2](1 + \varepsilon^s)^2}{(v_s u_t - u_s v_t)^2} + \frac{u_s^2 [Eu_t^2 + 2Fu_t v_t + Gv_t^2](1 + \varepsilon^t)^2}{(v_s u_t - u_s v_t)^2} \tag{28}$$

We minimize the strains $\varepsilon^s(u, v)$ and $\varepsilon^t(u, v)$ satisfying the condition that after adding these strains to a doubly curve surface along principal curvature directions, the surface maps to a planar shape on which a Gaussian curvature is zero. This results into a minimization problem. The constrained minimization problem is discretized by using finite difference method and trapezoidal rule of integration.

3.2 Determination of the Planar Developed Shape

Solving the non-linear minimization problem, we obtain the strains ε^s and ε^t at all grid points. The first fundamental form coefficient e, f, g of the planar developed shape is then obtained from Eqs. (26)–(28). For determination of planar coordinates

(X_{ij}, Y_{ij}) of the grid points at the corresponding planar development. These coordinates (X_{ij}, Y_{ij}) should satisfy the following equations at all grid points:

$$R_u \cdot R_u = \mathrm{e},\ R_u \cdot R_v = \mathrm{f},\ R_v \cdot R_v = \mathrm{g} \tag{29}$$

After discretizing the Eq. (29) using finite difference method (central difference for internal points and forward and backward difference for boundary points), we obtain a system of non-linear polynomial equations.

From stability point of view, the strain energy needs to be in minimal range for more stableness. So without calculating the distance directly, we can solve the following unconstrained minimization problem:

$$\sum_{i=1}^{N_g^u} \sum_{j=1}^{N_g^v} ((R_u \cdot R_u)_{at\,ij} - \mathrm{e}_{ij})^2 + ((R_u \cdot R_v)_{at\,ij} - \mathrm{f}_{ij})^2 + ((R_v \cdot R_v)_{at\,ij} - \mathrm{g}_{ij})^2 \tag{30}$$

The optimization problem can be solved by using quasi-newton method.

3.3 Heating Line Generation

Heating paths can be determined from information of direction of principal compressive strains (Clausen 2000). Heating along a line leads to bending along a direction perpendicular to that of the heating path. There is a combination of both bending and shrinkage in the heated region. When the heating paths are to be traced, the option is to be made between principal directions from either bending or shrinkage separately. The lines for bending are to be determined first, which is to be followed by shrinkage lines.

Heating patterns can be obtained based on the direction of principal bending of the target surface by the use of differential geometry.

3.3.1 Bending Paths

The target surface is given as a function, the principal bending vectors, $\overrightarrow{v_{1,2}}$, can be obtained from the Eigen value problem, i.e.

$$\left| \begin{bmatrix} L & M \\ M & N \end{bmatrix} - k \begin{bmatrix} E & F \\ F & G \end{bmatrix} \right| = 0 \tag{31}$$

where k is the Eigen value (principal curvature) and $\{E, F, G\}, \{L, M, N\}$ are the coefficients of first and second fundamental form respectively.

Once the Eigen values were obtained, which are basically the principal curvature, the corresponding Eigen vectors (principal curvature directions) can be evaluated from

$$\left[\begin{bmatrix} L & M \\ M & N \end{bmatrix} - k_1 \begin{bmatrix} E & F \\ F & G \end{bmatrix}\right] \begin{bmatrix} v_1 \\ v_2 \end{bmatrix} = \begin{bmatrix} 0 \\ 0 \end{bmatrix} \tag{32}$$

where k_1 is the largest principal curvature and $\vec{v}$ is the corresponding Eigen vector. The matrix is a singular matrix as the coefficient is equal to zero. It can be written as:

$$(L - k_1 E)v_1 + (M - k_1 F)v_2 = 0 \tag{33}$$

$$\text{Or } \overrightarrow{v_1} = S\left(-\left(\frac{M - k_1 F}{L - k_1 E}\right), 1\right), \quad \text{S} \in \text{R}. \tag{34}$$

If for instance, the surface, $\vec{x}$, is given as a function of x and y as $\vec{x} = x\vec{e}_1 + y\vec{e}_2 + f(x, y)\vec{e}_3$

Then the coefficients of the first and second fundamental form are simply:

$$E = 1 + (f_{,x})^2 \tag{35}$$

$$F = f_{,x} f_{,y} \tag{36}$$

$$G = 1 + (f_{,y})^2 \tag{37}$$

$$L = \frac{f_{,xx}}{\sqrt{1 + (f_{,x})^2 + (f_{,y})^2}} \tag{38}$$

$$M = \frac{f_{,xy}}{\sqrt{1 + (f_{,x})^2 + (f_{,y})^2}} \tag{39}$$

$$N = \frac{f_{,yy}}{\sqrt{1 + (f_{,x})^2 + (f_{,y})^2}} \tag{40}$$

where, x denotes differentiation with respect to x and, y denotes differentiation with respect to y.

When the vector field for the principal directions of curvature has been calculated, it must be traced into heating lines. Heating produces more bending perpendicular to heating directions than along it. Therefore bending paths can be traced perpendicular to the field.

3.3.2 Shrinkage Path

The largest shrinkage direction is not necessarily perpendicular to the heating path. Differential geometry can be used, provided that a function for the target surface is known. From the strain field of the elastic analysis the principal direction, θ_p and the principal strains $\varepsilon_{1,2}$ can be computed from the x and y components of the strain field, $\varepsilon_x, \varepsilon_y$ and ε_{xy}.

The z components of the strain are not needed as this is assumed to be zero in the principal direction is found from the plane strain formulation.

$$\theta_p = \frac{1}{2}\tan\left(\frac{\varepsilon_{xy}}{\varepsilon_x - \varepsilon_y}\right) \tag{41}$$

$$\varepsilon_{1,2} = \frac{1}{2}\left(\varepsilon_x + \varepsilon_y \pm \sqrt{\left(\varepsilon_x - \varepsilon_y\right)^2 + \varepsilon_{xy}^2}\right) \tag{42}$$

Equation (39) is may be the direction of either of the two perpendicular principal directions in the plane. As the direction for ε_1 is needed, ε_1 and ε_2 can be compared with (37) for finding the principal strains based on θ_p. The strains belong to the direction:

$$\varepsilon_{test} = \frac{1}{2}\left(\left(\varepsilon_x + \varepsilon_y\right) + \left(\varepsilon_x - \varepsilon_y\right)\cos 2\theta_p + \varepsilon_{xy}\sin 2\theta_p\right) \tag{43}$$

If the found ε_{test} is equal to ε_1 the angle θ_p used in (41) was in fact the wanted. Else it belongs to the other principal strain.

Previous to the tracing of strain directions into actual heating paths, requirement of distance between the heating lines must be determined. This can be explained in the following way: The displacement of heating line

$$d_L = \frac{S_y}{\varepsilon_1} w^p \tag{44}$$

where S_y, w^p, ε_1, d_L are shrinkage strain, width of the heating line, elastic analysis strain and distance between line of interest. Thus heating line can be traced.

4 Conclusions

From the above investigation the following conclusions can be made:

(i) A mathematical model has been proposed using differential geometry as a tool.
(ii) Detail step by step mathematical formulation have been configured for developing a doubly curve surface from a flat plate or blank.

(iii) A suitable step has been put forwarded for finding out desired optimum blank size.
(iv) Procedure for obtaining optimum heating line generation for the development of doubly curve surface has been suitably incorporated.

Overall a step by step problem solving procedure has been proposed mathematically, with the involvement of mechanical parameters for the process of laser line heating.

References

Shimizu, H. (1997). *A heating process algorithm for metal forming by a moving heat source.* Master's thesis, Masachusetts Institute of Technology, MA.

Liu. C., & Yao, Y. L. (2002). Optimal and robust design of laser forming process. *Journal of Manufacturing Processes*, *v4*(n1), 000–000.

Ueda, K., Murakawa, H., Rashwan, A. M., Okumoto, Y., & Kamichika, R. (1994a). Development of computer-aided process planning system for plate bending by line heating (report 1)—relation between final form of plate and inherent strain. *Journal of Ship Production*, *v10*(n1), 59–67.

Edwardson, S. P., Watkins, K. G., Dearden, G., & Magee, J. (2001). 3D laser forming of saddle shape. In Proceedings of LANE 2001.

Pressley, A. (2001). *Elementary Differential Geometry, Springer Undergraduate Mathematics Series ISSN 1615-2085*. London: Springer.

do Carmo, M. P. (1976). Differential Geometry of Curves and Surfaces. Englewood Cliffs, New Jersey: Prentice-Hall Inc..

Yu, G., Patrikalakis, N. M., & Takashi, M. (2000). Optimal development of doubly curved surfaces. *Computer Aided Geometric Design, 17*, 545–577.

Clausen, H. B. (2000). Plate forming by line heating. PhD thesis, Department of Naval Architecture and Offshore Engineering, Technical University of Denmark.

Cheng, J., Yao, Y. L. (2001). Cooling effects in multiscan laser forming. *Journal of Manufacturing Processes*, *3*(1), 60–72.

Jang, C. D., Moon, S. C. (1998). An algorithm to determine heating lines for plate forming by line heating method. *Journal of Ship Production*, *14*, 238–245.

Surface Alloying of Aluminum with Copper Using CO_2 Laser

Woldetinsay G. Jiru, Mamilla R. Sankar and Uday S. Dixit

Abstract Aluminium and its alloys have high demand in manufacturing and service industries due to their high specific strength. Addition of different metals like Cu, Mg, Ni, Cr, and Zn provides enhanced service life. In this work, commercially available 99 % pure aluminium was alloyed with copper powder of 10 μm particles size, which was melted by CO_2 laser. Three different methods were used for uniform placing of 95 % copper powder and 5 % aluminium powder on the aluminium substrate. The result was examined by Vickers hardness test. SEM and FESEM were used for studying surface and subsurface defects. Defect free aluminium alloy with improved microstructure and enhanced mechanical properties was obtained.

Keywords Surface alloying · CO_2 laser · Copper powder · Aluminium

1 Introduction

Most applications in real world require specific surface properties of components with good corrosion resistance, wear resistance and hardness. However, it is not easy to get alloys with such properties since they are expensive due to the cost of producing those parts. Investigations have been carried out to improve the properties and to reduce the cost of components made of Aluminium alloys. An important and recent technology is the laser surface alloying, which is a cost-effective technique to improve the surface properties of materials using laser beam

W.G. Jiru · M.R. Sankar · U.S. Dixit (✉)
Department of Mechanical Engineering, Indian Institute of Technology, Guwahati 781039, India
e-mail: uday@iitg.ac.in

W.G. Jiru
e-mail: woldetinsay@iitg.ac.in

M.R. Sankar
e-mail: evmrs@iitg.ac.in

S.N. Joshi and U.S. Dixit (eds.), *Lasers Based Manufacturing*, Topics in Mining, Metallurgy and Materials Engineering, DOI 10.1007/978-81-322-2352-8_7

heat (Hecht 2011). In this technique, the structure and physical characteristics at the surface get changed. Laser surface alloying utilizes a high power density laser beam to melt the metal coatings (alloying elements) and the surface of substrate material. This melting takes place in a very short period of time while bulk material remains cold and acts as heat sink for self-quenching and fast solidification phenomenon.

Laser surface alloying is similar to laser surface melting except that in the former, alloying materials are added into the melt pool in order to impart its properties as per requirements. It is also similar to laser surface cladding, but laser surface alloying requires more energy compared to cladding. The surface alloying improves corrosion resistance, oxidation and solidification resistance, wear resistance, mechanical properties, electrical & electronic properties, thermal insulation and aesthetic appearance. By improving the surface, it reduces the frictional energy loss.

2 Significances of Surface Alloying

Failure of engineering components due to some reasons such as mechanical, chemical or electrochemical interaction with the surrounding environment initiates at the surface due to the high intensity of external stress and environmental attack. To develop the resistance of materials, laser surface alloying is one method to modify the surface properties of a material without affecting its bulk property. The intimate contact between the melt and the solid substrate causes a very fast heat extraction during solidification resulting in very high cooling rates of the order of 10^5–10^8 K/s. The high cooling rates to which this surface layer is subjected result in the formation of different microstructures from bulk metal leading to improved surface properties (Dubourg et al. 2002).

Adding alloying elements into the melted pool makes them inter diffuse into the substrate material. Soon the solidification begins from the liquid-solid interface towards the direction of substrate material. It has been reported that the structure of alloyed zone depends on many properties like the particle size of the alloying material, power of laser irradiation and duration of irradiation time. Alloying brings changes in surface composition and microstructure of a material to give it certain desired properties. The benefits of rapid solidification by laser power for surface modification are as follows: fine grain structure due to fast solidification (Kannatey-Asibu 2009), reduced micro segregation and extended solubility of alloying elements.

3 Significance of (CO_2) Laser for Surface Alloying

Carbon dioxide laser is the most popular laser used in industrial material processing because of its high energy efficiency and high average optical power output. It uses N_2 gas for facilitating the pumping and population inversion process by providing molecule collisions with CO_2 gas.

In addition He gas is used for bringing the N_2 gas to ground state by from the lower excited state. CO_2 laser can be operated both in continuous and pulsed modes for material processing. In this work the continuous mode is used.

A CO_2 laser has the following characteristics (Dahotre 1998):

- Wave length of 10.6 μm. It is in the far infra-red spectrum thus not visible.
- It can be operated in continuous wave (CW) mode where the output power is at constant level, or pulse mode where the laser output power can be pulsed.
- The laser beam is well absorbed by organic materials and ceramics but poorly absorbed by metallic materials.
- Mirrors rather than optical fibers are used for beam delivery since normal glass or silica in optical fibers is opaque to the CO_2 laser beam.

4 Details of Experiments

In the present work, commercially pure Aluminium of 99 % purity was alloyed with copper powder of 10 μm particle size and melting was done by continuous CO_2 laser power. During alloying 50 % overlap is provided between consequent beads. After cooling, re-melting was carried out at orthogonal direction to alloyed layers.

Sample dimension used was 110 mm × 50 mm × 6 mm. Table 1 shows substrate material composition tested by analytical technique used for the elemental analysis in energy-dispersive X-ray spectroscopy (EDX). Figure 1 shows the schematic diagram of alloying used for this experiment.

In view of the importance of Aluminium for automobile, aerospace and many other sectors, this chapter investigates the surface alloying of aluminium by copper. Three methods of alloying have been used. The mechanical properties and microstructures of surface alloyed samples have been compared. Table 1 shows the substrate material composition.

Table 1 Chemical composition of substrate material

Element	Cu	Si	Fe	Al
Composition (wt%)	0.12	0.14	0.59	Balance

Fig. 1 Schematic of laser alloying

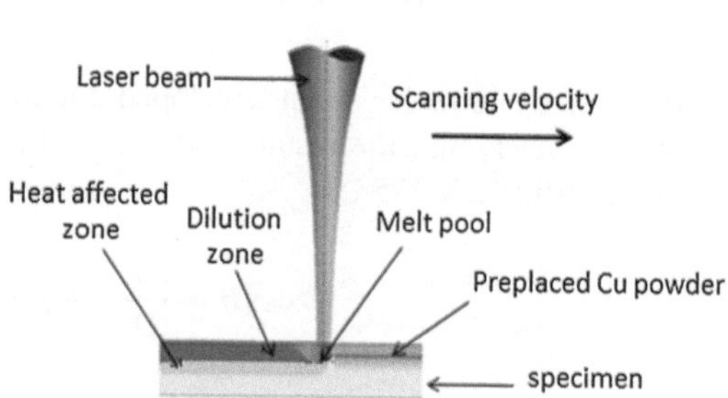

Table 2 Process parameters for surface alloying Al with Cu

S. No.	Process parameters for alloying				Process parameters for re-melting				Re-melting
	LP (kW)	SOD (mm)	SS (mm/min)	LSD (mm)	LP (kW)	SOD (mm)	SS (mm/min)	LSD (mm)	
S1	1.7	30	600	4.05	1.6	30	600	4.05	Yes
S2	1.7	30	500	4.05	1.6	30	600	4.05	
S3	1.7	30	500	4.05	1.6	40	800	5.39	
S4	1.7	40	500	5.39	1.6	40	800	5.39	
S5	1.7	30	400	4.05	–	–	–	–	No
S6	1.6	30	500	4.05	1.6	40	800	5.39	Yes
S7	2.0	40	400	5.39	–	–	–	–	No
S8	1.6	30	400	4.05	1.7	30	500	4.05	Yes
S9	1.8	30	400	4.05	1.6	40	800	5.39	
S10	1.7	30	600	4.05	1.7	30	500	4.05	
S11	1.8	30	500	4.05	1.5	40	400	5.39	
S12	1.7	40	600	5.39	–	–	–	–	No
S13	1.8	40	600	5.39	1.5	40	800	5.39	Yes
S14	1.6	30	600	4.05	1.6	40	800	5.39	
S15	1.7	30	400	4.05	1.7	30	500	4.05	

LP Laser power, *SOD* Stand of distance, *SS* Scanning speed, *LSD* Laser spot diameter

Three methods used for depositing the copper powder are as follows:

(a) Stirring of metal powder with Fevicol and then painting it on substrate to achieve 0.5 mm coating for samples 12, 13, 14, 15.
(b) First painting the substrate by Fevicol, then applying powder uniformly and finally compacting it to achieve total coating thickness of 1 mm for samples 4, 5, 6, and 9.
(c) First painting the substrate by Fevicol, then applying powder and blowing out excess powder to achieve total coating thickness of 0.5 mm for samples 1, 2, 3, 7, 8, 10 and 11. Table 2 shows the process parameters in different experiments.

5 Results and Discussions

5.1 Crystal Size and Lattice Strain

The crystal size and lattice strain developed due to thermal effect was examined for Sample 4 by XRD and the result is shown in Fig. 2. Strain was calculated from (Williamson and Hall 1953):

$$\beta \cos \theta = \frac{K\lambda}{D} + 4\eta \sin \theta \tag{1}$$

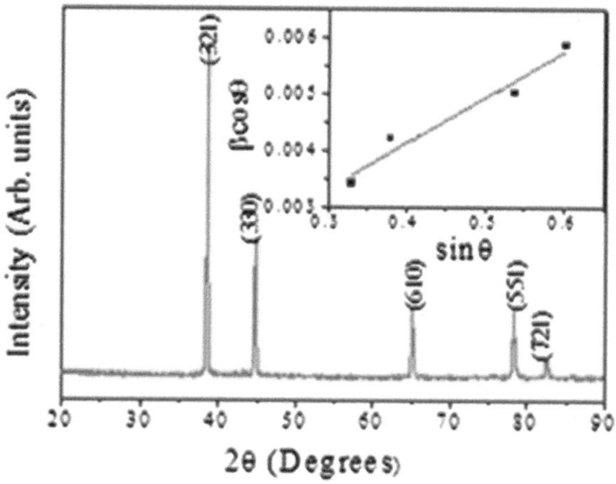

Fig. 2 X-ray diffraction pattern of Al_4Cu_9

where β is the full width at half maximum, K is a constant, λ is the X-ray wave length of Cuk_α radiation (λ = 1.5418 Å), D is the average crystal size and η is the lattice strain. The strain in the sample was 0.002 as observed in the inset of Fig. 2 and average size of crystallite is 152.06 nm. The crystal is cubic in structure as justified from the International Centre for Diffraction Data (ICDD)#: 02-1254, confirming the formation of Al_4Cu_9 phase.

5.2 Surface Roughness

Roughness plays an important role in determining how a real object will interact with its environment. Surfaces with roughness usually wear more quickly and have higher friction coefficients than smooth surfaces. The performance of a mechanical component highly depends on the condition of the surface finish, since irregularities in the surface may form nucleation sites for cracks or corrosion.

In the present work, the surfaces of the alloyed layers were tested without polishing the surface. Taylor Hobson make CCI Light, non-contact optical profilometer was used to check surface quality of alloyed samples. Figure 3a shows Sample 2, where the area of non-uniform coating resulted in void surface with larger peaks and valleys and Fig. 3b is Sample 9, with uniform deposited and alloyed surface with centre average roughness value (R_a) up to 1.9 μm.

5.3 Micro Hardness Analysis

Micro-hardness test was conducted on thickness side of the samples by Vickers hardness test using 500 gf and the results are plotted in Fig. 4. Maximum hardness

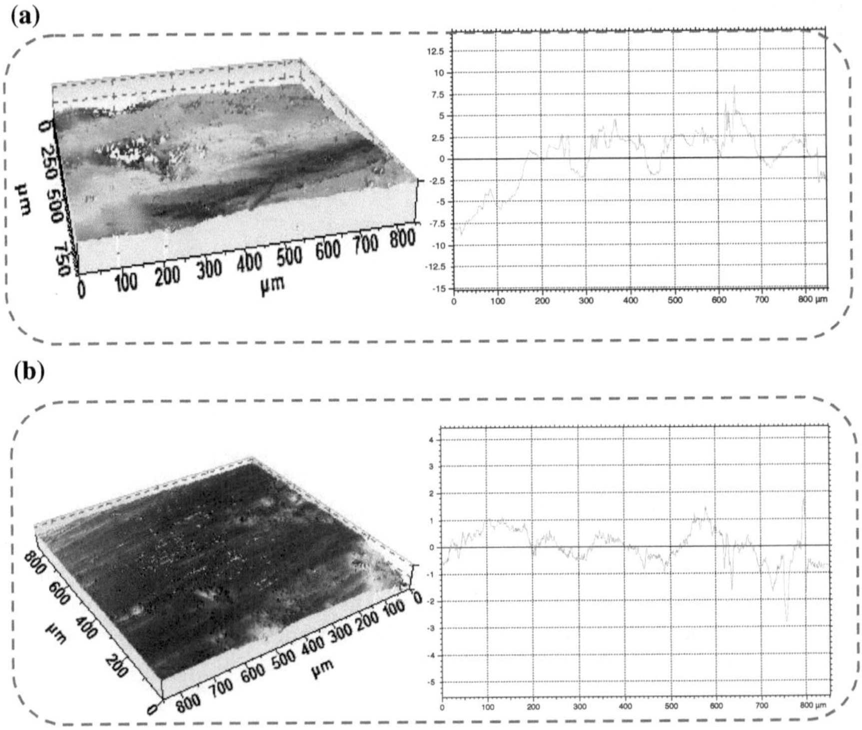

Fig. 3 Surface topography and corresponding line surface roughness of alloyed samples. **a** non uniform coated surface, **b** uniform coated surface

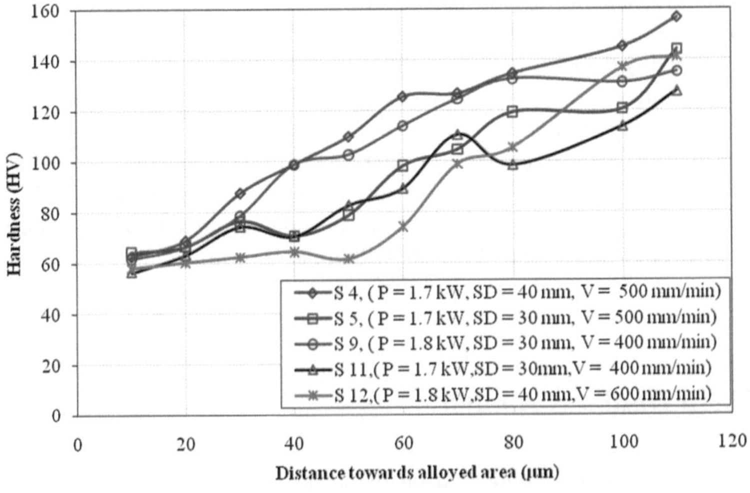

Fig. 4 Vickers hardness variation towards the alloyed area on the specimen

achieved is 156 *HV* at laser power of 1.7 kW laser beam diameter of 5.39 mm and scanning speed of 500 mm/min. The re-melting was done in orthogonal direction to the alloyed direction at laser power of 1.6 kW and scanning speed of 800 mm/min. The re-melting was done after alloying improved hardness (Pinto et al. 2003). Samples 4 and 9, surface alloyed with 1 mm deposition thickness of alloying element, resulted in improved hardness compared to other samples.

5.4 Microstructure and Morphology Analysis

A pure metal normally solidifies at a constant temperature with the inter-face between the solid and liquid media being planar for an alloy. Figure 5a for sample 4 and Fig. 5b for sample 9 show the morphological result at the solidified regions analysed by scanning electron microscopy (SEM). It shows the Cu growth in Al_4Cu_9 inter metallic phase with a fine microstructure region (Dubourg et al. 2002). After initial cellular growth from the bottom of the molten pool, the solidifying area developed lateral instability favourable for forming homogeneous structure.

Figure 6a shows uniformity of Cu growth into white Al primary grains and dark eutectic well distributed forming inter-metallic of Al_4Cu_9 phase. Figure 6b shows the grain boundary crack that contain appreciable amounts of copper, when the cooling rate is insufficient to prevent precipitation on grain boundaries at higher temperature and hence is susceptible to stress corrosion cracking (SCC) and micro-crack initiation (van Otterloo et al. 1995).

Atoms of Cu lower their free energy when they migrate to lattice defects such as grain boundaries, phase boundaries and dislocation. In Fig. 7a marked region 'A' shows un-melted region. A regular growth of binary Al_4Cu_9 eutectic morphology as

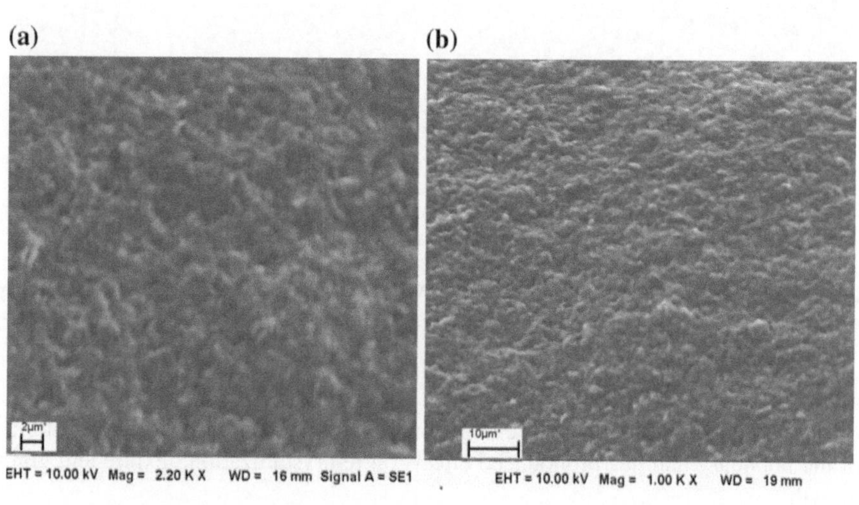

Fig. 5 Scanning electron microscopic for morphology test **a** sample 4 **b** sample 9

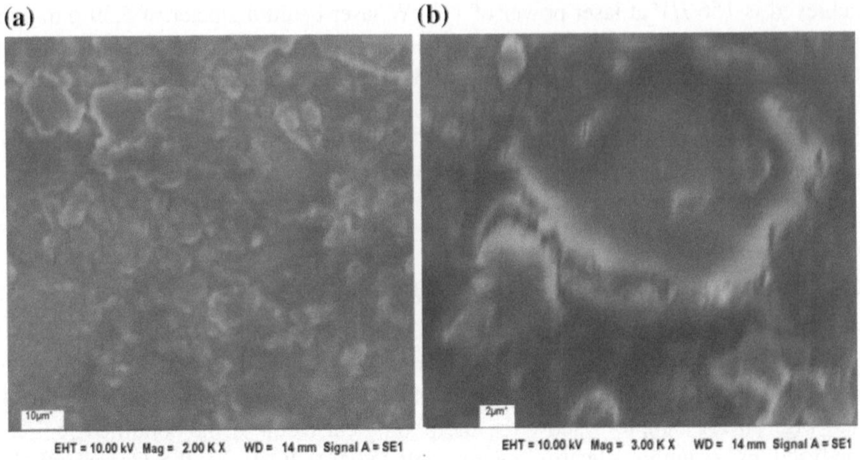

Fig. 6 SEM photograph of **a** sample 9 with laser alloying parameters $P = 1.8$ kW, SOD = 30 mm, $SS = 400$ mm/min and re-melting $P = 1.6$ kW, SOD = 40 mm and $SS = 800$ mm/min. **b** sample 13 with laser alloying parameters $P = 1.8$ kW, SOD = 40 mm, SS = 600 mm/min, re-melting = 1.5 kW, SOD = 40 mm, $SS = 800$ mm/min)

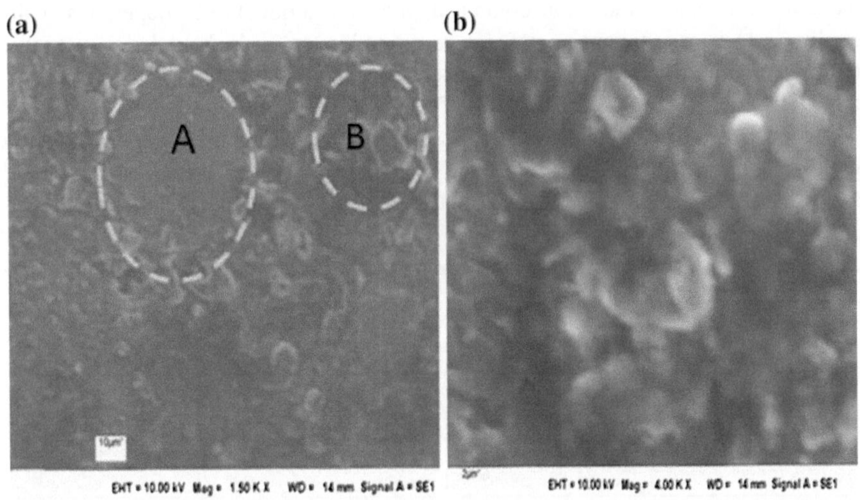

Fig. 7 SEM photograph of sample 5, $P = 1.7$ kW, SOD = 30 mm, SS = 400 mm, no re-melting **a** normal view **b** expanded view of region B

a result of uniform and fast cooling time resulted in fine microstructure as shown in Fig. 7b at higher magnification. Segregation to grain boundaries affects the mobility of the boundary and has pronounced effects on re-crystallization, texture and grain growth.

5.5 Porosity Challenges in Alloying

Porosity is not a usual problem in surface alloying like in other processes such as welding. However, if proper care is not taken it may cause material failure during service life. Porosity is the presence of gas pores or pockets within the fusion zone in the form of discontinuity formed by gas entrapment during solidification phenomenon. It occurs when gases such as oxygen, nitrogen and hydrogen are absorbed from atmosphere or developed when chemical reactions takes place between substrate and alloying elements during melting and solidification phenomenon.

The sources of these gases may also arises from contaminated substrate material or wetting conditions. For example, present experiments show the existence of gas pores or pockets within the alloyed area for samples first painted by adhesive followed by the copper powder deposition. Figure 8a shows a photograph and Fig. 8b a SEM photo. The voids are formed due to gas entrapment during solidification which are likely to occur when the rate of solidification is fast and there is no adequate time for the evolved gas to escape.

The laser parameter for alloying are P = 1.8 kW, SOD = 30 mm and SS = 500 mm/min. Re-melting laser parameters are P = 1.5 kW and stand of distance SOD = 40 mm and scanning velocity SS = 400 mm/min.

The quality of alloyed layers should be free of any cracks which may form under the subsurface layers due to some reasons. Powders of alloyed regions were tested by field emission scanning electron microscopy (FESEM) to check whether cracks and defects are formed at subsurface of alloyed regions. To check for it powders from alloyed layers were prepared and tested by the above mentioned machine for well alloyed layers selected. Figure 9 shows fractography results. No defects were observed.

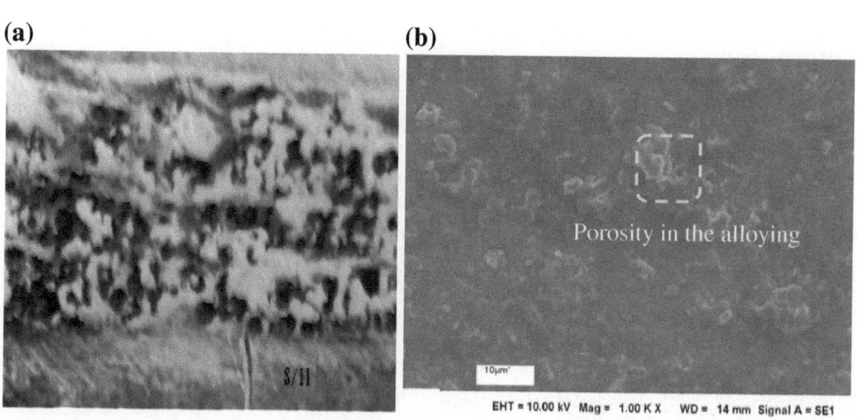

Fig. 8 Sample 11 **a** photograph, **b** SEM photo

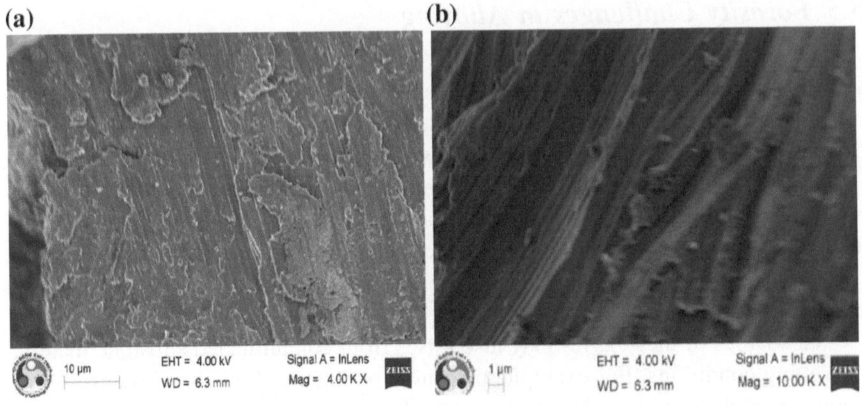

Fig. 9 Fractography tests by FESEM from powder of alloyed areas **a** sample 4, **b** sample 9

5.6 Conclusions

Surface alloying of commercially pure Aluminium by 10 μm copper powder was carried out. The following are the main observations:

- Uniformly deposited Cu powder by compacting method resulted in good surface alloyed layer with average surface roughness of 1.9 μm.
- For laser alloying at 1.7 kW and scanning speed of 500 mm/min and re-melting at lower power of 1.6 kW in orthogonal direction to alloyed layer resulted in improved micro hardness from 60 to 156 *HV*.
- Formation of fine microstructure with Al_4Cu_9 intermetallic phase developed due to fast solidification and uniform alloying of Cu powder on substrate material.

References

Dahotre, N. B. (Ed.). (1998). *Lasers in surface engineering (Vol. 1)*. USA: ASM International.

Dubourg, L., Pelletier, H., Vaissiere, D., Hlawka, F., & Cornet, A. (2002). Mechanical characterization of laser surface alloyed aluminium–copper systems. *Wear, 253*(9), 1077–1085.

Hecht, J. (2011). *Understanding lasers: An entry-level guide (Vol. 21)*. New York: Wiley.

Kannatey-Asibu, E, Jr. (2009). *Principles of laser materials processing (Vol. 4)*. New York: Wiley.

Pinto, M. A., Cheung, N., Ierardi, M. C. F., & Garcia, A. (2003). Microstructural and hardness investigation of an aluminum–copper alloy processed by laser surface melting. *Materials Characterization, 50*(2), 249–253.

van Otterloo, J. L. D. M., Bagnoli, D., & De Hosson, J. T. M. (1995). Enhanced mechanical properties of laser treated Al-Cu alloys: A microstructural analysis. *Acta Metallurgica et Materialia, 43*(7), 2649–2656.

Williamson, G. K., & Hall, W. H. (1953). X-ray line broadening from filed aluminium and wolfram. *Acta Metallurgica, 1*(1), 22–31.

Effect of Pulsed Nd:YAG Laser Parameters in Preplaced TiC Coating on Aluminium Substrate

Chinmaya Kumar Sahoo, Jageshwar Kumar Sahu and Manoj Masanta

Abstract In this chapter basic mechanism of laser surface modification technique has been discussed. Based on the process mechanism laser surface modification techniques are classified and discussed in brief. Specific advantages, disadvantages and applications of these processes are also explained. A brief review on laser surface modification of Aluminium substrate using different coating materials, surface modification of various engineering materials with TiC and surface modification of metallic substrate by utilizing pulse type low power laser have been presented. Finally, experimental details of TiC coating on pure Aluminium substrate using a pulsed Nd:YAG laser has been discussed. Effect of laser peak power and pulse overlapping on the Aluminium substrate by pre-placing TiC powder has been analyzed experimentally. Optical images of the cross-sectional view of the laser irradiated samples show successful formation of coating. Experimental results also show that, increase in laser peak power and pulse overlapping increases the coating thickness. High peak power results in removal of coating material but produces TiC mixed Aluminium layer at the top surface of substrate. Microhardness profile on the coating shows improvement in hardness up to 20 times than that of as received Aluminium substrate.

Keywords Pulsed Nd:YAG laser · TiC coating · Aluminium · Laser surface modification

C.K. Sahoo · J.K. Sahu · M. Masanta (✉)
Department of Mechanical Engineering, NIT, Rourkela, Rourkela 769008, India
e-mail: manoj.masanta@gmail.com

C.K. Sahoo
e-mail: chinmaya.cks@gmail.com

J.K. Sahu
e-mail: jageshwarkumarsahu@gmail.com

S.N. Joshi and U.S. Dixit (eds.), *Lasers Based Manufacturing*,
Topics in Mining, Metallurgy and Materials Engineering,
DOI 10.1007/978-81-322-2352-8_8

1 Introduction

Laser coating is an advance process in which surface properties of metallic components are improved by depositing materials with superior metallurgical properties i.e. higher hardness, wear resistance and corrosion resistance as compared to the base material, by using the heat source of laser beam. Laser coatings usually possess uniform composition, coating thickness and exhibit superior metallurgical bonding to the base material (Kadolkar and Dahotre 2003). In order to maintain the original properties of the coating materials, dilution of coating into the substrate should be minimum. Components coated with laser coating process produce extremely dense, crack-free and non-porous surfaces with high resistance against wear and corrosion (Vilar 1999). In addition to manufacturing of new component, laser coating or cladding process also used for repairing of various engineering component. Figure 1 shows the schematic diagram of a typical laser surface modification system which mainly consist of a laser source, beam delivery system with mirror and lens arrangement, CNC controlled 3-axis motion table and shielding gas supply.

2 Various Types of Laser Surface Modification Techniques

Laser surface modification is a broad area of laser material processing, basically used to change either microstructure or composition of substrate surface with or without addition of some new materials, to provide certain desired properties such as high hardness, wear resistance and corrosion resistance. Based on the working mechanism laser surface modification techniques are classified as follows:

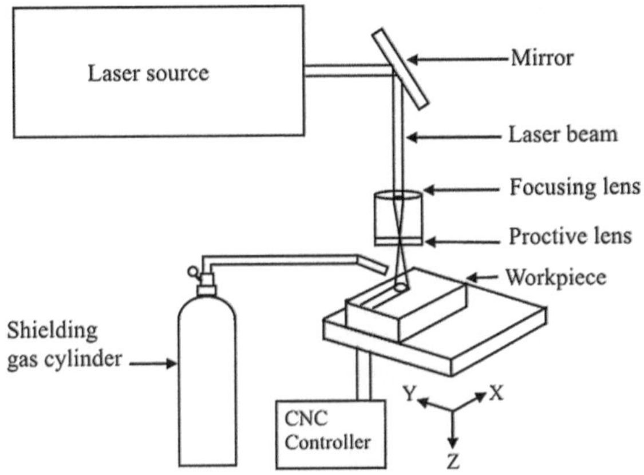

Fig. 1 Schematic diagram of a laser surface modification system

2.1 Laser Surface Modification Without Use of External Material

In this method of laser surface modification, no external material is used for changing the surface properties. By changing the microstructure of the surface through rapid heating and cooling or by melting a thin layer and subsequent rapid self-quenching, surface layer with enhanced mechanical properties are formed, which is harder than base material. Laser surface modification without use of any external material can be done by following ways.

2.1.1 Laser Surface Heat Treatment

In laser surface heat treatment, surface of base material is exposed to thermal cycle of rapid heating and cooling in such a way that, surface layers induce phase transformation. Due to laser heating and subsequent rapid cooling, structure of steel surface changes from austenitic to martensitic structure, which is harder than base material and possesses higher wear resistance. In laser surface heat treatment process material does not melt, transformation occur in the solid state only. Figure 2 shows the schematic of laser surface heat treatment process.

2.1.2 Laser Surface Melting and Re-solidification

In laser surface melting and re-solidification, a thin surface layer of base material is melted by high intensity laser beam, which subsequently solidifies due to mass effect of the bulk material (self-quenching), to form non-equilibrium structures. Figure 3 shows the schematic diagram of laser melting process, where temperature rise due to laser irradiation should be higher than the melting temperature of the material. Depending on the cooling rate, that is again depends on the laser power, interaction time of laser beam with material and its thermal properties, solidification of molten layer produces some metastable phases. These phase change in

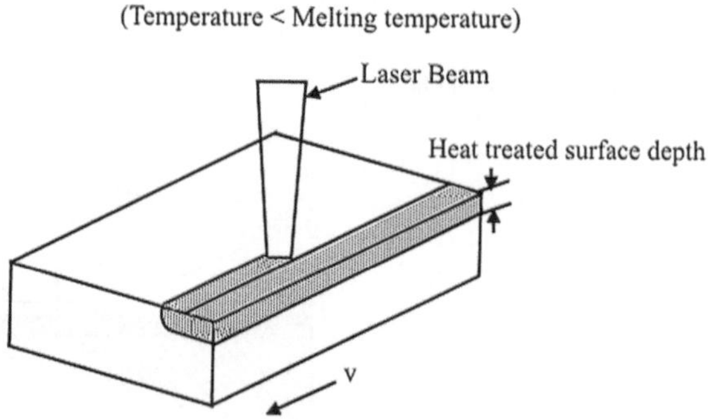

Fig. 2 Schematic of the laser surface heat treatment

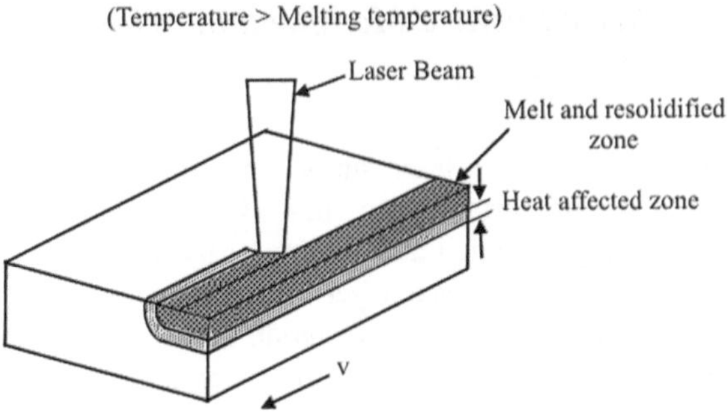

Fig. 3 Schematic of the laser surface melting and re-solidification

microstructure causes modified properties such as improved hardness, wear resistance and corrosion resistance of the base material surface.

2.1.3 Laser Shock Peening

Similar to peening by high velocity shots, laser beam can also be used to produce compressive stresses on the surface of base material for increasing its fatigue life and to improve the crack resistance property. Laser shock peening technique involves generation of tamped plasmas at the surface of component. High laser intensities (200 J/cm^2) with low pulse duration (30 ns) can generate shock pressure up to 10^5 atmospheres on the metal surface. The process is capable to produce compressive residual stress at a greater depth than that of conventional peening process without any significant change in surface finish. This deep residual stress is important for safety of critical items such as compressor and turbine blades. Figure 4 shows the schematic of laser shock peening process.

2.2 Laser Surface Modification by Using External Material

These processes can be distinguished either on the basis of 'degree of mixing' of coating material with substrate surface or based on the mode of supply of coating material to the substrate surface before or during the laser processing.

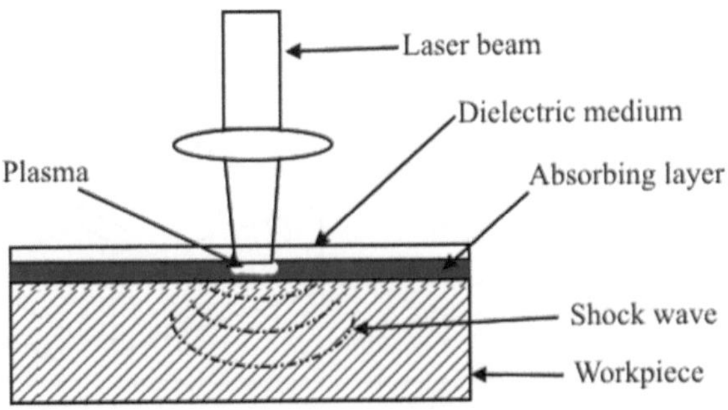

Fig. 4 Schematic of the laser shock peening process

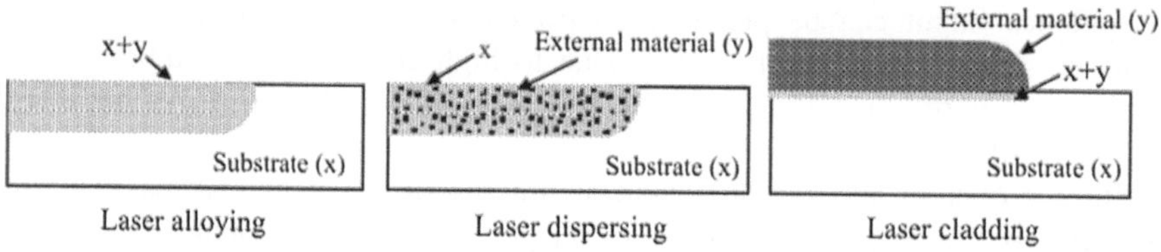

Fig. 5 Schematic representation of laser alloying, dispersing and cladding process

2.2.1 Based on Degree of Mixing of Coating Material with Substrate Surface

Laser alloying, laser dispersing and laser cladding are used as surface modification technique to form a coating layer by addition of external material with improved surface properties than the base material. Figure 5 shows the degree of mixing of coating material and the substrate for laser alloying, dispersing and cladding process.

Laser Alloying

In this process, melting of the substrate surface takes place due to scanning of laser beam while added elements (in the form of powder or pre-deposited layer) mix up with molten layer of substrate, to form an alloyed layer over the substrate surface. The alloyed layer shows improved mechanical properties in terms of hardness and wear resistance.

Laser Dispersing

In laser dispersing, hard ceramic particles are dispersed non-homogeneously in the matrix of base material. Here, coating material is non-soluble in the base material. Laser beam scan over the substrate to melt surface layer and simultaneously hard particles are blown towards the molten substrate, so that a layer of hard ceramic particle reinforced metal matrix composite can produces over the substrate surface.

Laser Cladding

In laser cladding a layer of different or same material is deposited over the substrate surface. Strong metallurgical bond between clad layer and substrate obtained due to slight dilution of the cladding into the substrate surface. The properties of the produced clad layer depend on the applied coating material and laser processing parameters. In comparison to conventional cladding process, in laser cladding overall heat input can be precisely controlled to make a layer with minimum dilution and low heat affected zone. Due to high cooling rate, resulting microstructures are fine and non-equilibrium, which shows high hardness and wear resistance of coated surface.

In case of laser alloying homogeneous mixing of the additional material with base material takes place and produces alloyed layer which contain properties of both the coating and substrate material or sometimes surface with improved properties. In laser dispersing process additional material (mainly ceramic particles) is dispersed non-uniformly in the matrix of base material which exhibit properties of both metal and ceramic particles. In case of laser cladding, layer of additional material at the top surface of the substrate takes place with better surface property as compared to the base material.

2.2.2 Based on the Mode of Supply of Coating Material

Preplaced Powder Method

This is considered to be a two-stage laser coating process. In first stage, substrate surface is covered with pre-placed powder to be coated. In next stage, laser beam scan over the pre-deposited coating layer, so that heat transfer takes place through the powder and melt the substrate surface along with powder and provide a metallurgical bond between coating and substrate surface. Uniform layer of the coating material on the substrate is achieved, which results homogeneous and crack-free coating with a strong metallurgical bond to the base material. This method is not suitable for complex geometric shape and also time consuming than the powder blowing/injection method. Figure 6 shows the schematic of laser coating process by preplaced powder method.

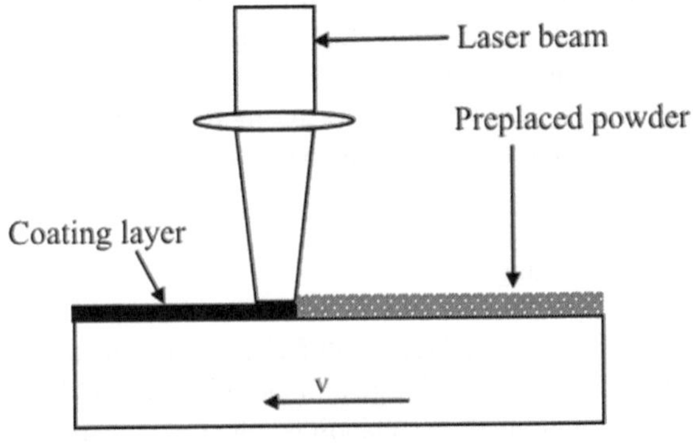

Fig. 6 Laser coating by preplaced powder method

Powder Injection Method

In this method, powder is injected into the path of laser beam through a nozzle with the help of an inert gas. The blown powder particles are partially melted by the laser beam. Again, the laser beam creates a small melt pool on the substrate surface that fully melts the blown powder. The melt pool after solidification produces a clad layer, which bonded with substrate by a minimum dilution. Powder can be injected towards the melt pool either by coaxial nozzle or by lateral ejection. By using coaxial nozzle, powder particles are rendered to the substrate through a coaxial system, where the laser beam and powder particles are feeding towards the substrate simultaneously. In lateral injection method, feed nozzle is positioned to the side of the laser beam. Figure 7 shows the schematic of laser coating by powder injection method (lateral injection).

Wire Feeding Method

Figure 8 shows the schematic of laser coating by wire feeding. In this method, clad material is supplied to the melt pool as a wire. It is a one-step process and efficiency

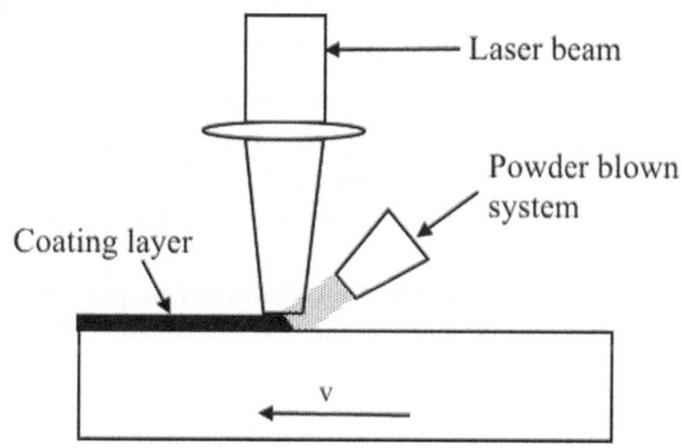

Fig. 7 Laser coating by powder injection method

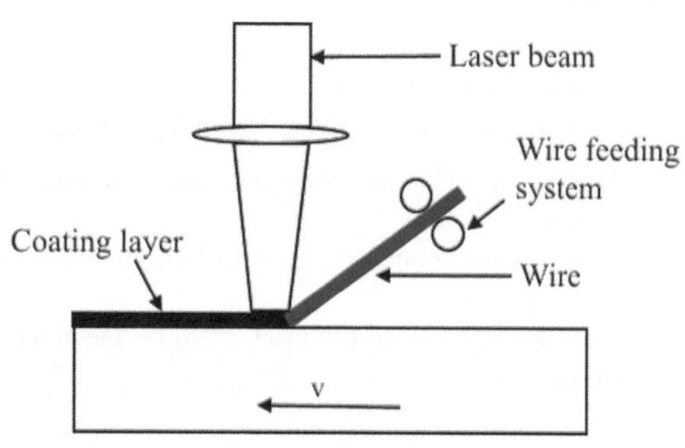

Fig. 8 Laser cladding by wire feeding method

of material usage is high compared to powder blown/injection method. However, major disadvantage of this process is that, laser energy is poorly absorbed by wire causes heat loss. The process is more efficient than the powder blown method, but extremely accurate alignment of the laser beam with the feed wire is required.

Among all these processes, preplaced powder method has advantages of controlling coating thickness properly and less amount of powder is required for depositing a predefined coating layer (Farnia et al. 2013).

3 Advantages, Limitations and Applications of Laser Coating Process

3.1 Advantages

High power lasers are emerging as efficient tools for the deposition of wear-resistant coatings. Laser surface coating technique is applicable on many engineering components with some simple arrangement. High cooling rates attainable in laser processing gives rise to extremely fine-grained structures and improved mechanical properties. Laser coating, in most cases ensures strong metallurgical bonding between coating and substrate material (Masanta et al. 2009).

Specific advantages of laser coating process are as follows:

(i) Low dilution between the coating and base material.
(ii) Excellent surface properties like high hardness, wear resistance and corrosion resistance due to refinement of microstructure.
(iii) Changes in properties of base material are minimum due to overall low heat input.
(iv) Controlled coating thickness could be possible.
(v) High productivity as the process is very fast.

3.2 Limitations

Although Laser coating process has several advantages, the process is limited in industrial application due to the following restrictions:

(i) High cost of laser setup and high running cost make the process very expensive.
(ii) Non-homogeneous energy distribution in the laser beam depending on the beam quality.
(iii) Poor absorptivity of the laser beam by metal surface reduces efficiency of the process.

3.3 Applications

Industrial components with high hardness, corrosion resistance and wear resistance properties are largely manufactured by deposition of hard layer using laser surface modification or coating process. Laser coating could be useful in the processes those require high productivity combined with high flexibility, without compromising the quality of the product. Uniform quality with a low heat input makes this process suitable for a broad range of applications in which minimum distortion is desirable.

Laser coating applications include spare part manufacturing, production of new components and maintenance and repairing of worn equipment and parts. Laser coatings are applied to produce surfaces resistant against adhesive, corrosive and abrasive wear, high temperature oxidation and wet corrosion. Typical applications of laser assisted surface engineering include: (Masanta 2010).

Automotive industry:

- Hardening and cladding of engine cam shafts and valve seats
- Hardening of cutting and bending edges on press tools involved in manufacture of automotive parts
- Surface modification for various tribo-components

Aerospace industry:

- Repairing of turbine blades by laser cladding
- Hardening of bearing surfaces on rotating shafts

Mining, power generation, oil and gas industry:

- Wear protection of drilling tools or hydraulic cylinders in mining industry
- Hardening of various parts of pumps, valves and tubular components
- Hard-facing of steam turbine blades
- Coating of long drilling tools for oil rigs by laser cladding

Other applications:

- Building up complex 3D geometries
- Coating of cutting tools and blades to protect them from wear

4 TiC Coating on Al Substrate Using Pulse Nd:YAG Laser

4.1 Laser Coating on Al Substrate

Aluminium has excellent property i.e. low specific weight, high corrosion resistance and thermal conductivity, which is desirable for component in aerospace and automobile industries. However, due to its low hardness and wear resistance

Table 1 Typical properties of aluminium

Property	Value
Density	2.70 g/cm^3
Melting point	660.32 °C
Boiling point	2470 °C
Thermal conductivity	237 W/m K
Thermal expansion coefficient	(25 °C) 23.1 × 10^{-6} K^{-1}
Modulus of elasticity	70 GPa
Hardness	~50 HV

application of Aluminium is sometimes restricted in specific application (Ravnikar et al. 2013). Table 1 shows the typical physical and mechanical properties of Aluminium. Improvement of surface properties of Aluminium is a big challenge by conventional heat treatment or alloying methods. Since, Aluminium does not show any allotropic transformation, conventional heat treatment process has fewer roles in increasing the hardness by forming harder phase like martensite (Fu and Batchelor 1998). Hence, alternate surface treatment like coating of harder material is necessary for improvement of the surface properties of Aluminium.

Several works have been done by various research groups to improve surface properties of Aluminium by laser surface modification technique. Das (1994) studied the effect of Ni alloying with pulsed Nd:YAG laser in the pure Aluminium by defocusing the beam diameter. Selvan et al. (2000) produced coating on Aluminium by alloying with an electrodeposited nickel layer by using a continuous wave (CW) CO_2 laser. A uniform coating was created on Aluminium by laser alloying process without any cavitation and porosity. This coating increases the hardness 20–27 times of base Aluminium. Uenishi and Kobayashi (1999) formed cladding of intermetallic compound Al_3Ti and its matrix composite layers on Aluminium surface by using continuous type CO_2 laser beam. During laser cladding, addition of TiB_2, TiC and SiC improved the wear property of Al substrate compare to Al_3Ti clad layer. Vreeling et al. (2001) investigated the formation of Al/SiC MMC layer on Aluminium substrate using a Nd:YAG laser by injection of SiC particles. Dubourg et al. (2005) studied the effect of Si and TiC reinforcement on the pure Aluminium by laser cladding of MMC layer. With SiC, TiC particle reinforcement both hardness and wear resistance of the coating improve significantly than Al/Si matrix. Majumdar et al. (2006a, b) studied development of a hard SiC dispersed composite layer and in situ TiB_2 dispersed Al matrix composite on Al substrate using a CW CO_2 laser. These composite layers improve the microhardness of the surface up to three times and wear-resistance of the composite surface significantly. Nath et al. (2012) studied laser surface alloying of Aluminium with WC + Co + NiCr through a continuous type Nd:YAG laser. Alloyed zone shows significant improvement in micro-hardness (up to 650 VHN) compared to Aluminium substrate (22 VHN). Improvement in fretting wear resistance was also observed due to presence of carbide particles in grain refined Aluminium substrate.

Table 2 Typical properties of TiC powder

Property	TiC
Density (g/cc)	4.93
Melting point (°C)	3067
Modulus of elasticity (GPa)	439
Heat conductivity (W/m °C)	28.9
Electrical resistivity (μΩ cm)	100
Thermal expansion coefficient (K^{-1})	8.15–9.45 × 10^{-6} (25–1500 °C)
Hardness (HV)	3000

4.2 Laser Coating with TiC

TiC is a promising coating material due to its high hardness, high melting point, thermal stability, wear resistance, low coefficient of friction (Katipelli et al. 2000). Therefore, TiC can be used as a suitable coating material for improving surface property of any metallic substrate. Table 2 shows the physical and mechanical properties of TiC.

Considerable numbers of works have been conducted by various research groups to improve the hardness, wear resistance and corrosion resistances of different structural materials with TiC or TiC reinforced metal matrix composite coating by laser surface engineering. Laser alloying of AISI 1045 steel were performed by TiC powder and TiC + Ni-alloy (Ariely et al. 1991; Ouyang et al. 1995). Both the coatings exhibit significant improvement in surface hardness (up to 1300 $HV_{0.2}$). It was also observed that, dissolution of TiC in the molten Fe produces dendrites upon re-solidification of the coating. To improve the wear resistance of tool steel laser surface alloying with 50 % TiC particles was performed which enhanced the hardness value of the coating layer up to 6 times than that of substrate material (Axén and Gahr 1992). A powder mixture of Ni alloy, titanium and crystalline graphite was used as the coating material on substrate of 5CrMnMo steel to produce a TiC reinforced composite coating which increased its hardness up to 1250 $HV_{0.2}$ (Wu 1999). Katipelli et al. (2000) studied the microstructure and wear behaviour of TiC on the 6061 Al alloy with high power continuous wave laser and found improvement in microhardness and dry sliding wear resistance compared to the substrate. Hardness values of H-13 steel increases up to 30 % by laser glazing and alloying with TiC (Jiang and Molian 2001). Tomida et al. (2001) shows that, hardness of Al–Mg alloy substrate was increased up to 500 HV by laser coating with TiC particle. The wear resistance of the produced coating was also improved up to six times than that of base material. Micro-hardness values of the laser clad layer with different pre-mixed composition of Mo/TiC are found 5–10 times higher than that of the as-received AA6061 Al-alloy (Chong et al. 2002). By using a mixture of TiC powder and 431 stainless steel, which undergo laser treatment, 65 % increase (724 HV) in hardness value compared to the deposit made with only 431 stainless steel powders (438 HV) was observed (Babu et al. 2006). Laser coating layer of TiC reinforced-H13 tool steel on AISI 4140 steel substrate exhibit hardness value in the range of 600–860 HV, which is much higher than that of substrate hardness (200–250 HV) (Jiang and Kovacevic 2007). Hardness of TiC reinforced iron based MMC coating was found 4 times greater than AISI 1030 steel substrate, when a preplaced layer of Ti, graphite and iron powder mixture was treated under laser beam (Emamian et al. 2010).

4.3 Laser Coating Using Pulsed Laser

Although laser surface modification normally done with continuous wave high power laser, due to specific advantages of pulse laser beam, low power pulse type lasers are also promising tools for laser surface modification. Pulsed laser offers overall low heat input and higher cooling rate due to its intermediate irradiation of laser beam, which causes less thermal distortion in the base metal compared to continuous wave laser (Sun et al. 2005). During pulse laser operation, energy is supplied through intermittent pulses which can be controlled by altering the pulse duration, pulse frequency and pulse energy. Such control of pulse parameters allows more precise regulation applied laser energy, which in turn create different thermal conditions to alter the microstructures and phase features of the laser processed zone (Farnia et al. 2013). Work of previous researchers show the improvement in surface property of the Aluminium alloy by adding suitable alloying elements using pulsed Nd:YAG laser (Das 1994; Gordani et al. 2008; Vaziri et al. 2009). Further, Fu et al. (1998) and Wendt et al. (2003) studied laser alloying of Aluminium alloy with Ni-Cr and Ti wire with the help of pulsed Nd: YAG laser. The alloyed zone shows high hardness and wear resistance compared to the substrate material. Maximum hardness was found at relatively low laser power, since higher power leads to dilution of the hard phases inside the Aluminium substrate. Yan et al. (2012, 2013) applied laser surface cladding technique to deposit (Ti, W) C reinforced composite coating on copper and Co-alloy/TiC/CaF_2 self-lubricating composite coatings on Cr-Zr-Cu alloy using a pulsed Nd:YAG laser. The reinforced MMC type coating exhibits higher wear resistance and lower friction coefficient than the substrate material.

From the study of these literatures, it is revealed that, considerable amount of work has been done in the field of laser surface modification on Aluminium using different coating materials. Again, TiC has been also used to improve surface properties of various substrate materials. However, there is a lack of study in the laser surface modification of pure Aluminium substrate with hard carbide material using pulse type laser. In this work, TiC layer deposited on pure Aluminium substrate with the help of a pulsed Nd:YAG laser. Objective of this work is to study the effect of peak power and pulse overlapping on the quality of deposited TiC coating on Aluminium substrate.

5 Experimental Planning Procedures

Aluminium plates of 30 × 50 × 3 mm^3 were used as substrate material for present experiment. At first, the plates were polished with 220 SiC emery paper to remove the oxide layer and unwanted material present in the surface. Samples were then cleaned and degreased with ethyl alcohol and acetone respectively. Specific amount of TiC powder were then mixed with acetone and organic binder to make a semi-solid solution. The semi-solid solution was then dispersed over the substrate surface to make a layer of TiC and dried in room temperature. In this way approximately 150 ± 10 μm thickness uniform powder layer was formed on Aluminium substrate.

A pulsed Nd:YAG laser (ALT-200) with maximum average power of 200 W, pulse duration of 0.5–20 ms and frequency of 20 Hz was used for laser irradiation. Pre-deposited samples were placed on a three-axis CNC table. Laser beam was focused on the sample surface by adjusting the height of the table. Laser beam diameter and pulse duration were fixed at 1.5 mm and 12 ms respectively prior to laser scanning. Experiments were conducted to identify the effect of peak power and pulse overlapping (OL). Laser pulse frequency was selected in such way that, the average power does not cross the laser machine limit (200 W). The average power (P_{avg}) was calculated considering the three parameters (P_{peak}, T_{on} and f) as shown in Eq. 1:

$$P_{avg} = P_{peak} \times T_{on} \times f \quad (1)$$

where,

P_{peak} = Peak power (W);
T_{on} = pulse duration (ms);
f = Frequency (Hz).

Table 3 shows the laser processing parameters used for this experiment. Gaussian shape pulse mode laser beam was used to scan the TiC pre-placed

Table 3 Experimental condition with variable laser-processing parameters

Expt. No.	Peak power (kW)	Pulse energy (J)	Frequency (Hz)	Pulse overlapping (%)	Scan speed (mm/s)	Average power (W)
1	1	12	14	60	8.4	168.00
2	1	12	14	70	6.3	168.00
3	1	12	14	80	4.2	168.00
4	1.5	18	9	60	5.4	162.00
5	1.5	18	9	70	4.05	162.00
6	1.5	18	9	80	2.7	162.00
7	2	24	7	60	4.2	168.00
8	2	24	7	70	3.15	168.00
9	2	24	7	80	2.1	168.00

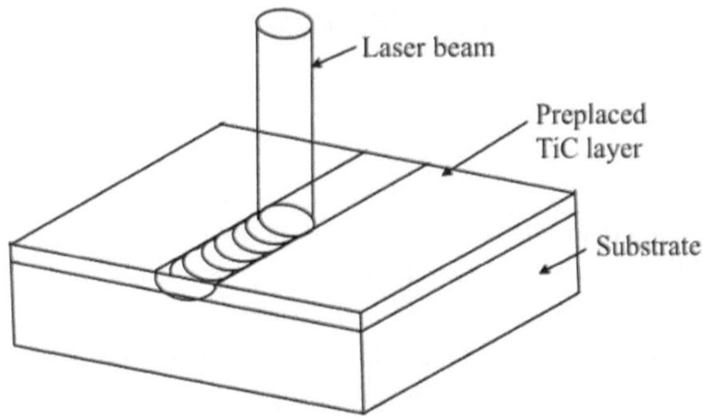

Fig. 9 Schematic diagram of pulsed laser coating process

Aluminium specimen by moving CNC table. Velocity of the workpiece was fixed to achieve a required pulse overlapping. Relation between laser scan velocity (v) beam diameter (d) and pulse overlapping (OL) is shown in Eq. 2, as described by Samant and Dahotre (2010).

$$v = f \times d \times (1 - OL/100) \tag{2}$$

Single track laser scanning was done to study the actual effect of each pulse and laser peak power. Distance between two tracks kept 3 mm to avoid heating effect of previous tracks. Argon protective atmosphere was used to avoid oxidation of substrate layer during laser processing of samples. Figure 9 shows schematic diagram of pulsed laser coating process.

After performing the laser scanning as per the parameters shown in Table 3, these laser-processed samples were cut perpendicularly to the laser scan direction with the help of wire-EDM and mounted with resin for metallurgical polishing. These mounted specimens were then polished with 600 and 1200 grade SiC polishing paper. Final polishing was done in diamond paste suspended polishing cloth. Optical microscope (AXIOCAM ERc 5s) images of the polished samples were used for study of the coating layer. Micro-hardness of the coating was taken at the cross-section of the coating layer and near surface region of the samples using a LECO micro-hardness tester (LM248AT) with 50 g load and 10 s time dwell.

6 Results and Discussion

6.1 Micro-hardness

Micro-hardness values of the coatings were measured at the polished cross section along a line from the top surface of the coating towards the depth of the substrate at an interval of 30 μm. Figure 10a–c shows the hardness profiles against depth of the coating from the top surface for the samples processed with different laser peak power i.e. 1, 1.5 and 2 kW respectively. From these graphs, variation in coating

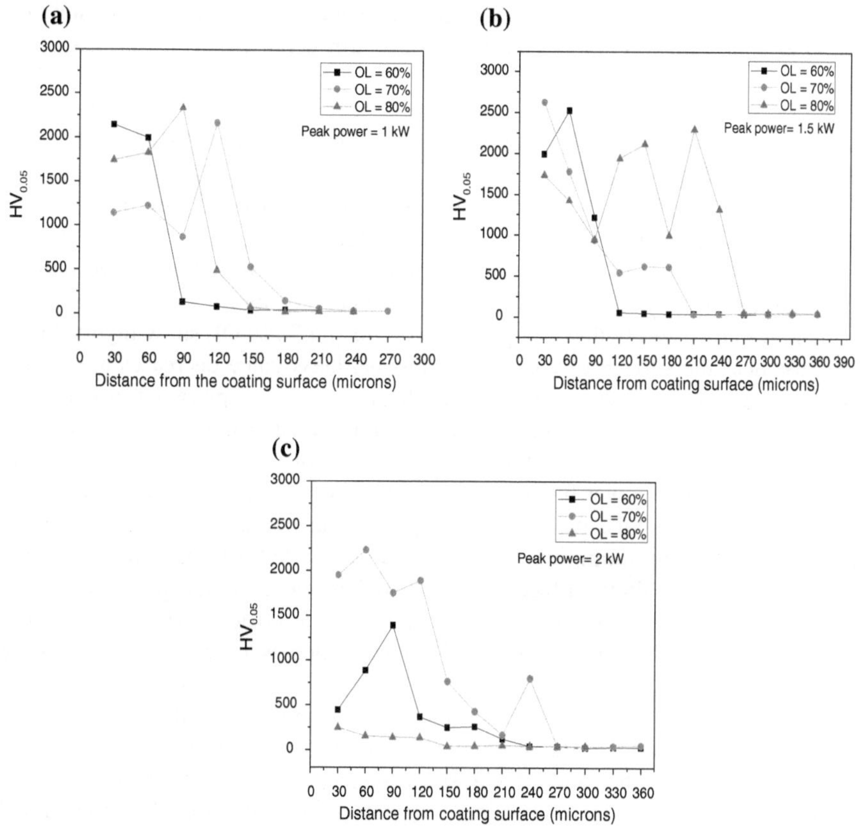

Fig. 10 Variation of hardness with distance from the coating surface, for samples treated with laser peak power **a** 1 kW **b** 1.5 kW and **c** 2 kW

thickness can also be observed due to difference of hardness value in Al substrate and TiC coating layer.

Figure 10a shows hardness profile for the laser processed sample with peak power 1 kW and different pulse overlapping (OL). High hardness value (2200 HV) observed near the coating layer and beyond a depth of 90 μm from surface of coating shows low hardness (35 HV) due to appearance of Aluminium surface. As the pulse overlapping (OL) increases from 60 to 70 % and 80 %, thickness of coating also increases. However, no specific trend on the increase of coating thickness at higher pulse overlapping (OL) condition i.e. 70 and 80 % has been observed.

Figure 10b shows the hardness value of coating with 1.5 kW peak power and different pulse overlapping condition. The profile shows that, for 60 % OL hardness value of the coating near the surface region is reasonably high (~2500 HV) and then this hardness value reduces to 35 HV, as the distance from the surface approaches to 120 μm. For 70 % OL, hardness value of coating gradually reduces from 2600 HV to a range of 600 HV at coating depth around 120–180 μm and then drops to approximately 35 HV at around 210 μm depth from the surface. Pulse overlapping of 80 % shows variation of hardness value from 2200 to 1000 HV for a coating depth around 240 μm and then drop drastically. Thus, it can be said that, thickness of carbide coating increases for increase in OL from 60 to 80 %. At higher OL, as the number of pulse per unit length increases, overall energy input and corresponding temperature of melt pool also increases. Due to low melting temperature, Al melted and dragged by capillary force to fill up void between the TiC particles (Chong et al. 2002), that results in better mixing of the preplaced powder with substrate material to form a uniform and thick coating.

Samples processed with 2 kW peak power, 60 and 70 % OL show high value of hardness near the surface of coating and this hardness value decreases toward the substrate surface as observed in Fig. 10c. It may be due to presence of deposited TiC at the upper portion of coating and TiC reinforced Aluminium matrix at the bottom portion of the coating. Laser processed sample with 80 % OL and 2 kW peak power shows a lower hardness value (250 HV) near coating surface compare to the 60 and 70 % OL. At higher peak power and overlapping condition better mixing of TiC with Aluminium took place and a diluted TiC-Al composite layer produced on Aluminium surface. Again at higher energy input, partially vaporization of preplaced TiC reduces percentage of TiC in coated surface and accordingly reduces the overall hardness. This phenomenon is further discussed in the next sub-section.

Figure 11 shows the effect of laser peak power with 80 % OL on variation of hardness from the top surface of the coating. From the graph, it is observed that, samples processed with 1 and 1.5 kW show higher hardness value than the sample processed with 2 kW laser power. At low peak power condition, TiC melted and deposited over the Al substrate, which exhibit relatively high hardness. In case of high peak power due to high energy density, possibility of evaporation of preplaced

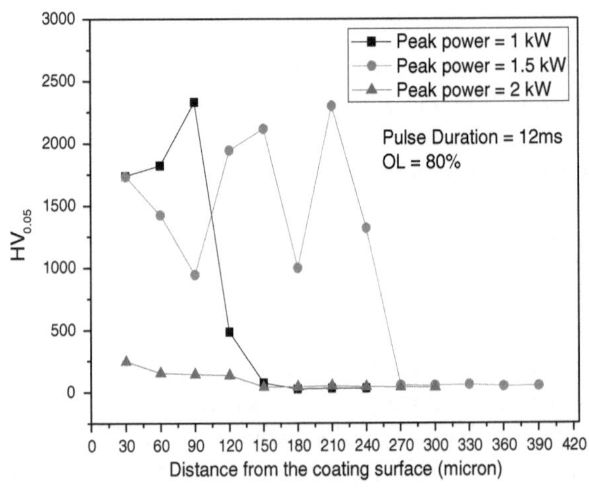

Fig. 11 Comparison of hardness variation with distance from the coating surface for sample processed at different laser peak power

TiC powder layer is more and at higher temperature remaining TiC mix with molten Aluminium substrate and produces a diluted composite layer, which shows lower hardness value than the TiC layer deposited at lower peak power condition.

6.2 Microstructure Analysis

Optical images of laser processed samples show the successful formation of TiC coating on Aluminium substrate. Figure 12a–c show the effect of laser processing parameter on the formation of TiC coating layer on Al substrate. Figure 12a, b show the formation of TiC coating on the Al substrate which is not uniform in centre area and near the edges. This may be due to Gaussian heat distribution of laser irradiation during scanning with pulse Nd:YAG laser. Sample processed with 1 kW peak power and 80 % OL (Fig. 12a) shows relatively lower coating thickness than the sample processed with 1.5 kW peak power and 80 % OL. At low peak power heat input in the processed area is relatively low, which melted lower amount of pre-deposited TiC layer, as well as dilution of TiC with Aluminium substrate is less compared to the coating deposited at higher peak power (1.5 kW) and pulse

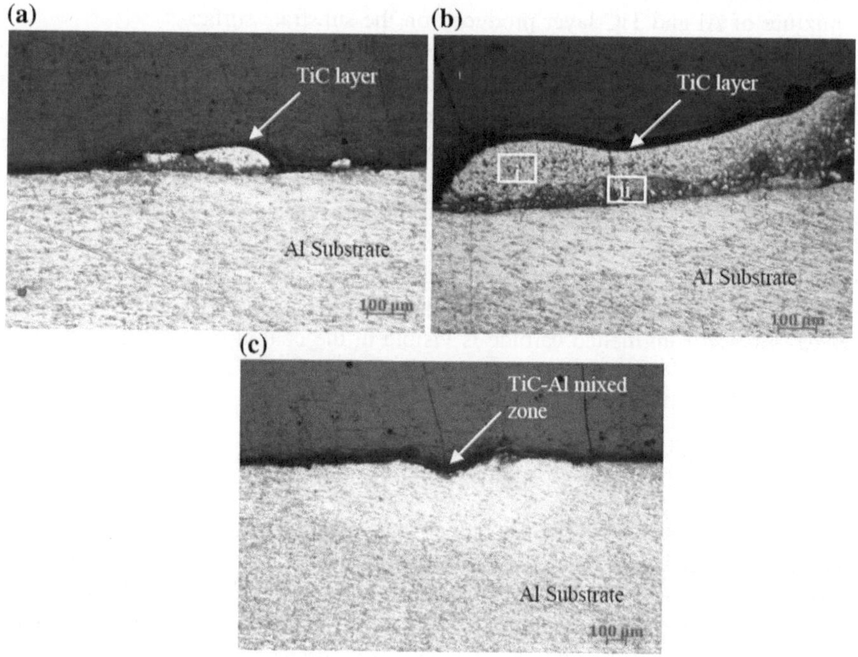

Fig. 12 Optical images of the cross-section of coatings processed with **a** laser power = 1.0 kW, overlapping = 80 % **b** laser power = 1.5 kW, overlapping = 80 % **c** laser power = 2 kW, overlapping = 80 %

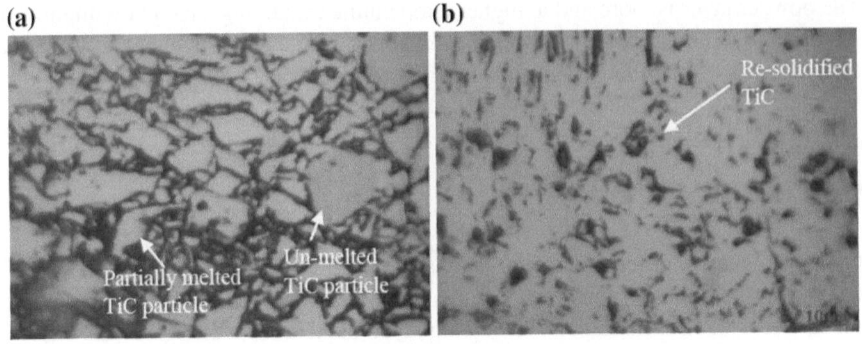

Fig. 13 High magnified optical images of the **a** upper portion of coating corresponding to mark 'i' **b** lower portion of coating corresponding to marked 'ii' for Fig. 12b

overlapping condition. For the samples processed with 2 kW peak power and 80 % OL (Fig. 12c) no separate layer of TiC is observed, however at the surface layer of the substrate a different shaded zone can be visible. Higher energy input due to high peak power may remove the maximum amount of preplaced TiC powder along with some amount of base material as vapour. Again, due to mixing of remaining amount of TiC powder with Aluminium substrate, at relatively higher temperature, a mixture of Al and TiC layer produced on the substrate surface.

Figure 13 shows the high magnified view of marked portions corresponding to Fig. 12b i.e. coating processed with 1.5 kW, 80 % overlapping. From this image it is observed that, at the upper portion of the coating (Fig. 13a corresponding to zone (i) in Fig. 12b) some bulk and partially melted TiC deposited after laser processing and less amount of Al matrix is present. However, at the lower portion of the coating (Fig. 13b corresponding to zone (ii) in Fig. 12b), amount of TiC is relatively less and present in the form of reinforcement in Aluminium matrix. Here, TiC melted and re-solidified after laser processing. As melting temperature of TiC (3067 °C) is higher than the vaporization temperature of Aluminium substrate (2470 °C), some un-melted carbide is visible in the coating.

7 Conclusion

TiC coating has been successfully deposited on Aluminium substrate by using pulsed Nd:YAG laser. The deposited coating shows significant improvement in hardness value compared to the Aluminium substrate. Effect of peak power and overlapping factor on the produced coating can be observed clearly. From the experimental results following conclusion can be drawn:

1. Higher hardness and low coating thickness have been observed for coating deposited with low peak power and low pulse overlapping condition (1 kW and 60 % OL).
2. Thickness of TiC coating increases for increase in overlapping from 60 to 80 %. At higher pulse overlapping, overall energy input and corresponding temperature of melt pool increases, which cause better mixing of the preplaced TiC powder with Al substrate to form a thicker coating.
3. At high peak power (2 kW) and higher overlapping (80 %), no separate layer of TiC is observed; however, mixed zone of TiC and Aluminium has been formed at the substrate surface.

With the possibility of further improvement of the coating following works could be done in future as extension of the present study:

1. Wear test for the coated samples can be done by ball on disc or adhesive type wear testing.
2. Continuous type high power lasers can be use to produce the same coating on Aluminium alloy.
3. As TiC individually is not a suitable coating material on Al substrate due to mismatch in properties, addition of some binder materials like Ni, Cr with TiC coating on Al substrate can be performed to improve the quality of the coating.

References

Ariely, S., Shen, J., Bamberger, M., Dausiger, F., & Hugel, H. (1991). Laser surface alloying of steel with TiC. *Surface and Coatings Technology, 45*(1–3), 403–408.

Axén, N., & Gahr, K. H. Z. (1992). Abrasive wear of TiC-steel composite clad layers on tool steel. *Wear, 157*, 189–201.

Babu, S. S., Kelly, S. M., Murugananth, M., & Martukanitz, R. P. (2006). Reactive gas shielding during laser surface alloying for production of hard coatings. *Surface and Coatings Technology, 200*, 2663–2671.

Chong, P. H., Man, H. C., & Yue, T. M. (2002). Laser fabrication of Mo-TiC MMC on AA6061 Aluminium alloy surface. *Surface and Coatings Technology, 154*, 268–275.

Das, D. K. (1994). Surface roughness created by laser surface alloying of aluminium with nickel. *Surface and Coatings Technology, 64*, 11–15.

Das, D. K., Prasad, K. S., & Paradkar, A. G. (1994). Evolution of microstructure in laser surface alloying of Aluminium with nickel. *Materials Science and Engineering A, 174*, 75–84.

Dubourg, L., Ursescu, D., Hlawka, F., & Cornet, A. (2005). Laser cladding of MMC coatings on aluminium substrate: Influence of composition and microstructure on mechanical properties. *Wear, 258*, 1745–1754.

Emamian, A., Corbin, S. F., & Khajepour, A. (2010). Effect of laser cladding process parameters on clad quality and in-situ formed microstructure of Fe–TiC composite coatings. *Surface and Coatings Technology, 205*, 2007–2015.

Farnia, A., Ghaini, F. M., & Sabbaghzadeh, J. (2013). Effects of pulse duration and overlapping factor on melting ratio in preplaced pulsed Nd:YAG laser cladding. *Optics and Lasers in Engineering, 51*, 69–76.

Fu, Y., & Batchelor, A. W. (1998). Laser alloying of Aluminium alloy AA 6061 with Ni and Cr. Part II. The effect of laser alloying on the fretting wear resistance. *Surface and Coatings Technology, 102*, 119–126.

Fu, Y., Bathelor, A. W., Gu, Y., Khor, K. A., & Xing, H. (1998). Laser alloying of Aluminium alloy AA 6061 with Ni and Cr. Part 1: Optimization of processing parameters by X-ray imaging. *Surface and Coatings Technology, 99*, 287–294.

Gordani, G. R., Razavi, R. S., Hashemi, S. H., & Isfahani, A. R. N. (2008). Laser surface alloying of an electroless Ni–P coating with Al-356 substrate. *Optics and Lasers in Engineering, 46*, 550–557.

Jiang, W., & Molian, P. P. (2001). Nanocrystalline TiC powder alloying and glazing of H13 steel using a CO_2 laser for improved life of die-casting dies. *Surface and Coatings Technology, 135*, 139–149.

Jiang, W. H., & Kovacevic, R. (2007). Laser deposited TiC/H13 tool steel composite coatings and their erosion resistance. *Journal of Materials Processing Technology, 186*, 331–338.

Kadolkar, P., & Dahotre, N. B. (2003). Effect of processing parameters on the cohesive strength of laser surface engineered ceramic coatings on Aluminium alloys. *Materials Science and Engineering A, 342*, 183–191.

Katipelli, L. R., Agarwal, A., & Dahotre, N. B. (2000). Laser surface engineered TiC coating on 6061 Al alloy: Microstructure and wear. *Applied Surface Science, 153*, 65–78.

Majumdar, J. D., Chandra, B. R., Nath, A. K., & Manna, I. (2006a). Compositionally graded SiC dispersed metal matrix composite coating on Al by laser surface engineering. *Materials Science and Engineering A, 433*, 241–250.

Majumdar, J. D., Chandra, B. R., Nath, A. K., & Manna, I. (2006b). In situ dispersion of titanium boride on aluminium by laser composite surfacing for improved wear resistance. *Surface and Coatings Technology, 201*, 1236–1242.

Masanta, M. (2010), Characterization and performance evaluation of ceramic composite coatings synthesized by laser cladding of pre-placed precursor on steel substrate. *Ph.D. Thesis*, IIT Kharagpur, Kharagpur.

Masanta, M., Ganesh, P., Kaul, R., Nath, A. K., & Choudhury, A. R. (2009). Development of a hard nano-structured multi-component ceramic coating by laser cladding. *Materials Science and Engineering A, 508*, 134–140.

Nath, S., Pityana, S., & Majumdar, J. D. (2012). Laser surface alloying of aluminium with WC + Co + NiCr for improved wear resistance. *Surface and Coatings Technology, 206*, 3333–3341.

Ouyang, J. H., Pei, Y. T., Lei, T. C., & Zhou, Y. (1995). Tribological behaviour of laser clad TiC_p composite coating. *Wear, 185*, 167–172.

Ravnikar, D., Dahotre, N. B., & Grum, J. (2013). Laser coating of aluminium alloy EN AW 6082-T651 with TiB_2 and TiC: Microstructure and mechanical properties. *Applied Surface Science, 282*, 914–922.

Samant, A. N., & Dahotre, N. B. (2010). Three-dimensional laser machining of structural ceramics. *Journal of Manufacturing Processes, 12*, 1–7.

Selvan, J. S., Soundararajan, G., & Subramanian, K. (2000). Laser alloying of aluminium with electrodeposited nickel: optimisation of plating thickness and processing parameters. *Surface and Coatings Technology, 124*, 117–127.

Sun, S., Durandet, Y., & Brandt, M. (2005). Parametric investigation of pulsed Nd:YAG laser cladding of stellite 6 on stainless steel. *Surface and Coatings Technology, 194*, 225–231.

Tomida, S., Nakata, K., Saji, S., & Kubo, T. (2001). Formation of metal matrix composite layer on Aluminium alloy with TiC-Cu powder by laser surface alloying process. *Surface and Coatings Technology, 142*, 585–589.

Uenishi, K., & Kobayashi, K. F. (1999). Formation of surface layer based on Al_3Ti on Aluminium by laser cladding and its compatibility with ceramics. *Intermetallics, 7*, 553–559.

Vaziri, S. A., Shahverdi, H. R., Torkamany, M. J., & Shabestari, S. G. (2009). Effect of laser parameters on properties of surface-alloyed Al substrate with Ni. *Optics and Lasers in Engineering, 47*, 971–975.

Vilar, R. (1999). Laser alloying and laser cladding. *Materials Science Forum, 301*, 229–252.
Vreeling, J. A., Pei, Y. T., Wind, B., Ocelı́k, V., Hosson, J. T. M. D. (2001). Formation of γ-Al_2O_3 in reaction coatings produced with lasers. *Scripta Materials, 44*, 643–649.
Wendt, U., Settegast, S., & Grodrian, I. U. (2003). Laser alloying of aluminium with titanium wire. *Journal of Materials Science Letters, 22*, 1319–1322.
Wu, X. (1999). In situ formation by laser cladding of a TiC composite coating with a gradient distribution. *Surface and Coatings Technology, 115*, 111–115.
Yan, H., Zhang, P., Yu, Z., Li, C., & Li, R. (2012). Development and characterization of laser surface cladding (Ti, W) C reinforced Ni–30Cu alloy composite coating on copper. *Optics and Laser Technology, 44*, 1351–1358.
Yan, H., Zhang, J., Zhang, P., Yu, Z., Li, C., Xu, P., & Lu, Y. (2013). Laser cladding of Co-based alloy/TiC/CaF2 self-lubricating composite coatings on copper for continuous casting mold. *Surface and Coatings Technology, 232*, 362–369.

Finite Element Simulation of Laser Cladding for Tool Steel Repair

Santanu Paul, Ramesh Singh and Wenyi Yan

Abstract Laser cladding is a coating technique, wherein several layers of clad materials are deposited over a substrate so as to enhance the physical properties of the work-piece such as wear resistance, corrosion resistance etc. Strong interfacial bond with minimum dilution between the material layers is a pre-requisite of the process. This technique also finds widespread applications in repair and restoration of aerospace, naval, automobile components. A thermomechanical finite element models is developed wherein the Gaussian moving heat source is modelled along with element birth and death technique to simulate powder injection laser cladding of CPM9V over H13 tool steel, which is extensively used for repair of dies. The present work focuses on predicting the clad geometry and other clad characteristics such as the heat affected zone, dilution region and the subsequent residual stress evolution. It is expected that this knowledge can be used for repair of structures subjected to cyclic thermomechanical loads.

Keywords Finite element model · Gaussian laser heat source · Element birth technique · Laser cladding

S. Paul (✉)
IIT B-Monash Research Academy, Mumbai 400076, India
e-mail: santanupaul@iitb.ac.in

R. Singh
Indian Institute of Technology, Mumbai 400076, India
e-mail: ramesh@me.iitb.ac.in

W. Yan
Monash University, Clayton VIC3800, Australia
e-mail: wenyi.yan@monash.edu

S.N. Joshi and U.S. Dixit (eds.), *Lasers Based Manufacturing*,
Topics in Mining, Metallurgy and Materials Engineering,
DOI 10.1007/978-81-322-2352-8_9

1 Introduction

The high cost components used in the automobile/aerospace industry are sometimes operated beyond their original design life during, which they are subjected to cyclic/repeated thermomechanical loading thereby causing fatigue, corrosion and wear (Liu et al. 2011). Such age-related problems during service are common initiators of failures that cause high-performance and high-value components to be rendered useless (Pinkerton et al. 2008). Therefore, repair/restoration of such worn out or damaged high cost components used in the automobile/aerospace industry becomes beneficial as it can drastically reduce the overall cost multiple times by further extending the service life of these components (Wang et al. 2002).

Alternatively, the introduction of new forming materials like High Strength Steels (HSS), such as H13 tool steels (used mainly in the automotive sector) have introduced new challenges in tool manufacturing and repair of dies/moulds to the cold and hot shaping industry. These HSS are extremely aggressive for tools and dies thereby forcing the die sector to use new powder metallurgical tool steels, with an excellent combination of toughness, hardness and wear resistance for cutting, deep-drawing and bending dies (Leunda and Soriano 2011). A group of high vanadium-containing tool steels (such as CPM9, 10 and 15V), produced via powder metallurgy have proven to be successful in achieving high yield strength with high elongation and considerable work hardening along with excellent wear resistance during forming of HSS (Wang et al. 2006).

The moulds and dies used in hot and cold working industry are subjected to cyclic/repeated thermomechanical loading and thereby undergo wear and other localized damage. This calls for a repair process that does not induce tensile residual stresses. Traditional thermal spraying, Tungsten Inert Gas, Gas Metal Arc welding (GMAW) or High Velocity Oxygen Fuel (HVOF) techniques cannot be used effectively for powder metallurgical steels due to thermal damage and the process inaccuracies. On the other hand, these dies have very complex 3-D shape and very precise contoured deposition is required for the repair of such components. Moreover, repair being a localized process requires smaller beam size and precision. Therefore, it is imperative that the substrate properties should not deteriorate due to the heat affected zone and/or dilution. In this regard, a laser-based deposition technology such as laser cladding is a promising technique, in the remanufacturing industry, as it is characterized by localized and rapid fusion of materials. As a result, a relatively narrow Heat Affected Zone (HAZ) is generated.

Laser cladding is a material deposition technique in which the metallic materials in powdered form are supplied into a laser generated heat spot, by means of a carrier gas, where the material melts and forms a melt pool, which quickly solidifies into metal layers. As the metal powder passes through the laser beam, it is melted and deposited in the melt pool created by the laser beam on the metal substrate. By completely fusing the feedstock material, metal powders are directly transformed into fully dense solid objects composed of metallurgical bonded tracks of material that require no final finishing. In addition, laser cladded material exhibits mechanical properties equivalent to those of similar alloys in the wrought condition. Partial overlapping of individual tracks in a suitable pattern produces a continuous layer of material. By overlapping such layers, fully dense and metallurgically sound 3D objects are generated (Costa and Vilar 2009).

The aim of laser surface cladding is to deposit a coating material, with required properties, on to different metallic substrates in such a way as to produce a good metallurgical bonding with the substrate to improve surface properties such as wear resistance, corrosion resistance, and high-temperature oxidation resistance. It is considered as a strategic technique, since it can yield surface layers that, compared to other hard facing techniques, have superior properties in terms of homogeneity, hardness and microstructure. Compared to the conventional repair methods, laser cladding process yields minimum dilution of the cladded layer by the elements from the substrate or vice versa thereby resulting in a very small HAZ (Hu et al. 1998a, b; Mc Daniels et al. 2008). The relatively small HAZ therefore results in tiny deformation and stress along with high dimensional accuracy and integrity of the final products (Wang et al. 2002; Mc Daniels et al. 2008).

Conventional methods such as TIG/GMAW though relatively easy to apply, produce a lot of heat, thereby causing high residual stresses, resulting in distortion heat-related effects in the base metal (Tusek and Ivancic 2004). On the other hand alternative techniques such as plasma transferred arc (PTA) welding (Su et al. 1997) and electron beam (EB) welding (Henderson et al. 2004), although gives very precise heat flux, require complex and expensive apparatus. The high velocity oxy-fuel (HVOF) thermal spraying technique (Tan et al. 1999) procedure which finds widespread applications in many industries produces less component distortion than with TIG welding and it has many advantages over plasma spraying, including deposition of a thicker and lower-porosity coating. However, tight control of depth and spread of deposited material is not possible, which necessitates machining for finishing stage. In comparison, laser cladding due to its localized heat affected zone, flexibility and precise control over the deposition area, produces lower residual stress than from TIG welding based repair. The process can induce desirable compressive residual stresses at the surface (Grum and Slabe 2003; Moat et al. 2007). The physical and corrosion properties of the clad material can be difficult to predict because it undergoes a repeated heating–cooling cycle (Pinkerton et al. 2006), but in many cases clad properties are superior to those of the parent material (Majumdar et al. 2005).

In addition to the surface coating applications, due to a localized heat affected zone, flexibility and precision, powder deposition by laser cladding has a vast scope in repair of worn out aerospace structures, gears and dies/moulds for various manufacturing operations with little distortion and intermixing as compared with plasma powder/TIG build-up welding and thermal spraying techniques (Steen 2003). Repair of power station turbine blades (Brandt et al. 2009), engine valve seats (Kawasaki et al. 1992) and other components using LC have been reported in literature (Liu et al. 2011). For laser cladding of high-alloy tool steels such as H13 tool steels, cracking often occurs in the coating as a result of thermal and/or phase-transformation stresses thereby restricting the applications of these tool steels in laser cladding. Pre-heating of the matrix has been used to eliminate cracking of the coating (Zhang et al. 2001). However, this reduces the cooling rate and, thereby, affecting the microstructure of the cladding. On the other hand, a group of high vanadium-containing tool steels (such as CPM9, 10 and 15V), produced for powder metallurgy application, capable of producing high yield strength with high elongation and considerable work hardening along with excellent wear resistance (Wang et al. 2006) can be used as cladding materials for anti-wear applications. In fact, studies have been reported on laser cladding with CPM10V (Hu et al. 1998a, b) and other high vanadium tool steels (Zhang et al. 1999).

Residual stresses are produced in the parts produced by laser powder deposition techniques such as laser cladding due to the thermal history dependence phenomena in such processes. Residual stresses in clad material could affect the component's resistance to corrosion and fatigue cracks due to high thermal stress concentration (Sun et al. 2012). Therefore, the control of residual stresses plays a significant role in determining the mechanical performance of the fabricated parts which can be conveniently analysed by using modelling techniques. Although the interaction between certain phase transformations and the stress field are known and have been studied and modelled by researchers working on other heat treatment processes, such interaction has only been analysed briefly in the context of laser powder deposition (Griffith et al. 1998; Ghosh and Choi 2005, 2006, 2007).

One of the pre-requisites of laser cladding process is to keep dilution to a minimum to minimize the mixing between the clad layer and the substrate in order to maintain the properties of the baseline material (Steen 2003). However, high dilution allows stronger bonding between the clad and base material and in some case may have beneficial properties (Schneider 1998). Therefore the weakest point in a laser cladded component is the clad/HAZ interface due to inconsistent dilution/fusion (Mc Daniels 2008; Schneider 1998; Pinkerton et al. 2008). In the HAZ, the substrate material is heated to a temperature below the melting temperature and cooled at a lower rate than the coating surface. This trend can lead to microstructural changes in the HAZ that are difficult to control and could have a detrimental effect on the mechanical properties of the part (Mc Daniels et al. 2008). Cracks in laser-welded, high-strength, low-alloy steels formed near the borderline of the fusion line and the HAZ have been reported by Onoro and Ranninger (1997) with fatigue resistance minimum in the HAZ near the fusion line (Lee et al. 2000).

Finite element modelling is an appropriate tool to predict the temperature field, heat affected zone (HAZ), dilution zone and residual stress developed, so as to predict the clad quality and to develop optimum and successful cladding conditions. Previous efforts in developing sequential thermomechanical models provide temperature profile and cooling rate to predict the microstructure of the substrate (Wang et al. 2006; Picasso et al. 1994; Huan et al. 2006). In addition to this a few studies reported in the literature have also considered molten metal flow and phase transition (Wang et al. 2006). However, the finite elements models to evaluate the temperature profile for powder injection technique have been developed with constant convection co-efficient which may introduce errors (Liu et al. 2011; Wang et al. 2002; Shi and Qianchu).

Most modelling efforts focus their attention on phenomena occurring during the deposition of a single track of material or the build-up of thin wall geometry by overlapping several single pass layers. Not only does the thin wall geometry represent the simplest case of multilayer laser powder deposition, it is also the one that requires least effort to create a numerical representation of the problem (Amon et al. 1998) and the one that (potentially) requires less computation time and disk storage space.

Sequential thermomechanical analysis of laser cladding process in ANSYS® has been performed to obtain temperature profile for both planar and curved mesh of clad profile (Chen and Xue 2010; Deus and Mazumder 2006; Zhang et al. 2008, 2011; Plati et al. 2006). The temperature values obtained from transient thermal analysis are used as input for obtaining longitudinal and shear residual stresses for the thermomechanical analysis (Chen and Xue 2010; Deus and Mazumder 2006; Zhang et al. 2008, 2011; Plati et al. 2006) for various material systems, such as, copper on aluminium (Crespo et al.), Stellite on austenitic stainless steel AISI 304 (Suarez et al. 2010) and Monel on Ni-based alloys (Chunhua et al. 2012).

As noted previously, most of the laser cladding work reported is for non-powder metallurgical materials which are not very applicable for die repairs. However, thermomechanical model to investigate the HAZ, dilution zone and residual stress developed and thereby predict the clad quality have been reported in literature (Chunhua et al. 2012) for deposition of CPM9V on H13 tool steel. As a whole the finite element models available in literature (Plati et al. 2006; Paul et al. 2014) simulate the addition of new clad elements to the substrate using the element birth technique with the clad elements being deposited at the solidus temperature of the substrate, and therefore lack the ability to predict the clad geometry for the process. Consequently, this work is focused on the development of a 3-D coupled thermomechanical finite element modelling in ABAQUS® for laser cladding of CPM9V (crucible steel) on H13 tool steel. The addition of new elements to the substrate is simulated by using the element birth technique and the heat load addition is simulated by writing a user defined subroutine DFLUX in ABAQUS®. The clad geometry, clad dilution, heat affected zone and the residual stresses have been predicted from the model and compared with the experimental results (Paul et al. 2014).

2 Process Modelling

2.1 Physical Description of Process

In this work, the *powder injection* mode of laser cladding is considered. In this technique of laser cladding the powdered material is injected from a nozzle and is deposited over the base material in the presence of a laser beam. The description of the process is elaborated schematically in Fig. 1.

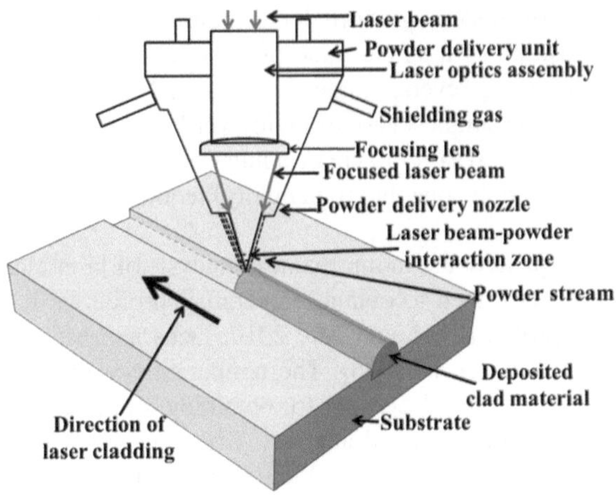

Fig. 1 Schematic of powder injection laser cladding technique

Table 1 Chemical composition of H13 tool steel (wt%)

C	Cr	Mn	Mo	Si	S	V
0.39	5.2	0.4	1.4	1.1	0.003	0.95

Data source Chen and Xue (2010)

Table 2 Chemical composition of CPM9V (wt%)

C	Cr	Fe	Mo	V
1.2	5.250	82.54	1.3	9.10

Data source Chen and Xue (2010)

In this work the substrate employed is H13 tool steel and clad material is vanadium carbide steel, CPM9V, extensively used for repair of damaged/worn out dies. Chemical composition of H13 tool steel is provided in Table 1 and the chemical composition of CPM9V is listed out in Table 2.

The actual cladding with dimension of substrate as 125 mm × 105 mm × 15 mm over which clad was deposited at length of 40 mm has been obtained from literature (Crespo et al.). Gaussian laser heat source is employed with beam diameter of 3 mm and power of 2000–3800 W. Powder particles was spherical in shape with size of 44–104 μm. In the current work a coupled thermomechanical finite element models of powder injection technique of laser cladding with Gaussian moving heat source is developed. The coupled thermomechanical model developed predicts the clad quality for laser power of 1700 W, feed rate of 5 g/min, scanning speed of 200 mm/min and beam diameter of 3 mm.

2.2 Model Assumptions

The following assumptions are made about the process for developing the coupled thermomechanical model:

- Stationary frame of reference has been attached with the laser beam, considering that the dimensions of the work-piece are large compared to those of the molten pool.
- Heat transfer in the process is assumed to occur without any internal heat generation and the variation of heat transfer co-efficient with temperature is neglected.
- Gaussian moving heat source with linear decrease of heat input with penetration depth is assumed.
- Fluid flow in the melt pool and its subsequent effect on the heat transfer co-efficient is neglected.
- Body force and surface traction are neglected in thermomechanical analysis and only the loading due to transient thermal field (Gaussian moving heat source) on the body is considered.
- For residual stress analysis elastic perfectly plastic behaviour with temperature dependent yield stress, with no work hardening and prior strain history effects are assumed.
- Creep (time dependent deformation) effects are also neglected.

2.3 Governing Equations

In a coupled thermomechanical finite element analysis of laser cladding the values of nodal temperature obtained after thermal analysis of Gaussian moving heat laser source are used as input to calculate the mechanical response in particular the residual stresses developed. Transient heat conduction equation is used as the basic governing equation for thermal analysis which is given as:

$$\frac{\partial}{\partial x}\left(k\frac{\partial T}{\partial x}\right)+\frac{\partial}{\partial y}\left(k\frac{\partial T}{\partial y}\right)+\frac{\partial}{\partial z}\left(k\frac{\partial T}{\partial z}\right)+\mathrm{Q}=\rho\mathrm{C_p}\frac{\partial T}{\partial \mathrm{t}}+\rho U_x C_p\frac{\partial T}{\partial x} \quad (1)$$

where ρ, C_p and k refers to density, specific heat and thermal conductivity respectively of material; T and t refer to the temperature and time variables respectively.The thermal stresses in a clad-substrate system are developed due to the presence of relatively high thermal gradients between dissimilar materials. The relatively high thermal gradient between dissimilar materials in general develops due to the fact that, the clad and substrate materials have different coefficients of thermal expansion and different reference temperatures. As laser cladding is a much localized process therefore, during a short interval of time, a thin layer of the substrate is heated up to the melting point, and a molten layer of the clad material is deposited on it. Thereafter, the heat is conducted into the substrate and the clad solidifies and starts shrinking due to thermal contraction, whereas the substrate first expands and later contracts according to the local thermal cycle. Hence, the residual stresses are evaluated as a function of time, when after heating the work-piece is left to cool under normal conditions and stress is analysed after infinite time period. Therefore, in the thermomechanical analysis of the process the only loading on the system due to the transient thermal field is considered, as any external loading, body force and surface traction are neglected. Thus, the strain–displacement relation is given by:

$$\in_{ij} = \frac{du_i}{dx_j} + \frac{du_j}{dx_i} \tag{2}$$

where σ_{ij} is stress tensor, n_{ij} is exterior normal to surface, u_1 is displacement and $\in_{ij}$ is total strain tensor, where the total strain $\in_{ij}$ is given by:

$$\in_{ij} = \in_{ij}^{el} + \in_{ij}^{p} + \in_{ij}^{th} \tag{3}$$

where $\in_{ij}^{el}$ represents elastic strain, $\in_{ij}^{p}$ represents plastic strain and $\in_{ij}^{th}$ is the thermal strain. The stress and elastic strains are connected through elastic moduli or stiffness tensor C_{ijkl}. For isotropic material stiffness tensor is function of Young's modulus, E and Poisson's ratio, υ given by:

$$\sigma_{ij} = C_{ijkl}\left(\in_{ij} - \in_{ij}^{th} - \in_{ij}^{p}\right) \tag{4}$$

The thermal strains are given by:

$$\in_{ij}^{th}(T) = \alpha_T(T)\left(T - T_{ref}\right) - \alpha_T(T_0)\left(T_0 - T_{ref}\right) \tag{5}$$

where α_T is coefficient of thermal expansion, T_{ref} is reference temperature (i.e., starting temperature from which coefficient of thermal expansion is obtained) and T_0 is initial temperature.

Besides these the following initial and boundary conditions are also satisfied:

$$T(x, y, 0) = T_0 \tag{6}$$

$$T(x, y, \infty) = T_0 \tag{7}$$

2.4 Numerical Formulation

For simulation a half symmetric model with dimension 6 mm × 6 mm × 6 mm has been developed. Coupled analysis for thermal and mechanical phenomena is considered. To evaluate the effect of thermal stresses produced during the process 8-node coupled temperature-displacements, C3D8RHT reduced integration elements are used. Mesh size of 54 μm × 100 μm × 100 μm has been considered across the cross-section of the clad, whereas in the substrate 46 μm × 100 μm × 100 μm. Figure 2 presents a pictorial representation of the model geometry along with the dimensions and meshed geometry. Total number of node is 27,510 and total number of elements is 24,820.

In powder injection technique, the continuous addition of material or mass on to the substrate is modelled by representing geometry of finite elements that change over time so as to simulate the powdered nature of the material. This is achieved by means of successive discrete activation of new set of elements into the computational domain or geometry using element birth and death feature as illustrated by Fig. 3. Furthermore, as the heat source moves forward the elements activated in the previous set are deactivated so as to initialize the cooling process.

Along with the element birth–death technique used to simulate the effect of powder deposition, the Gaussian moving heat source is also modelled using the DFLUX subroutine in ABAQUS®. The Gaussian moving heat source with linear decrease of heat input with penetration depth is given by (Shanmugam et al. 2013):

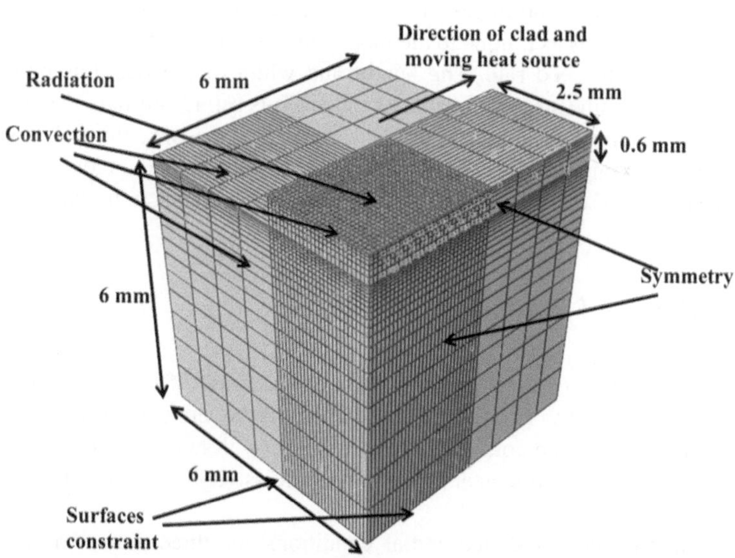

Fig. 2 Geometry of the model and mesh design along with the loading and boundary conditions

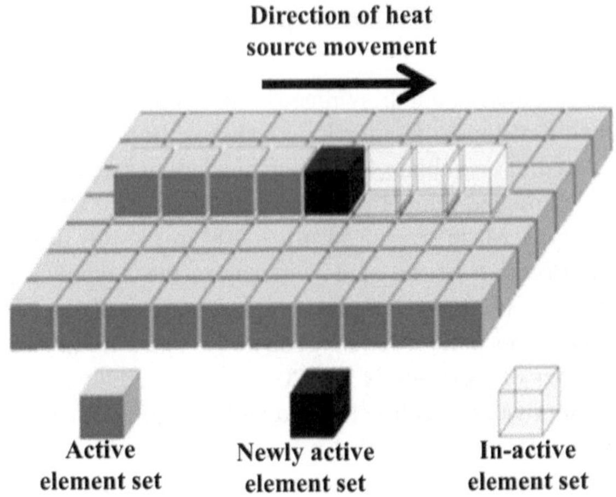

Fig. 3 Powder deposition modelled by element birth technique

$$Q(r,z) = \frac{Q_L}{\pi r_0^2 d} \exp\left[\left(1 - \frac{r^2}{r_0^2}\right)\left(1 - \frac{z}{d}\right)\right] \tag{8}$$

where Q_L is the laser power, r_0 is the beam radius, d is the heat penetration depth and r the instantaneous laser beam location given by:

$$r^2 + x^2 + y^2 + z^2 \tag{9}$$

where x, y, z are the co-ordinates of the laser heat source at time t.

While applying this technique to the model, initial volume of powder deposited is calculated using the feed rate. The height and width of powder thus obtained is deposited in each time step, and value of time step is set accordingly. Thereafter as the Gaussian moving heat source scans the surface, the clad and a thin substrate layer is melted and a strong interfacial bond is formed. The elements then lose heat mainly by conduction into work-piece and heat loss is also by convection and radiation.

2.5 Loading and Boundary Conditions

Gaussian distribution of laser power with linear decrease of heat input with penetration depth given by Eq. (8) is considered to be emitted from the source. Boundary conditions like convective heat transfer (with heat transfer coefficient of 15 $Wm^{-2} K^{-1}$) and radiative heat transfer (with emissivity 0.3) results in heat losses from surfaces.

Subsequently, the associated initial conditions for three-dimensional thermal analysis are given by Eq. (6). The addition of clad to the substrate is modelled using element birth-death technique as depicted in Fig. 3. The reactivation of the cladding elements is followed by the release of latent heat during solidification of the clad. After the activation of all the cladding elements, external heat injection is continued as the heat source moves forward, but the thermal analysis is continued until the system reached a steady state.

The rest of the work piece surfaces are open to atmosphere and are subjected to convective and radiative heat losses as depicted in Fig. 2. The above boundary condition is stated mathematically as:

$$K_n \frac{dT}{dn} + h(T - T_0) + \sigma \in \left(T^4 - T_0^4\right) = 0 \quad (10)$$

where n denotes the direction normal to surface, k_n refers to thermal conductivity; h, $\in$, σ and T_0 refers to surface heat transfer co-efficient, emissivity, Stefan-Boltzmann constant and ambient temperature respectively. The first term represents heat loss due to conduction from the surface whose unit normal is n. The second and third term refers to convection and radiation heat losses from the surface of the work-piece.

For the mechanical analysis external loading is not considered and to prevent the rigid body motion the nodes on the base are fully constrained to prevent elemental motion. The application of thermal load as Gaussian moving heat flux and the mechanical boundary conditions showing the surfaces which are fully constrained are illustrated in Fig. 3.

2.6 Material Properties

Temperature dependent thermo-physical properties are considered for both clad and substrate as these properties changes with temperature. Along with thermal properties like thermal conductivity, specific heat and latent heat, mechanical properties namely co-efficient of thermal expansion, young's modulus, poison's ratio and yield strength are provided as input. Stress and strain fields are dependent on evolution of plastic strains so kinematic hardening in addition to Von Misses yield criteria is assumed which is valid for clad, interface and the substrate region.

The yield strength as a function of temperature, decreases exponentially with temperature and tends to zero as the nodal temperature approaches the liquidus temperature. Accordingly, the "anneal temperature" feature in ABAQUS® is used which resets stress and strain values above molten temperature to zero. Also a very low value of Young's modulus is considered to make the melting zone a stress free zone.

Initially zero stress is considered in the material and then analysis is performed for residual stress by considering elastic perfectly plastic behaviour. The properties of CPM9V (clad) are listed in Table 3 and the thermo-physical properties of H-13 (substrate) are listed in Table 4.

Table 3 Thermo-physical properties of CPM9V mixture (Chen and Xue 2010)

Temperature (K)	Conductivity(W m^{-1} K^{-1})	Expansion (K^{-1})	Density (kg m^{-3})
300	20.48	1.102×10^{-5}	7455
373	21.6	1.105×10^{-5}	–
573	25.25	1.141×10^{-5}	–
820	26.08	1.186×10^{-5}	–

Other material properties of CPM9V used in the simulation are as follows (Chen and Xue 2010)
Melting temperature: 1773 K
Young's modulus: 221 GPa
Poisson's ratio: 0.28
Yield stress: 1600 MPa

Table 4 Thermo-physical properties of H13 tool steel (Chen and Xue 2010)

Temperature (K)	Conductivity (W m^{-1} K^{-1})	Expansion (K^{-1})	Density (kg m^{-3})
310	25	1.09×10^{-5}	7600
400	–	1.1×10^{-5}	–
500	26.3	1.15×10^{-5}	–
810	28	1.24×10^{-5}	–
900	–	1.31×10^{-5}	–
1000	30	–	–
1200	32	–	–
1500	35	–	–

Other material properties of H-13 used in the simulation are as follows (Chen and Xue 2010)
Melting temperature: 1730 K
Young's modulus: 210 GPa
Poisson's ratio: 0.3
Yield stress: 1400 MPa

3 Results and Discussion

The coupled thermomechanical analysis of cladding process is based on the fact that the temperature field obtained from the thermal analysis serves as the basis for prediction of clad geometry, the clad dilution and HAZ. Furthermore, for mechanical analysis as no external loading is considered apart from the application of thermal load as moving heat flux the prediction of thermal field for residual stress analysis becomes imperative.

3.1 Temperature Field

Figure 4a shows the contour plot of nodal temperature as the clad is being deposited. The red portion depicts the burnt away clad materials for a laser power of

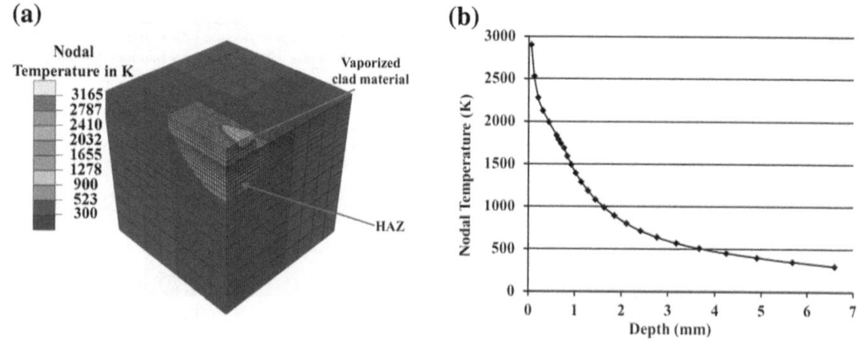

Fig. 4 **a** Contour of nodal temperature, **b** Variation of nodal temperature with depth

1700 W, feed rate of 5 g/min and scanning speed of 200 mm/min. The variation of nodal temperature along the cross-section of the clad when the laser beam is directly over the clad surface is shown in Fig. 4b. The finite element models developed calculates the clad height by eliminating the elements which have exceeded the vaporization temperature of the powder (CPM9V) from the computational domain. The comparison of the clad height as predicted by model shows a variation of 14 and 18 % variation for the prediction of clad width, from that of the experimental data (Plati et al. 2006). Therefore it is evident that the model is able to predict the clad geometry.

3.2 Results in Dilution and Heat Affected Zone

Dilution is the contamination of cladding with the substrate material which is detrimental to the clad quality. Therefore it is a pre-requisite to obtain clad with minimal dilution. However, it is very difficult to eliminate dilution but can be significantly reduced by selecting optimum parameters. For finite element analysis it is assumed that the part of the substrate that has melted typically results in dilution. Therefore, the molten depth of substitute can be used as a good estimate of dilution depth. Figure 5a compares the dilution value for the current model with that actually obtained through experiment (Paul et al. 2014). Figure 5a also compares the dilution as obtained for a coupled finite element models available in literature with thermal boundary condition (Paul et al. 2014). The current model predicts dilution with a variation of 25 % from the experimental which shows that the model is able to capture dilution that occurs during the process.

In the present work, the region between substrate melting temperature and 1000 K is considered as the Heat Affected Zone (HAZ) as illustrated by Fig. 4a. Figure 5b compares the HAZ value for the model and that actually obtained through

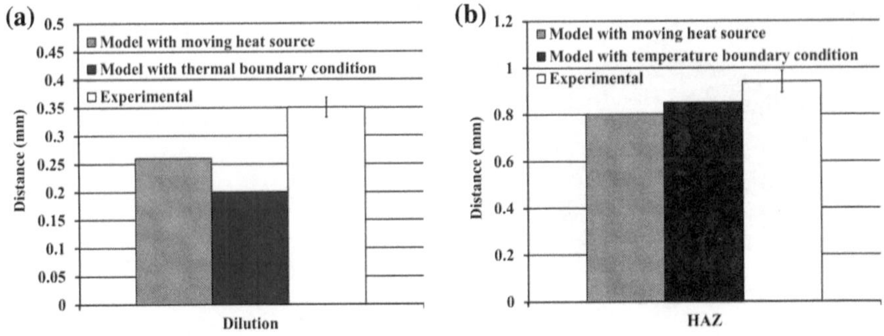

Fig. 5 Comparison of **a** dilution zone, **b** HAZ

experiments (Paul et al. 2014). The current model is able to predict the HAZ with a variation of 15 %. Thus it can be concluded that through the model both dilution and heat affected zone are being captured.

3.3 Residual Stress Analysis

As discussed in previous section in a clad-substrate system the presence of relatively high thermal gradients between dissimilar materials leads to the development of thermal stresses because of the presence of different co-efficient of thermal expansion and different reference temperatures for the clad and substrate materials. Due to laser cladding being a localized heating application during a short time interval, a thin layer of substrate is heated up to the liquidus temperature and a molten layer of the clad material is deposited on it. Thereafter, the heat is conducted into the substrate and the clad solidifies and starts shrinking due to thermal contraction, whereas the substrate first expands and later contracts according to the local thermal cycle which leads to the development of residual stresses.

The residual stress contour in the longitudinal direction, i.e., along the cladding direction is shown in Fig. 6a. The model predicts compressive stresses on the surface of the clad which originates to counter the tensile stresses induced at the interface because of high thermal contraction. The interface region contracts rapidly due to the rapid heat transfer to the relatively cooler substrate underneath whose dimensions do not change significantly, thereby inducing tensile stresses as the substrate.

Experimental values of longitudinal stresses were obtained through X-ray diffraction (XRD) technique (Paul et al. 2014) using a X-ray beam of diameter 500 μm. Figure 6b compares the model average values which are typically the nodal values averaged over an area of 500 μm. The model predicts a variation of 12 % in the residual stress in the longitudinal direction on the clad-substrate

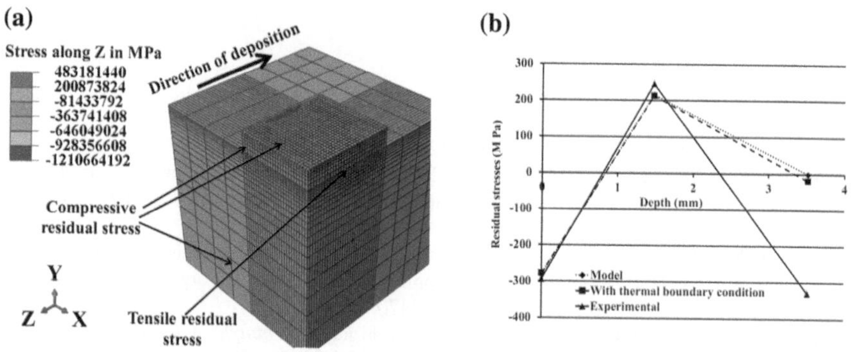

Fig. 6 **a** contour of residual stress along longitudinal direction, **b** comparison of residual stress along the direction of clad

interface. However, the development of compressive residual stress in the clad highlights the potential application of the process for in situ repair applications for die and aerospace components.

4 Conclusions

A coupled thermomechanical finite element models of powder injection laser cladding technique has been developed and the model results have been compared with the available experimental results [43]. Following are the main conclusions drawn from the present work:

- The thermomechanical model for powder injection laser cladding process with moving Gaussian heat source is able to predict the clad geometry with prediction errors lying between 14 and 18 %.
- Thermal analysis of the process can predict the temperature profile which can be used to estimate the extent of dilution and heat affected zones. The prediction errors lie between 15 and 25 %.
- The longitudinal residual stress has been characterized for clad, interface and the substrate regions. The nodal average values for the model over an area of 500 μm qualitatively predicted similar nature with reasonable variation.
- The longitudinal stress in the interface region is tensile in nature which can be attributed to the rapid heat transfer and higher thermal contraction at the interface.
- The analysis of the process also highlighted the application of the technique for repair and restoration applications of various high cost components due to evolution of compressive stresses on the surface of the clad.

References

Amon, C. H., Beuth, J. L., Weiss, L. E., Merz, R., & Prinz, F. B. (1998). Shape deposition manufacturing with micro-casting: Processing, thermal and mechanical issues. *Journal of Manufacturing Science and Engineering, 120*, 656–665.

Brandt, M., Sun, S., Alam, N., Bendeich, P., & Bishop, A. (2009). Laser cladding repair of turbine blades in power plants: From research to commercialization. *International Heat Treatment and Surface Engineering, 3*, 105–114.

Chen, J., & Xue, L. (2010). Laser cladding of wear resistant CPM9V tool steel on hardened H13 substrate for potential automotive tooling applications. In *Materials Science and Technology 2010 Conference and Exhibition*, (pp. 2459–2470).

Costa, L., & Vilar, R. (2009). Laser powder deposition. *Rapid Prototyping Journal, 15*, 264–279.

Da Sun, S., Liu, Q., Brandt, M., Janardhana, M., Clark, G. (2012). Microstructure and mechanical properties of laser cladding repair of AISI 4340 steel, 28th International Congress of the Aeronautical Sciences.

Deus, A., Mazumder, J. (2006). Three-dimensional finite element models for the calculation of temperature and residual stress fields in laser cladding. In *Laser Materials Processing Conference, ICALEO 2006 Congress Proceedings* (pp. 496–505).

Ghosh, S., & Choi, J. (2005). Three-dimensional transient finite element analysis for residual stresses in the laser aided direct metal/material deposition process. *Journal of Laser Applications, 17*, 144–158.

Ghosh, S., & Choi, J. (2006). Modeling and experimental verification of transient/residual stresses and microstructure formation in multi-layer laser aided DMD process. *Journal of Heat Transfer, 128*, 662–679.

Ghosh, S., & Choi, J. (2007). Deposition pattern based thermal stresses in single-layer laser aided direct material deposition process. *Journal of Manufacturing Science and Engineering, 129*, 319–332.

Griffith, M. L., Schlienger, M. E., Harwell, L. D., Oliver, M. S., Baldwin, M. D., Ensz, M. T., et al. (1998). Thermal behavior in the LENS process. In D. Bourell, J. Beaman, R. Crawford, H. Marcus, & J. Barlow (Eds.), *Paper presented at Solid Freeform Fabrication Symposium*. Austin, TX: University of Texas at Austin.

Grum, J., & Slabe, J. M. (2003). A comparison of tool-repair methods using CO_2 laser surfacing and arc surfacing. *Applied Surface Science, 208–209*, 424–431.

Henderson, M. B., Arrell, D., Larsson, R., Heobel, M., & Marchant, G. (2004). Nickel based superalloy welding practices for industrial gas turbine applications. *Science and Technology Welding and Joining, 9*, 13–21.

Hu, Y. P., Chen, C. W., & Mukherjee, K. (1998a). Development of a new laser cladding process for manufacturing cutting and stamping dies. *Journal of Materials Science, 33*, 1287–1292.

Hu, Y., Chen, C., & Mukherjee, K. (1998b). Innovative laser-aided manufacturing of patterned stamping and cutting dies: Processing parameters. *Materials and Manufacturing Processes, 13*, 369–387.

Kawasaki, M., Takase, K., Kato, S., Nakagawa, M. & Mori, K. (1992). Development of engine valve seats directly deposited onto aluminium cylinder head by laser cladding process, SAE Technical Paper Series, SAE Paper No. 920571. *SAE International, Warrendale, PA*, pp. 1–15.

Lee, H. K., Kim, K. S., & Kim, C. M. (2000). *Engineering Fracture Mechanics, 66*, 403–419.

Leunda, J., & Soriano, C. (2011). Laser cladding of vanadium-carbide tool steels for die repairs. *Proceedings of the Sixth International WLT Conference on Lasers in Manufacturing, physics procedia, 12*, 345–352.

Liu, Q., Janardhana, M., Hinton, B., Brandt, M., & Sharp, K. (2011). Laser cladding as a potential repair technology for damaged aircraft components. *International Journal of Structural Integrity, 2*, 314–331.

Majumdar, J. D., Pinkerton, A., Liu, Z., Manna, I., & Li, L. (2005). Mechanical and electrochemical properties of multiple-layer diode laser cladding of 316L stainless steel. *Applied Surface Science, 247*, 373–377.

Mc Daniels, R. L., White, S. A., Liaw, K., Chen, L., McCay, M. H., Liaw, P. K. (2008). Effects of a laser surface processing induced heat-affected zone on the fatigue behavior of AISI 4340 steel. *Materials Science and Engineering: A, 485*, 500–507.

Moat, R., Pinkerton, A. J., Hughes, D. J., Li, L., Preuss, M., and Withers, P. J. (2007). Stress distributions in multilayer laser deposited Waspaloy parts measured using neutron diffraction. In *Proceedings of 26th International Congress on Applications of Lasers and Electro- optics (ICALEO).* Orlando, California, CD.

Onoro, J., & Ranninger, C. (1997). Fatigue behavior of laser welds of high-strength low-alloy steels. *Journal of Material Process and Technology, 68*, 68–70.

Paul, S., Ashraf, K., Singh, R. (2014). Residual stress modeling of powder injection laser surface cladding for die repair applications. In *Proceedings of the ASME 2014 International Manufacturing Science and Engineering Conference MSEC2014.* Detroit, Michigan, USA.

Picasso, M., Marsden, C.F., Wagniib.RE J.-D., Frenk, A., Rappaz, M. (1994). A Simple but realistic model for laser cladding, metallurgical and materials transactions B *25B*, p. 281.

Pinkerton, A., Wang, W., Li, L. (2008). Component repair using laser direct metal deposition. In *Proceedings of the Institution of Mechanical Engineers, Part B: Journal of Engineering Manufacture 222*, 827–836.

Pinkerton, A. J., Karadge, M., Syed, W. U. H., & Li, L. (2006). Thermal and microstructural aspects of the laser direct metal deposition of Waspaloy. *Journal of Laser Applications, 18*, 216–226.

Plati, A., Tan, J., Golosnoy, I., Persoons, R., Acker, K., & Clyne, T. (2006). Residual stress generation during laser cladding of steel with a particulate metal matrix composite. *Advance Engineering Materials, 8*, 619–624.

Qi, H., Mazumder, J., Ki, H. (2006). Numerical simulation of heat transfer and fluid flow in coaxial laser cladding process for direct metal deposition. *Journal of Applied Physics, 100*, 024903.

Schneider, M. F. (1998). *Laser cladding with powder; Effect of some machining parameters on clad properties. Ph.D Thesis,* University of Twente, Enschede, The Netherlands.

Shanmugam, N. S., Buvanashekaran, G., Sankaranarayanasamy, K. (2013). Some studies on temperature distribution modelling of laser butt welding of AISI 304 stainless steel sheets. *World Academy of Science, Engineering and Technology 7.*

Steen, W. M. (2003). *Laser material processing* (3rd ed.). London: Springer.

Su, C. Y., Chou, C. P., Wu, B. C., Lih, W. C. (1997). Plasma transferred arc repair welding of the nickel-base superalloy IN-738LC. *Journal of Materials Engineering and Performance 6*, 619–627.

Suarez, A., Amado, J., Tobar, M., Yanez, A., Fraga, E., & Peel, M. (2010). Study of residual stresses generated inside laser cladded plates using FEM and diffraction of synchrotron radiation. *Surface and Coatings Technology, 204*, 1983–1988.

Tan, J. C., Looney, L., & Hashmi, M. S. J. (1999). Component repair using HVOF thermal spraying. *Journal of Materials Processing Technology, 92–93*, 203–208.

Tusek, J., Ivancic, R. (2004). Computer-aided analysis of repair welding of stamping tools. *Z. fu¨r Metallkunde, 95*, 8–13.

Wang, J., Prakash S., Joshi Y., Liou F. (2002). Laser Aided Part Repair-A Review. In *Solid Freeform Fabrication Proceedings*, pp. 57–64.

Wang, S.-H., Chen, J.-Y., & Xue, L. (2006). A study of the abrasive wear behavior of laser-clad tool steel coatings. *Surface and Coatings Technology, 200*, 3446–3458.

Zhang, C. H., Hao, Y. X., Qi, L., Hu, F., Zhang, S., & Wang, M. C. (2012). Preparation of Ni-Base alloy coatings on monel alloy by laser cladding. *Advanced Materials Research, 472–475*, 313–316.

Zhang, P., Ma, L., Yuan, J., Yin, X., & Cai, Z. (2008). The finite element simulation research on stress-strain field of laser cladding. *Journal of Engineering Materials, 373–374*, 322–325.

Zhang, Y., Yuan, X., & Zeng, X. (1999). *ICALEO 1999: Laser materials processing conference* (p. 241). USA: San Diego.

Zhang, Y., Zeng, X., & Yuan, X. (2001). *ICALEO 2001: Applications of lasers and electro-optics* (p. 577). USA: Jacksonville.

Zheng, L., Xie, W., & Li, Y. (2011). The numerical simulation on the temperature field of laser cladding. *Journal of Engineering Materials, 467–469*, 1372–1376.

Excimer Laser Micromachining and its Applications

James Jacob, P. Shanmugavelu, R. Balasubramaniam and Ramesh K. Singh

Abstract Non-conventional, non-contact type advanced machining process like laser based micro machining process is widely used in modern industries for producing components with geometrically complex profiles. Though laser based micro machining of polymer, by and large, is a cold ablation process, photo thermal process associated with the laser heating may affect the surface characteristics. This chapter starts with an introduction to various excimer laser sources and proceeds to micromachining application areas with specific reference to polymers. The study of excimer laser micromachining in different gaseous media, conducted by the authors is elaborated further. This study was conducted to ascertain the impact of purging with gases such as air, argon, nitrogen, helium and hydrogen during the laser ablation process. A negative photo resist, E-1020 obtained from M/s Cadmosil Chemical Pvt. Ltd, India was studied using 248 nm KrF excimer laser. The effect of gas purging on the ablation rate and surface characteristics of the polymer was studied. Amongst the gases used, hydrogen gas showed distinct results with respect to ablation rate and surface characteristics. It has been observed that hydrogen gas has enhanced both the ablation rate and the surface quality significantly. The role of hydrogen gas in enhancing the laser ablation rate may be attributed to the possible involvement of hydrogen gas in the laser assisted chemical reaction with polymer. Eximer laser micromachining process is characterized by a number of process parameters that determines efficiency, economy and quality of the whole process. In this chapter, the details of the experiment along with the results and observations, with areas for future study have been presented.

J. Jacob (✉) · P. Shanmugavelu · R. Balasubramaniam
Bhabha Atomic Research Centre, Mumbai 400085, India
e-mail: jamesmankottil@gmail.com; jamesm@barc.gov.in

P. Shanmugavelu
e-mail: psvelu@barc.gov.in

R. Balasubramaniam
e-mail: cwsbalu@barc.gov.in

R.K. Singh
Indian Institute of Technology, Bombay, Mumbai, India
e-mail: ramesh@me.iitb.ac.in

S.N. Joshi and U.S. Dixit (eds.), *Lasers Based Manufacturing*,
Topics in Mining, Metallurgy and Materials Engineering,
DOI 10.1007/978-81-322-2352-8_10

Keywords Laser micromachining · Excimer laser · Micromachining applications · Photo resist · Assist gas

1 Introduction

Many Non-conventional, non-contact type of advanced micro machining methods are developed and widely being employed in modern industries for making components with geometrically complex profile. These non-conventional, non-contact advanced machining methods include electro discharge machining, energy beam machining (Laser, electron, ion and plasma), electrochemical machining and chemical machining processes. Each of these processes has its own limitations with regard to the type of material and applications. Laser beam is a unique energy source, which can, by and large, remove materials by photo-thermal processes. This process is employed for the machining of variety of materials, such as metals, ceramics, composites and glasses. The ability of Laser beam machining, to form complex patterns with close tolerances in wide variety of materials, has opened up brighter avenues to wider range of applications. Miniaturization is an ongoing trend in modern technologies. Industries like semi-conductors, biotechnology, micro-electronics, Opto-electro mechanical systems, telecommunication, entertainment, instrumentation, optical data memory/storage, and remote sensing and imaging and ever increasing number of medical applications require sub-micrometer features in various materials. Material processing requirements of high precision, of the order of micrometers or even sub micrometers, at high speed, has become the norm for many of these high tech industries. The combination of high resolution, accuracy, flexibility and high speed has resulted in the advancement of micro-engineering in the recent past. For example, the ability to drill smaller holes down to ~ 1 μm diameter, is an important technological challenge in many industries that manufacture high-tech products. While high-power solid state lasers are fundamental to large scale industrial production, excimer laser on the other hand has revolutionised the manufacturing industry with high precision, easy 3D structuring and less stringent production requirements. Industrial applications such as the drilling of ink-jet printer nozzles (Rowan 1995), production of sensors (Gifford and Bartnik 1998). and the manufacture of display panels now routinely use excimer laser micro machining in production environments. Consequent to the rapid improvement in the speed and memory capacities of the ICs (Integrated circuits), the requirement for enhancement in the packaging density has multiplied. The multi-layer printed circuit boards (PCBs) which consists of multi-layer sandwiches of conductor-insulator-conductor, require electrical connection between layers, made possible by drilling small holes through the dielectric layer (Gower 2000). Typically the requirement is to drill a large number of around 100 μm diameter holes at around 500 μm pitch, which represent around 30 % of the overall cost of the PCB. The combination of high-resolution, accuracy, speed and flexibility is allowing excimer laser micromachining to gain acceptance in many high technology industries. A brief outline on excimer based micromachining of different polymeric materials for various industrial applications is briefly illustrated in this paper.

2 Types of Excimer Laser

Lasers can be classified based on a number of factors e.g. active medium (solid, liquid and gas), output power (low, medium and high power lasers), excitation method (electrical, optical and chemical), operating mode (continuous wave, pulsed mode and Q-switched output mode), efficiency and applications. excimer lasers, which fall under the category of gas lasers, are the most powerful lasers with short (of the order nano seconds) pulse rates and durations, whose wavelengths fall in the UV or DUV range of the electromagnetic spectrum. The word Excimer comes from, Excited dimer (the lasing medium of this class of lasers), which refers to a class of diatomic molecules formed by the chemical combination of two identical constituents, in their excited state. For example, in Xe_2 laser, the Xenon gas was used as the lasing medium. In fact the word Excimer is a misnomer, as far as most of the current industrial excimer lasers are concerned. The most commonly used excimer lasers like ArF, KrF, XeCl, XeF etc., should have been called 'Exciplex' Lasers (rather than Excimer Lasers), as they are complexes of Rare Gas Halides (RGH) in the excited states. Excimer describes a diatomic molecule that is chemically bound in its electronically excited upper state, but is repulsive or only weakly bound in its lower ground state. In the normal course, rare gas atoms won't be able to react chemically with any other atom to form a molecule. In that sense, Rare-gas halides are peculiar molecules that emit laser light on an unusual type of electronic transition. The two atoms are bound only when the molecule is in an excited state. When the molecule drops to the ground state, which is the lower laser level, the molecule falls apart. That produces a population inversion in a rather unusual way—there can't be any molecules in the lower laser level because they are not bound together. Figure 1 shows the energy levels of a typical rare gas halide as a function of the spacing between the two atoms in the molecule. R represents the rare gas and H the halide. The dip in the excited-state curve shows where the molecules are metastable. The absence of a dip in the ground-state curve indicates that the molecules fall apart. When the molecule is excited, the energy is at a minimum when the two atoms are at a certain distance apart, trapped in a potential well. However, in the "ground-state," with the lowest possible energy, there is no bonding energy to hold the two atoms together and the molecule fall apart, as shown in the lower curve. This reflects something we know from elementary chemistry, namely, rare gases do not like to form compounds, even with elements as highly reactive as halogens.

A simplified molecular transition in KrF excimer laser is depicted in Fig. 2. A typical KrF excimer laser gas mixture is a Kr:F2:Ne blend with neon constituting most of the volume. The neon acts as a third body collision partner in the formation of the excited KrF* molecule. The voltages and currents required for excimer laser operation test the limits of electronic technology. Consequently, excimer lasers are

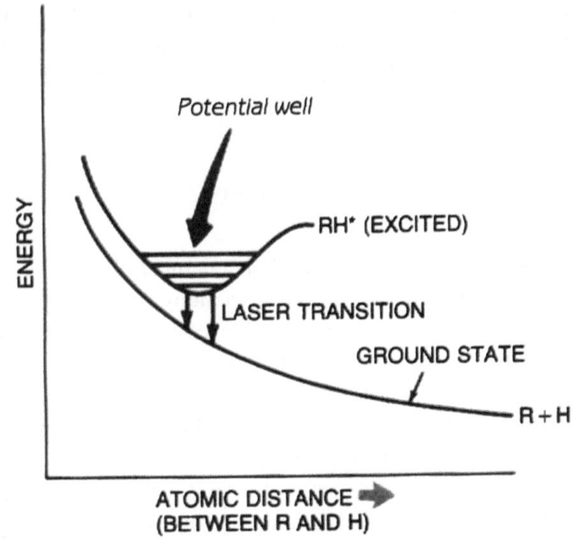

Fig. 1 Internal energy of a rare-gas halide molecule in excited and ground states

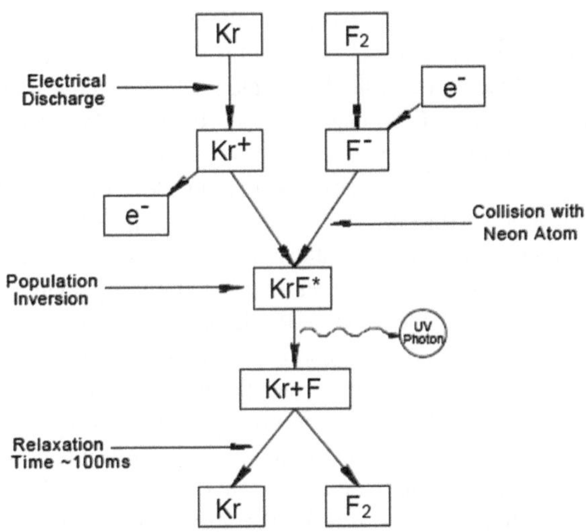

Fig. 2 Molecular transition in the KrF excimer laser

more complex than other types of lasers. They require more maintenance and are more expensive to maintain.

Most excimer lasers are capable of using any of the six gas mixtures available, but it is usually not advisable to mix fluorine and chlorine in the same laser in industrial situations.

Each gas mixture has its own characteristic wavelength and application areas as shown in Table 1.

Excimer lasers are excited by passing a short, intense electrical pulse through a mixture of gases containing the desired rare gas and halogen. Normally, 90 % or

Table 1 Photon energies and application areas of various excimer laser sources

Active medium	Wave length in nm	Photon energy in eV	Advantages/application areas	Disadvantages
Molecular fluorine (F_2)	157	7.9	Suitable for machining high resolution features, even on ceramics with very good ablated surface quality	The laser beam path must be isolated from the ambient O_2 molecules to prevent the formation of ozone by absorption of laser energy by O_2
Argon fluoride (ArF)	193	6.42	Suitable for ablating glass and ceramic materials with high machining rate and good surface quality owing to the combination of lasing efficiency.	The beam path needs to be either in vacuum or in an inert gas purged environment to prevent photon absorption by oxygen
Krypton fluoride (KrF)	248	5.0	Best candidate for polymer ablation owing to the combination of average power, lasing efficiency, absorption characteristics and operating cost	
Xenon chloride (XeCl)	308	4.02	Very long gas life in the laser cavity. Used mostly in marking applications	Larger absorption depth in polymers than KrF laser, leading to higher ablation depth, but lower average powers generate a lower overall machining rate
Xenon fluoride (XeF)	351	3.53	Useful for pulse annealing and laser assisted etching owing to less energy intensity	Less attractive even for polymers owing to lower absorption

(Data source from: Tseng 2007)

more of the mixture is a buffer rare gas (typical helium or neon) that does not take part in the reaction. The mixture also contains a small percentage of the rare gas (argon, krypton, or xenon) that becomes part of the excimer molecule, and a smaller fraction of the molecules that supply the needed halogen atoms. The halogen atoms may come from halogen molecules, such as F_2, Cl_2, or Br_2, or from molecules that contain halogens, such as nitrogen trifluoride (NF3). The advantage of avoiding pure halogens is that they are very reactive and hence may result in the corrosion of the material of construction of the laser tube. Electrons in the discharge transfer energy to the laser gas, breaking up halogen molecules and causing formation of electronically excited molecules like xenon fluoride (XeF*, the * means excited). The reactions involved are very complex and depend on the type of gases. The molecules remain excited for about 10 ns, then drop to the ground state and dissociate. The molecular kinetics as well as the duration of the driving electrical pulses limit the laser operation to pulses lasting tens or hundreds of nanoseconds. The energies involved are large, due to the high gain of the lasing medium, and the output is at ultraviolet wavelengths. Excimer-laser repetition rates are more dependent on the power supply than on the gas. The principal limitation is the speed of high-voltage switches. The highest repetition rates are around 1000 Hz, but more typical values are tens to few hundreds of hertz. Pulse energies range from about 10 mJ to a few joules, and differ somewhat among gases, with KrF and XeCl generally being more energetic. Average power—the product of the pulse energy times the repetition rate —can reach a couple of hundred watts, although lower values are more common. Note that in general the pulse energy tends to decrease with repetition rate.

As mentioned above, excimer lasers are, generally produced by a fast electrical discharge in a high-pressure mixture of rare gas (krypton, argon or xenon) and a halogen gas. Since each of these binary mixtures has a characteristic emission wavelength, the excimer laser is normally designated by the name of the mixture. As a result, the wavelength of an excimer laser can be altered, as indicated in the Table 1, by changing the gas mixture with associated modifications of mirrors to obtain maximum output. The excimer plasma tube operates at a very high pressure of the order of 3.5 times atmospheric pressure. This high pressure results in high electrical resistance to the passage of current through the gas. The gas is therefore, first pre-ionized to facilitate a smooth, homogeneous discharge. The main discharge, up to 40 kV is applied across two parallel electrodes, which run through the entire length of the plasma tube. The resulting plasma contains very high concentration of excited dimers resulting in a laser with high gain and thus high energy per pulse of the order of 3–10 eV. As in other high-gain pulsed lasers, the discharge in an excimer laser is perpendicular to the length of the tube. The excimer laser tubes must resist attack by the highly corrosive halogens in the laser gas. Excimer laser tubes are filled with the laser gas mixture then sealed and operates for a certain number of shots until the gas needed to be replaced. The tube's total volume is much larger (typically 100–1000 times) than the volume where the discharge excites laser action. Often, the gas is passed through a recycling system that helps regenerate the proper gas mixture and extend the life of the gas fill. The laser's pulse energy drops with time, until the spent gas must be pumped out of the chamber and replaced. The number of shots depend on the gas, and can be many millions of shots for gases, such as xenon chloride. Although that number of shots may sound impressive, a simple multiplication will show that at high repetition rates it doesn't amount to very long operation time. A 200-Hz laser generates 720,000 pulses an hour. Thus, gas supply is part and parcel of any excimer laser setup. Although other lasers such as YAG and CO_2 lasers are also extensively used in High Density Interconnection (HDI) technology, the excimer laser ablation is indispensable when it comes to 'fine' finish micro- and nano-fabrications. This is particularly true for hard and delicate materials. This is largely due to its short wavelength, short pulse duration, and of course its pulse energy allowing for what is generally termed as a 'cold ablation' process. The excimer laser also excels others in its ability to 'mask-project' patterns, using stencil or metal-on quartz masks on to a sample with a minimal Heat Affected Zone (HAZ). The minimal HAZ is due to the short pulse duration/interaction time between the laser beam and the material.

Nevertheless, picosecond and femtosecond lasers are now available today. These classes of lasers are designed to further reduce the HAZ. They are also characterized by higher etch rate, strong absorption by the material, improved surface roughness and lower ablation thresholds. These desirable features of the excimer laser have attracted and favoured its use not only for polymers but also with other materials such as ceramics, glasses and silicon which are often hard to machine.

3 Laser Interaction with Polymers

A laser beam striking a material can be reflected, transmitted, or absorbed. In practice, all three occur to some degree. In order for laser micromachining to be practical, the laser beam must be absorbed by the material. Reflected and transmitted light are lost during processing. So only the absorbed light is able to do the material removal in the machining process. The transmitted portion of the power will normally be absorbed within a certain depth of the surface. As the material is heated by the absorbed portion of the transmitted radiation, material vapor, and sputtered material can be ejected from the surface into the oncoming beam. This causes further interaction where coupling of the beam into the material is affected by scattering off, by the ejected material or by shielding of the incoming laser radiation from the surface with a plasma plume, which is created by the beam in the ejected vapor. In addition to creating emission of the material from the surface, the absorbed energy may also create thermal, mechanical, and phase changes which propagate into the material. When the laser beam strikes the material, the photon energy is transferred to the material and subsequently converted to other forms of energy depending on the material. With metals, this is transferred to the mobile electrons which results in the heat energy that can cause vapourisation and disintegration of the metal. However, with non-metals, the energy can either be converted to chemical energy required for bond-breaking or heat energy for vaporization. These two possibilities depend on the type of material, its bond energy and the wavelength of the laser or more precisely the photon energy. Essentially, there are two common mechanisms (Garrison and Srinivasan 1985) for laser material interactions, which can occur at varying degrees while processing a material.

- Thermal (photothermal or pyrolytic): This is an electronic absorption in which the photon energy is used to heat up the material to be processed and thus part of the material is removed as a result of molecule vaporization, such as in CO_2 laser cutting.
- Athermal (photochemical or photolytic): This is a photochemical process whereby the material is ablated by direct breaking of molecular bonds when hit by photons (energy) of the incident beam. In principle, this is only possible if the photon energy is equal or greater than the bond energy of the molecules of the material to be processed. During this process, a particular area of the surface of the material is removed with minimum (or without any, theoretically) thermal

damage to the surrounding material. This process is generally called ablation, though photothermal processes are also referred to as ablation. Ablation is generally used in reference to polymer and/or soft materials, but laser ablation is also possible with other materials such as ceramic and glass. However higher fluences are required in their case.

The etch rate—the amount of material removed per pulse—is mainly a function of the photon energy and the material being processed. The ablation threshold is the point at which the applied energy density is enough to cause ablation either photolytic or pyrolytic (Fig. 3). The ablation threshold varies from polymer to polymer depending on the nature and strength of the bonds in the polymer and also on laser wavelength (Table 2). An ablation threshold can be obtained from a plot of etch rate against a logarithmic scale of fluence at zero ablation rate, as indicated in the Fig. 4 below.

Table 3 shows a list of common bonds in polymers with their respective bond energies, which need to be broken during any laser ablation, regardless of the nature of the mechanism. For photochemical ablation, the laser wavelength has to be carefully chosen such that the photon energy obtained from the laser is equal or greater than the bond energy of the polymer to be processed. When working at fluence levels below the Ablation threshold, no ablation/micromachining is expected to occur. However, the chemical properties of the materials are subject to certain changes. Furthermore, operating at well above the threshold may leads to increase in the heat-affected zone (HAZ) and debris deposition. The former is due to

Fig. 3 The principle of laser ablation

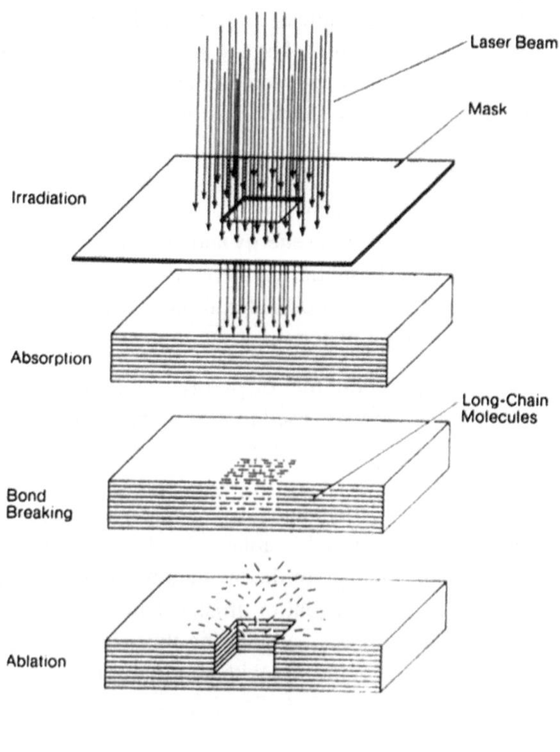

Table 2 Ablation threshold fluence and depths for some selected materials

Material	Excimer laser source/wavelength(nm) [nm]	Fluence (J/cm^2)	Ablation depth/pulse (μm)
Polycarbonate (PC)	KrF/248	4.0	0.4
Polyester (PES)	KrF/248	4.0	0.8
Polyethylene (PE)	KrF/248	3.7	1.0
	ArF/193	6.0	4.0
Polyimide (PI)	XeCl/308	0.3	0.1
	KrF/248	0.7	0.3
Alumina	ArF/193	45	0.06
	KrF/248	45	0.19
	XeCl/308	25	0.17
Zirkonia	KrF/248	10	0.12
Boron nitride	ArF/193	20	0.15
Silicon nitride	KrF/248	10	0.18
Silicon carbide	KrF/248	10	0.13
Piezoelectric ceramics	KrF/248	5.0	0.05
(PZT)	XeCl/308	5.4	0.20

(Data source from Basting and Marowsky 2005)

high energy while the latter is, as a result of bombarding the ejected materials. It may be noted that intense bombardment of ejected particles above the ablation zone may result in the reduction of the micromachining rate/the ablation rate. This is because the ejected materials might absorb fractions of the incoming beam thus reducing the effective fluence at the zone of ablation. Wavelength is one of the factors that determine the thresholds of ablation. For example, the ablation threshold for PMMA (Poly Methyl Meth Acrylate) is ~150 mJ/cm^2 at 193 nm and ~500 mJ/cm^2 at 248 nm—this is a 3-time increase in value between the two wavelengths. The rule-of-thumb for laser ablation of polymers is to have lower threshold fluences for ablation at shorter wavelengths. The pulse duration is another important factor. Ultra short laser pulses can minimize a photothermal process during the laser process so that the higher quality structures might be obtained. Especially, melting effects (recast and debris) around the machining areas can be minimized by ultrashort pulsed laser.

As mentioned earlier, a key concept in excimer laser machining is the fluence, F, which is defined as the energy per area per pulse delivered onto the sample surface, with the unit of [J/cm^2]. Invariably, polymers display a threshold fluence as indicated in Fig. 4, which must be exceeded for ablation to take place. The threshold fluence is often in the order of few hundred mJ/cm^2, depending on the polymer and on the excimer wavelength. Most polymers obey Beer's law of attenuation. From Table 3, it is clear that, KrF laser which is the most commonly used excimer laser source for polymer micromachining, is able to break many types of polymeric bond, as its photon energy is 5 eV. Ideal photochemical processes are unlikely to occur, and some thermal effects usually accompany ablation processes.

Fig. 4 Ablation rate as a function of laser fluence

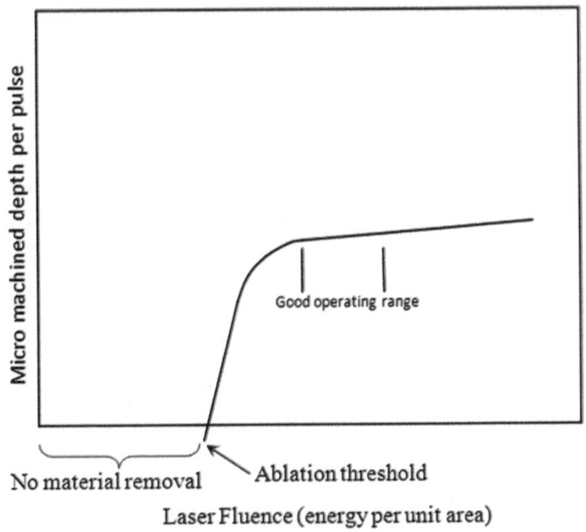

Table 3 Bond dissociation energy

Type of bond	Bond energy
O–O	1.47
N–N	1.69
C–N	3.16
C–C	3.6
H–N	4.03
C–H	4.29
N=N	4.33
H–H	4.51
O–H	4.81
O=O	5.16

(Data source from Lide 2009)

4 Excimer Laser Machining Process

Excimer laser can create structures on substrate by two means either by direct writing or by using mask patterns as shown in Fig. 5. The poor beam quality factor of the excimer laser beam makes it unsuitable for focusing on a small area, as the minimum spot size is directly proportional to the beam quality factor. The relatively large beam size of ~25 mm × ~10 mm is able to satisfy the requirement for projection printing. Hence, the technique of mask projection is commonly used in a large number of applications. With a mask and de-magnifying optics, laser projection printing allows parallel production of a large array of de-magnified microstructures simultaneously, to increase the throughput (Tseng 2007). In the other major fabrication mode viz, direct writing, the microstructure is constructed by direct bombardment of a series of small spots using a focused laser beam

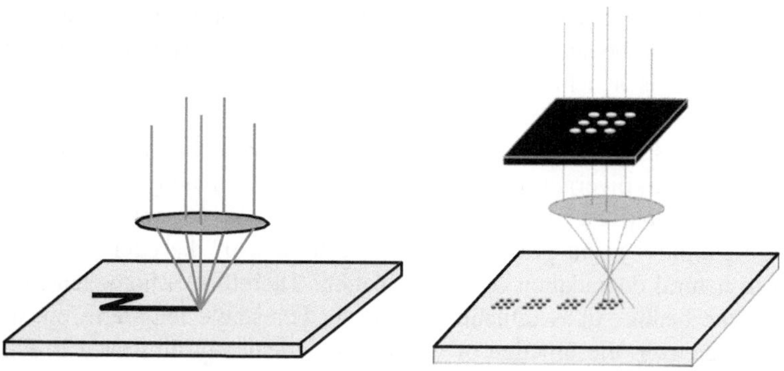

Fig. 5 Direct writing and mask projection techniques

scanning a specific pattern over the target. Since direct writing is operated in a series mode, its throughput is much lower than that of projection printing. Consequently, excimer lasers are normally used for projection printing to machine microstructures and are fundamentally different from most other industrial lasers, such as the solid-state Nd:YAG and CO_2 lasers, which are generally employed in direct writing during machining.

As shown in Fig. 6, a basic laser micromachining workstation consists of an excimer laser source, beam shutter, attenuator, a set of steering mirrors, Beam

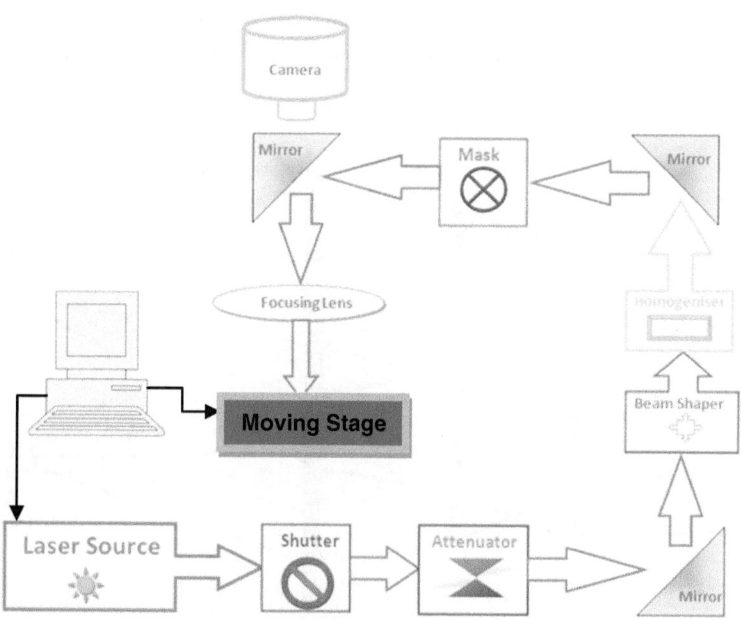

Fig. 6 Typical excimer laser micro machining setup

Shaper, Beam Homogenizer, Mask, Projection Lens, motorized stage and camera; the whole system being mounted on an optical table. The movement of the motorized stage is synchronized with the Laser pulse firing from the Excimer source using the computerized control system.

Excimer lasers exhibit relatively poor beam quality, including high-order mode structures and high divergence. Also, the beam profile is mainly determined by the discharge conditions and the laser resonator optics. With continuous operation, discharge conditions and gas quality, the cleanliness of the cavity windows are subject to natural degradation or contamination. Therefore, a homogeneous beam profile at the beam exit is difficult to achieve. For stable long-term operation, a beam homogenizer, the function of which is schematically illustrated in Fig. 7, is frequently used to compensate and average these influences. A typical fly's eye homogenizer can convert a non-uniform profile of an excimer laser beam into a homogeneous beam with higher than 95 % uniformity in the energy density.

After the homogenizer the laser beam is passed through a projection mask and projected onto the target surface by a de-magnifying objective/projection lens. As indicated in Tables 1 and 3, the photon energies of F_2 and ArF lasers are higher than the oxygen binding energy, 5.1 eV. As a result, the F_2 and ArF beams can be absorbed by O_2 in the air, generating ozone. To prevent the generation of hazardous ozone gases and to have efficient propagation without unnecessary energy absorption, the laser beam passage should be encapsulated, and flushed by an inert gas or drawn into a vacuum. In projection printing, the mask is one of the most critical components. In basic excimer laser micromachining, two types of binary masks, stencil and chrome-on quartz (or more generally, metal-on-quartz), are normally used to define the regions to be patterned. In stencil masks, patterns are etched on metal foils. This type of mask has a long lifetime with excellent contrast

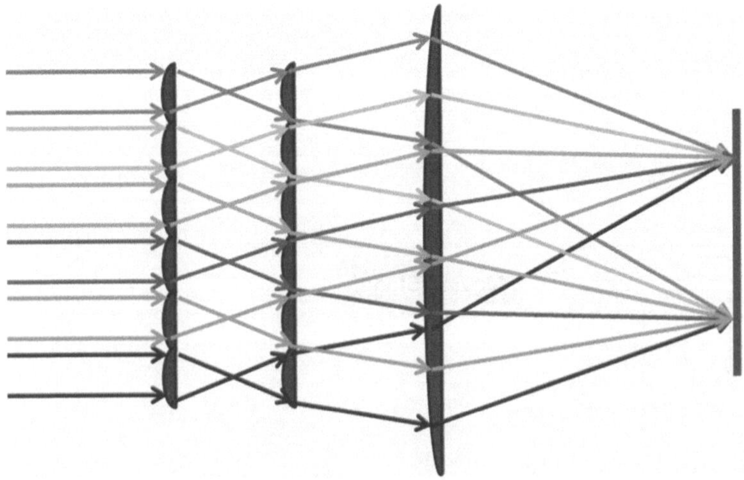

Fig. 7 Ray diagram of typical beam homogenizer

because it not only exhibits high resolution or small feature sizes, but also withstands high-energy laser irradiation.

Successive laser pulses cause material ablation/removal, thus the controlled local micromachining of the part, without any adverse thermal effect associated with continuous wave, high power laser applications. Because of the short pulse duration, of the order of just nanoseconds, there is not much time for the heat to diffuse out of the irradiated or micromachining-area into the bulk material of the work piece, resulting in very good surface characteristics, without the adverse heat affected zone. Excimer lasers allow high-intensity energy to be absorbed in a very thin surface layer, for the effective removal of material from irradiated target area. The machining mechanism of polymers, by excimer lasers is usually referred to as ablation as mentioned above. Here, the material explodes into vapour and is ejected in the form of species such as atoms, molecules, ions, and clusters due to the interaction of the intense laser pulse with the material. To ablate a high precision and high-quality structure at micrometer scales, the interaction between the laser and material must be well understood and controlled.

The complex microstructure is created through movement of the work piece or mask or both. The type of laser used, the selection of polymeric material, environment under which the ablation is carried out, parameters of the laser beam etc. are some of the vital inputs to be considered to have optimal laser ablation for manufacturing micro features with excimer laser. The mask projection method offers great flexibility and there are numerous features which can be utilized when considering micro-machining applications. Few of them are listed below.

Mask dimensions: The projection lens usually de-magnifies the mask pattern onto the work piece and so the features on the mask need not be as small as the microstructures to be produced. Typically, de-magnifications of ×4, ×10, ×12, ×16 or ×20 are used so the mask does not have to be made of ultra-high resolution features, thereby reducing its complexity and cost of manufacture.

Mask Damage: Due to the de-magnification used, the energy density of the laser beam is much lower at the mask than at the sample mounted on the work stage. This reduces the risk of damage to the mask with intense laser radiation, and increases the mask lifetime as well.

Separation of Mask and Work-piece: Because the mask and work piece are not in close proximity, the mask does not get exposed to any debris or get deteriorated due to particulate damage from the sample ablation.

Independent Control: Mask projection allows independent control of the motion of the mask and workpiece and this allows many different processing techniques to be used depending on the desired micro-engineering application.

Resolution and Depth of Focus: The smallest feature resolution R which can be obtained is fundamentally governed by the projection lens being used and can be expressed as $R \propto k1\lambda/NA$, where λ is the laser wavelength and NA is the numerical aperture of the optical system and k1 is a constant which depends on the sample material and processing conditions. The depth of focus (DoF) is similarly given by $DoF \propto k_2\lambda/NA^2$ where k_2 is another optical/process-related constant.

5 Different Polymers Used for Industrial Micromachining

Nearly all engineering materials, including plastics, glass, metals, semiconductors, composites, and biological materials can be processed by excimer lasers. Polymers with high absorption (See Table 4 below) at 248 nm, such as PET and PI are better suited for machining than materials of low (PS, PMMA) or negligible (PTFE, PE, PP) absorption. Generally, the machining depth for high-absorption polymers is smaller than for low-absorption polymers, because the penetration of the photons is limited to a smaller interaction volume. Even polymers which show negligible absorption such as PE, PP and Teflon (PTFE) can be machined to some extent. This is due to nonlinear absorption (multi-photon processes), but the structures invariably display poor machining quality.

The high intensity of the laser will not only cause ablation, but also the formation of a plume of plasma above the interaction zone. The plasma plume has at least two important effects on the ablation process. The plume is formed on a timescale comparable to that of the laser pulse itself. Thus, some of the incoming radiation may interact with the early stages of the plume, before reaching the job surface. The extent of the plume may be several mm, depending on the fluence and the material, often orders of magnitude more than the size of the structure that is ablated. When the plume is extinguished, the material in it could be re-deposited along the ablated structure, often as visible debris. For some applications this is not a concern, whereas particular contaminants may be detrimental in some other cases. Consequently, excimer laser machining may often be followed by a cleaning procedure to remove the debris. Some control is offered by carrying out the ablation in inert atmospheres, such as He or N_2. Debris is often found in the form of carbon dust, which is readily wiped off, from the surface.

The etching process of polymeric material with a pulsed UV Laser is named as Ablative Photo Decomposition (APD) by Srinivasan and Mayne-Banton (1982). The ablation process is characterized by pulse-by-pulse removal of small amount of material at submicron thickness, with minimal damage to the surrounding areas. A photon of short-wavelength laser has energy ranging from 4 to 10 eV, which covers the bond energy of most of the organic polymeric materials. The bonding energies of different types of covalent bonds of organic polymeric materials are shown in Table 3.

Table 4 Absorption coefficients and penetration depths of selected polymers at 248 nm

Polymer	Absorption, α (μm^{-1})	Penetration depth (nm)
PMMA	0.0063	158×10^3
PI	22	45
PET	16	62.5
PC	1	1 × 103

(Data Source from: Chen et al. 2005; Jackson et al. 1995; Meijer 2004; Pfleging 2006; Yung et al. 2000; Zakariyah 2010; Zeng et al., 2003)

Numerous models have been proposed to explain the observed etching behaviour of various polymeric materials by Garrison and Srinivasan (1985). These models invariably assume the ablation to be either photochemical or photo-thermal process. In photochemical process, the material absorbs photons, leading to bond breaking. In photo- thermal process, the deposited energy is converted to heat, leading to thermal decomposition. It has been generally accepted that ablation proceeds by a combination of these two mechanisms. Experimental work by Garrison and Srinivasan (1985) showed that photochemistry plays a less role for wavelengths greater than 193 nm. A photon of short-wavelength laser has energy ranging from 4 to 10 eV, which covers the bond energy of most of the organic polymeric materials. The bonding energies of different types of covalent bonds of organic polymeric materials are shown in Table 3. It is evident that a photon of 248 nm KrF excimer laser having energy of 5.0 eV, is sufficient to break most of the covalent bonds of organic polymeric materials. As a result, Excimer lasers are used in the first step of pre-process pattern printing method for metals. Even though the photochemical action dominates over the photo thermal process, photo thermal process still has a significant role in the whole polymeric ablation process, which may have detrimental effects on the ablated surface characteristics. Similar to the case of conventional mechanical machining process, where liquid coolant is used to reduce tool wear out resulting in improved surface finish of the work piece, the effect of using coolants during the process of laser ablation were extensively studied by researchers. The thermal effect was also extensively studied by researchers on metals and ceramic materials during the machining process using high energy pulsed Nd:YAG laser (Li and Achara 2004). Formation of recast layer in metals due to thermal effect was reported in such cases. Use of inert gases to minimize the thermal effect was also reported (Rao et al. 2005). Researchers have also studied (Issa et al. 2008; Sinkovics et al. 2010) a thermal model based on transient heat conduction model, using finite element methods by irradiating CO_2 laser to create micro channels on the surface of soda-lime glass.

6 Applications of Excimer Laser

There are a lot of applications using laser machining introduced because of its own advantages. Few of the important area of applications are cited here.

6.1 Fabrication of Microfluidic System

A microchannel is a tiny channel to deliver liquid for certain functionality in microsystems. It may be used in various applications, ranging from applications in manufacturing industries, such as micromolding and printer nozzles, to biomedical devices for medical diagnostics, such as LOC (lab-on-a-chip) or μTAS (micro total analysis system). Microchannels are mainly classified as either open channels or closed channels. An open channel, in general, has an open top part; a closed channel does not have any open parts. The closed microchannel demands more complicated serial fabrication strategy. A microchannel is generally created by micro/nano fabrication. However, other methods such as micromilling are also used, according to the size of the system. Currently, wet/dry etching is the most popular method to fabricate microchannels; in particular, deep reactive ion etching (DRIE) can be used to create the channels, giving a high aspect ratio compared to other methods, and various materials, ranging from silicon to glass, are etched using this method. However, DRIE requires additional masking processes using organic layers, photoresists or metals. Additionally, it is very hard to create channels with different depths on the same system. Soft lithography, including micro contact printing (μCP) and a few different micro molding methods, can also be used to fabricate the micro channel (Rosqvist and Johansson 2002). It creates microstructures by replicating prefabricated mold structures using certain materials, such as a molten polymer. Thus, it is relatively low-cost and suitable for mass production. However, the range of materials that can be used for soft lithography processes are limited because of the nature of the process. The methods introduced above need serial fabrication steps, not a single process. Meanwhile, excimer laser micromachining can create micro channels directly on various substrates (Shah et al. 2008; Zheng et al. 2006). The resolution of structures created by laser micromachining is mainly dependent upon the wavelength of laser. It offers several advantages compared to other fabrication methods. First, it is easy to implement by controlling a few parameters. Second, it does not require complicated serial processes and a restricted working environment such as masking and coating processes in a clean room facility. In Laser Micromachining process to fabricate micro channels, the geometry of a structure can be easily controlled with simple tool path codes, which are essentially the ones used for traditional CNC machining. It enables the change of geometries of micro channels with ease, based on the design, simply by using a different tool path. In addition to above benefits, it can prevent contamination during the process as it is a non-contact process. Although the laser method demands a long machining time to fabricate a structure with a high aspect ratio and a large area (however, it can be controlled by laser specifications such as laser energy and frequency), it is a good method to create a microstructure.

Kim (2010) has successfully developed excimer laser micromachining process for fabrication of micro channel, using three different materials such as silicon, PE and PMMA Their primary objective was to apply excimer laser micromachining to fabricate microfluidic devices. By optimizing machining parameters, the simple reservoir and micro channels were fabricated by these researchers and microfluidic experiments were performed using three different liquids (water, ethanol and glycerine) on the created microchannels.

6.2 Fabrication of Micro Lens Array

Micro-lens and micro-lens array are important elements for many applications on optical data storage, digital display, and optical communication. Although many conventional methods are widely used for fabrication of micro-lens array, a common problem shared by all these methods is that the micro-lens surface profile is not accurately controllable. Chiu and Lee (2011) demonstrated a unique method for fabricating aspheric micro-lens array based on a KrF 248 nm excimer laser micromachining with precise surface profile control. They have developed a planetary scanning method which can fabricate axially symmetric 3D microstructures with an arbitrary profile based on the sample rotation and revolution approaches. In this planetary contour scanning method, a metal mask screen was combined with x–y movable stage to solve the problem of outskirt laser machining. Their result of SEM image of single aspheric micro-lens fabricated by planetary scanning machining shows a very smooth surface profile. Detailed comparison between the machined surface profiles with their designed and desired counterparts showed that the machined surface is smooth and well matched to the desired surface profile. This method demonstrated superior results over other currently existing micromachining or micro fabrication methods in terms of profile flexibility, versatility, and profile accuracy.

7 Excimer Laser Micromachining Under Gaseous Environment

Though laser based micro machining of polymer, by and large, is a cold ablation process, photo thermal process associated with the laser heating may affect the surface characteristics. Limited information is available on the use of gaseous coolants during laser ablation of polymeric materials. The authors have conducted a study on establishing the effect of certain gaseous environment on the ablation rate of polymer during excimer laser ablation. Attempts have been made by the authors to study whether the usage of gaseous cooling media in the laser ablation process has any meaningful bearing on the surface characteristics of the work piece. Five different gaseous media viz, hydrogen, helium, argon, nitrogen and air were selected for the study. Among these gases, hydrogen and helium possess higher thermal conductivity than the other three gaseous media used in the experiment. The details of the experiments along with the results and observations are presented here.

A circular mask is used to get a final beam size of 5 mm on the polymeric surface to be ablated.

A gas handling setup was fabricated for admitting gas at a predefined flow rate over the surface of polymer during the laser ablation process. This set up consists of a cylindrical steel chamber made up of Stainless Steel 304 grade material having a volume of about 100 ml with a gas inlet and outlet for gas admittance at predefined flow rates. A calcium fluoride window of 2 in. dia. is fixed over ablation chamber, normal to laser beam axis Stainless steel coupons (2 in. dia × 2 mm thick) are used as metal substrate, over which negative photo resist (E-1020 obtained from M/s Cadmosil Chemical Pvt. Ltd, India) was coated to about 100 micron thickness by standard spin coating technique. The coating was air dried and subsequently cured at 110 °C for about 1 h under vacuum at 10^{-2} mbar absolute pressure. This is done to achieve uniform coating thickness of the polymer which is free from entrapped gases in the polymeric material. The coated coupon was placed inside the steel chamber and mounted on the X-Y translation stage of laser ablation system. The photo-resist was exposed to 2000 pulses of the laser under various experimental conditions. Under a pre-set condition where the laser pulse energy, number of pulses and pulse repetition rate are fixed to a value, the ablation process was performed under continuous flow of the gases. The gases used for study include hydrogen, helium, nitrogen, argon and air. In all cases, the laser fluence is set to a value which is above the threshold value for the polymeric material. The mass ablated (the amount of polymeric material removed for a fixed no. of pulses) was computed as a function of gas flow rate at different laser power levels. Gas flow rate was varied in the range of 0.2–2 l/min. See Fig. 8, which shows the variation of Ablated mass with gas flow rate, under different gaseous blanketing environment. Though there is no drastic change in ablation rate with change in flow rate for each of the gases, the variation is significant among the gases as shown in Fig. 9. The variation in ablation rate follows a trend of Air ≈ Nirogen < Argon < Helium <<Hydrogen. With respect to air, hydrogen gas showed 5.2 times higher ablation rate, while helium showed 2.4 times higher values. Our observations could be well explained as detailed below. Laser ablation rate, in general, is sum of the rates attributed to photochemical and photo-thermal process. Depending upon the conditions in which the ablation process is conducted, the extent of contribution by photo chemical (PC) and photo-thermal (PT) process may vary. Above the threshold limit of laser energy, photo-thermal process could be contributing significantly to the net ablation rate.

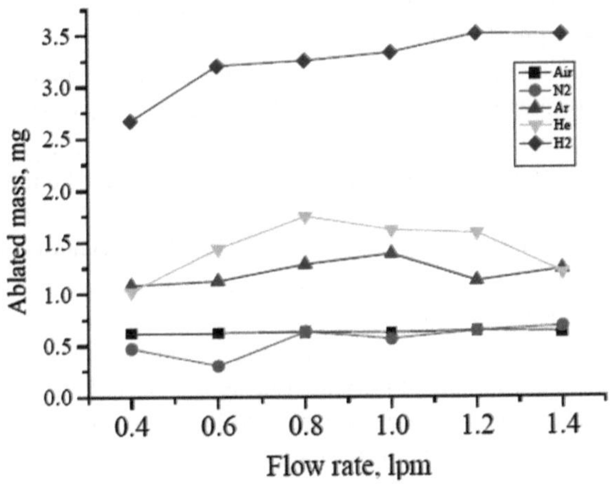

Fig. 8 Variation of mass ablated in different gaseous environment for a laser energy of 20 mJ

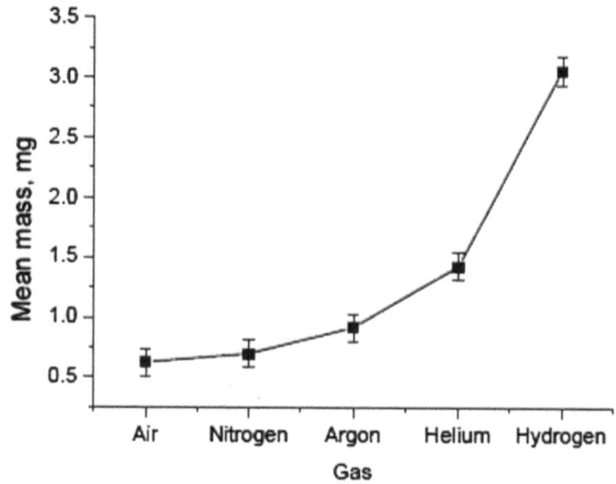

Fig. 9 Variation of mean mass ablated for different gas environment

Table 5 Thermal conductivity of gases

Gas	Thermal conductivity at 300 K (W/mK × 10^{-3})
Argon	17.9
Nitrogen	26
Air	26.2
Oxygen	26.6
Helium	156.7
Hydrogen	186.9

(Data source from: Lide 2009)

Based on the thermal conductivity values of gases shown in Table 5, hydrogen and helium gases should act as a better coolant compared to other gases to minimize the thermal effects associated with laser ablation process. It is observed that laser ablation of polymeric material using excimer laser under the cover of air, nitrogen, argon and helium, produced similar surface characteristics with insignificant impact on ablation rate. Owing to its higher thermal conductivity, helium gas showed higher thermal quenching effect compared to Argon, Nitrogen and air media. On the contrary, hydrogen showed different behaviour compared to other gaseous media studied. Despite its higher thermal conductivity compared to helium, hydrogen gas showed remarkable increase in ablation rate, thus suggesting to be an assisting-chemical agent during the laser ablation of polymeric material. Its involvement in the chemical assisted laser ablation of polymeric materials provides many beneficial effects. It increases the ablation rate. It improves the quality of etched surface, which is desirable for laser micro machining. It surprisingly suppresses the thermo chemical burning of the polymer with oxygen, thereby eliminating the deposit of black colored carbonaceous soot over ablated surface. Work is in progress to examine its reaction kinetics, characterization of machined surface with different polymeric materials and optimize the ablation parameters to derive the desirable surface characteristics.

8 Conclusions

Excimer laser based micro machining methods have been under continuous development for the last few decades. Now it is possible to create extremely tiny and advanced structures for various applications. It has realized a lot of things beyond imagination. It has been developed in various applications, ranging from medical surgery to semiconductor industry due to its unique advantages. It provides several advantages compared to other micro fabrication methods. Excimer laser micromachining process is characterized by a number of process parameters that determines efficiency, economy and quality of the whole process. Researchers are trying to optimize the process through experimental and modelling techniques. The two extreme application areas such as machining of thick materials and machining of micro-parts need considerable research work.

Acknowledgments The authors wish to express their sincere thanks to Mr. Abhishek Singh and Mr. Rajashekara Biradar, for their assistance in carrying out the experiments.

References

Basting, D., & Marowsky, G. (2005). *Excimer laser technology*. Berlin: Springer.

Chen, Y.-M., Yang, C.-L., Cheng, Y.-L., Chen, H.-H., Chen, Y.-C., Chu, Y. & Hsieh, T.-E., (2005). 10 Gbps multi-mode waveguide for optical interconnect. In *Proceedings of the 55th electronic components and technology conference* (Vol. 2, pp. 1739–1743).

Chiu, C. C., & Lee, Y. C. (2011). Fabricating of aspheric micro-lens array by excimer laser micromachining. *Optics and Lasers in Engineering, 49*(9), 1232–1237.

Dupont, A., Caminat, P., & Bournot, P. (1995). Enhancement of material ablation using 248, 308, 532, 1064 nm laser pulse with a water film on the treated surface. *Journal of Applied Physics, 78*(3), 2022–2028.

Garrison, B. J., & Srinivasan, R. (1985). Laser ablation of organic polymers: microscopic models for photochemical and thermal processes. *Journal of Applied Physics, 57*, 2909–2914.

Gifford, R. S., & Bartnik, D. J. (1998). Using optical sensors to measure arterial blood gases. *Optical and Photonics News, 9*, 27.

Gower, M. C. (2000). Industrial applications of laser micromachining. *Optics Express, 7*(2), 56–67.

Issa, A., Brabazon, D., & Hashmi, M. S. J. (2008). 3D transient thermal modelling of laser microchannel fabrication in lime-soda glass. *Journal of Materials Processing Technology, 207*, 307–314.

Jackson, S. R., Matheringham, P. E., & Dyer, P. E. (1995). Eximer laser ablation of Nd:YAG and Nd:glass. *Applied Surface Science, 86*, 223–227.

Kim, E. S. (2010). *Ph.D. thesis*, Raleigh, North Carolina State University.

Li, L., & Achara, A. (2004). Chemical assisted laser machining for the minimization of recast and heat affected zone. *Annals of CIRP, 53*, 175–178.

Lide, D. R. (Ed). (2009), *CRC handbook of chemistry and physics* (89th ed.). Boca Raton, FL: CRC Press/Taylor & Francis.

Meijer, J. (2004). Laser beam machining (LBM), state of the art and new opportunities. *Journal of Materials Processing Technology, 149*, 2–17.

Pfleging, W., Przybylski M., & Bruckner H.J. (2006). Excimer laser material processing—state of the art and new approaches in microsystem technology. In *Proceedings of SPIE—the international society for optical engineering,* (Vol. 6107, pp. 61070G1 61070G15).

Rao, B. T., Kaul, R., Tiwari, P., & Nath, A. K. (2005). Inert gas cutting of titanium sheet with pulsed mode CO_2 cutting. *Optics and Lasers in Engineering, 43*, 1330–1348.

Rosqvist, T., & Johansson, S. (2002). Soft micromolding and lamination of piezoceramic thick films. *Sensors and Actuators A, 97–98*, 512–519.

Rowan C. (1995). Excimer lasers drill precise holes with higher yields. *Laser Focus World, 31*(8), 81–83.

Shah, L., Kam D. H., & Mazumder, J. (2008). Fabrication of microfluidic networks using a high power femtosecond fiber laser. *Proceedings of SPIE, 7005*, 70050E-1–70050E-5.

Sinkovics, B., Gordon, P., & Harsányi, G. (2010). Computer modelling of the laser ablation of polymers. *Applied Thermal Engineering, 30*(2010), 2492–2498.

Srinivasan, R., & Mayne-Banton, V. (1982). Self-developing photoetchining of poly (ethylene tere phthalate) films by far-ultraviolet excimer laser radiation. *Applied Physics Letters, 41*, 576–578.

Tseng, A. A. (2007). Recent developments on microablation of glass materials using excimer lasers. *Optics and Lasers in Engineering, 45*, 975–992.

Yung, W. K. C., Liu, J. S., Man, H. C., & Yue, T. M. (2000). 355 nm Nd:YAG laser ablation of polyimide and its thermal effect. *Journal of Materials Processing Technology, 101*, 306–311.

Zakariyah, S.S. (2010). Laser ablation of polymer waveguide and embedded mirror for optically-enabled printed circuit boards (OEPCB). *Ph.D. thesis,* Loughborough University, UK.

Zeng, D. W., Yung, K. C., & Xie, C. S. (2003). UV Nd:YAG laser ablation of copper: chemical states in both crater and halo studied by XPS. *Applied Surface Science, 217*, 170–180.

Zheng, H. Y., Liu, H., Wan, S., Lim, G. C., Nikumb, S., & Chen, Q. (2006). Ultrashort pulse laser micromachined microchannels and their application in an optical switch. *International Journal of Advanced Manufacturing Technology, 27*, 925–929.

Laser Induced Micromachining and Preliminary Experiments on Manufacturing of Micro-channel on Mild Steel

Sanasam Sunderlal Singh, S.N. Joshi and Alika Khare

Abstract Laser-Induced Micromachining (LIMM) is a non-conventional machining process in which a high power laser beam is focused over the work surface in order to remove the material selectively. The material is removed through melting, evaporation and plasma formation. LIMM offers better machining efficiency than other non-conventional machining processes in terms of the machining rate, efficient debris removal, better surface morphology and capability to machine wide range of materials irrespective of its hardness and electrical conductivity. Further precision can be obtained by confining the laser induced plasma plume from the surface using an external magnetic field or electric field. This chapter deals with a brief introduction and basic principle of laser induced micromachining. This is followed by a discussion on various configurations of LIMM. Further, a review on numerical and experimental studies of the laser induced micromachining is presented. At the end of the chapter, preliminary experimental work on laser induced micromachining of mild steel has been reported.

Keywords Laser induced micromachining · Micro channel · Numerical modeling of LIMM

1 Introduction

To fulfill the need and requirement of modern society in terms of MEMS devices, the demand for miniaturized features like fuel injection nozzles in automobiles, micro holes in electrical printed circuit board and textile industries, nano level

S.S. Singh · S.N. Joshi (✉)
Department of Mechanical Engineering, Indian Institute of Technology Guwahati, Guwahati 781039, Assam, India
e-mail: snj@iitg.ernet.in

A. Khare
Department of Physics, Indian Institute of Technology Guwahati, Guwahati 781039, Assam, India

S.N. Joshi and U.S. Dixit (eds.), *Lasers Based Manufacturing*, Topics in Mining, Metallurgy and Materials Engineering, DOI 10.1007/978-81-322-2352-8_11

surface finish on complex geometry, micro fluidic channels have been tremendously increasing. This has lead to the development of micromachining such as laser micromachining, ultrasonic micromachining, micro electrical discharge machining, electrochemical micromachining, etc. Micromachining indicates machining in the range of 1–999 μm. In recent years, laser beam micromachining (LIMM) has emerged as one of the dominant non-contact processing tools in the fabrication of micro-components. LIMM can remove material in very small amount (in micron range), while traditional machining removes material in macro scale. Application of lasers have advantages such as narrow cut, minimum heat affected zone (HAZ), better accuracy, smooth and flat edges, minimal deformation of the workpiece and fast cutting of intricate profile cut and fast adaptation to changes in manufacturing programs.

1.1 Laser Induced Micromachining

Laser, an acronym of light amplification by stimulated emission of radiation, is a device that emits light through the process of stimulated emission. It comprises of three main components namely the lasing medium (active medium), pump (means of exciting the lasing medium into its appropriate higher energy states) and optical feedback system or optical cavity comprising of two mirrors as shown in Fig. 1 (Silfvest 2004). One of the mirrors is made partially reflecting making the laser beam to come out of the cavity for useful purpose.

One can classify laser based on the lasing medium: Gas laser, Liquid laser, Solid state laser, Semiconductor, Free electron laser etc. Laser can be operated either in continuous wave (CW) mode or in pulsed mode. The main properties of laser are: high brightness, monochromatic, low divergence and high spatial and temporal coherence.

Laser Induced Micromachining (LIMM) can be applied to any material while traditional machining processes have to choose compatible tools for materials with different strength or hardness. It is difficult to machine hard material or brittle material such as ceramics by traditional machining processes. However, laser

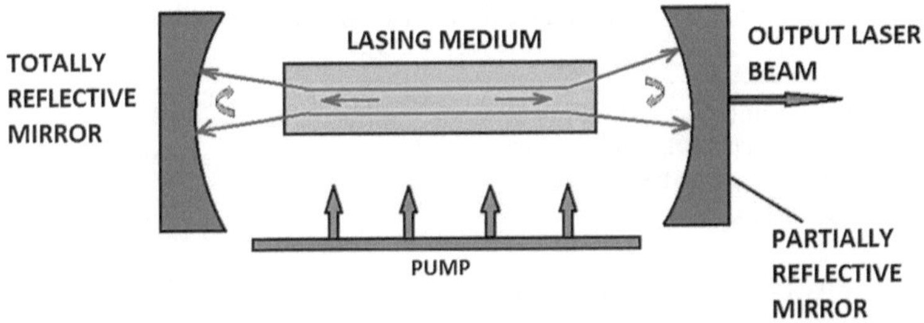

Fig. 1 Schematic diagram of laser system

machining can easily machine such materials with high precision. In LIMM, laser beam is focused by a lens on the work surface, removing a small portion of the material by melting and vaporization. The process of removing the material is known as ablation. It is generally accepted that two phenomena happen during the ablation process i.e. photochemical ablation and photo-thermal ablation. Photochemical ablation involves the absorption of photons leading directly to bond breaking in the material without any intermediate heat dissipation. It uses UV photons having energies 3–7 eV (Johan 2004), which is used to break the chemical bonds directly. Alternatively, multi-photon process can be utilized with the longer wavelength laser. This phenomenon is generally used for machining of polymers. In photo-thermal ablation, deposited energy is converted into heat and decomposes the material surface in the focal volume. The processes involved are: heating, melting, ablation of the surfaces etc. The focal intensity of laser influences the vaporization rate and high vaporization rate causes a shockwave that can reach pressure of more than 50 kbars (Fabbro et al. 1990). The high pressure created during the melting and superheating of liquid at the end of laser pulse ejects the material within the focal volume at high speed.

Laser can also be utilized to cut the work piece of any material in desired geometrical shapes. It can even cut the work pieces which are normally difficult to cut by other machining processes. CO_2 laser and Nd:YAG laser are the most commonly used in cutting as these can deliver the required high powers for high speed cutting. UV lasers are mostly used for thin layer cuttings or organic material cutting. Plastics, polymers, composites, wood, rubber, papers, stones and crystals have been successfully machined by infrared lasers (Caiazzo et al. 2005; Lau et al. 1990). Some materials such as composites and gemstones can be readily cut with high quality using lasers, while they may be difficult to cut using other techniques. UV lasers can even cut polymers with negligible heat affected zone, because the photon energy is comparable to the bonding energy of the material dominated by photochemical process over photo-thermal (Znotins et al. 1987).

Drilling is one of the simplest and successful applications of industrial lasers. Drilling of holes in ceramic, silicon and polymer substrates is widely utilized in electronics industry. While, laser drilling of metals is used for fabrication of cooling channels in air turbine blades, tiny orifices for nozzles, etc. Holes less than 0.25 mm diameter can be drilled using laser with ease; especially for hard and brittle materials such as ceramics and gemstones (Forget et al. 1988). Large holes can be drilled by trepanning, i.e., by scanning the beam around the circumference of a circle.

Laser marking, scribing and texturing are important applications of laser micromachining besides cutting and drilling. In these applications, a very thin layer of material is ablated and a mark or pattern is formed (Qi et al. 2003). Permanent laser markings of almost all materials with high precision have considerable advantages over other techniques. It doesn't require any sample preparation or post treatment. Applications of laser markings range from shallow to deep-marking of hard and corrosion-resistant materials.

The general parameters which control laser machining are laser wavelength, beam shape, focal intensity of the laser on to the surface, pulse width, pulse repetition rate and working environment (vacuum/inert/air/liquid). Laser intensity is the main input process parameter that directly influences machining features such as material removal or HAZ. Generally, longer the wavelength, lower the absorption, hence decrease in material removal (Johan 2004). The output parameters are Material Removal Rate (MRR), machining geometry (kerf width, hole diameter, taper), surface quality (surface roughness, surface morphology), metallurgical characteristics (recast layer, heat affected zone, dross inclusion) and mechanical properties (hardness, strength).

1.1.1 Process Mechanism of Laser Induced Micromachining

When laser beam is focused on the target material, fraction of the energy is reflected and part of it is absorbed, the absorbed energy directly ablates the material by local heating in the focal region as depicted in Fig. 2.

The fraction of the incident power that is reflected from the surface depends on the polarization and angle of incidence of the laser as well as the reflectivity of the working medium (William and Mazumder 2010). Further, part of the incident power that is being absorbed by the work material propagates through the electron subsystem and then gets transferred to the lattice there by transferring to the ambient target material within the focal volume and its surrounding.

Based on the intensity of the laser, the surface undergoes melting, vaporization, plasma formation etc. At very high intensity, surface starts to vaporize before a significant melting depth of molten material is formed. The material gets removed through laser ablation. For laser pulse durations longer than microseconds, hydrodynamic ablation is the dominating phenomenon over the direct ablation. During laser cutting and drilling operations, there exists a time scale which allows the surface to heat to the vaporization temperature and remain there for some time. During the initial phase, the ablated particle surrounding the medium expands. This leads to the formation of recoil pressure which thereby helps in expulsion of the

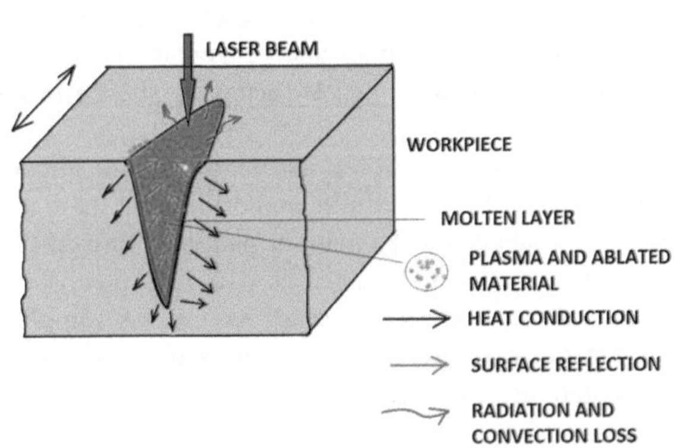

Fig. 2 Laser irradiation showing different phenomena involved

molten material from the irradiated region. In addition to this, there exist strong temperature gradients in the radial direction in the molten material due to the beam profile i.e. the center of the molten material is usually hotter than the outer. Thus, recoil pressure and temperature gradient drive the molten material out in the form of liquid droplets. The combination of vaporization and hydrodynamic ablation results into drilling or cutting the material in the focal region.

When the laser intensity is higher than that required to vaporize the work material, the vaporized material gets ionized. This leads to the formation of plasma at the vicinity of the work surface. With the formation of plasma, laser energy transmitted to the material depends on the density of the plasma. Below a certain critical value of the plasma density, laser energy is transmitted to the work material. When the critical value of the plasma density is reached, further transmission of laser energy onto the work material is cut off due to the absorption or reflection of the incident laser beam by laser induced plasma. Shock waves are generated when the above formed plasma expands, which is the basis for laser shock processing to improve residual stress of metal components (Sano et al. 1997). The sequence of processes in material removing via high power laser is depicted in Fig. 3.

Laser-induced micromachining can be performed under vacuum, in presence of low-high pressure gases and in liquid. The process of LIMM can be improved by placing the work piece in presence of magnetic field which controls the expansion of plasma. Figure 4a shows the schematic of LIMM under vacuum. Figure 4b that of in presence of liquid/air and Fig. 4c depicts the laser Induced Micromachining in presence of magnetic field. The process in LIMM can be improved using these careful choices of surrounding environment (Fabbro et al. 1990).

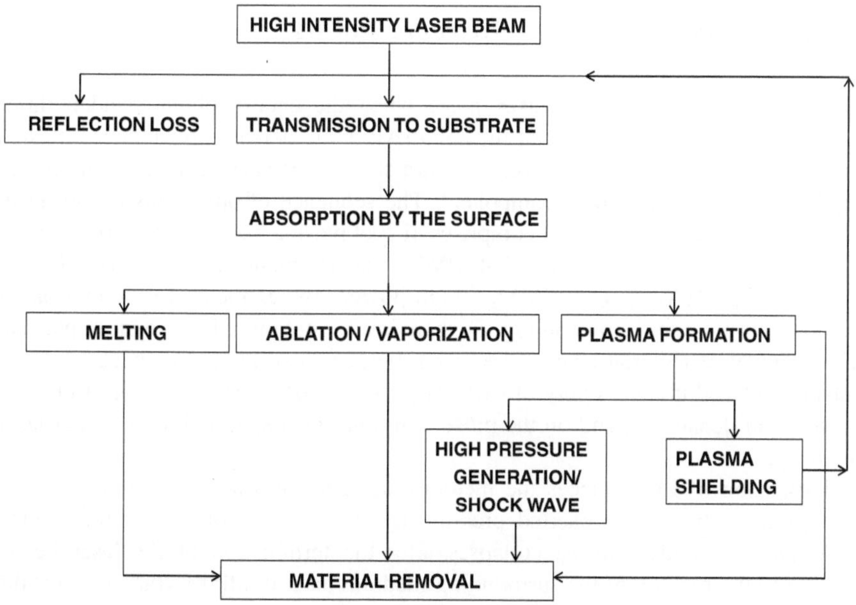

Fig. 3 Principle of LIMM

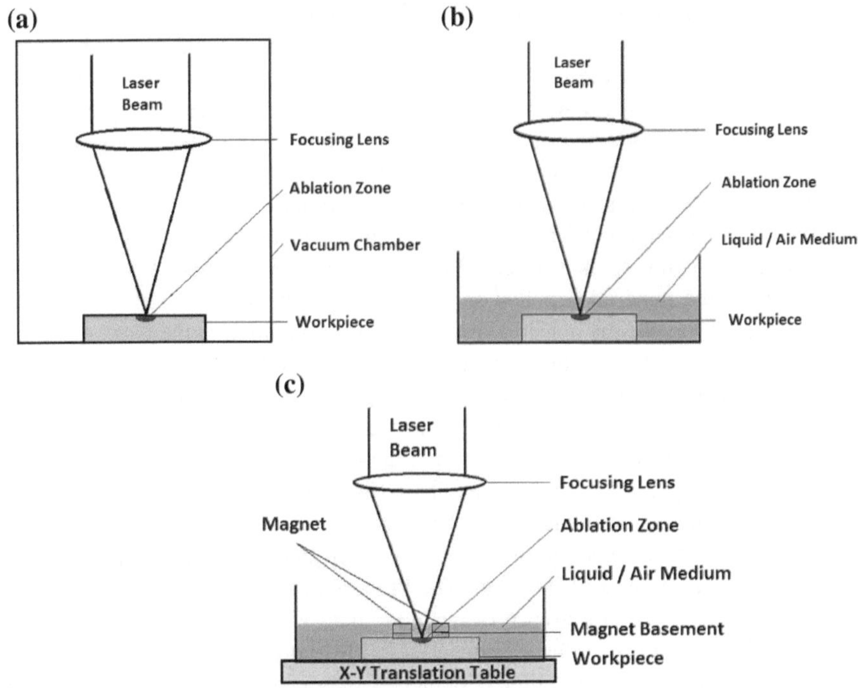

Fig. 4 **a** Laser induced micromachining in vacuum. **b** Laser induced micromachining in liquid/gas. **c** Laser induced micromachining process in magnetic field

1.2 Laser Induced Plasma Micromachining

Laser induced plasma (LIP) takes place when the energy of the incident laser exceeds the ablation threshold of the solid (Ali et al. 2014). Laser induced plasma micromachining (LIPMM) utilizes the ultra-short laser beam generated plasma in liquid medium to machine a component. The sequence of operations occurring in LIPMM is depicted in Fig. 5. It comprises of spot plasma based LIPMM process (S-LIPMM), Line-LIPMM process (L-LIPMM) and magnetically controlled LIPMM process (MC-LIPMM). In LIPMM, a high power laser is focused in a transparent liquid or air medium (shown in Fig. 5), there is a formation of laser induced plasma. The removal of material takes place when the generated plasma interacts with the substrate placed in the vicinity of LIP (Pallav et al. 2014). The duration of plasma, its size and density depend on the pulse duration, focal spot and the incident laser intensity.

Expansion of LIP replaces the medium and generates a large pressure on the workpiece surface. This exerted pressure by the plasma holds back the molten material. Eventually, plasma collapses after the termination of the laser beam, leading to the rushing of the surrounding medium back to fill the void. This results

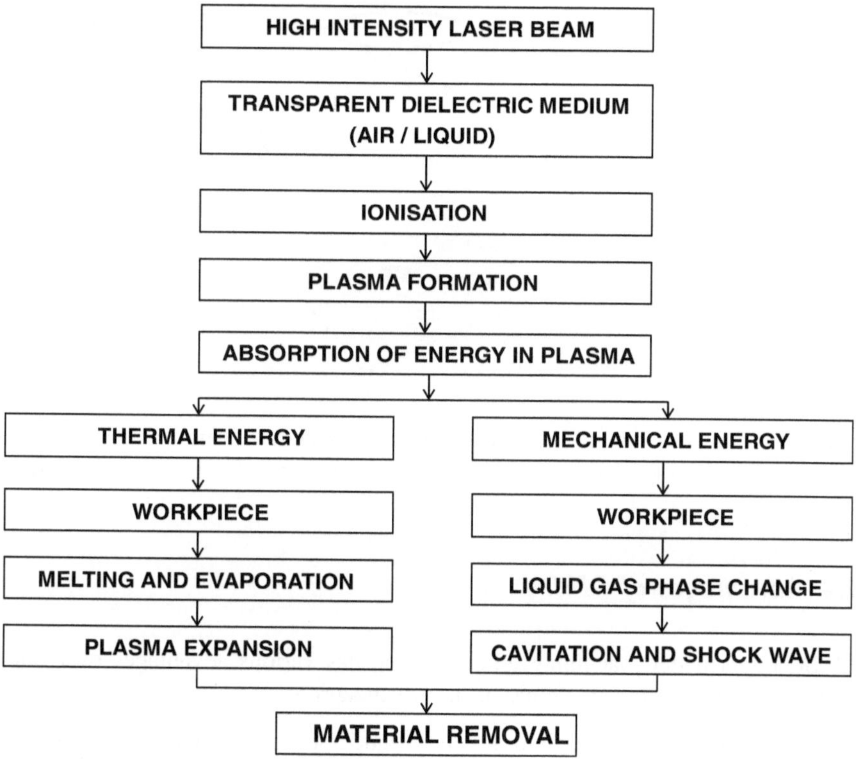

Fig. 5 Laser induced plasma micromachining

in an instantaneous expulsion of the material from the workpiece surface. Thus a crater is generated on the workpiece surface corresponding to every shot of LIP.

Micro-channel can be fabricated if the workpiece is moved with respect to the plasma in the focal region of the surface at a certain feed-rate in a preprogrammed motion. Overlapping of the craters in succession writes the micro channel (Pallav et al. 2014).

1.3 Laser Induced Plasma Assisted Ablation (LIPAA)

It utilizes the laser generated plasma as a by-product during laser ablation of a metallic target placed below a transparent workpiece (Hanada et al. 2006; Kadan et al. 2005; Zhang et al. 1999) as shown in the schematic of Fig. 6. In this process, laser beam is made to pass through a glass substrate which is transparent. Then the transmitted beam is focused on a solid target (typically a metal) placed behind the substrate so that the target is ablated resulting in plasma generation. The generated

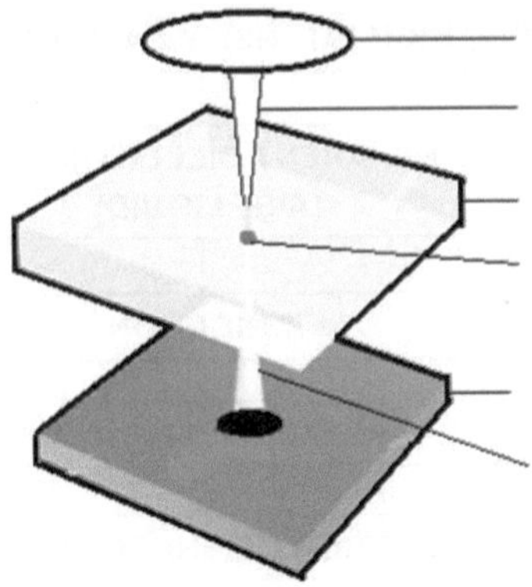

Fig. 6 Laser induced plasma assisted ablation

plasma is then used to machine the transparent material. This process has been widely applied for micromachining of various transparent hard and soft materials.

1.4 Advantages of LIMM

LIMM offers better machining efficiency and has various advantages over other micromachining processes. Some of the advantages are:

- Laser induced micromachining offers wide range of machining capabilities irrespective of the properties of the material.
- The issues of tool related problems such as tool wear, manufacturing of tool and cutting force are not required in laser induced micromachining.
- Greater process flexibility can be achieved by automation of laser system such as CNC incorporated laser system.
- The material removal rate is comparatively higher as compared to EDM, USM, EBM, etc.
- There is reduction in Heat Affected Zone (HAZ) due to very thin re-solidified layer.
- Machining can be performed with high precision in any geometry.

1.5 Limitations of LIMM

In spite of the various advantages of LIMM, it also has some unavoidable limitations.

- The machining depth is limited by the Rayleigh length.
- Heat affected Zone cannot be avoided completely in laser induced micromachining since the main material removal mechanism is thermal in nature.
- It is difficult to machine material possessing high reflectivity.

2 Literature Review on Laser Induced Micromachining

In view of the importance of laser induced micromachining (LIMM) in manufacturing of MEMS devices, micro-fluidic devices, relevant literatures on LIMM have been referred in this section. The following subsections deal with the papers related to modeling of temperature profile, HAZ and geometric features such as groove shape, width and depth of cut. Further, experimental findings towards the enhancement of laser induced micromachining are also discussed.

2.1 *Numerical Studies of LIMM*

Literature reports various numerical studies on LIMM mainly using finite element method (FEM) and finite difference method (FDM). Un-Chul and Francis (1972) developed a 1-D laser machining model that uses continuous, distributed and moving heat source for describing the temperature profile and thermal stress propagation for laser drilled holes in high purity fired-alumina ceramic substrate. Roy and Modest (1993) used moving CW gaussian beam profile to model evaporative laser machining of silicon nitride by incorporating temperature dependent properties such as specific heat and absorption coefficient using finite difference method. Pietro and Yao (1995) have developed a 3-D transient heat transfer model for the prediction of kerf geometry. Yu (1997) carried out a numerical study on the laser drilling and cutting process of copper plate using Ansys 5.0/ED which had incorporated with Ansys Parametric Design Language (APDL) to monitor and impose the changing boundary and loading conditions. The laser beam was modeled as a heat source moving on the workpiece in a prescribed manner.

Prusa et al. (1999) had developed a numerical model for the calculation of heat conduction losses, cutting speed and temperature distribution in HAZ in CO_2 laser cutting of thick materials. Meung and Jingwei (2001) introduced a finite element method for simulation of pulsed laser-cutting process in air to study amount of material removed and smoothness of groove shape with the gaussian pulsed laser beam, model in two dimensional unsteady heat conduction. Numerical simulation of LIMM was derived considering only the solid-liquid phase change using finite element method by Khalil and Sreenivasan (2005). Here, the enthalpy formulation was implemented in the Finite Element Method (FEM) based model.

Effect of multiple reflection, recoil pressure were incorporated in the numerical simulation of 1-D laser machining for alumina in addition to consideration of temperature dependent thermal conductivity and specific heat by using COMSOL multi-physics software (Samant and Dahotre 2009). Numerical simulation was also reported on laser machining of Carbon Fiber Reinforced Plastic composite material by Negarestani et al. (2009). A 3D finite element model was developed for predicting the transient temperature field, cut profile and heat affected zone. For the first time, the model employed the heterogeneous mesh for the composite with anisotropic material properties.

The thermal and optical properties of the target material change during the process of laser ablation. These temperature dependent material properties of the target material were taken into consideration for FEM model to predict the ablation depth and temperature distribution in pulsed laser ablation (Nikhil et al. 2010). The effect of plasma shielding on the incident laser flux was also considered in the model. Moscicki et al. (2011) developed a model which describes both the target heating, formation of the plasma and expansion. The model employed conservation of mass, momentum and energy which are solved by using Fluent™ software package.

Finite element modeling technique was employed to study the laser machining in air and water, in order to understand the underlying mechanism of underwater crack free machining. In this investigation, FEM based ANSYS™ software was employed to study the temperature distribution. The difference in heat convection coefficient was considered as the varying parameter in numerical simulation for laser machining in air and water (Yan et al. 2011).

2.2 Experimental Studies on LIMM

Investigations have been carried out on machining of glass matrix composite comprising SiC fibres in a borosilicate glass matrix. It has been reported that the material removal primarily depends on the pulse energy rather than the pulse duration (Ian et al. 1998). The effect of cutting parameters on the cut quality of ultra low carbon steel thin sheets has also been studied considering heat affected zone (HAZ) as the output parameter. It was found that HAZ increases with the increase in laser power and decreases with increase in scanning speed and gas pressure (Hanadi et al. 2008). Rajaram et al. (2003) reported that laser power plays a major role on the kerf width while feed rate played a minor role for experiments conducted on 4130 steel using CO_2 laser. Decreasing power and increasing feed rate generally led to a decrease in kerf width and HAZ. An increasing feed rate generally increases surface roughness and striation frequency. Comparison of CO_2 laser cutting with Continuous Wave (CW) and pulsed mode laser operation in presence of different gases namely argon, helium and nitrogen were carried out on titanium sheet (Rao et al. 2005). Use of helium as shear gas results in narrow HAZ and low dross as compared to that of Ar owing to high heat convection and shear stress provided by He gas. However, the laser cut edge produced in presence of He gas results in waviness of the cut. During the pulsed mode laser cutting of austenitic stainless steel sheet, rough surface and incomplete cutting has been observed on increasing the speed due to pulse separation and insufficient overlapping. While with continuous wave mode using Nd:YAG laser, increasing the speed with increase in power results in smoother surface, better quality and smaller kerf width (Ghany and Newishy 2005).

The cutting quality of laser machining depends on many input process parameters such as wavelength, pulse energy, pulse duration, optical and thermal properties of the material, composition, pressure of assist gas etc. Therefore, optimization of these process parameters is required to obtain desired process performance. Sharma et al. (2010) employed Taguchi L_{27} orthogonal array to optimize the input parameters for both straight and curved cut profile using Nd: YAG laser for cutting of Nickel based super alloy thin sheet. It was observed that the optimum input parameter levels for curved cut profiles are entirely different from straight cut profiles except kerf width. Optimization of the multi performance for Nd:YAG laser cutting of nickel based superalloys has been carried out using grey relational analysis for quality characteristics of average kerf taper and average surface roughness (Sharma and Yadava 2011). It was observed that cutting speed was the most significant cutting parameter for minimizing the value of average kerf taper and average surface roughness. However the repetition rate, pulse width and back ground oxygen pressure were found to be less significant parameters.

The cross section of a drilled silicon nitride in air and water as reported by Noboru et al. (1988) using an Nd:YAG pulsed laser is shown in Fig. 7. In the air

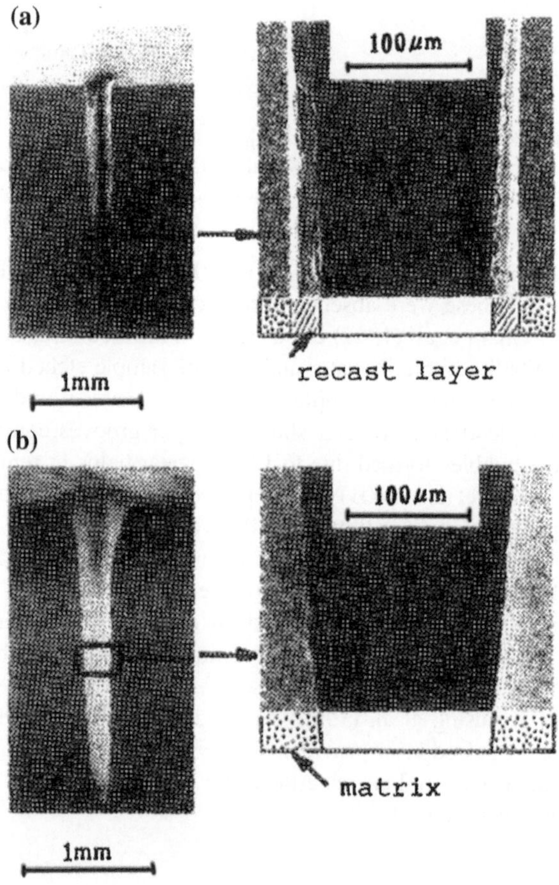

Fig. 7 Cross section of a silicon nitride ceramics drilled in **a** air **b** water with Q-switched power (pulse duration of 100 ns, laser energy of 5 mJ/pulse, Q-switching frequency of 1 kHz, processing time of 60 s, laser power density of 7.1×10^7 W/ cm^2) (Noboru et al. (1988), with permission)

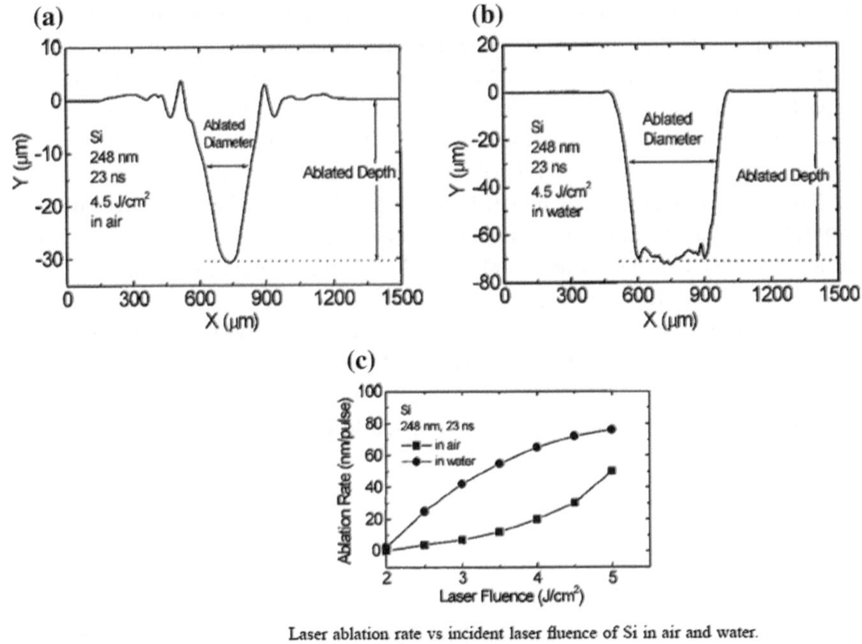

Fig. 8 **a** Profile of the ablated region of Si after 1000 pulse irradiation in air. **b** Profile of the ablated region of Si after 1000 pulse irradiation in water. **c** Laser ablation versus incident laser fluence of Si in air and water (Zhu et al. (2001), with permission)

ambient, a recast layer of about 20 μm thickness and micro-cracks were observed while these were absent in presence of water as back ground medium.

Ohara et al. (1997) did a study on infrared laser etching of micro coil pattern of Al bulk sample in water and air. The sample etched in air showed debris along the scanned line. The sample etched under water was devoid of any such debris. The sample etched in water showed deeper grooves compared to that of air. Probably the bubbles formed due to LIP in water helps in removing the debris.

Zhu et al. (2001) also reported that the laser ablation rate was enhanced in presence of water film above the workpiece. A KrF excimer laser with a wavelength of 248 nm and pulse duration around 23 ns was used for ablation of silicon sample. Figure 8 shows the profile of the ablated region in air and water. It shows that ablation rate in air in around 30 μm (Fig. 8a), whereas that of under water was 70 μm (Fig. 8b). It was also demonstrated that the ablation rate increases with the laser fluence for both the cases.

Kruusing et al. (1999) reported that machining in water of NdFeB magnet by short-pulsed laser provides clean grooves while machining in air results in grooves filled with molten material using Nd:YAG laser of 180 ns, 1.8 mJ, 1 kHz with varying feed rate.

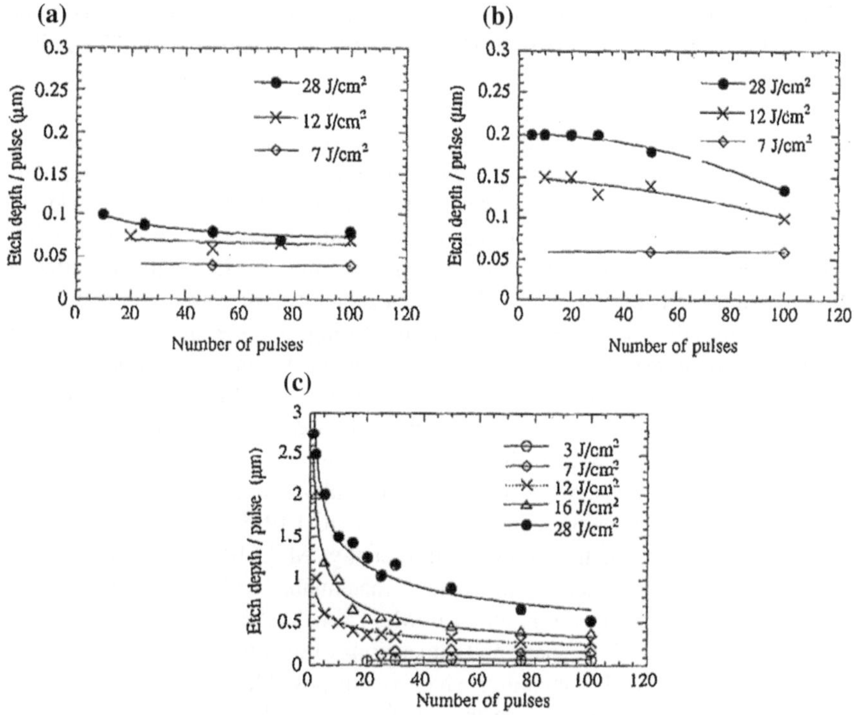

Fig. 9 **a** In the ambient atmosphere only. **b** With the addition of the argon gaseous cover. **c** With the addition of a water film on the treated surface (Dupont et al. (1995), with permission)

Dupont et al. (1995) carried out an experimental study on pulsed laser processing of stainless-steel alloy in ambient air, argon gas and also under a flowing water film on the material surface at atmospheric pressure. Figure 9 shows the variation of the etch depth per pulse as a function of number of pulses at the 532 nm laser wavelength. Figure 9a shows the effect of increase in number of pulses on the etch depth for different laser fluences, and that of in presence of argon in Fig. 9b. Figure 9c shows the effect in presence of water film on the treated surface. The etch depth was found to be maximum in presence of water film and decreases with the number of pulses at higher fluence.

Yuan-Jen et al. (2012) demonstrated the use of magnetic field in laser micromachining on highly reflective material Al6061 for drilling of micro hole. With the use of static magnetic field, the machining process could be enhanced by many folds in depth and the inlet diameter is reduced by 42 %. It was also observed that the roundness of the inlet was improved. The second harmonic of Nd:YAG laser at

532 nm wavelength, pulsed mode with maximum energy of 270 mJ, pulsed width of around 6 ns and frequency of 50 Hz were used in the experiments. The NdFeB permanent magnets with different arrangements were also tried in this work.

Zhang et al. (1999) performed micro-grating operation on workpieces of optically transparent materials: fused quartz and Pyrex glass using LIPAA. Kadan et al. (2005) also performed LIPAA for micro-marking and relieve grating in sapphire, silica and glass work material. Further, Hanada et al. (2006) demonstrated that crack free marking and color marking of glass material is possible using LIPAA.

Pallav and Ehmann (2010) introduced a new process, Laser Induced Plasma Micromachining (LIPMM) in which plasma generated in a transparent working medium is directly used to perform micromachining. Rajiv et al. (2013) extended this work, and observed that it can machine a wider variety of materials as compared to direct laser ablation. It was demonstrated that Line-LIPMM can be used for machining a line instead of a spot by optically manipulating the shape of the plasma. L-LIPMM can reduce the time for machining micro-textures over large areas by about six times, while retaining the multi-materials capability of Spot-LIPMM. Further, magnetic control-LIPMM (MC-LIPMM) was developed to manipulate the shape of the plasma. Additionally, MC-LIPMM can increase the length and reduce the width of the channel machinable with line plasma as well. Since no additional optics are required in MC-LIPMM the results showed that in-process manipulation of the shape and dimensions of the machined feature might be possible in MC-LIPMM, something which is not possible in laser ablation as the optics is required to be changed in the middle of a machining cycle.

Sarah and Ishan (2014) carried out an experimental study to show the feasibility of manipulating or controlling the plasma plume during laser machining using the magnetic field around the plasma plume in order to maximize the aspect ratio of the machined spot. Ultra short pulsed laser of Nd-YVO_4 solid state, 8 ps pulse duration at 532 nm wavelength and having Gaussian profile is used for the study. Polyamide (PA66) and aluminium alloy (A 5052) were used as substrates. Kerosene and distilled water was used as dielectric media.

Comparison of the different variation of LIPMM based on working parameters, dielectric and workpiece is tabulated in Table 1.

During laser ablation, formation of plasma occurs which comprises of electrons, ions and neutrals. This laser induced plasma absorbed the laser energy further. The energy is absorbed by means of photo-ionization and inverse Bremsstrahlung leading to increase in temperature and pressure. This can block laser radiation completely on the work piece (Moscicki et al. 2011). However, many researchers have been working on this aspect to machine different work materials including transparent and highly reflective materials by utilizing the generated plasma.

Table 1 Summary for variation of LIPMM

Sl. No.	Variation of LIPMM	Working parameters	Dielectric	Workpiece	Comparison	Remarks
1	S-LIPMM Rajiv et. al. (2013)	$Nd\text{-}YVO_4$ pulsed laser, 8 ps pulse duration, 0.12 W, 10 kHz	Distilled water	Semitransparent polycarbonate, opaque ABS polymer sheets, transparent alumina ceramic, polished silicon wafer, opaque sialon	Same condition followed for direct ablation	Transparent alumina could not be machined while sialon was machinable with depth of 6 μm in direct ablation while 18 μm was obtained using S-LIPMM
	L-LIPMM Rajiv et. al. (2013)		Kerosene, distilled water, EDM oil, mineral oil	AA5052	Same experimental conditions with varying dielectric	Kerosene—deposition on the sides and insides of the channel due to carbonization; Distilled water—no carbonization and comparable machined depth to kerosene EDM oil and mineral oil—no deposition, much lower machined depth; L-LIPMM has higher productivity than S-LIPMM and has 84% improved machining time
	MC-LIPMM Rajiv et. al. (2013)		Distilled water	AA5052	MC-LIPMM compared with S-LIPMM and L-LIPMM	MC-LIPMM results in longer channel with more uniform depth and width across the length of the channel; 200% increase in channel length

(continued)

Table 1 (continued)

Sl. No.	Variation of LIPMM	Working parameters	Dielectric	Workpiece	Comparison	Remarks
2	MC-LIPMM Sarah and Ishan (2014)	Nd-YVO_4 pulsed laser, 8 ps pulse duration, 0.13 W, 10 kHz	Kerosene, distilled water	PA 66; AA5052	Experiments conducted with varying magnetic configuration (8, 12, 1 neodymium magnet with 4400 Gauss surface field) and its location	Difference in aspect ratio observed; Only HAZ and no machining with LIPMM at same setting and focal point; Increase in aspect ratio upto 6 using external magnetic field
3	LIPMM Saxena et al. (2014)	Nd-YVO_4 pulsed laser, 8 ps pulse duration, 10–50 kHz, pulse energy 6 μJ	NaCl solution in distilled water, Fresh water	AA5052	Influence of the presence of salinity in dielectric	100% increase in depth at an optimum salinity of 2–4 g/100 ml

3 Preliminary Experimentation on Laser Induced Micromachining

In order to observe the effect of the sample scan speed on the surface roughness, laser micro channels were fabricated by focusing a second harmonic (532 nm) of a Q-switched Nd:YAG laser having pulse duration of a 10 ns, repetition rate of 10 Hz, focused to a size of 8.31×10^{-3} cm^2 on to a work sample in ambient environment. The work sample used for the experimentation is mild steel of dimension 30 × 10 × 3 mm. The sample is finely polished by mounting on a double disc rotating polishing machine. The polished work sample is then irradiated with second harmonic of a Q-switched Nd:YAG laser over a scanning length of 15 mm. The LIMM was performed at two different scan speeds. The experimental details are as listed in Table 2.

After irradiation with nanosecond laser in a single pass, formation of microchannel has been observed. Figure 10 shows the pictorial view and profilometer image of the microchannel formed as observed by non-contact high precision

Table 2 Experimentation details

Experimental conditions		
Parameters	Sample 1	Sample 2
Energy per pulse	30 mJ	30 mJ
Focal length of the lens	150 mm	150 mm
Length of traverse	15 mm	15 mm
Speed	1800 step/s (540 μm/s)	900 step/s (270 μm/s)

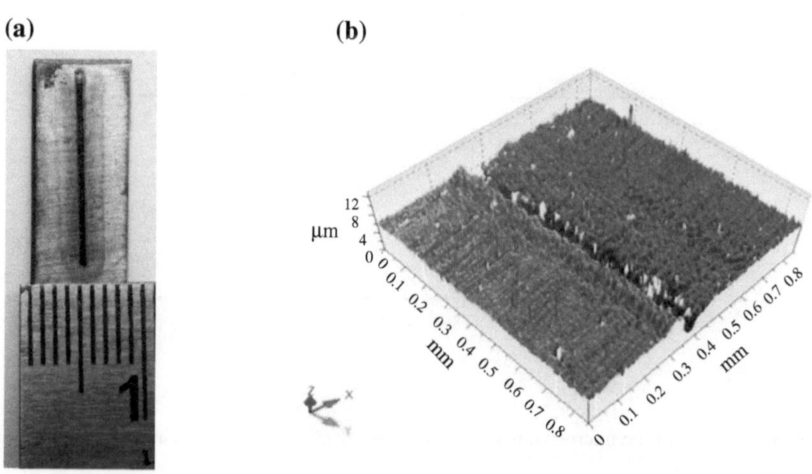

Fig. 10 **a** Pictorial view **b** Profilometer image of the microchannel for sample 1

profilometer for sample 1. It is observed from the figure that the formed channel is uniform throughout the traversed length. The width and the depth of the channel were being measured from the generated profilometer. For this, channel width and the depth of the cut were measured at four different sections along the channel length viz. section A-A, B-B, C-C and D-D as shown in Fig. 11. The sections were taken at equal intervals.

Figures 12 and 13 show the scan profile of the microchannel at scan speed of 540 and 270 μm/s respectively recorded by Profilometer.

The experimental results are listed in Table 3. The average value of the channel width was 114.7 and 146.5 μm for scan speed of 540 and 270 μm/s respectively. The average value of channel depth was 4.4 and 3.2 μm respectively. The decrease in the channel width with increase in scan speed is due to the reduction in

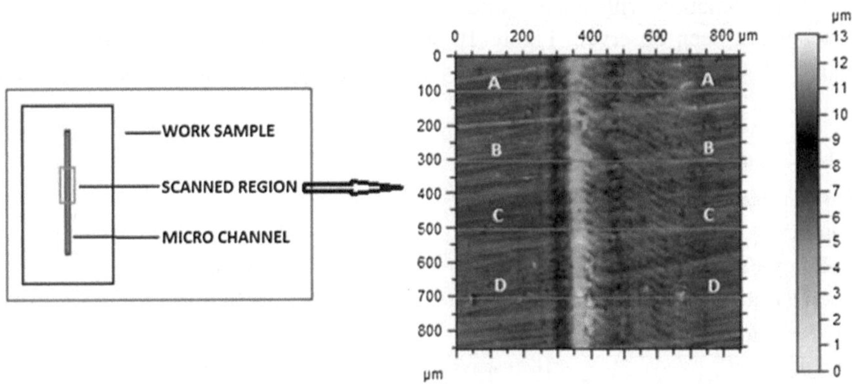

Fig. 11 Scanned area and roughness measured at different sections of the microchannel for sample1

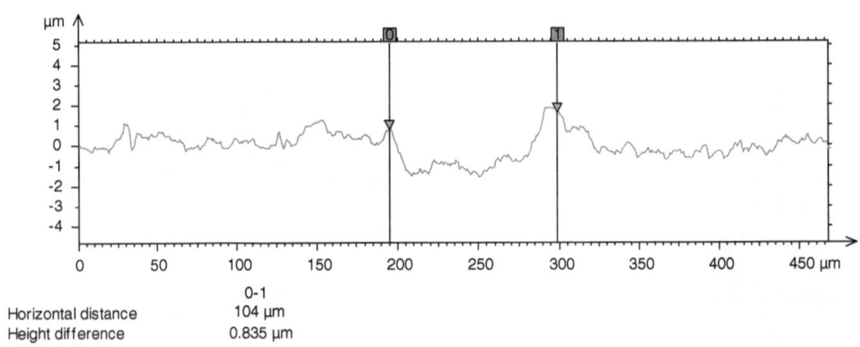

Fig. 12 Scan profile of the microchannel at scanning speed of 540 μm/s showing the width of the microchannel for sample 1

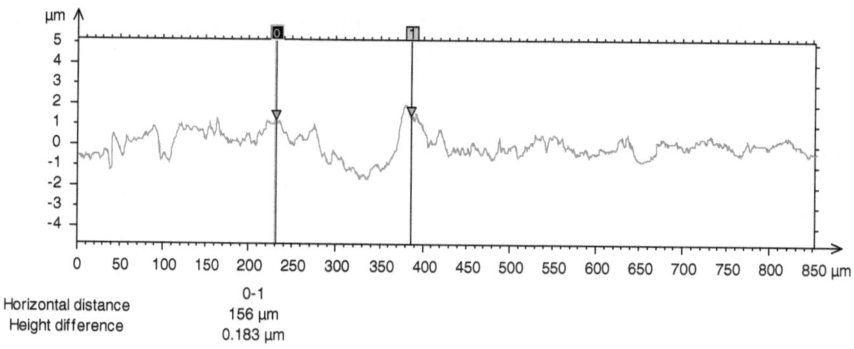

Fig. 13 Scan profile of the microchannel at scanning speed of 270 µm/s showing the width of the microchannel for sample 2

Table 3 Experimental results

Sample	Traverse speed (µm/s)	Section	Width (µm)	Depth (µm)
Sample 1	540	A-A	134	3.93
		B-B	121	4.6
		C-C	104	3.04
		D-D	100	5.94
Sample 2	270	A-A	141	3.51
		B-B	156	2.81
		C-C	121	4.01
		D-D	168	2.46

interaction time of the laser beam with the work sample at a given location. The decrease in the depth at lower scan speed could be due to the formation of recast layer in the channel.

4 Summary

After an in-depth study of the available literature related to LIMM, it has been observed that many attempts have been made in modeling of LIMM towards complete phenomenon. Initially, modeling of LIMM has been developed considering only the direct vaporization of the substrates. Later, modeling of the phenomenon considering multiphase viz. liquid, gas and plasma has also been carried out. Simultaneously, various time dependent physical properties of the material and the process are being incorporated in the trend. Above all of these, it is found that a very scant amount of work has been done on modeling of LIMM in liquid medium and in presence of magnetic field and electric field. Therefore, this area of research is still open for further advancement and to improve the understanding of the process and modeling so as to match the experimental observations and improve the precision.

The quality of laser machined component depends on many input process parameters viz; wavelength, repetition rate, pulse energy, pulse duration, optical and thermal properties of the material, composition and pressure of assist gas etc. Therefore, optimization of these process parameters is required to obtain the desired performance. Varying the machining medium has a major impact on the quality of the machined component. Various theories have been reported to study this effect. The performance of LIMM can further be enhanced with the application of the magnetic field.

References

Ali, A. K. H., Kadhim, A. A., & Mohammed, R. A. (2014). Investigation of some physical parameters of laser induced copper plasma. *Asian Journal of Applied Sciences, 2*, 151–157.

Caiazzo, F., Curcio, F., Daurelio, G., & Minutolo, F. M. C. (2005). Laser cutting of different polymeric plastics (PE, PP and PC) by a CO_2 laser beam. *Journal of Materials Processing Technology, 159*, 279–285.

Di Pietro, P., & Yao, Y. L. (1995). A numerical investigation into cutting front mobility in CO_2 laser cutting. *International Journal of Machine Tools and Manufacture, 35*(5), 673–688.

Dupont, A., Caminat, P., & Bournot, P. (1995). Enhancement of material ablation using 248, 308, 532, 1064 nm laser pulse with a water film on the treated surface. *Journal of Applied Physics, 78*(3), 2022–2028.

Fabbro, R., Fournier, J., Ballard, P., Devaux, D., & Virmont, J. (1990). Physical study of laser-produced plasma in confined geometry. *Journal of Applied Physics, 68*, 775–784.

Forget, P., Jeandin, M., Lechervy, P., & Varela, D. (1988). Laser drilling of a superalloy coated with ceramic. *The metallurgical society* (pp. 553–562).

Ghany, K. A., & Newishy, M. (2005). Cutting of 1.2 mm thick austenitic stainless steel sheet using pulsed and CW Nd:YAG laser. *Journal of Materials Processing Technology, 168*, 438–447.

Hanada, Y., Sugioka, K., & Midorikawa, K. (2006). Laser-induced plasma-assisted ablation (LIPAA): Fundamental and industrial applications (Vol. 6261, pp. 626111-1-15). In: *Proceedings of SPIE*

Hanadi, G. S., Mohy, S. M., Yehya, B., & Wafaa, A. A. (2008). CW Nd:YAG laser cutting of ultra low carbon steel thin sheets using O_2 assist gas. *Journal of Materials Processing Technology, 196*, 64–72.

Ian, P. T., Tony, P. H., & Ian, P. R. (1998). Nd-YAG laser machining of SiC fibre/borosilicate glass composites. Part I. Optimisation of laser pulse parameters. *Composites Part A, 29A*, 947–954.

Johan, M. (2004). Laser beam machining (LBM), state of the art and new opportunities. *Journal of Materials Processing Technology, 149*, 2–17.

Kadan, V. M., Blonsky, I. V., Salnikov, V. O., & Orieshko, E. V. (2005). Effect of laser induced plasma in machining of transparent materials. *Proceedings of SPIE 5715, Micromachining and Microfabrication Process Technology X, 130*, 130–137.

Khalil, A. A. I., & Sreenivasan, N. (2005). Study of experimental and numerical simulation of laser ablation in stainless steel. *Laser Physics Letters, 2*, 445–451.

Kruusing, A., Uusimaki, A., Petretis, Br, & Makarova, O. (1999). Micromachining of magnetic materials'. *Sensors and Actuators A, 74*(1–3), 45–51.

Lau, W. S., Lee, W. B., & Pang, S. Q. (1990). Pulsed Nd: YAG laser cutting of carbon fibre composite materials. *CIRP Annals—Manufacturing Technology, 39*, 179–182.

Meung, J. K., & Jingwei, Z. (2001). Finite element analysis of evaporative cutting with a moving high energy pulsed laser. *Applied Mathematical Modelling, 25*, 203–220.

Moscicki, T., Hoffman, J., & Szymanski, Z. (2011). Modelling of plasma formation during nanosecond laser ablation. *19th Polish National Fluid Dynamics Conference (KKMP), Poznan, 63*, 99–116.

Negarestani, R., Sundar, M., Sheikh, M. A., Mativenga, P., Li, L., Li, Z. L., et al. (2009). Numerical simulation of laser machining of carbon-fibre-reinforced composites. *Proceedings of IMechE, 224*, 1017–1662.

Nikhil, A. V., Upendra, V. B., & Suhas, S. J. (2010). A finite element model to predict the ablation depth in pulsed laser ablation. *Thin Solid Films, 5*(19), 1421–1430.

Noboru, M., Shuichi, I., Yasutomo, F., & Ken, I. (1988). Pulsed laser processing of ceramics in water. *Applied Physics Letters, 52*, 1965–1966.

Ohara, J., Nagakubo, M., Kawahara, N., & Hattori, T. (1997). High aspect ratio etching by infrared laser induced micro bubbles. In: *Proceedings of the IEEE Tenth Annual International Workshop on Micro Electro Mechanical System* (pp. 175–179), New York: IEEE.

Pallav, K., & Ehmann, K. F. (2010). Feasibility of laser induced plasma micromachining (LIP-MM). *Precision Assembly Technologies and Systems, 315*, 73–80

Pallav, K., Han, P., Ramkumar, J., Nagahanumaiah, & Ehmann, K. F. (2014), Comparative assessment of the laser induced plasma micromachining and the micro-EDM processes. *Journal of Manufacturing Science and Engineering, 136*, 011001-1-16.

Prusa, J. M., Venkitachalam, G., & Molian, P. A. (1999). Estimation of heat conduction losses in laser cutting. *International Journal of Machine Tools and Manufacture, 39*, 431–458.

Qi, J., Wang, K. L., & Zhu, Y. M. (2003). A study on the laser marking process of stainless steel. *Journal of Materials Processing Technology, 139*, 273–276.

Rajaram, N., Sheikh-Ahmad, J., & Cheraghi, S. H. (2003). CO_2 laser cut quality of 4130 steel. *International Journal of Machine Tools and Manufacture, 43*, 351–358.

Rajiv, M., Ishan, S., Kornel, E., & Jian, C. (2013). Laser-induced plasma micromachining (LIPMM) for enhanced productivity and flexibility in laser based micromachining processes. *CIRP Annals—Manufacturing Technology, 6*, 211–214.

Rao, B. T., Kaul, R., Tiwari, P., & Nath, A. K. (2005). Inert gas cutting of titanium sheet with pulsed mode CO_2 laser. *Optics and Lasers in Engineering, 43*, 1330–1348.

Roy, S., & Modest, M. F. (1993). CW laser machining of hard ceramics-I. Effects of three-dimensional conduction, variable properties and various laser parameters. *International Journal of Heat and Mass Transfer, 36*, 3515–3528.

Samant, A. N., & Dahotre, N. B. (2009). Laser machining of structural ceramics—A review. *Journal of the European Ceramic Society, 29*, 969–993.

Sano, Y., Mukai, N., Okazaki, K., & Obata, M. (1997). Residual stress improvement in metal surface by underwater laser irradiation. *Nuclear Instruments and Methods in Physics Research B, 121*, 432–436.

Sarah, W., & Ishan, S. (2014). A preliminary study on the effect of external magnetic fields on laser-induced plasma micromachining (LIPMM). *Manufacturing Letters*. doi:http://dx.doi.org/10.1016/j.mfglet.2014.02.003.

Saxena, I., Ehmann, K., & Cao, J. (2014). Productivity enhancement in laser induced plasma micromachining by altering the salinity of the dielectric media. *ICOMM—2014*, (Vol. 93).

Sharma, A., & Yadava, V. (2011). Optimization of cut quality characteristics during Nd:YAG laser straight cutiing of Ni-based superalloy thin sheet using grey relational analysis with entropy measurement. *Materials and Manufacturing Processes, 26*, 1522–1529.

Sharma, A., Yadava, V., & Rao, R. (2010). Optimization of kerf quality characteristics during Nd: YAG laser cutting of nickel based superalloy sheet for straight and curved cut profiles. *Optics and Lasers in Engineering, 48*, 915–925.

Silfvast, W. T. (2004). *Laser fundamentals* (2nd ed.). Cambridge: Cambridge University Press.

Un-Chul, P., & Francis, P. G. (1972). Thermal analysis of laser drilling processes. *IEEE Journal of Quantum Electronics, 8*, 112–119.

William, M. S., & Mazumder, J. (2010). *Laser material processing* (4th ed.). Berlin: Springer.

Yan, Y., Li, L., Sezer, K., Wang, W., Whitehead, D., Ji, L., et al. (2011). CO_2 laser underwater machining of deep cavities in alumina. *Journal of the European Ceramic Society, 31*, 2793–2807.

Yu, L. M. (1997). Three-dimensional finite element modelling of laser cutting. *Journal of Materials Processing Technology, 63*, 637–639.

Yuan-Jen, C., Chia-Lung, K., & Nai-Yu, W. (2012). Magnetic assisted laser micromachining for highly reflective metals. *Journal of Laser Micro/nanoengineering, 7*, 254–259.

Zhang, J., Sugioka, K., & Midorikawa, K. (1999). Micromachining of glass materials by laser-induced plasma assisted ablation (LIPAA) using a conventional nanosecond laser. *Proceedings SPIE conference on Laser Applications in Microelectronics and Optoelectronic Manufacturing IV, 3618*, 363–369.

Zhu, S., Lu, Y. F., Hong, M. H., & Chen, X. Y. (2001). Laser ablation of solid substrates in water and in ambient air. *Journal of Applied Physics, 89*(4), 2400–2403.

Znotins, T. A., Poulin, D., & Reird, J. (1987). Excimer lasers: An emerging technology in materials processing. *Laser Focus, 23*, 54–70.

Fabrication of Micro Lens Array by Excimer Laser Micromachining

Syed Nadeem Akhtar, Shashank Sharma and J. Ramkumar

Abstract Micro Lens arrays are widely used in optical devices such as photo-sensors, digital projectors, photovoltaic cells, 3D imaging etc. These have traditionally been fabricated by photolithography, moulding and embossing, reactive ion etching and electroforming. These processes are wet processes and require expensive setup and running cost. A novel method is presented in this work that allows fabrication of micro lens array using excimer laser micromachining. The fabrication has been done using mask projection with work piece scanning. A KrF excimer laser has been used to micro machine lenses on a poly (methyl methacrylate) substrate. The surface profile of the lens array is measured and then related to the laser-material coupling and the energy of the laser pulses. Using this method, it is possible to fabricate micro lenses down to a diameter of 5 μm over a considerably large area.

Keywords Excimer laser · Lens array · Ablation rate · Micromachining

1 Introduction

Researchers had begun to investigate the mechanism of laser ablation in the early 1960s with tools like high speed photography and atomic force microscopy. Ready (1965), for example had put in efforts to study, both experimentally and analytically, the laser ablation using classical heat transfer and thermodynamic principles. The observations started out with basic measurements of depth of ablation and understanding its correspondence with analytically calculated data. Since then, the study of laser ablation mechanisms, for various materials, has progressed to the extent that advanced techniques like time-resolved spectroscopy and molecular dynamics simulations are able to observe and predict the escape velocity of ablated

S.N. Akhtar (✉) · S. Sharma · J. Ramkumar
Department of Mechanical Engineering, Indian Institute of Technology Kanpur, Kanpur 208016, India
e-mail: snadeem@iitk.ac.in

S.N. Joshi and U.S. Dixit (eds.), *Lasers Based Manufacturing*, Topics in Mining, Metallurgy and Materials Engineering, DOI 10.1007/978-81-322-2352-8_12

fragments, their composition and time duration of ablation (Dijkkamp et al. 1987; Srinivasan et al. 1986, 1995).

This paper reviews the observations and findings relating to the mechanism of ablation of polymers when irradiated by nanosecond UV lasers. Excimer lasers are a part of this category and have been considered for this review. The paper first introduces the basic parameters of lasers that play a significant role in the ablation process. Definition of key terminologies related to the process and properties of materials are then discussed. This is followed by a discussion on mechanisms active during the excimer laser ablation of polymers and the various pathways and products as a result of this ablation.

2 Parameters of Laser Radiation

A laser radiation is characterized by certain key parameters. Concerning lasers used in machining of materials, these parameters are the wavelength of radiation, the nature of radiation (continuous or pulsed), the duration of the pulse and the quality of the beam. Wavelength (λ) determines the energy carried by individual photons in the laser beam and hence the kind of interaction these photons will have with the material. Large wavelength lasers, like CO_2 lasers ($\lambda = 10.6\ \mu m$), mainly cause ablation by a thermal process because the photons do not carry enough energy to cause any other effect. Short wavelength lasers, like F_2 lasers ($\lambda = 157$ nm), mainly cause ablation by a photolytic process wherein the photons are directly involved in breaking of bonds within the material. A description of the various types of excimer lasers, their wavelengths and photon energies is given in Table 1. Since we are concerned with UV lasers ($\lambda < 400$ nm), photolytic ablation process will dominate the discussion.

3 Key Terminologies in Laser Material Interactions

When a laser radiation is incident on a material most of the photons get absorbed by the electrons present in the outermost shell of atoms in the top skin layer. The thickness of the skin layer, also called the optical penetration depth (δ), is $2/\alpha$ where

Table 1 Excimer lasers and the associated wavelengths and photon energies

Laser	Wavelength (nm)	Photon energy (eV)
XeF	351	3.53
XeCl	308	4.03
KrF	248	5.00
KrCl	222	5.50
ArF	193	6.42
F_2	157	7.43

From Meijer (2004), reprinted with permission

'α' is the absorption coefficient of the material for the particular wavelength. The optical penetration depth for polymers is about 200 nm. The time it takes for the photons to get absorbed is called the photon absorption time (t_a) and is of the order of 1 fs. These photons cause the electrons to get excited. The time it takes for the excited electrons to relax to their previous state is called the electron relaxation time (or electron cooling time, t_e) and is of the order of 1 ps. After 1 ps of irradiation, the electrons relax and pass on the energy to the lattice in the form of heat. This heat diffuses into the material at the rate determined by the thermal diffusivity (κ) and thermal conductivity (k) of the material (Meijer 2004; Choudhary 2012; Meijer et al. 2002). The depth through which the heat diffuses into the material in a certain time 't' is called the thermal diffusion depth (d) and is given by $d = \sqrt{(4\kappa t)}$.

The pulse duration is a property of the laser radiation. However, the pulses are classified as short or ultra short based on whether the optical penetration depth is more or less than the thermal diffusion depth (in time 't_e'), respectively. Generally speaking, pulses with duration less than 1 ps are called ultra short pulses. An often used term in laser machining is 'fluence'. Fluence is defined as the irradiated energy per pulse per unit area of the material and its units are J/cm^2. Ablation is noticeable only above a certain value of fluence (depending on the material) and this value is called the threshold fluence (F_{th}). Researchers have over time realized that instead of fluence (that makes no reference to the time in which the energy is delivered), intensity better describes the effect of radiation (Srinivasan et al. 1995; Jensen 2004). Intensity, then, is defined as average power per pulse per unit area and its units are W/cm^2. Several parameters affect the mechanism and extent of laser ablation in polymers, and some of those are mentioned in Table 2 for some common polymers.

4 Mechanism of Laser Ablation of Polymers

For ablation of polymers by nanosecond UV laser, the process is a combination of thermal (pyrolytic) and photo (photolytic) ablation. Since the pulse duration here is much larger than that of ultra short pulses, non-linear processes like multi-photon absorption do not occur (Srinivasan et al. 1986). Pyrolytic ablation starts with

Table 2 Some common polymers and their absorption coefficients and thermal diffusivities

Polymer	Absorption coefficient (α, μm^{-1}) at 248 nm	Thermal diffusivity (κ, m^2/s)
Polycarbonate	1	0.144×10^{-6}
Polypropylene	–	0.096×10^{-6}
Polyimide	22	0.122×10^{-6}
PMMA	0.0063	0.115×10^{-6}

From Meijer (2004), Jensen (2004), Tsutsumi and Kiyotsukuri (1988), reprinted with permission absorption of the incident photons and electronic excitation of the material. It however proceeds via random distribution of this energy amongst the various degrees of freedom of the constituents (molecules) which in turn leads to heating of the material. The heated material ablates via both melt/vaporization and thermal degradation (bond-scission leading to fragmentation) routes. Both these processes happen in the ground state of the constituents (Dyer 2003). This is significantly observed for higher wavelength (low energy photons) and longer pulse lasers ($t_p \geq 1$ μs). The two basic reasons for pyrolytic ablation are that, firstly, the energy from the low energy photons is transferred to the electrons and then to the lattice in the form of heat energy. The photons do not carry enough energy to cause any other effect, e.g. breaking the covalent bonds in the material. Secondly, the duration of irradiation (t_p) is significantly larger than the electron relaxation time (t_e) and the energy from the photons begins to travel into the material during the irradiation, leading to heating of the vicinity of the area being irradiated. Pyrolytic ablation often leads to a heat affected zone (HAZ) around the irradiated area, thermal degradation and evidence of melt flow. These effects, however, are generally not significantly observed in case of ablation of polymers with nanosecond lasers.

Photolytic ablation is a photo-chemical process that involves bond breaking and formation with the aid of incident photons. Polymers are composed of elements (like C, O, H, N, F, Cl etc.) joined together by covalent bonds. The incident photons, if sufficiently energetic, can modify the bond structure within the material leading to bond-scission or cross-linking effects. The intensity of radiation is so high in nanosecond UV lasers that the numbers of bonds that are broken are several orders of magnitude higher than those that are formed. This, however, requires that the incident photons carry sufficient energy to break the covalent bonds. Typical bond energies of some common covalent bonds are given in Table 3 that may be compared to the energy of photons as mentioned in Table 1. Upon comparison it can be inferred that UV lasers ablate majorly through photolytic ablation owing to the high energy of the photons, although this will be further discussed in the following section. The incoming photons break the covalent bonds in the long polymer chains resulting in generation of several smaller chains present within the same volume (Meijer 2004; Jensen 2004). This fragmentation or decomposition of the polymer happens when the reactants are in electronically excited state (Dyer 2003). When the number of chains in a unit volume exceeds a certain value, meaning that when the bond density in that volume decreases below a critical value, the smaller chains explode out in the form of vapor and small particles. This explosion leads to expelling of the material and ablation.

Table 3 Bonds and their bond energies

Bonds	Bond energy (eV)
Si–Si, Cl–Cl	1.8–3.0
C–N, C–C	3.0–3.5
C–O, C–H	4.5–4.9
C=C, C=O	7.0–8.0

From Meijer (2004), reprinted with permission

5 Analysis and Observations

5.1 Nature of Ablation

Pure photolytic ablation, also called cold ablation, is unlikely with nanosecond lasers. Several researchers have observed signatures of both pyrolytic and photolytic ablation in polymers. Dijkkamp et al. (1987) observed through time resolved reflectivity measurements that the ablation of poly-methylmethacrylate (PMMA) and poly (3-butenyltrimethylsilane sulfone) (PBTMSS) was entirely due to pyrolytic ablation with KrF excimer laser. Feng et al. studied the ablation of polypropylene (PP) with KrF excimer laser. They observed that even with fluence near the ablation threshold of the polymer, ablation was observable on the surface. Although the ablation was due to expelling of fragments generated after polymer chain scissions, the scission was caused not directly by the photons but by the heating of the polymer surface by the radiation. A rim of solidified polymer was observed around the ablated crater suggesting that polymer had melted and flown out of the surface and then resolidified around the crater (see Fig. 1). The observation that the rim increased in size with increase in fluence and the number of pulses suggests accumulation of heat with higher fluence and number of pulses. Dyer (2003) observes that polymers that are thermally robust (like polyimide) have been confirmed to ablate by extensive fragmentation, indicating that photolytic ablation resulting in covalent bond scission was the active mechanism.

Relative importances of the two ablation mechanisms have also been studied by researchers. Feng et al. (2000) found that photo ablation dominates at very low or very high fluence. For the fluence range generally used in ablation pyrolytic ablation was found to be dominant. However, since laser ablation employs irradiation with several pulses, a cumulative effect of both processes is always observed. Jensen suggested that the relative dominance can be checked by comparing the values of the optical and the thermal penetration depths of a material for a particular wavelength and pulse duration.

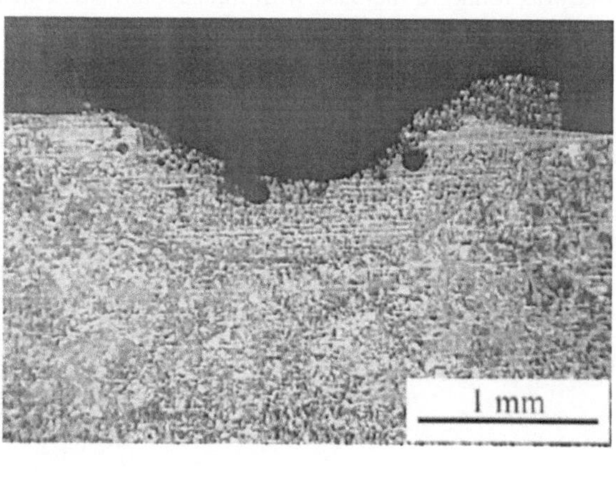

Fig. 1 The cross sectional micrograph of an ablated crater in PP showing the rim formation (from Feng et al. (2000), reprinted with permission)

5.2 Plume Interaction During the Laser Pulse

Pulses of KrF excimer laser, for example, last for about 20 ns. Immediately after a pulse hits the material, a high temperature plasma plume is formed that contains gaseous components, small fragments and radicals. This plume is continuously fed by the fragments being expelled out from the material surface during the pulse. Since the plume is developed almost immediately after the pulse hits the material, a significant portion of the pulse strikes the plume while on its way to the material surface. The plume (i.e. the species present in the plume) are known to significantly absorb the incoming radiation and letting only a small portion of it through. Then the effective absorption coefficient of the polymer is considerably reduced and leads to a lower extent of ablation (etch rate). For a highly absorbing plume, low fluence results in efficient ablation of the polymer surface and vice versa. Using a high fluence, in this case, leads to heating of the plume and hence to undesirable thermal effects (Dyer 2003).

5.3 Ablation Rate and Threshold

Ablation rate is one of the most studied parameters in laser machining of polymers. The ablation rate is also called the etch rate and is defined as the depth (ablated) per pulse of laser. It is measured either in nm or in μm and is an average quantity. It has been always observed that the threshold fluence must be exceeded in order for ablation to set in. This is required across the entire band of wavelengths. For sub-100 ns pulse duration lasers, the fluence threshold is independent of the pulse duration. This, however, is not true for longer duration pulses (Dyer 2003). It is still a matter of investigation and debate as to whether a true ablation threshold exists for a material. The authors of this paper have measured the etch depth and ablation rate of biaxially-oriented polyethylene terephthalate (boPET) when irradiated with KrF excimer laser (Akhtar 2013). The measurements are shown in Fig. 2. The ablation rate was found to increase with fluence and was constant up to a certain number of pulses for the same fluence. Etching of polymers has been observed in the vicinity (lower) of ablation thresholds. In their study of ablation of polyimide using excimer lasers, Kuper et al. (1993) found smooth curves for etch rate versus fluence, suggesting the absence of any kind of ablation threshold. A sharp threshold has, however, been observed with ArF (193 nm) laser while machining polyimide. A clearer description of polymer behaviour at high temperatures or energy densities is difficult at this time due to the absence of experimental data regarding this process. In the domain of photolytic ablation, if a minimum of 'n' bonds per unit volume of the material need to be broken to initiate ablation, the threshold fluence, F_{th}, can be expressed as

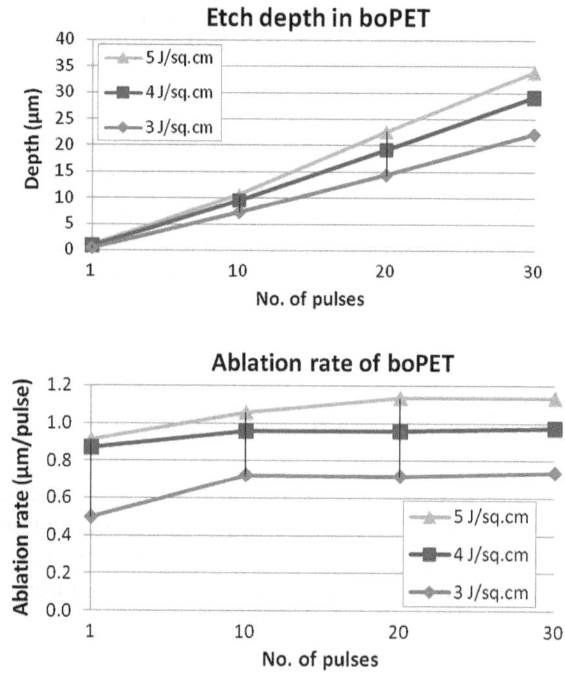

Fig. 2 Ablation depth (*top*) and ablation rate (*bottom*) plotted as a function of the number of pulses for various fluences during machining of boPET by KrF excimer laser [from Akhtar et al. (2013)]

$$F_{th} = \frac{nhv}{\eta\alpha(1 - R)} \tag{1}$$

where η is the quantum efficiency of bond-scission, hv is the photon energy and R is the surface reflectivity. The quantum efficiency of bond scission increases with decrease in wavelength. Hence η for ArF laser is more than that for KrF laser. The η for KrF, however, can be increased by increasing the temperature so that the photochemical processes are accelerated (Srinivasan et al. 1986). In the domain of pyrolytic ablation,

$$F_{th} = \frac{C(T_{th} - T_s)}{\alpha(1 - R)} \tag{2}$$

where 'C' is the volumetric specific heat capacity of the material, T_{th} is the minimum temperature required to be reached to initiate thermal degradation of the material, and T_s is the surface temperature of the material (Dyer 2003). The intensity of radiation at a depth 'y' into the material, $I(y)$, follows the Beer's law of attenuation, $I(y) = I(0)e^{-\alpha y}$, where $I(0)$ is the intensity at the surface. The etch rate (depth per pulse, 'd') shows a logarithmic dependence on the fluence and follows the relation, $d = \alpha^{-1}\ ln(F/F_{th})$, where α is experimentally determined. The etch depth varies linearly with the log of fluence for values of fluence ranging between F_{th} and a certain upper limit. For a constant fluence, the etch depth varies linearly with the number of pulses. Decrease of etch rate at high fluence is due to secondary absorption of incoming photons by the expelled products (present in the plume) and distribution of the irradiation over the walls of the ablated crater. Also, low energy pulses do not break enough number of C–C bonds and hence lead to small etch depths. Moderate energy pulses are able to break enough number of C–C bonds that leads to efficient ablation and comparatively greater etch depth. High energy pulses break a large number of C–C bonds but over the same polymer mass (albeit into smaller polycarbon fragments that explode out) leading to almost the same etch depth. Hence beyond a particular pulse energy (fluence) the efficiency of photolytic ablation levels off (Dayal et al. 2013).

5.4 Mask Projection Technique

Excimer lasers are widely used with mask projection. The technique involves use of a mask, through which the laser beam passes before hitting the work piece, as shown in Fig. 3. A beam delivery section in the laser machine serves to shape the beam before allowing it to reach the mask. It comprises a number of optics to shape, collimate, homogenize and focus the beam.

A typical focusing arrangement is shown in Fig. 4. When a collimated beam passes through a double convex lens, it is converged on to a plane, called the *focal plane* of the lens. The point of convergence, also called the *focal point*, is in fact, not a point, but a spot. The spot, with the minimum radius of w_o, is called the *beam waist*. The minimum possible spot size, to which a beam can be practically focused, is called the *diffraction limited spot size*. Hence, while using mask projection, it is generally not possible to focus a beam to a size smaller than its diffraction limited spot size.

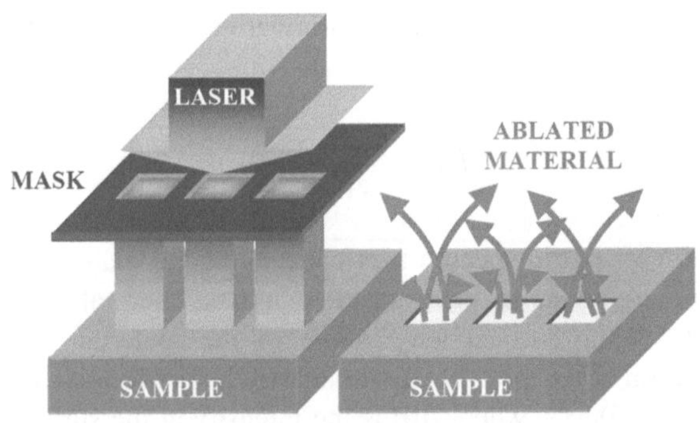

Fig. 3 Schematic of a mask projection setup, showing the incident laser beam, passing through a mask with *three square holes*, before falling on the work piece and ablating a similar pattern on it [Reprinted with permission, Rizvi (1999)]

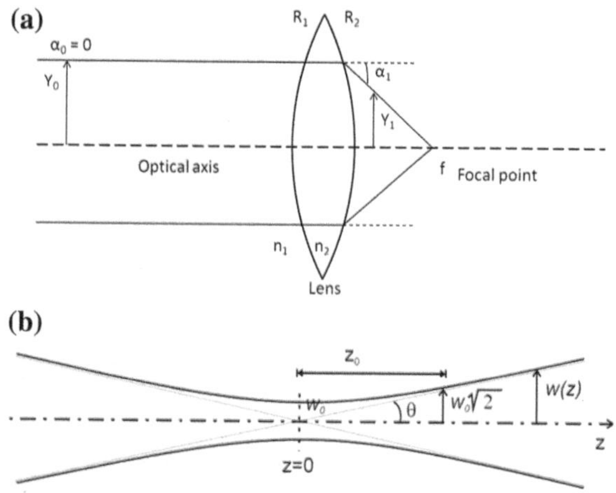

Fig. 4 Schematic diagrams of a simple optical system showing the convergence of a beam onto a focal plane, with a detailed description of the beam waist. **a** A focusing arrangement with a double convex lens. 'f' is the focal point, lying on the optical axis 'z'. **b** Enlarged view of the focal region, showing the beam waist with a radius of 'w_o', and Rayleigh range 'z_o'

Lasers with smaller wavelengths can be focused to smaller spots, and vice versa. Hence, it is advantageous to use lasers with small wavelengths (typically in the ultraviolet range) while fabricating micrometer sized features. The spot radius at the beam waist, w_o, is expressed as,

$$w_0 = \frac{4\lambda f}{2\pi y_0} \tag{3}$$

The Rayliegh range, z_o, is the distance along the optical axis, on either side of the beam waist, at which the beam diameter reaches 1.414 times its value at the beam waist. It is expressed as

$$z_0 = \frac{\pi w_0^2 M}{\lambda} \tag{4}$$

In the expressions above, λ is the wavelength of the laser, M is the beam quality factor, and f is the focal length of the lens. In a laser machining setup with mask projection, the workpiece is kept at the image plane of the mask, with the required placement accuracy of the order of the Rayleigh range. If the workpiece is not properly positioned, a blurred image of the mask features is obtained on the workpiece.

Mask projection employs a mask, which is generally a screen with transparent and opaque portions. The laser beam is allowed to pass through the transparent portion, and the shape of the transparent portion is machined/imprinted on the workpiece by virtue of its ablation. An example of a mask projection setup is shown in Fig. 5.

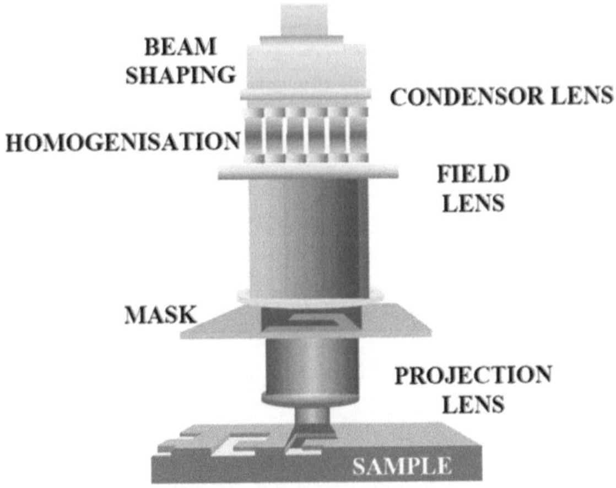

Fig. 5 Schematic of a mask projection system [Reprinted with permission, Rizvi (1999)]

The mask projection technique is a contrast to the direct write technique, and offers several advantages over the latter. The mask and the optics, being away from the work piece, do not suffer any damage due to ablation. Many setups allow movement of the mask as well as the work piece. The combination of mask and work piece scanning provides the ability to produce several complicated features, like arrays of micro features, nested holes, micro-channels with different kinds of cross sections etc. Since an image of the mask is used to machine micro features, it is possible to de-magnify the image and machine features that are much smaller than those on the mask. Demagnifications of 2X, 5X and 10X are widely used. The demagnification allows use of much lower fluences on the mask, thereby saving the mask from laser induced damage.

Several kinds of masks are used in the industry. Binary masks are the most widely used and easiest to fabricate. These generally contain either opaque elements deposited on a transparent substrate, or a pattern cut out on an opaque sheet of polymer or metal. These are similar to lithographic masks, but are more resistant to damage, primarily caused due to laser. Gray scale masks are another kind of masks that have varying transmissivity over the cross section. The smoothly varying transmissivity allows fabrication of smooth profiles on the workpiece. These are however much more difficult to fabricate and require calibration of transmissivity to produce desired patterns. Phase shift masks employ interference between portions of a beam passing through two separate regions of the mask, each with a different refractive index. Fine patterns with high contrast and resolution can be produced on the workpiece using such masks, but the masks require complex fabrication. Since binary masks are easy to fabricate, these are mostly used for producing micro features on polymer surface. In the following section we shall present a method to fabricate micro lens array with the help of a binary mask.

6 Fabrication of Micro Lens Array

In recent years the fabrication of micro lenses has gained attention of researchers. These micro lenses have found use in a plethora of components such as CCD arrays (Gale et al. 1997), digital projectors, 3D imaging (Hess 1912), and integral photography (Lippmann 1908). Moreover, current research indicates that array of micro lenses has the ability to act as concentrators for high efficiency photovoltaics (Karp et al. 2010). They are also used to couple light to optical fibres (Cohen et al. 1974). Therefore, the use of micro lenses in compact high performance micro optical devices has become essential. That is why, in the past decade, researchers have emphasized on deriving novel methods to fabricate lenses with lens diameter ranging from a few to several hundred micrometers.

Thus, there is a growing need to establish a cost effective, less complex and efficient method to fabricate micro-lenses. For the past two decades, researchers have been exploring various methods to fabricate refractive micro lenses. Some research was based on glass based lenses, while the study on polymer based lenses has opened a whole new window of opportunities. The photo-resist reflow method (Popovic et al. 1988), ultraviolet curing of polymer (Okamoto et al. 1999), LIGA method (Sankur et al. 1995), micro jet technique (MacFarlane et al. 1994) and micro moulding or hot embossing method (Ong et al. 2002) are some such new techniques which are used to fabricate micro lenses. However, these methods suffer from poor surface quality of features that cannot be improved. Hence, the use of these methods is restricted to some extent. Excimer laser micromachining has an important role in the fabrication of polymer based micro lenses as it interacts via photo-chemical mechanism or the cold ablation mechanism and therefore the surface quality can be controlled accurately (Brannon 1990; Dyer et al. 1985). The material is removed by laser ablation while the shape of the pattern is controlled by mask projection of laser source and the motion of a micro positioning stage.

For fabricating a 3-dimensional micro structure, one can use a photo mask which modulates the spatial laser intensity on the work piece, producing features of varying depth (Tien et al. 2003). Another method is known as contour mask scanning (Zimmer et al. 2000), in which contour or mask opening is synchronized with sample movement to achieve depth variation.

Lee et al. (2005) have successfully fabricated axis symmetric micro lenses using a new approach termed as "planetary mask contour scanning method". In this method the mask revolves as well as rotates at the same time. The underlying principle is to create a probability distribution of laser intensity with the help of a self spinning mask which revolves around the sample producing axis symmetric feature. Authors have used 0.5 mm polycarbonate samples. According to the principle, the motion of the sample stage and mask revolution should be synchronized with laser firing sequence and before machining the mask centre should be in alignment with the sample stage. Lee et al. (2007) used this method to fabricate axis symmetric micro lenses of 200 μm aperture with high surface quality, i.e. surface roughness ranged from 3 to 6 nm.

Although several methods have been proposed over the years but still a systematic study and derivation of less complex and accurate method is required. In this paper we present a systematic study of fabrication process of micro lenses via use of a contour mask and work piece scanning.

6.1 Experimental Setup

A 248 nm KrF Excimer laser (Coherent Variolas Compex Pro 205F) is used for machining the micro lens array. The machine can deliver pulses of energy up to 750 mJ with 20 ns pulse width. The energy of the pulse can be varied by changing the discharge voltage or by manually tuning the attenuator. A pair of 8 × 8 fixed array of insect eye lenses is used to create a square field of 20 mm × 20 mm with homogeneous top hat beam profile at the mask plane. The setup for mask projection is shown in Fig. 6. The beam transmitted across the mask is imaged on the work piece through an imaging lens which has a demagnification of 10X. Optical microscopes and proprietary image analysis software have been used to capture and analyze the images. The work piece is an 8 mm thick piece of poly(methyl methacrylate) (also known as PMMA).

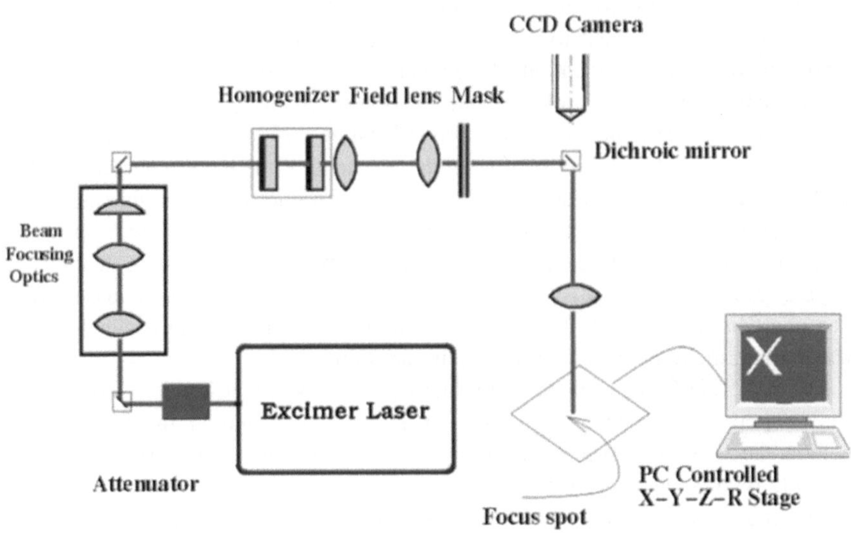

Fig. 6 Schematic of the experimental setup [from Dayal et al. (2013), reprinted with permission]

Table 4 Machining parameters and their values used in the experiments

Pulse energy (mJ)	176
Pulse repetition rate (Hz)	5
Scanning speed (μm/s)	8.3, 16.7, 25.0, 33.3

6.2 Experimental Procedure

The fabrication of micro lens array proceeds through generation of masks in two stages, followed by the fabrication of the lenses. The first stage mask used was a 30 mm × 30 mm piece of aluminium foil inside which the desired cross-sectional profile of the lenses was cut, at a scale of 100X. This mask was used to further machine the same profile, albeit at a 10 times smaller scale, on a polymer sheet. The polymer mask was finally used to fabricate the lenses on the PMMA work piece. The work piece was kept on a micro-positioning XYZ stage, and scanned along the X and Y axes to generate the lens array.

The machining parameters used for the experiments are mentioned in Table 4. The pulse energy was fixed at 176 mJ and the pulse repetition rate was fixed at 5 Hz. A set of experiments was conducted by varying the scanning speed of the work piece. The variation in scanning speed causes a variation in the number of pulses falling at a particular spot, and thereby varies the depth to which the feature gets machined.

7 Result and Discussion

7.1 Analysis of Masks

The first stage mask was fabricated in a piece of aluminium foil, and is shown in Fig. 7. The profile of the mask was designed in a way that the exposed length is largest at the center and it reduces, by second order, towards the periphery. This allows the manufacture of two halves of a lens, with peaks at the periphery and valley in the middle. Note that the region that has a larger exposed length sees more number of pulses per spot, and hence experiences greater machining depth.

The maximum exposed lengths in the first and second stage masks are 18.095 and 1.579 mm respectively, which is a reduction of about 11.5X. Clearly, an overcut of about 12.7 % is observed, which is expected in a laser machining process. The radius of the curved profile, has been reduced 10 times, as can be seen from Fig. 7.

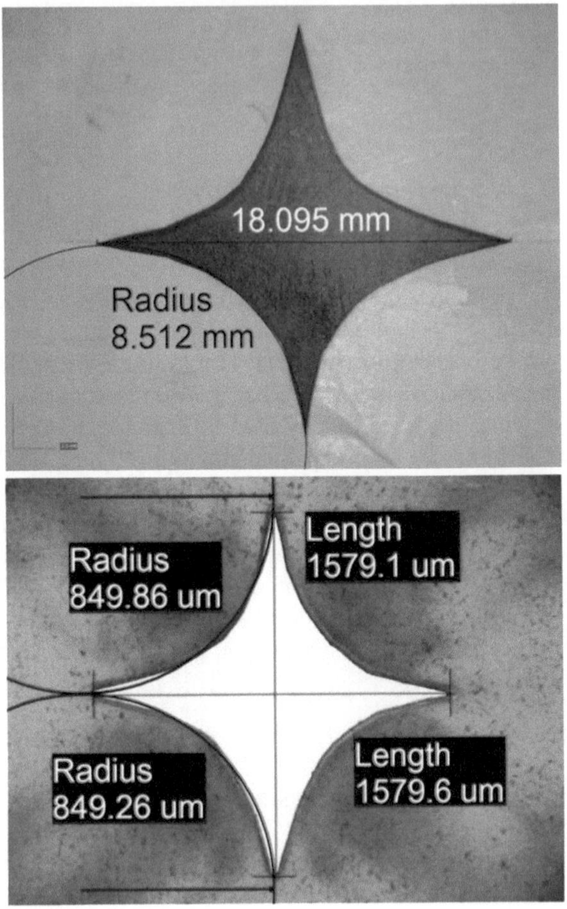

Fig. 7 First (*top*) and second (*bottom*) stage masks

7.2 Analysis of Micro Lens Profiles

The cross-sectional micrographs of the various lenses are shown in Fig. 8. Arrays of 3 × 3 lenses have been fabricated and their profiles are measured and compared with the theoretical profile expected from machining in PMMA. Note that it is easy to extend the size of the array to 10 × 10 or 100 × 100. The dimensions shown on the top of each figure (e.g. see Fig. 8a) show the depth to which the entire array has been machined. This depth is compensated for when comparing these profiles with the theoretical profile.

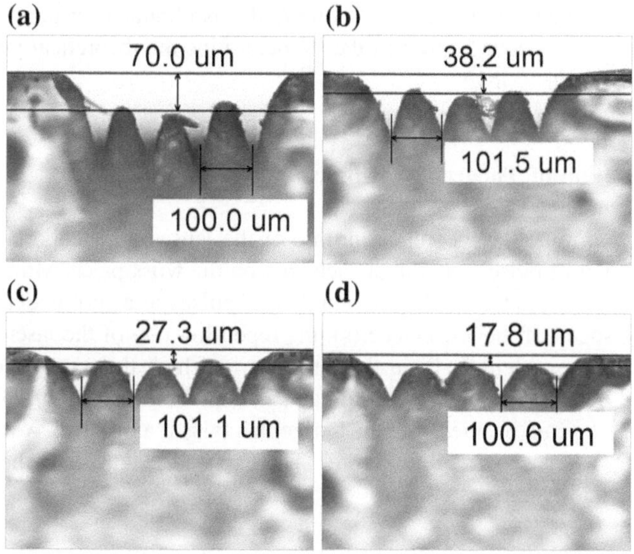

Fig. 8 Cross-sectional micrographs of lenses machined at various scanning speeds, **a** 8.3 μm/s, **b** 16.7 μm/s, **c** 25.0 μm/s and **d** 33.3 μm/s, with 176 mJ pulses and at 5 Hz. The adjacent lenses have been fabricated after indexing the work piece by 100 μm

7.2.1 Measurement of Ablation Rate of PMMA

Basic experiments were conducted on PMMA to measure the depth of ablation. Depths were measured after machining with several pulses of laser at three different pulse energies (120, 175 and 225 mJ). It is important that these measurements are done with several pulses so that incubation effects of photon absorption are compensated for, and the ablation depth per pulse reaches a steady value. Figure 9 shows the plot of ablation depth per pulse at the three different pulse energies, for varying number of pulses. As is widely reported, the etch depth per pulse varies

Fig. 9 Ablation rate of PMMA versus fluence

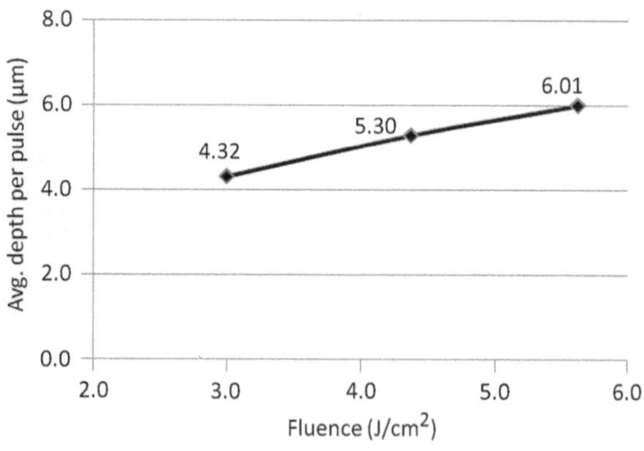

linearly with the fluence. A value of 5.3 μm/pulse is obtained for the pulse energy used (175 mJ). This value is used further to determine the theoretical profile of the lenses machined on PMMA.

7.2.2 Calculation of Theoretical Profile

In order to determine the theoretical profile of the lenses, it is important to determine the number of pulses that fall at each spot on the work piece. More number of pulses generates a greater depth. The number of pulses at a spot is determined by the scanning speed of the workpiece (s), the repetition rate of the laser (r) and the feature size (exposed length, l), in the direction in which the work piece is being scanned. In Fig. 10, the origin of coordinates is fixed at the point 'O.' The vertical distance to the bottom profile, from the horizontal axis, at a distance 'X' from 'O,' is calculated (P_0 P_1, as shown in the figure). The exposed length, L, is then, twice of this distance, and is expressed as

$$L = 2\left(R - \sqrt{R^2 - X^2}\right), \tag{5}$$

If the distance moved by the stage in 1 s is 's', then 's' units of length see 'r' pulses in a second. Hence, the number of pulses per spot will be the number of pulses falling per 'L' units of length, and is expressed as Eq. 6. The ideal depth of ablation is, then, given by,

$$pulses\,per\,spot = \frac{rL}{s}. \tag{6}$$

$$\text{Ideal depth} = \text{Pulses per spot} \times \text{ablation rate}. \tag{7}$$

The ideal depth of ablation helps to generate the theoretical profile of the lens. The theoretical profiles for all four cases of scanning speed are shown in Fig. 11.

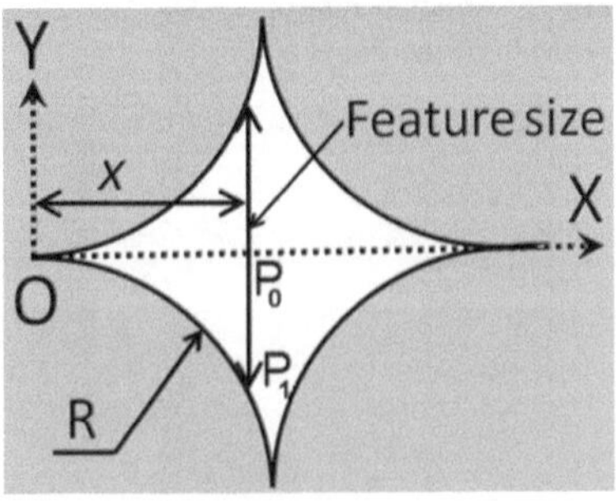

Fig. 10 Schematic of the procedure to calculate the feature size (exposed length)

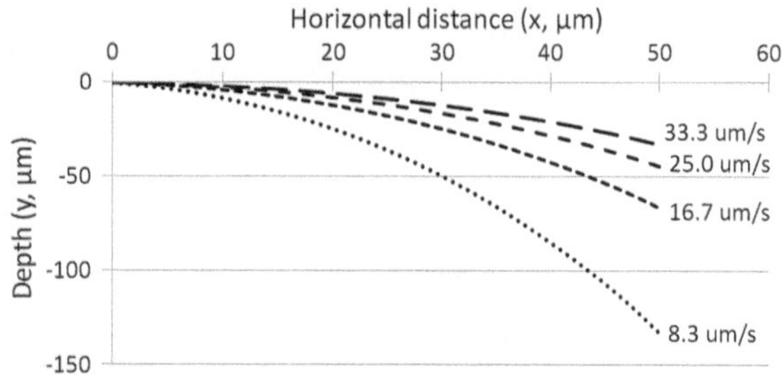

Fig. 11 Theoretical profiles of lenses machined at different scanning speeds

7.2.3 Micro Lens Profile—Experimental Results

Figure 12 shows the experimentally determined profiles of micro lenses in comparison with the theoretically predicted ones. All features show conformance to the theoretical profile to a very high degree. It can be noted that lenses machined at higher scanning speeds have lower machined depths. This is because lesser number of pulses fall per spot at a higher speed. Figure 13 shows the surface profiles of the machined lenses for the case of scanning speed of 33.3 μm/s.

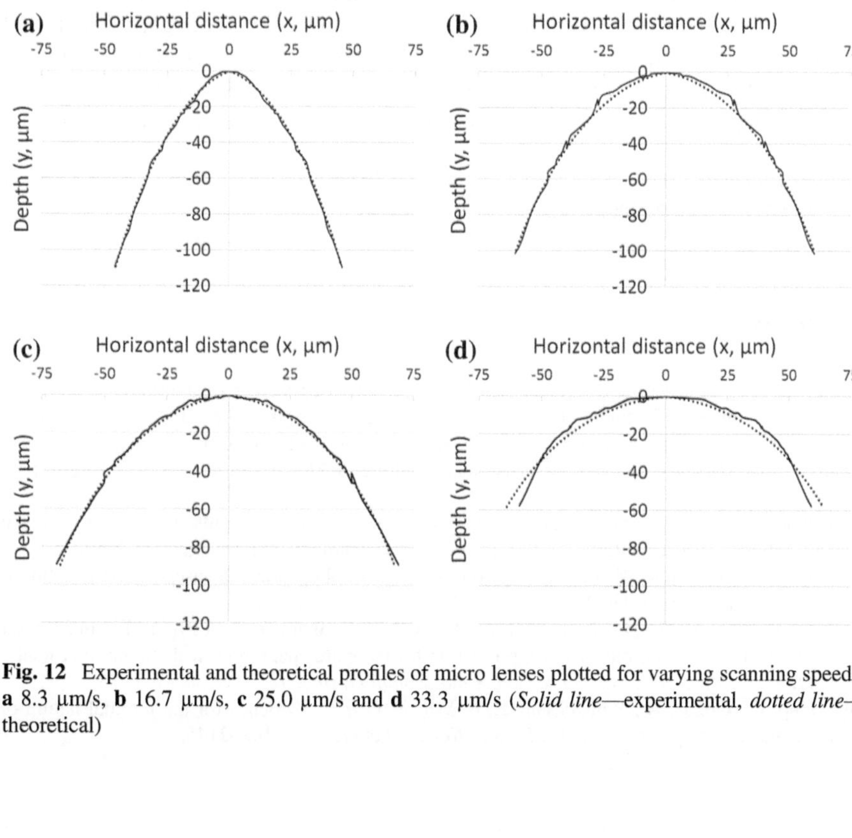

Fig. 12 Experimental and theoretical profiles of micro lenses plotted for varying scanning speeds, **a** 8.3 μm/s, **b** 16.7 μm/s, **c** 25.0 μm/s and **d** 33.3 μm/s (*Solid line*—experimental, *dotted line*—theoretical)

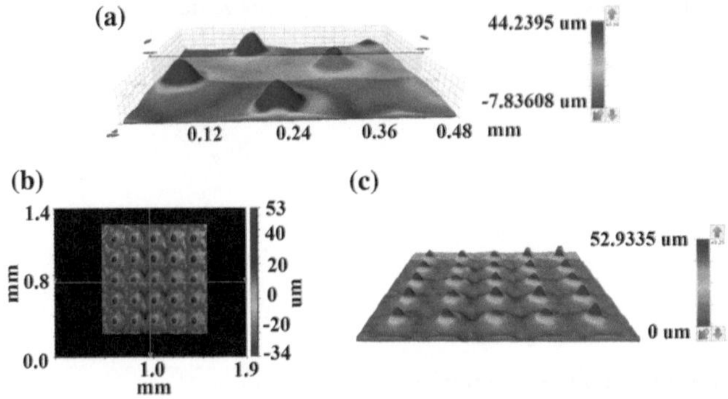

Fig. 13 3D Optical profilometery images of 5 × 5 array of micro lenses fabricated at the scanning speed of 33.3 μm/s. **a** Side view (closeup) of the micro lenses. **b** Micro lens array, fabricated over 1 mm × 1mm, when viewed from the top. **c** Side view of the micro lens array

8 Conclusions

Excimer laser micromachining has provided a new method to fabricate micro lens arrays. Therefore, it is important to study the various fabrication methods and their effects on feature quality. The effect of scanning speed and mask pattern is reviewed in this paper. At moderate scanning speeds the experimental profile agrees with the theoretical feature accurately. The number of pulses per spot is high at low scanning speeds. Hence, lenses are machined to a greater depth at low scanning speeds.

Acknowledgments The authors thank Dr. Nadeem Rizvi and ASME, SPIE and Elsevier for permissions to use some figures and tables in this chapter.

References

Akhtar, S. N., Ramakrishna, S. A., & Ramkumar, J. (2013). Excimer laser micromachining on polymers under different atmospheres and at different length scales. In *8th International Conference on Micro Manufacturing (ICOMM)*, March 25–28, Victoria, Canada.

Brannon, J. H. (1990). Excimer-laser ablation and etching. *IEEE Circuits and Devices Magazine, 6* (5), 18–24.

Choudhary, H. (2012). Numerical simulation and experiments of nano pulsed UV laser on metal and polymer (Master's thesis, Indian Institute of Technology Kanpur, Kanpur, India).

Cohen, L., & Schneider, M. (1974). Microlenses for coupling junction lasers to optical fibers. *Applied Optics, 13*(1), 89–94.

Dayal, G., Akhtar, S. N., Ramakrishna, S. A., & Ramkumar, J. (2013). Excimer laser micromachining using binary mask projection for large area patterning with single micrometer features. *ASME Journal of Micro and Nano Manufacturing, 1*(3), 031002-1–7.

Dijkkamp, D., Gozdz, A., Venkatesan, T., & Wu, X. (1987). Evidence for the thermal nature of laser-induced polymer ablation. *Physical Review Letters, 58*, 2142–2145.

Dyer, P. (2003). Excimer laser polymer ablation: Twenty years on. *Applied Physics A, 77*, 167–173.

Dyer, P. E., & Sidhu, J. (1985). Excimer laser ablation and thermal coupling efficiency to polymer films. *Journal of Applied Physics, 57*(4), 1420–1422.

Feng, Y., Liu, Z., & Yi, X. (2000). Co-occurrence of photochemical and thermal effects during laser polymer ablation via a 248-nm excimer laser. *Applied Surface Science, 156*, 177–182.

Gale, M. T., Pedersen, J., Schütz, H., Povel, H., Gandorfer, A., Steiner, P., & Bernasconi, P. N. (1997). Active alignment of replicated microlens arrays on a charge coupled device imager. *Optical Engineering, 36*(5), 1510–1517.

Hess, W. (1912). Improved manufacture of stereoscopic pictures. UK Patent 13, 034.

Jensen, M. (2004). Laser micromachining of polymers (Ph.D. thesis, Technical University of Denmark, Denmark).

Karp, J. H., Tremblay, E. J., & Ford, J. E. (2010). Planar micro-optic solar concentrator. *Optics Express, 18* (2), 1122–1133.

Kűper, S., Brannon, J., & Brannon, K. (1993). Threshold behavior in polyimide photoablation: Single-shot rate measurements and surface-temperature modeling. *Applied Physics A, 56*, 43–50.

Lee, Y. C., & Wu, C. Y. (2007). Excimer laser micromachining of aspheric micro lenses with precise surface profile control and optimal focusing capability. *Optics and Lasers in Engineering, 45*, 116–125.

Lee, Y. C., Chen, C. M., & Wu, C. Y. (2005). A new excimer laser micromachining method for 3D microstructures with continuous surface profiles. *Sensors Actuators A, 117*(2), 349–355.

Lippmann, G. (1908). Epreuves reversibles. Photographies integrales. *Comptes Rendus, 146*, 446–451.

MacFarlane, D. L, Narayan, V., Tatum, J. A, Cox, W. R., Chen, T., & Hayes, D. J. (1994). Microjet fabrication of microlens array. *IEEE Photonics Technology Letters*, *6*(9), 1112–1114.

Meijer, J. (2004). Laser beam machining (LBM), state of the art and new opportunities. *Journal of Materials Processing Technology, 149*, 2–17.

Meijer, J., Du, K., Gillner, A., Hoffmann, D., Kovalenko, V., Masuzawa, T., et al. (2002). Laser machining by short and ultrashort pulses, state of the art and new opportunities in the age of the photons. *CIRP Annals-Manufacturing Technology, 51*, 531–550.

Okamoto, T., Mori, M., Karasawa, T., Hayakawa, S., Seo, I., & Sato, H. (1999). Ultraviolet-cured polymer microlens arrays. *Applied Optics, 138*, 2991–2996.

Ong, N. S., Koh, Y. H., & Fu, Y. Q. (2002). Microlens array produced using hot embossing process. *Microelectronic Engineering, 60*(3–4), 365–379.

Popovic, Z. D., Sprague, R. A., & Connell, G. A. (1988). Technique for monolithic fabrication of microlens arrays. *Applied Optics, 27*(7), 1281–1284.

Ready, J. (1965). Effects due to absorption of laser radiation. *Journal of Applied Physics, 36*, 462–468.

Rizvi, N. H. (1999). Production of novel 3D microstructures using excimer laser mask projection techniques. *Proceedings of SPIE, 3680*, 546–552.

Sankur, H. O., Motamedi, E., Hall, R., Gunning, W. J., & Khoshnevisan, M. (1995). Fabrication of refractive microlens arrays. *Proceedings of SPIE, 2383*, 179–183.

Srinivasan, R., Hall, R., Loehle, W., Wilson, W., & Allbee, D. (1995). Chemical transformations of the polyimide Kapton brought about by ultraviolet laser radiation. *Journal of Applied Physics, 78*, 4881–4887.

Srinivasan, R., Braren, B., Dreyfus, R., Hadel, L., & Seeger, D. (1986). Mechanism of the ultraviolet laser ablation of polymethyl methacrylate at 193 and 248 nm: laser-induced fluorescence analysis, chemical analysis, and doping studies. *Optical Society of America B, 3* (5), 785–791.

Tien, C. H., Chien, Y. E., Chiu, Y., & Shieh, H. D. (2003). Microlens array fabricated by excimer laser micromachining with gray-tone photolithography. *Japanese Journal of Applied Physics, 42*(Part 1, No. 3), 1280–1283.

Tsutsumi, N., & Kiyotsukuri, T. (1988). Measurement of thermal diffusivity for polymer film by flash radiometry. *Applied Physics Letters, 52*, 442–444.

Zimmer, K., Braun, A., & Bigl, F. (2000). Combination of different processing methods for the fabrication of 3D polymer structures by excimer laser machining. *Applied Surface Science, 154–155*, 601–604.

Studies on CO_2 Laser Micromachining on PMMA to Fabricate Micro Channel for Microfluidic Applications

Rishi Kant, Ankur Gupta and S. Bhattacharya

Abstract Microfluidic devices are highly commonplace in the field of biomedical technology, point of care diagnostics and chemical analysis. The rapid and low cost manufacturing of these devices have always been a challenge. CO_2 laser micromachining has played an important role in micro-machining of devices at a scale similar to the microfluidic devices although it renders the machined surfaces with high surface roughness. The chapter reports an initiative to do process optimization of laser micromachining technique for producing smooth machined surfaces in the micro scale devices. The chapter discusses the impact of process parameters like raster speed, laser power, print resolution etc. and its optimization using two target functions of dimensional precision and surface roughness on micro-channels made in PMMA (Poly methyl metha acrylate) substrates. The laser machined PMMA samples are analyzed using 3D-profilometry and Field emission scanning electron microscope (FESEM) for surface quality and dimensional precision. To investigate optimum process parameters of CO_2 laser for fabricating the micro-channel on PMMA with dimensional accuracy and good surface quality, Analysis of variance (ANOVA) and regression analysis is conducted. It is found that optimum surface roughness of this process is around 7.1 μm at the optimum value of the process parameters 7.5 mm/s (50 % of maximum machine limit) raster speed, 17.9 W (51 % of maximum machine limit) laser power and 1200 DPI (100 % of maximum machine limit) printing resolution. The static contact angle of the micro-machined surface has also been observed for analyzing the amenability of these channels to flow of water like fluids for micro-fluidic applications. The chapter also covers a

R. Kant (✉) · A. Gupta
Microsystems Fabrication Laboratory, IIT, Kanpur 208016, India
e-mail: dsrishikant@gmail.com

A. Gupta
e-mail: drankur@iitk.ac.in

R. Kant · A. Gupta · S. Bhattacharya
Mechanical Engineering Department, IIT, Kanpur 208016, India
e-mail: bhattacs@iitk.ac.in

S.N. Joshi and U.S. Dixit (eds.), *Lasers Based Manufacturing*,
Topics in Mining, Metallurgy and Materials Engineering,
DOI 10.1007/978-81-322-2352-8_13

review of work done by various researchers in which they developed different methodology for successful manufacturing of microfluidic devices by employing CO_2 laser micromachining.

Keywords PMMA · CO_2 laser · Micromachining · Contact angle · Microfluidics

1 Introduction

Polymeric materials are utilized for fabricating microfluidic devices as applied to diagnostics and detection principally due to their ease in fabrication, biocompatibility, optical clarity and easy conformability to features of up to micro/nano-meter scale. Microfluidic devices made in polymeric devices find wide applications in the health care sector. The high level of interest and participation of the industry in microfluidic research and development illustrate the commercial value of these devices for important and diverse applications. Thus microfluidics is poised to become one of the most dynamic segments of the MEMS/NEMS technology.

Microfluidic devices explore the envisioned applications in confined space of several micron ranges. This micron sized space further can act as a micro-channel, micro-mixer, micro pumps, micro valves and several other active or passive devices which are commonly fabricated using expensive advanced lithography processes or other processes borrowed from the microelectronic industry. Lasers are already widely used to do laser lithography and sometimes 3-D lithography although their efficacy in micro-machining of microfluidic devices is not very widely explored. Laser and some other non-traditional machining processes are being increasingly explored for a high throughput fabrication of microfluidic devices. Among all these processes the laser machining process holds a lot of promises due to its flexibility towards parameter selection, its applicability on a wide range of materials and its rapidity of manufacture. We have successfully demonstrated the utility of CO_2 laser in micromachining of Poly Methyl Methacrylate (PMMA) devices. However, the post machined surface obtained through this process is highly non-uniform and poses a variety of restrictions on the devices regarding the flow of soft biological matter through the devices so fabricated. For example, channels used for detection and counting of mammalian cells may need a very low level of roughness to prevent lysis of the analytes. Also, high level of roughness sometimes induces local eddies and vortices very close to the rough surface leading to nonspecific adsorption of soft biological species and hinders the process of re-usability of the devices and very often gives out false positive signals in detection protocols. In addition cell adhesion, non-selective protein adsorption, particle/cell aggregation, bubble generation, optical signal detection etc. is range of problems posed by rough surfaces associated with microfluidic devices.

In earlier work Lawrence and Li (2001), Nayak et al. (2008), Heng et al. (2006), Davim et al. (2008a, b), Snakenborg et al. (2004) and a variety of other authors have investigated the roughness of CO_2 laser machined samples of PMMA and have been able to obtain roughness in the range of few microns (Lawrence and Li 2001). While applying such methodology to fabricate the features and structures of micro-fluidic devices we were highly limited in efficacy of the devices to handle soft biological fluids by virtue of the channel roughness (Kant et al. 2013). The range of roughness that has been reported in earlier work is not enough for the specific kind of microfluidic applications as discussed and therefore we were enabled to look into the process optimization aspects so that we could achieve less than 10 microns roughness and we have been quite successful and have produced 7 microns average roughness and a high level of dimensional precision by repeated process optimization and also making a hybrid step to perform surface levelling.

In the past, to improve cutting quality (Davim et al. 2008a, b) and minimization of bulges (Li et al. 2009) during machining has been carried out by various researchers. Chung and Lin (2011) reported a microchannel fabrication strategy which reduces bulges and feature size using foil assisted CO_2 laser micromachining. Their technique is able to reduce the defects such as bulges formation, clogging, resolidification and splashing phenomena in CO_2 laser micromachining of PMMA. They could reduce channel width up to 75 μm or less using this process. Li et al. (2009) developed a technique "two times of laser cutting" to eliminate the bulges near the wall of micro channel fabricated by CO_2 micromachining. They also proposed a relationship between laser process parameter and bulge height which formed during machining. They concluded that if ratio of laser power and scanning velocity of laser beam increases bulge formation height increases. Davim et al. (2008a, b) studied the cutting quality of PMMA substrate during CO_2 micromachining. They proposed that Heat affected zone (HAZ) increases with increase in laser power and decreases with increase in cutting velocity while the surface roughness increases as laser power decreases and increases of cutting velocity. Huang et al. (2010) proposed preheating of PMMA substrate at 85 °C during machining to reduce surface roughness of the chamber sidewall. However it is also required to optimize process parameter for laser ablation on PMMA which is being presented in the current work.

1.1 Theoretical Background

The first CO_2 laser was developed at Bell Labs, United States by Patel in 1964 which produced a light beam with an infrared long wavelength band at around 9.4–10.6324 μm range. We have explored the common laser heating process assuming a circular well focussed beam on the surface and tried to infer from a commonly used one dimensional heat equation to calculate the needed beam intensity to produce a thermal ablation in PMMA surface.

The laser power density 'I' varies as it propagates inside PMMA (in 'Z' direction) and follows the Beer–Lambert law which is as

$$I = I_0 \mathrm{e}^{-\mu z} \tag{1}$$

where I_0 is initial Power density of laser beam, μ is spectral linear absorption coefficient (cm^{-1}) and z is depth. Figure 1 shows a figure schematic of the circular lasing action on the PMMA substrate.

Yuan and Das (2007) reported a mathematical model for predicting the depth and shape of micro machined channels on PMMA substrate. The depth of channel is predicted by balancing laser irradiation energy to sum of conduction energy within PMMA and energy required to break the polymer chain (i.e. conversion into monomers). The model predicts the depth of melting temperature as H(y) which is the depth of ablation of the channel at a location 'y'. As the spot size is assumed to be circular the 'x' coordinate automatically gets defined once the 'y' is explicitly stated.

The mathematical model is based on following assumptions:

(i) The laser beam intensity distribution has been assumed to be Gaussian type in three dimensional space as,

$$I(x, y, z) = \frac{P}{\pi w^2(z)} \mathrm{e}^{-\left[\left(x^2+y^2\right)/w^2(z)\right]} \tag{2}$$

$$w^2(z) = R^2\left[1 + \left(\lambda z/2\pi R^2\right)^2\right] \tag{3}$$

where P is power of laser, $w(z)$ is radius of laser beam at distance z from focal waist, λ is wavelength and R is radius of laser beam.
Assuming $\lambda z/2\pi R^2 \ll 1$ makes Eq. (3) of constant radius laser beam

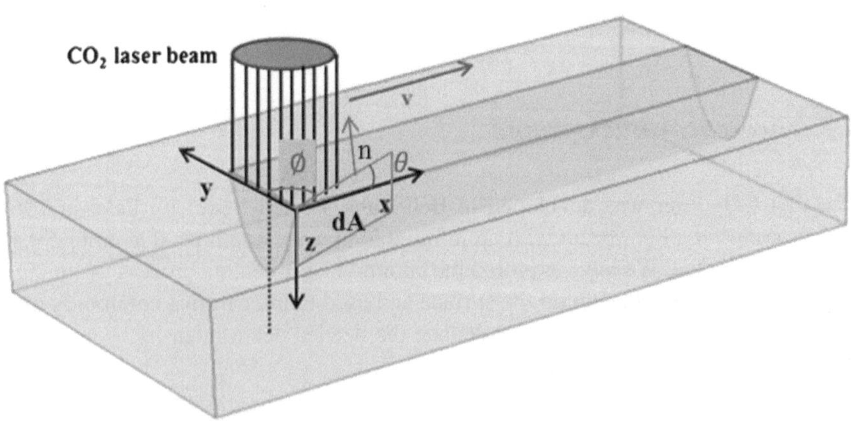

Fig. 1 CO_2 laser beam travelling with velocity v in +x direction

(ii) Energy losses due to conduction and radiation is negligible.
(iii) Converted monomers from PMMA will affect laser beam.

Heat conduction in PMMA substrate can be expressed as

$$\frac{\partial^2 T}{\partial x^2}+\frac{\partial^2 T}{\partial y^2}+\frac{\partial^2 T}{\partial z^2}-\frac{1}{\alpha}\frac{\partial T}{\partial t}=\frac{1}{k} \tag{4}$$

where α is thermal diffusivity, k is thermal conductivity and c_p is the specific heat.

Since, $\alpha = k/c_p$where T is the temperature at point (x, y, z) and at time t.

Due to high heat absorption coefficient of PMMA, Eq. (4) can be written by considering the direction of propagation of the laser beam (Dinger et al. 1993).

Choi and Chryssolouris (1995) proposed that three dimensional micro channel can be divided into finite surface elements which can be expressed by inclination θ in x direction and ϕ in y direction.

So the temperature distribution at surface can be achieved by solving Eq. (4) which is as

$$T(t)|_{z=0}=\frac{2I}{k}\sqrt{\frac{\alpha t}{\pi}} \tag{5}$$

If we consider energy balance for each element than input laser irradiation energy is equal to sum of energy conducted by surface elements and energy which make element into volatile products.

$$E_{laser}\, dxdy = E_{conduction}\, dA + E_{decompostion}\, dxdy \tag{6}$$

By geometry

$$dA=\sqrt{1+\tan^2\theta+\tan^2\ \phi}\, dx\, dy \tag{7}$$

The laser input energy density can be expressed as

$$E_{laser}=\int_{-\infty}^{+\infty} I(x,y,z)\frac{dx}{v} \tag{8}$$

Putting the value of $I(x, y, z)$ from Eq. (2)

$$E_{laser}=\int_{-\infty}^{+\infty}\frac{P}{\pi v w^2(z)}\mathrm{e}^{-\left[\left(x^2+y^2\right)/w^2(z)\right]}dx \tag{9}$$

On evaluation of above integral

$$E_{laser}=\frac{aP}{\sqrt{\pi}Rv}*\mathrm{e}^{-\frac{y^2}{R^2}} \tag{10}$$

The energy required to decompose PMMA into volatile MMA can be expressed as

$$E_{decomposition}\, dxdy = \rho LD(y)dxdy \tag{11}$$

Since laser beam is moving with v velocity in x direction, for this heat transfer can be expressed as

$$\nabla^2 T = \frac{\partial^2 T}{\partial n^2} = \frac{v}{\alpha}\frac{\partial T}{\partial x} \tag{12}$$

Since

$$\frac{\partial T}{\partial n} = \sqrt{\left[\left({}^{\partial T}/_{\partial x}\right)^2 + \left({}^{\partial T}/_{\partial y}\right)^2 + \left({}^{\partial T}/_{\partial z}\right)^2\right]} \tag{13}$$

Substituting above equation into

$$\frac{\partial^2 T}{\partial n^2} = \frac{v \tan\theta}{\alpha\sqrt{1 + \tan^2\theta + \tan^2\phi}}\frac{\partial T}{\partial n} \tag{14}$$

The energy which goes into conduction can be calculated as

$$E_{conduction}\, dA = \int_{-\infty}^{+\infty}\left(-k\left(\left(\frac{\partial T}{\partial n}\right)|_{n=0}\frac{dx}{v}\right)dA\right. \tag{15}$$

Keeping the following boundary condition in Eq. (15)

$$\begin{aligned} T &= T_v \quad \text{at} \quad n = 0 \\ T &= T_0 \quad \text{at} \quad n = \infty \\ \left(\frac{\partial T}{\partial n}\right)|_{n=0} &= \frac{v \tan\theta}{\alpha\sqrt{1 + \tan^2\theta + \tan^2\phi}}(T_v - T_0) \end{aligned} \tag{16}$$

Now

$$E_{conduction}\, dA = \left[\int_{-\infty}^{+\infty} \rho\, c_p (T_v - T_0) \tan\theta dx\right] dxdy$$

$$E_{conduction}\, dA = \rho\, c_p (T_v - T_0) D(y) dxdy \tag{17}$$

where

P is the power of the laser beam
R is radius of laser beam
V is velocity with which laser beam travels in x direction
a is the absorptance of PMMA at the CO_2 laser wavelength of 10.6 μm
ρ is the density of PMMA
L is the latent heat of decomposition from PMMA to MMA
T_v, T_0 is decomposition and room temperature respectively.

The depth of channel is predicted by balancing laser irradiation energy to sum of conduction energy within PMMA and energy required to break the polymer chain (i.e. conversion into monomers).

Using Eqs. (6), (10), (11) and (17) and solving we get following

$$H(y) = \frac{1}{\sqrt{\pi}} \frac{aP}{Rv\rho\left[L + c_p\left(T_v - T_0\right)\right]} e^{-\frac{y^2}{R^2}} \tag{18}$$

If $H_0 = \frac{1}{\sqrt{\pi}} \frac{aP}{Rv\rho\left[L + c_p(T_v - T_0)\right]}$

Equation (18) gives a Gaussian profile for the depth of micro channel

$$H(y) = H_0 e^{-\frac{y^2}{R^2}} \tag{19}$$

Equation (19) is valid only for higher laser power at which entire underlying area of PMMA could be ablated. For low power, laser energy will not be sufficient to melt and vaporize underlying material so radius of beam (R) is not constant.

For this case depth of micro channel can be predicted as

$$H(y) = H_0 e^{-\frac{y^2}{y_{th}^2}} \tag{20}$$

where

$$y_{th} = R\sqrt{\ln \frac{2P}{\pi R v F_{th}}} \tag{21}$$

where

$$F_{th}\ (\text{threshold fluence for ablation}) = I_s t_s \tag{22}$$

where I_s is given laser power density and t_s is laser irradiation time.

1.2 Mechanism of Material Removal During CO_2 Laser Micromachining of PMMA

The molecular structure of PMMA is shown in Fig. 2 which is a type of thermoplastic. The material removal takes place by heating, melting and decomposition steps.

The decomposition of PMMA into volatile monomers (MMA) takes place at 360 °C. The decomposition reaction has been very well understood by Lippert et al. (1999) and Fig. 3 illustrates the decomposition pathway of the PMMA.

2 Experimental Set-Up

We have performed all our Laser micromachining using EPILOG WIN32 laser machine (M/S Epilog laser, USA) in laboratory conditions. All machining is done at (25 °C) room temperature condition and a positive pressure environment. The laser that we use has a maximum power of 35 W and has a total working bed size of 2 ft × 1 ft. A design of the device that needs to be laser machined is drafted using a computer assisted drawing package (Coral Draw) which is used to drive the various stepper motors with the lasing stage and also the working stage. The working stage is capable of movement in the '*z*' direction and the lasing stage can operate in the raster (engraving) and vector (deep cutting) modes in the x-y plane. Figure 4 shows schematic representation of laser machining set up.

2.1 Micromachining of PMMA Substrate

PMMA is highly absorptive at the infra-red region (wavelength = 10.6 μm). Laser machining operates on the physics of photon to phonon conversion where beam matter interaction results in rapid molecular vibrations leading to generation of thermal energy. The local intensity posed by the laser is so intense (being in a very

Fig. 2 PMMA molecule's structure

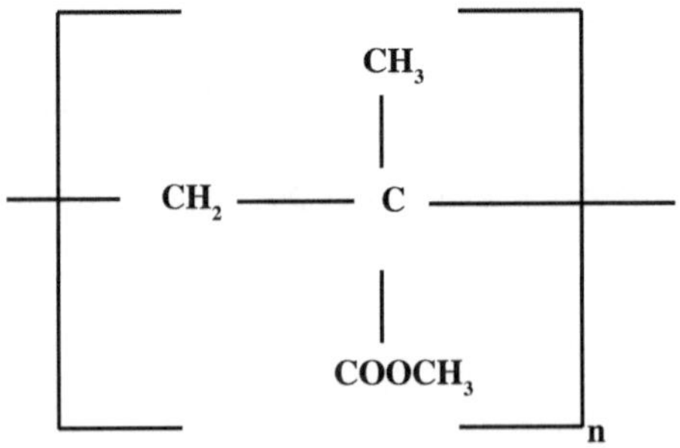

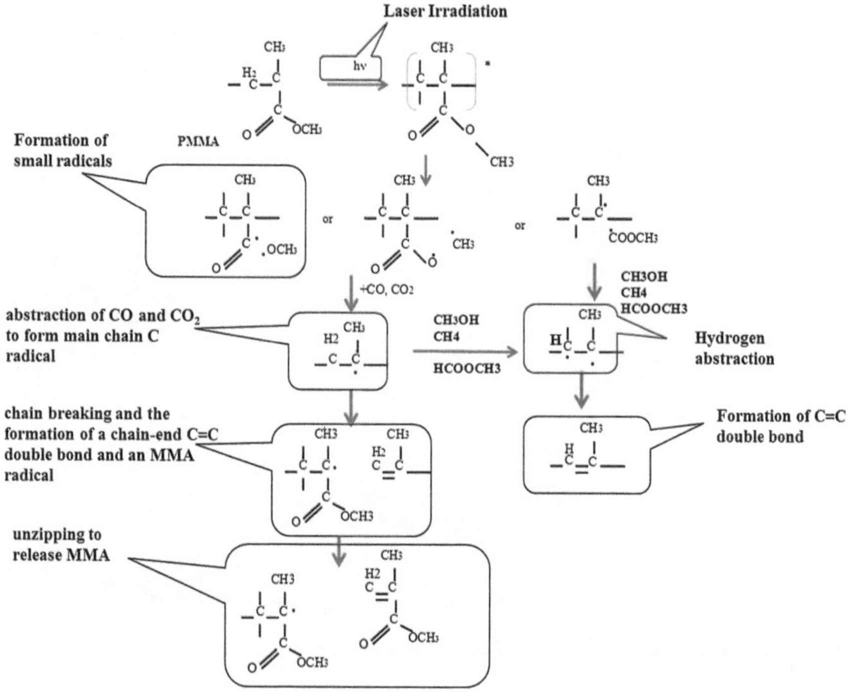

Fig. 3 Photolysis of PMMA's ester side chain

small zone) that it results in energy that is sufficient to melt and vaporize the PMMA material. The machining parameters for the laser are highly dependent on the overall feature size and correct choice of parameters leads to good dimensional accuracy. The various combinations of parameters resulting into different surface finish levels can be easily achieved using Laser processes (Nayak et al. 2008; Heng et al. 2006).

We have conducted around 30 trials with a combination of multiple parameters designed by Design Expert software. Although several trials did not scribe PMMA substrate, hence its surface roughness is not reported here and is denoted by N/A. Table 1 shows the various combinations that are proposed and the measured.

2.2 *Characterization of Micro-machined PMMA Substrate*

The surface roughness characterization is performed on 3D-optical profilemeter (NanoMap-D, AEP Technology, USA). Various channel widths varying from 10–100 microns have been evaluated for roughness. The interferometer has optical resolution in nanometer range. The minimum area that is scanned by the system is $0.064516\ mm^2$. The interferogram is used to reconstruct the topology of the surface

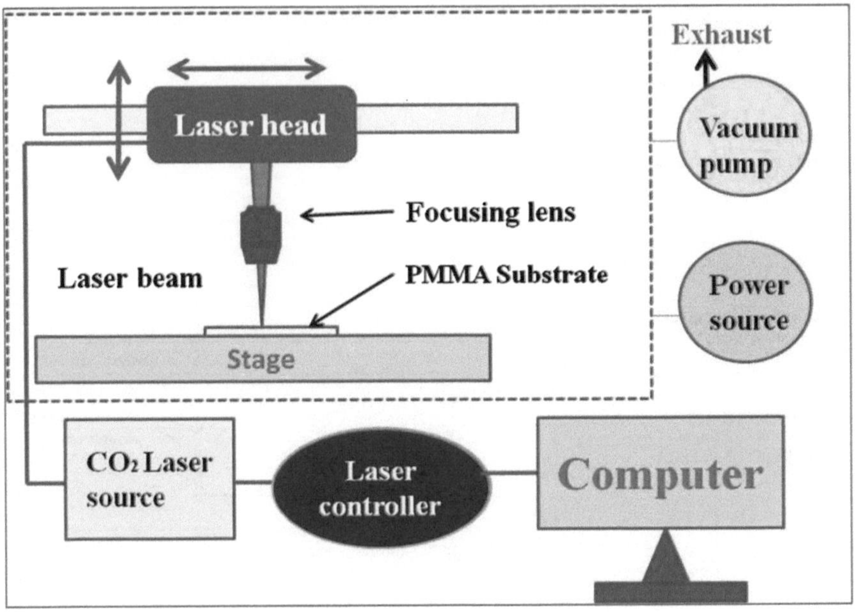

Fig. 4 Schematic diagram of CO_2 laser micromachining process

being scanned and the 3-D imaging is performed across the different micro-channel widths. The average roughness 'Ra' is reported to vary with the combination of different machining parameters as illustrated in Fig. 5a–j.

3 Optimization of CO_2 Laser Process Parameters for Smooth Surface

Optimization of laser machining parameters as discussed above has been performed with design of experiments (DOE) in which a central composite design (CCD) is used to fit a model. An optimum solution is extracted from the different machining parameters, including raster speed, resolution and power. Optimization was first carried out with respect to laser power and printing resolution, then with respect to speed and printing resolution and finally for speed and power. Figure 6a–c shows the output of this optimization process as contour plots. As can be seen in Fig. 6a the contour corresponding to a roughness level of 7.0647 μm starts at a base power of around 50 % and the value 7.0647 μm happens while printing at a resolution of 1200 DPI. We could not really explore a resolution more than 1200 DPI as we were limited by the machine specifications. There are of course more precise electro-mechanical stages particularly those used in scan jet printers which may be able to print at a resolution of as high as 5000 DPI and it intuitively makes sense that if the

Table 1 Details of designed parameters and measured roughness values

Resolution (DPI)	Power (%)	Speed (%)	Roughness (μm)
600	50	51	11.113
75	10	100	N/A*
1200	10	10	N/A
600	50	51	11.113
75	10	10	N/A
600	10	51	N/A
600	50	51	11.113
600	50	51	11.113
75	10	10	N/A
1200	100	10	22.003
75	100	100	N/A
600	50	51	11.113
1200	100	100	8.7607
600	50	51	11.113
600	50	100	14.461
600	50	10	N/A
75	100	10	N/A
600	50	51	11.113
75	100	100	N/A
1200	10	100	N/A
600	100	51	22.014
75	10	100	N/A
1200	10	100	N/A
75	50	51	N/A
1200	50	51	7.0647
1200	10	10	N/A
75	100	10	N/A
600	50	51	11.113
1200	100	10	N/A
1200	100	100	8.7607

Where N/A* value not available, as laser did not scrib the PMMA substrate

resolving power of the lasing head is higher than it would be able to handle higher power values while sacrificing the roughness level not too much. Similarly from Fig. 6b the minimum roughness attainable at 1200 DPI resolution limit of the system is corresponding to a speed that is 50 % of the maximum limit of the system if a similar roughness level were to be contained in Fig. 6a earlier for a combinatorial of the power as well. This gets further validated by Fig. 6c which reports a contour of 7.0647 μm average roughness corresponding to a power value of 51 % of the maximum limit of the system while maintaining the 1200 DPI level.

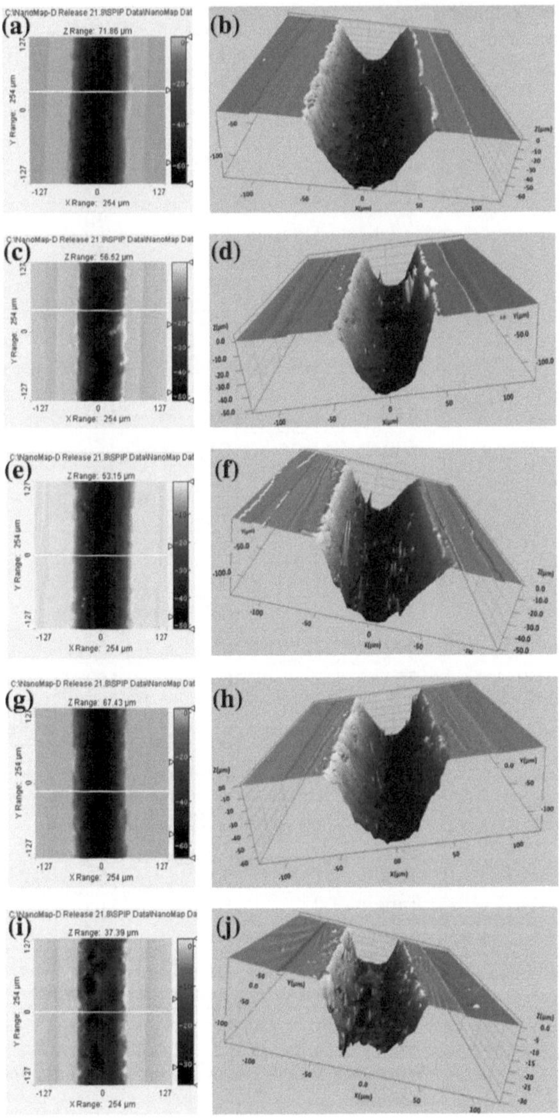

Fig. 5 **a**, **b** Plan view of the reconstructed image of the microchannel/3-Dimensional view (cutting parameters 1200 DPI, 17.5 W power and 7.65 mm/s rastering speed, Ra = 7.0647 μm). **c**, **d** Plan view of the reconstructed image of the microchannel/3-Dimensional view (cutting parameters 1200 DPI, 35 W power and 15 mm/s rastering speed, Ra = 8.7607 μm). **e**, **f** Plan view of the reconstructed image of the microchannel/3-Dimensional view (cutting parameters 600 DPI, 17.5 W power and 7.65 mm/s rastering speed, Ra = 11.113 μm). **g**, **h** Plan view of the reconstructed image of the microchannel/3-Dimensional view (cutting parameters 600 DPI, 17.5 W power and 15 mm/s rastering speed, Ra = 22.0114 μm). **i**, **j** Plan view of the reconstructed image of the microchannel/3-Dimensional view (cutting parameters 600 DPI, 17.5 W power and 15 mm/s rastering speed, Ra = 14.463 μm)

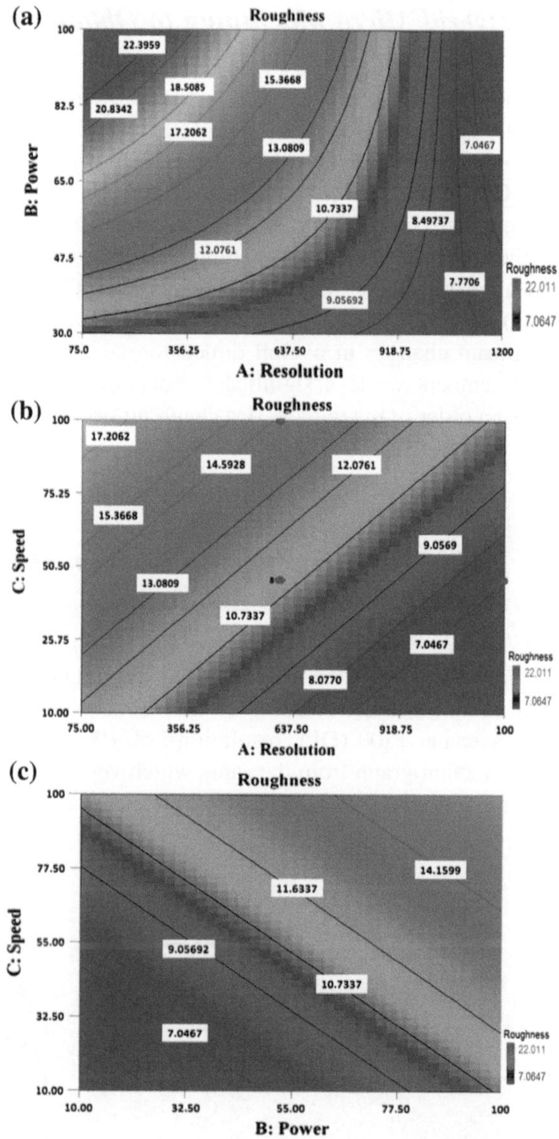

Fig. 6 **a** Contour plot of power and resolution for roughness. **b** Contour plot of speed and resolution for roughness. **c** Contour plot of power and speed for roughness

Therefore, we find out the optimized average surface roughness as approximately 7.06 microns and also through analysis report the speed to be 50 % of machine limit and power to be 51 % of the machine limit.

3.1 Hybrid Micro Machining to Obtain Smooth Surface

Hybrid machining strategy has been incorporated with a first step of lasing the PMMA surface followed by a wet etching of the Laser scribbed PMMA micro-channel. The wet etching is carried out by using a mixture of Toluene (M/s S D Fine-Chem limited, Mumbai, India, CAS no. 108-88-3) and Methanol (M/s Loba Chemie Pvt. Ltd., Mumbai, India, CAS no. 67-56-1) in a Volume ratio of 1:4. Further samples from laser etching step are immersed totally within this solution and ultrasonicated at 42 kHz frequency (M/s Citizen Scale (I) Pvt. Ltd., Mumbai, India) inside a laminar flow hood for 15 min at room temperature. There were no significant changes in overall dimensions including depth and width of the channels/chambers while a significant reduction in the average roughness values to almost a order of magnitude. (Ra changing from 7–22 μm to 270–860 nm) as can be seen in Fig. 7.

3.2 Field Emission Scanning Electron Microscope (FESEM) Imaging of PMMA Substrate

We have also conducted field emission scanning electron microscopy (FESEM)on a laser micro-machined samples which was laser exposed using optimized machining parameter i.e. 1200 (DPI, resolution), 50 (% power) and 51 (% speed). Figure 8 shows a micrograph from the same which very clearly shows the impression left by melting and recondensation of PMMA material during machining.

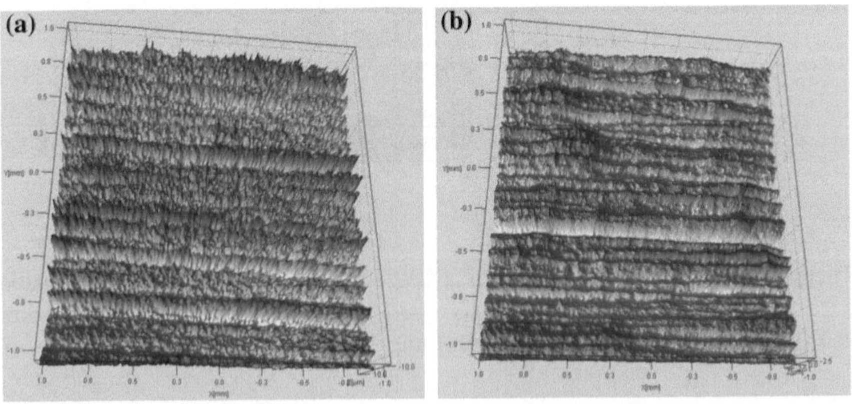

Fig. 7 **a** Reconstructed surface image of a post laser scribbed PMMA surface using optical profilometer (Ra value of 7 microns). **b** Reconstructed surface image of a post wet etched PMMA surface (Ra value of 860 nm)

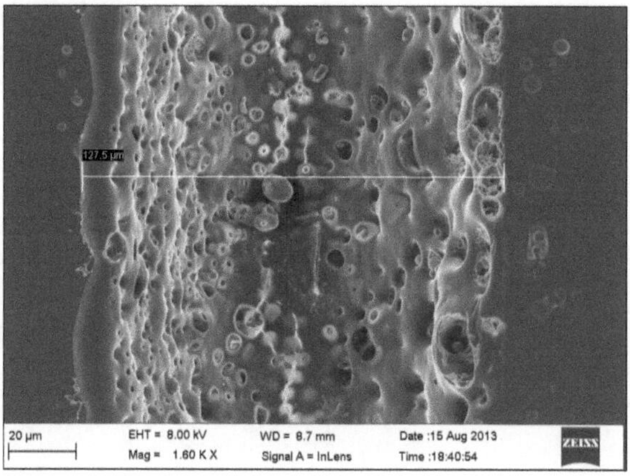

Fig. 8 FESEM image of micromachined PMMA substrate

4 Wettability Measurements for Microchannels

One of the main difficulties faced in working with polymeric platforms in microfluidics is the inability of the polymeric devices to drain out the fluids entrapped along embankments (Kant et al. 2013). This is more of a problem in particularly devices which actively control microscale flows like micro-pumps or valves. Therefore, it becomes highly desirable in applications involving devices whose one or more dimensions are more than 100 microns to have some kind of repulsion to water based which most of the times find useful as carrier fluids in microfluidic biodevices, Therefore the wettability characteristics become absolutely important in many situations and it is worthwhile to investigate the micro-machined channels in terms of the surface wettability. On doing such characterization we have found the surfaces whose roughness goes down to an average of 7.0647 µm owing to the lasing process and also to the Hybrid step for leveling of the surface that the static contact angle as measured on a Goniometer (M/S Dataphysics Instruments,

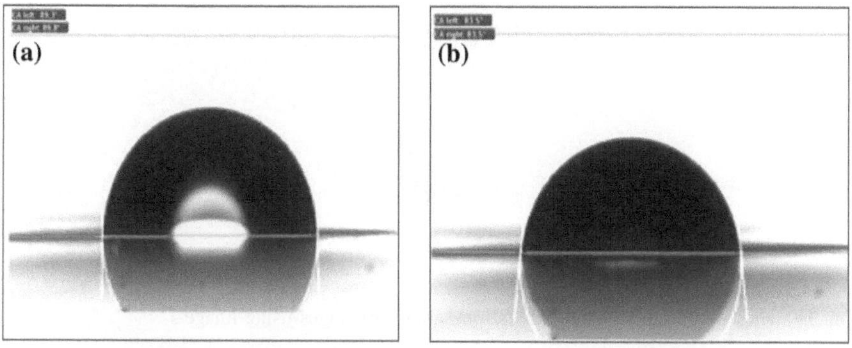

Fig. 9 **a** Contact angle 89.3° with water. **b** Contact angle 83.5° with glycerol

Germany) is quite high of the order of 83–89° with fluids like water and glycerol (see Fig. 9). Thus the micro-machined surfaces can be considered as quite hydrophobic and therefore would find applicability in microfluidic device realization.

S. No.	Original image from UV Transilluminator	Processed image using Image J
1	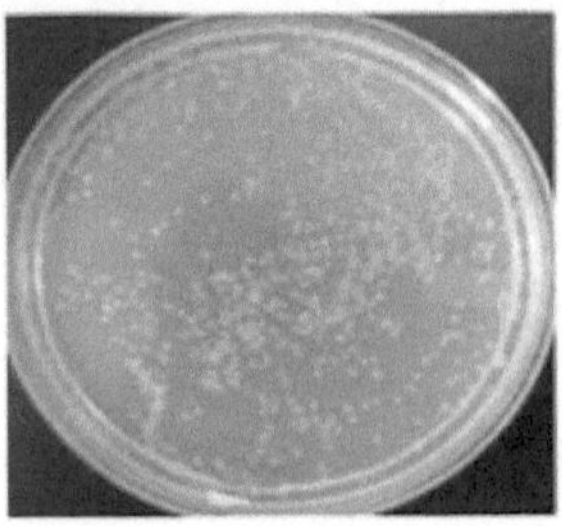	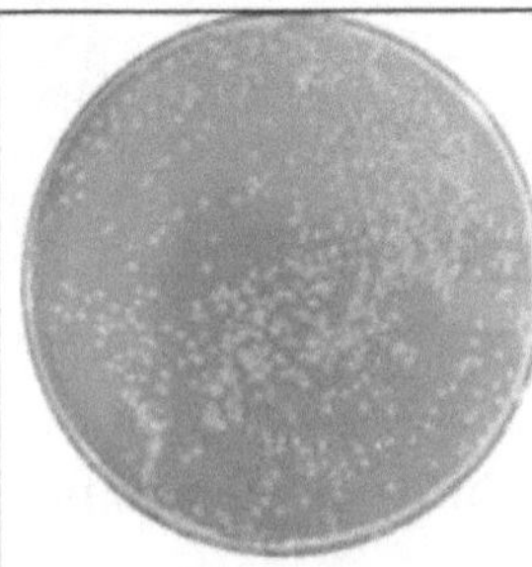 No of colonies 402
2	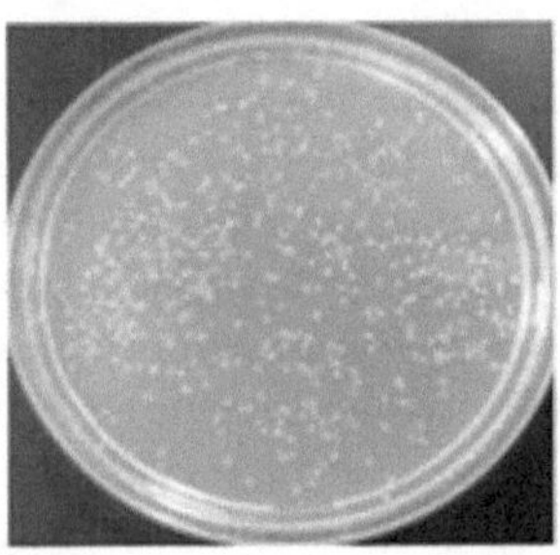	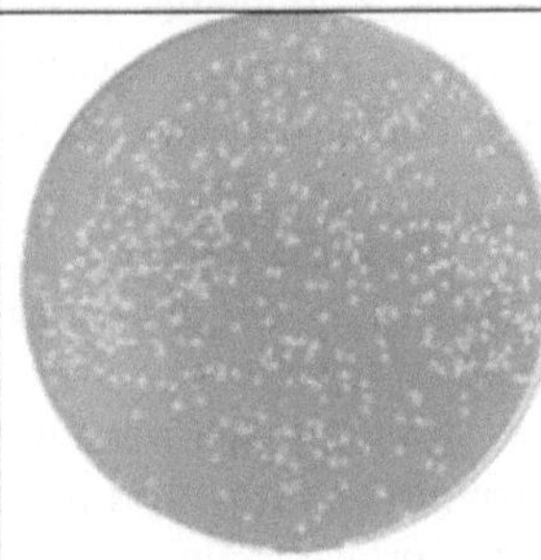 No of colonies 398
3	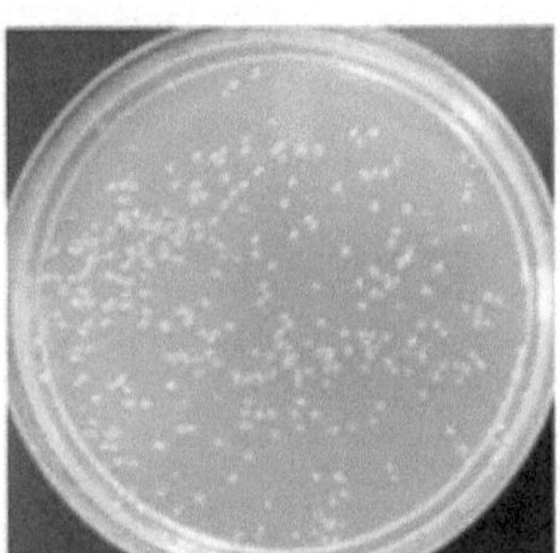	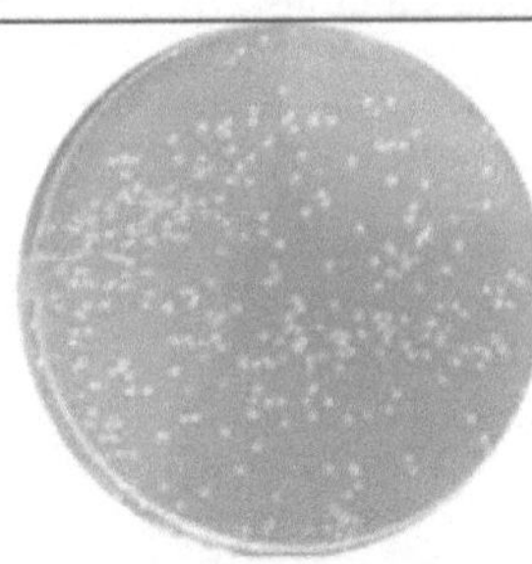 No of colonies 386

Fig. 10 The counting of cells at pre and post transport stages using image-J

4.1 *Bacterial Cell Viability Studies on Surface Obtained by Hybrid Micromachining*

We have also tested the PMMA surface achieved by hybrid micromachining for micro-organisms viability. We have developed piezo based micro pump for transporting bacterial cell and various experiments are conducted. The initial sample diluted to a concentration of 10^3 cfu/ml was plated using brain heart infusion (BHI) agar. The culture was diluted, plated, using transilluminator (M/s Bangalore genie) and cell counting is done using Image-J software. The initial sample recorded a count of 402 colonies and then recorded the viable colonies to be within 386–398 for three subsequent passes through the micro pumping architecture using the piezo actuator to transport the sample solution as shown in Fig. 10.

5 Conclusion

PMMA micromachining using CO_2 laser is presented and channel depth profile, reaction mechanism of material removal etc. are reviewed by thorough literature survey. We further demonstrate the basic thematic of this work by looking at the various optimization modalities which can normally be used for such a machining process in context of microfluidic device realization particularly in case of those devices which are widely used for handling devices for soft biological materials, We have optimized CO_2 laser process machining parameter (raster speed, power and resolution) to achieve smooth surface for microfluidic applications. We have further coupled this to a hybrid leveling chemical process where the overall roughness can be taken down to 7.04 μm average roughness at optimized lasing parameters with a resolution of 1200 DPI an average power of 50 % of maximum power amounting to 17.5 W and rastering speed of 51 % of machine scale amounting to 7.65 mm/s. The wettability of optimized smoothen surface is also measured using goniometer and found out to be 89° with water and 83.5° with glycerol and further analysis of viability of bacterial cells are being performed to test the claims made related to the ability of microfluidic devices to handle biological soft matter.

References

Choi, W. C., & Chryssolouris, G. (1995). Analysis of the laser grooving and cutting processes. *Journal of Applied Physics, 28*, 873–878.

Chung, C. K., & Lin, S. L. (2011). On the fabrication of minimizing bulges and reducing the feature dimensions of microchannels using novel CO_2 laser micromachining. *Journal of Micromechanics and Microengineering, 21*(06), 5023.

Davim, J. P., Barricas, N., Marta, C., & Oliveira, C. (2008a). Some experimental studies on CO_2 laser cutting quality of polymeric materials. *Journal of Materials Processing Technology, 198*, 99–104.

Davim, J. P., Oliveira, C., Barricas, N., & Conceição, M. (2008b). Evaluation of cutting quality of PMMA using CO_2 lasers. *International Journal of Advanced Manufacturing Technology, 35*, 875–879.

Dinger, C., Sterkenburgh, T., Holler, T., & Franke, H. (1993). *Nonconducting Photopolymers and Applications* (pp. 278–287). San Diego, Bellingham: SPIE.

Heng, Q., Tao, C., & Zho, T. (2006). Surface roughness analysis and improvement of micro-fluidic channel with excimer laser. *Microfluidics and Nanofluidics, 2*, 357–360.

Huang, Y., Liu, S., Yang, W., & Yu, C. (2010). Surface roughness analysis and improvement of PMMA-based microfluidic chip chambers by CO_2 laser cutting. *Applied Surface Science, 256*, 1675–1678.

Kant, R., Singh, H., Nayak, M., & Bhattacharya, S. (2013). Optimization of design and characterization of a novel micro-pumping system with peristaltic motion. *Microsystem Technologies, 19*, 563–575.

Lawrence, J., & Li, L. (2001). Modification of the wettability characteristics of polymethyl methacrylate (PMMA) by means of CO_2, Nd: YAG, excimer and high power diode laser radiation. *Materials Science and Engineering A, 303*, 142–149.

Li, J. M., Liu, C., & Zhu, L. Y. (2009). The formation and elimination of polymer bulges in CO_2 laser microfabrication. *Journal of Materials Processing Technology, 209*, 4814–4821.

Lippert, T., Webb, R. L., Langford, S. C., & Dickinson, J. T. (1999). Dopant induced ablation of poly(methyl methacrylate) at 308 nm. *Journal of Applied Physics, 85*(3), 1838–1847.

Nayak, N. C., Lam, Y. C., Yue, C. Y., & Sinha, A. T. (2008). CO_2-laser micromachining of PMMA: The effect of polymer molecular weight. *Journal of Micromechanics and Micro Engineering, 18*(09), 5020.

Patel, C. K. N. (1964). Continuous-wave laser action on vibrational-rotational transitions of CO_2. *Physical Review, 136*, A1187.

Snakenborg, D., Klank, H., & Kutter, J. P. (2004). Microstructure fabrication with a CO_2 laser system. *Journal of Micromechanics and Micro Engineering, 14*, 182–189.

Yuan, D., & Das, S. (2007). Experimental and theoretical analysis of direct-write laser micromachining of polymethyl methacrylate by CO_2 laser ablation. *Journal of Applied Physics, 101*, 024901.

Energy Based Analysis of Laser Microchanneling Process on Polymethyl Methacrylate (PMMA)

Shashi Prakash and Subrata Kumar

Abstract CO_2 laser micromachining provides low cost machining solution for fabrication of three dimensional microfluidic channels on poly-methyl-methacrylate (PMMA). In this research work CO_2 laser microchanneling process has been analyzed from the first principle. Considering the Gaussian distribution of laser beam, an energy based model has been proposed to predict the microchannel depth and channel profile. For fabricating microfluidic devices, PMMA has emerged as a cheap alternative to many other costly materials like silicon, quartz etc. Its material properties like absorptivity and thermal properties have been investigated. In order to physically verify the proposed model, experiments have been performed on a 3 mm thick PMMA sheet and actual and predicted results have been compared. Simultaneous TGA/DSC tests have been conducted to determine various thermal properties of PMMA. Since thermal conductivity of the PMMA is very low, the conduction loss has been neglected while developing the model. The proposed model successfully predicts the channel depth and profile without much loss of accuracy. energy based analysis has been found to be simple yet powerful method to predict the channel dimensions for low thermal conductivity materials.

Keywords CO_2 laser beam machining · PMMA · Microchanneling · Energy analysis

1 Introduction

The use of lasers in micro manufacturing is growing rapidly. The usual advantages of using laser in micromachining are its simplicity in operation and fast production rate. CO_2 laser machining is a noncontact and heat based material removal process

S. Prakash (✉) · S. Kumar
Mechanical Engineering Department, Indian Institute of Technology Patna, Patna, India
e-mail: spasthana@iitp.ac.in

S. Kumar
e-mail: subrata@iitp.ac.in

S.N. Joshi and U.S. Dixit (eds.), *Lasers Based Manufacturing*,
Topics in Mining, Metallurgy and Materials Engineering,
DOI 10.1007/978-81-322-2352-8_14

where lower surface roughness and higher dimensional tolerance can be achieved using low beam diameter and highly concentrated beam energy (Madic et al. 2012). Laser micro-machining is a material processing technique that employs lasers to induce managed vaporization to provide required micro scale geometrical shape and dimensional ablation. Despite the facts that laser machining is a technically complicated process, research work has allowed the fabrication of precise, smooth and clean components at high speed. Lasers do not employ traditional tool-fixture setups, which makes the whole process lot easier as well as less time consuming. With different wavelength ranges of available laser systems, one can possibly machine all kinds of materials. Moreover, apart from cutting, lasers can also do many other jobs like welding, drilling and forming. Owing to this, modern industries are now replacing the conventional machines with laser machining system. This also results in getting lower cost, much higher productivity and better surface quality (Zhou et al. 2004).

Miniaturization of analytical tools and devices has been an ongoing trend to improve the performance of analytical tools. Microdevices make many different types of analysis including chromatography, electrophoresis and DNA analysis possible. Such microdevices reduce sample consumption, cost, time to results and also yield better performance and portability. Microchannels are part of several microfluidic devices, microelectronics based devices and micro-electro-mechanical systems (MEMS). Microchannels are used in various bio-analytical devices, DNA synthesis, electrophoresis etc. The size specifications of these microchannels vary in different types of applications. Microchannels are generally defined as channels having any of the sizes in micron range (1–999 μm). Microchannels are generally fabricated with non-conventional manufacturing techniques unlike channels of higher size which can be fabricated using conventional fabrication techniques. Microchannels with high aspect ratios are used in most of the devices (Costano-Alvarez et al. 2008). However, microchannels with low aspect ratios are also used in several applications like particle separation devices, DNA analysis etc. (Russom et al. 2009).

In earlier times, silicon, glass and quartz have been the most preferred materials for microfluidic devices because of some of the favorable properties like optical transparency, high mechanical strength, high melting point, and well defined surface characteristics. However, producing such devices on these substrates has always been a costly affair. Producing microchannels on such substrates involve typically longer manufacturing processes like lithography, embossing and etching (Iliescu et al. 2012). The use of polymeric materials for microfluidic devices is a recent trend. Polymers offer several advantages over other materials in terms of lower cost, lighter weight and easier manufacturing. Thus they can be made portable and disposable (Qi et al. 2009).

In recent times, poly-methyl-methacrylate (PMMA) has been evolved as one of the most preferred polymer for many microfluidic devices (Prakash et al. 2014). PMMA possesses high absorbance in mid-infrared zone which makes it particularly suitable for CO_2 laser (Malek 2006). It has low heat capacity and low heat conduction. Therefore, all the energy transferred to material is utilized in rapid vaporization (Prakash et al. 2013). PMMA also offers high degree of optical transparency which makes it perfectly suitable for many microfluidic devices. PMMA has been especially useful in applications where high temperature is not involved because of its low thermal stability.

CO_2 laser ablation has been investigated by various authors over the years. Helebrant et al. (1993) developed one dimensional steady state heat conduction equation for CO_2 laser irradiation of glass. Julia et al. (1998) utilized the energy balance method for CO_2 laser ablation of biological tissues. They have also considered the effects of temperature on optical properties. Laser micromachining of PMMA has also been studied by Klank et al. (2002). They utilized the CO_2 laser for rapid fabrication of microfluidic structures on PMMA Snakenborg et al. (2004) analyzed the laser micro-machining process on PMMA using 1-dimensional heat conduction equation. Xiang et al. (2006) used the three dimensional transient heat conduction equations to model the CO_2 laser cutting of PMMA. They used finite element method to solve the model. Yuan and Das (2007) developed analytical model for low power CO_2 laser ablation of PMMA. The model included heat conduction as well as energy loss in decomposition of PMMA. Shulepov et al. (2010) proposed a thermo-chemical model for CO_2 laser ablation of PMMA and polymide. They studied the phenomenon of plasma generation with CO_2 laser irradiation. The model considers the plasma screening effect. They observed the presence of plasma pipe in CO_2 laser cutting at high intensities. However, most of the models employ computer simulation based numerical analysis to determine the channel profile and depth.

In this research work, CO_2 laser based microchanneling process has been analyzed based on simple energy balance method. Energy based modeling also gives physical insight into the laser micromachining process. Channel profile and depth have been determined utilizing this model. Implementation of finite element analysis or other numerical analysis based modeling is difficult as well as time consuming when compared to energy based modeling. This energy based model is simple in nature and can readily be applied in laser micromachining processes without much loss of accuracy for most of the polymers. This energy based modeling is especially suitable for materials having very low thermal conductivity.

2 Determination of Material Properties

PMMA is one of the most widely used amorphous thermoplastic materials. Because of its high optical transparency it is widely used in place of glass. It is also known by many other names like Plexiglass, Lucite, Perspex and Acrylite. PMMA is generally produced by polymerization process. Most of the properties of PMMA vary product to product because of large numbers of commercial compositions which are copolymers with co-monomers other than methyl methacrylate. The chemical structure of PMMA has been given in Fig. 1.

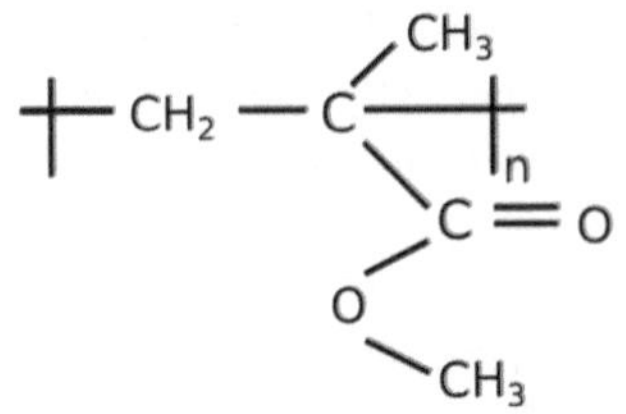

Fig. 1 Chemical structure of poly-methyl-methacrylate (PMMA)

PMMA behaves differently with different kinds of lasers according to their wavelength. Transparent PMMA generally transmits a major part of laser beam of near IR wavelength (1.06 μm). However, it shows a good absorption in ultraviolet and mid infrared range (10.6 μm). In order to determine the absorptivity of used PMMA specimen for CO_2 laser wavelength i.e. 10.6 μm, far-infrared (FIR) spectroscopic analysis has been performed using Shimazdu IRAffinity-1 spectrophotometry.

In this test light beams covering infrared spectrum are produced. These are based on the fact that the molecules of particular substance absorb specific frequencies. Two separate tests for determining absorptivity and reflectivity have been performed. The experiments have been performed under normal atmospheric conditions. These tests give the absorptivity and reflectivity values of used PMMA over a wide range of wavelengths. In this experiment cast acrylic (PMMA) of 3 mm thickness has been used. The transmissivity and reflectivity results are shown in Fig. 2. The reflectivity and transmissivity corresponding to 10.6 μm i.e. 943 cm^{-1} have been found to be 4.67 and 0.31 % respectively. Therefore it can be concluded that the used PMMA specimen absorbs nearly 95 % of CO_2 laser radiation. In other words, it can be assumed that 95 % of total CO_2 laser power is utilized in material processing, while remaining 5 % is lost. This amount is still very high when compared to laser interactions with many other materials. Therefore, with very little

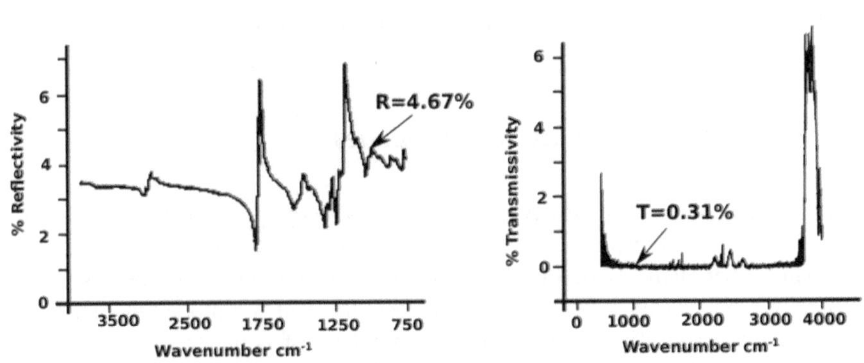

Fig. 2 Reflectivity and transmissivity test of PMMA

loss of total beam power, CO_2 laser can be assumed to be perfectly suitable for PMMA.

Determination of thermal properties is most important for understanding of CO_2 laser interaction of PMMA. In order to determine its various thermal properties, Simultaneous Thermogravitymetric analysis/Differential scanning calorimetry (TGA/DSC) test have been performed. The tests have been performed using STA 6000 (Perkin-Elmer).

Table 1 lists the thermal properties obtained using TGA/DSC tests. The heating of the sample has been performed at temperature increment rate of 10 °C per minute. An initial weight of 32 mg of the sample has been taken. TGA test reveals the thermal decomposition behavior of PMMA. TGA curve has been shown in Fig. 3 in black dashed line. PMMA starts turning into rubbery state at 110 °C. It starts melting at around 160 °C. However, both the processes do not involve significant latent heat. Sample starts vaporizing at 220 °C. At 377 °C, the highest rate of decomposition takes place. The sample gets completely vaporized at 393 °C. DSC curve has been shown in red in color. Area under the DSC curve represents the enthalpy of vaporization process. Some of the material properties have been taken from the available literature and been mentioned in this table. In this modeling, it has been assumed that material properties do not change with temperature.

In this research work, commercial CO_2 laser, (VLS 3.60, Universal Laser System Inc., USA), has been used (Fig. 4). The laser is quasi-continuous (QCW) wave in nature. The system is air-cooled and no other cooling aid has been used during the experiments. Laser head can move in X and Y direction while bed moves in Z-direction. The system is fully CNC controlled and runs through graphics based software like AutoCAD, Corel draw etc. The focal length of the focusing lens is 50 mm. Further details of CO_2 laser system has been provided in Table 2. The pulse frequency of this system can be varied by adjusting the speed and PPI (pulse per inch) parameter. The output power of the laser system remains constant over the time during "ON" period. Figure 5 shows the power time variation of the laser system. Laser beam can be focused to a very thin spot. Laser beam diameter has been determined on a 1 mm thick PMMA sheet as described by Powell (1998).

Table 1 Thermal properties obtained from simultaneous TGA/DSC

Property	Symbol	Value
Glass transition temperature	T_g	110 °C
Melting temperature	T_m	165 °C
Thermal decomposition starts at	T_{ds}	230 °C
Thermal decomposition ends at	T_{de}	393 °C
Specific heat	c_p	1.466 kJ/kg-K
Enthalpy of vaporisation	ΔH_v	2757 kJ/kg
Material density (Radice et al. 2012)	ρ	1070 kg/m^3
Thermal conductivity (Radice et al. 2012)	k	0.19 W/mK

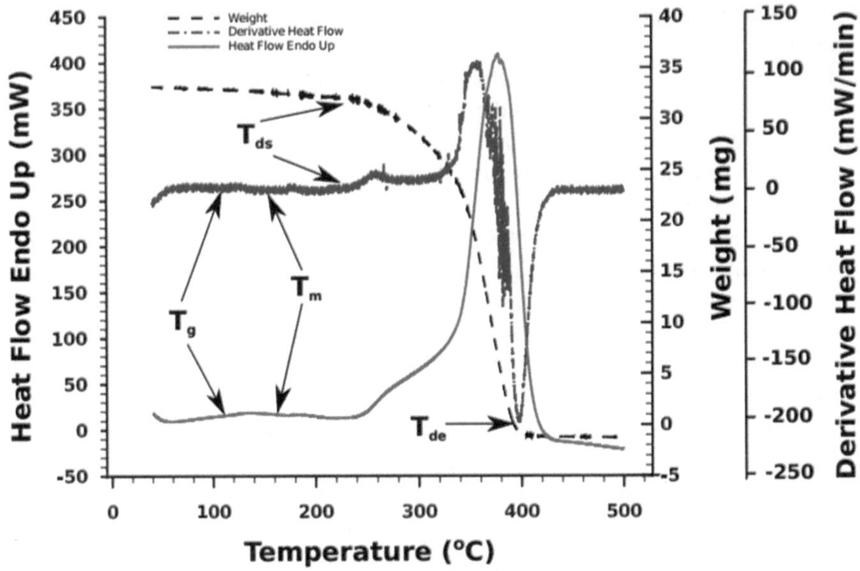

Fig. 3 Simultaneous TGA/DSC test results of PMMA

Fig. 4 Universal CO_2 laser systems

Table 2 CO_2 Laser system specification

Class type	IV
Max power	60 W
Wavelength	10.6 μm
Output beam shape	Gaussian
M^2 value	<1.5
Mode of operation	RF excited ON/OFF
Maximum working area	24″ × 12″

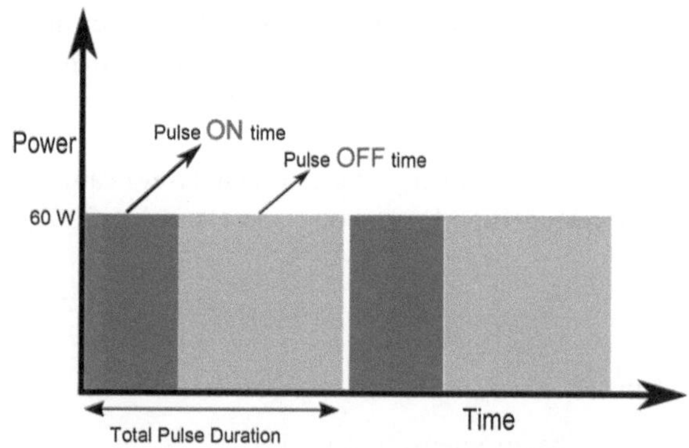

Fig. 5 Power-time variations for CO_2 laser

Average beam diameter has been found to be 237 μm. Also, beam periphery has been found to be very closely approximating the circle. Laser system can be allowed to work in two different modes namely raster mode and vector mode. In this work, all experiments have been performed in vector cutting mode of the laser system. All the experiments have been performed at focus point of the lens resulting in minimum beam spot diameter.

3 Energy Based Modeling of CO_2 Laser Microchanneling of PMMA

The thermal conductivity of PMMA is very low. As such the energy loss due to thermal conduction is also very small. Therefore, neglecting the conduction heat loss does not affect the modeling process to a significant extent. Also the heated zone is extremely small in a microchanneling process. The maximum temperature rise is also not significantly high. These factors make the convection and radiation loss to be negligible. Material properties inevitably change with temperature. However, since the vaporization temperature of the PMMA is small, these changes may also be assumed to be insignificant for making the modeling simpler and easy to understand. As the laser beam vaporizes the material, in the way they also interact with the vaporizing particles. Some part of the incoming beam energy gets absorbed by such particles. However, since the amount of vaporizing material is very small, the energy absorption in this process can also be assumed to be very small. Therefore, on the basis of the above discussion, following assumptions have been made in this energy based modeling;

1. There is no conduction heat loss into the substrate.
2. There is no convection and radiation loss.
3. The material properties do not change and remains constant throughout the process.

Let the laser beam moves through the center line in x-direction and at the center line y = 0. Figure 6 shows the schematic diagram of laser beam movement on PMMA. Since the beam is near perfect Gaussian, the cross-section of the produced microchannel is also nearly Gaussian.

The optical intensity of a Gaussian laser beam at any location (x, y) can be given by following Eq. (1) (Dahotre and Harimkar 2007):

$$I(x, y) = I_0 e^{-2\left(\frac{x^2+y^2}{w^2}\right)} \tag{1}$$

where,

I = laser beam intensity at (x, y),
I_0 = laser beam peak intensity $= \frac{2P}{\pi w^2}$;
P is average laser power
w = beam spot radius.

Let us consider an infinitesimally small area $dA = dx \times dy$ within the beam diameter (Fig. 6). P is the average laser power supplied for the time dt. Let α is the absorptivity of the material for CO_2 laser beam wavelength. The laser beam travels with a scanning speed U (Fig. 7).

Since the conduction and radiation energy losses are negligible and PMMA quickly vaporizes without undergoing significant melting, it can be assumed that all the energy given by laser to PMMA is utilized into material removal.

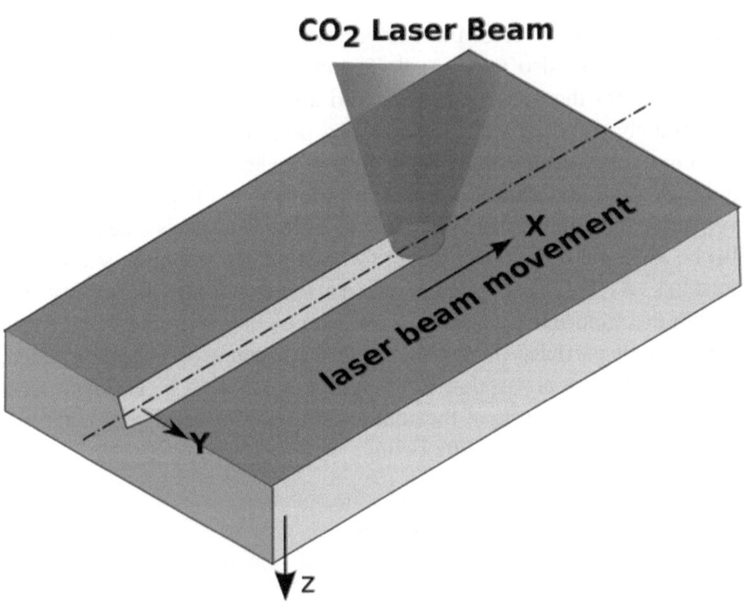

Fig. 6 Schematic of laser microchanneling process on PMMA

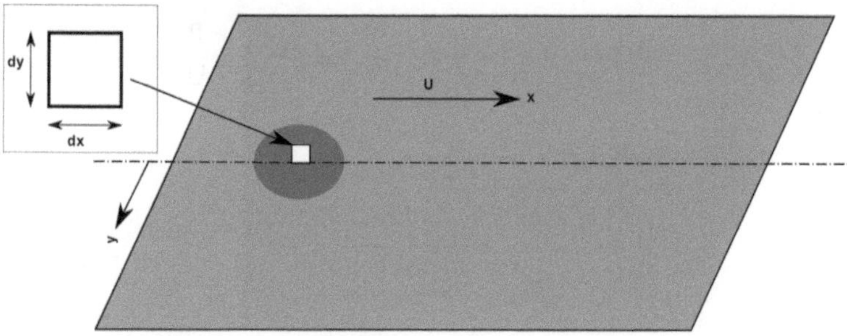

Fig. 7 Focused laser beam moving in x-direction

Energy transfer from laser to the small area *dA* can be computed as follows: Amount of laser energy transferred to area *dA*;

$$E_{laser} = \alpha \times I_0 e^{-2\left(\frac{x^2+y^2}{w^2}\right)} \times dA \times dt = \alpha \times I_0 e^{-2\left(\frac{x^2+y^2}{w^2}\right)} \times dA \times \frac{dx}{U} \quad (2)$$

Energy required to vaporize the small volume can be computed as per Eq. (3):

$$E_{vaporization} = m(c_p \Delta T + H_L) = \rho \times dA \times dz\ (c_p \Delta T + H_L) \quad (3)$$

where,

m = Mass of the vaporized volume,
ρ = mass density of PMMA
dz = depth of vaporized volume
ΔT = Temperature change (from room temperature to vaporization temperature),
H_L = Latent heat of vaporization

Equating Eqs. (2) and (3),

$$E_{laser} = E_{vaporization}$$

$$\alpha \times I_0 e^{-2\left(\frac{x^2+y^2}{w^2}\right)} \times dA \times \frac{dx}{U} = \rho \times dA \times dz\ (c_p \Delta T + H_L)$$

$$dz = \alpha \times I_0 e^{-2\left(\frac{x^2+y^2}{w^2}\right)} \times \frac{dx}{U} \times \frac{1}{\rho\,(c_p \Delta T + H_L)}$$

Applying the limits;

$$\int_0^Z dz = \int_{-\infty}^{+\infty} I_0 e^{-2\left(\frac{x^2+y^2}{w^2}\right)} \times \frac{dx}{U} \times \frac{\alpha}{\rho(c_p \Delta T + H_L)}$$

$$Z = \frac{\alpha\sqrt{2}}{w\rho\sqrt{\pi}(c_p\Delta T + H_L)} \times e^{-2\left(\frac{y^2}{w^2}\right)} \times \frac{P}{U} \quad (4)$$

At the centerline i.e. at y = 0, $Z = Z_{max}$

$$Z_{max} = \frac{\alpha\sqrt{2}}{w\rho\sqrt{\pi}(c_p\Delta T + H_L)} \times \frac{P}{U} \quad (5)$$

$$Z = Z_{max} e^{-2\left(\frac{y^2}{w^2}\right)} \quad (6)$$

Equation (5) can be used to determine the maximum depth of the microchannel, i.e. depth at the center. From Eq. (5) it can be interpreted that maximum depth is directly proportional to the laser input power. However, the maximum channel depth decreases when scanning speed increases. The channel depth along the radius i.e. along the y axis can be determined by using Eq. (6).

In order to physically verify the developed model experiments have been performed with three distinct levels of power and scanning speed. Average power and scanning speed have been varied at three different levels. Total nine experiments have been performed based on full factorial design as depicted in Table 3. The output characteristics i.e. microchannel depth has been measured using 3-D Olympus microscope. The depth at the center of the microchannel has been termed as microchannel depth. V-Shape of the microchannel has been obtained due to Gaussian nature of laser beam. Figure 8 shows the output characteristics of the produced microchannels.

The ends have been ultrasonically cleaned and polished before measurement. The depth has been found to be consistent across the length of the microchannel. Bulging at the top corners of the microchannels are also visible in Fig. 8. It has also been observed that larger is the energy density, larger is the amount of bulging and splashing due to larger amount of resolidification. Top view of the microchannels reveals that the channels are straight and clean. Some part of the channel

Table 3 Input and output parameters of CO_2 microchanneling process on PMMA

Experiment no.	Power (P)	Scanning speed (U)	Kerf width	Maximum depth
1	2	7.5	255.86	582.12
2	3	7.5	271.21	850
3	4	5	301.91	1553.19
4	3	5	287.84	1233.79
5	3	10	259.7	678.03
6	2	5	273.77	847.43
7	2	10	237.95	471.37
8	4	7.5	276.33	1063.79
9	4	10	267.37	855.15

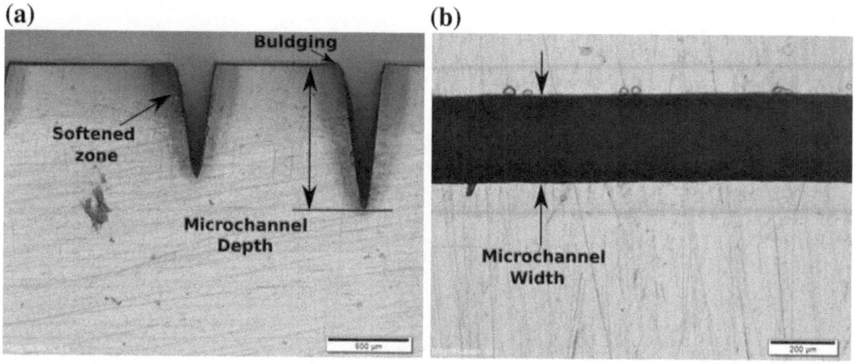

Fig. 8 **a** Cross-sectional view. **b** *Top* view of fabricated microchannel

surrounding becomes softer than the rest of the part after the microchanneling process and therefore has been termed as softened zone. The softened zone can be distinguished by a different color on the optical micrograph image as shown in Fig. 8. The softened zone can also be termed as heat affected zone (HAZ) in this context. The largest width of the softened zone takes place at the top most surface and reduces thereafter. This can be attributed to the fact that heat radiates radially after striking the PMMA surface. Since, highest amount of energy strikes the top-most surface, and reduces thereafter, the largest width of HAZ takes place at the top most surface. No burrs and charring phenomenon has been observed around the microchannels in this process. All the microchannels have been found to be clean and straight. The experiments have been performed on transparent 3 mm thick PMMA sheet.

The depth and profile of the microchannels have also been evaluated based on Eqs. (5) and (6). Table 4 enlists the comparison between actual values and calculated values. Percentage error in prediction of maximum depth has been calculated using following equation.

Table 4 Percentage error calculation

Experiment no.	Experimental depth value	Predicted depth	Error (%)
1	582.12	578.1333	0.68485
2	850	867.2	2.02352
3	1553.19	1734.4	11.6669
4	1233.79	1300.8	5.43123
5	678.03	650.4	4.07504
6	847.43	867.2	2.33293
7	471.37	433.6	8.01281
8	1063.79	1156.267	8.69313
9	855.15	867.2	1.40911

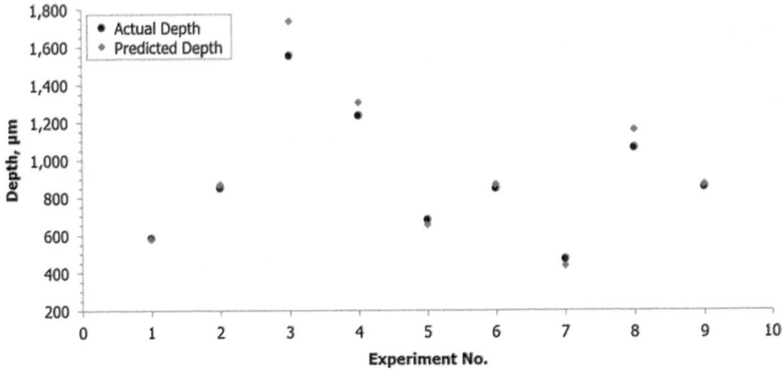

Fig. 9 Graph showing relative errors in actual and predicted values

$$\text{Error} = \left|\frac{\text{Experimental result} - \text{Predicted result}}{\text{Experimental result}}\right| \times 100\,\%$$

Figure 9 shows the relative differences in actual and predicted values of microchannel depth for all the experiments.

4 Material Removal Mechanism

CO_2 laser ablation is purely thermal ablation in which material removal takes place purely due to vaporization through melting. PMMA strongly absorbs the CO_2 laser and vaporizes immediately as soon as the average laser power crosses certain threshold value. Glass transition temperature of the PMMA has been found to be 110 °C. As soon as temperature of the substrate material reaches to 110 °C, the material starts to turn into glassy rubbery state. Once the temperature of the part crosses glass transition temperature, it may not return to its original condition. The refractivity and some of the surface properties change. The swelling at the surface can be found in the region close to microchannel. This particular region where temperature crossed the glass transition temperature but did not reach to decomposition temperature is termed as softened zone. Since the thermal conductivity of the PMMA is very low, heat gets accumulated into very small region near to microchannel. The formation of softened zone or heat affected zone can be attributed to its low thermal conductivity. Lower conductivity causes localized accumulation of heat in this zone. PMMA involves very little amount of melting at around 160 °C. However, melting does not involve significant amount of latent heat. Therefore, material removal process can be assumed to involve direct sublimation of solid PMMA into vapors without melting. Due to this direct phase change process, the chemical degradation is at minimum. Therefore microchannel edges are of superior quality than many other cutting processes. It has also been observed during simultaneous TGA/DSC test of PMMA sample. As the temperature of the PMMA reaches to 230 °C, the decomposition of PMMA starts. Although decomposition temperature of the PMMA is not sharp, it takes place very rapidly. PMMA fully decomposes at 393 °C. The highest rate of decomposition takes place at 377 °C. Under the incident laser beam, the temperature of the surface immediately rises to its vaporization temperature and PMMA $(C_5O_2H_8)_n$ decomposes into monomer MMA (methyl-meth-acrylate) units, carbon dioxides (CO_2), carbon monoxides (CO) and water $(H_2O)_{10}$. The unzipping of various chemical chains leads to formation of monomers. For a radically polymerized PMMA, following two kinds of processes have been proposed. In the first one, the depolymerization process starts with the vinyl end groups. While the second depolymerization process is initiated at saturated C–C bonds. Monomers are highly volatile in nature. Therefore, very little amount of melting takes place. This is one of the reasons that CO_2 laser ablation of PMMA does not involves redeposition of materials. The cut is very much clean with very little or no defects. However a little charring and burning effects are quite visible at the edges forming the bulging zone. The only undesirable effect is softened zone and bulging of the material at the adjacent surfaces. All the byproducts are perfectly volatile in nature and do not cause any further reaction on the PMMA substrate, leaving behind a clean structure. Due to presence of carbon content, a little amount of charring takes place at the edges. The cross sectional shape of the microchannels is of near Gaussian shape. The Gaussian shape occurs due to the Gaussian nature of the laser beam. The microchannels with lower depths follow Gaussian nature more closely than channels with larger depths. However, in most of the cases the cross-section very much resembles to alphabet "V". This shape of the microchannel also signifies the shape of the beam. Therefore it can be concluded that laser beam shape resembles to the shape of a cone with a pointed crest. The top view of the microchannel is a clean channel structure. Both edges are perfectly parallel to each other.

5 Results and Discussion

All of the experiments have been performed in order to have a wide range of different aspect ratios. The aspect ratio varies from 2 to 5 approximately. Developed model has been tested on all these wide ranges of aspect ratio. The model has been successfully able to predict the maximum depth and profile of the microchannels as shown in Fig. 10 (not to scale). Figure 10 shows the relative differences in actual channel profile and predicted channel profiles. The predicted channel profiles have been drawn with the help of MATLAB software. The background figure in grey shows the actual channel width while the white colored curve shows the predicted channel profile. The minimum error in predicting the maximum depth is 0.68 % while the maximum error is around 11.66 %.

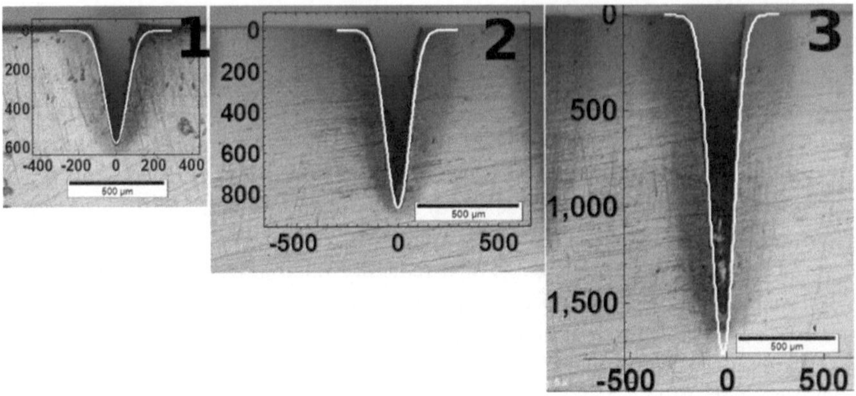

Fig. 10 Actual and predicted microchannel profiles

6 Conclusions

CO_2 laser has been used successfully to create microchannels on PMMA. The etched microchannels are clean with negligible amount of burrs around the channel. The developed model utilizes fundamental physical aspects of thermodynamics. The material properties of PMMA have been determined utilizing spectroscopy and simultaneous TGA/DSC. The developed model presents an excellent agreement between actual and predicted channel profiles. The successful modeling requires the accurate values of material properties. The developed model can be used directly to predict channel depth and profiles. The model has been found to predict the dimensions accurately over a large zone of dimensions. Computational approaches, though more accurate, consume time in order to predict the microchannel profiles and depth. Energy based analysis presents an easy solution for CO_2 laser machining of PMMA.

Acknowledgments The authors gratefully acknowledge the financial support from Department of Science and technology (DST), Govt. of India for providing INSPIRE fellowship to one of the author for carrying out this research.

References

Castaño-Álvarez, M., Pozo Ayuso, D. F., Garcia Granda, M., Fernández-Abedul, M. T., Rodriguez Garcia, J., & Costa-Garcia, A. (2008). Critical points in the fabrication of microfluidic devices on glass substrates. *Sensors and Actuators B: Chemical, 130*, 436–448.

Dahotre, N. B., & Harimkar, S. P. (2007). *Laser Fabrication and Machining of Materials* (1st ed.). Springer, London.

Helebrant, A., Buerhop, C., & Weissmann, R. (1993). Mathematical modelling of temperature distribution during CO_2 laser irradiation of glass. *Glass Technology, 34*(4), 154–158.

Iliescu, C., Taylor, H., Avram, M., Miao, J., & Franssila, S. (2012). A practical guide for the fabrication of microfluidic devices using glass and silicon. *Biomicrofluidics, 6*(1), 016505.

Julia, J. E., Aboites, V., & Casillas, M. A. (1998). CO_2 laser interaction with biological tissue. *Instrumentation and Development, 3*(10).

Klank, H., Kutter, J. P., & Geschke, O. (2002). CO_2-laser micromachining and back-end processing for rapid production of PMMA-based microfluidic systems. *Lab on a Chip, 2*(4), 242–246.

Madic, M., Radovanovic, M., & Nedic, B. (2012). Correlation between surface roughness characteristics in CO_2 laser cutting of mild steel. *Tribology in Industry, 34*(4), 232–238.

Malek, C. G. K. (2006). Laser processing for bio-microfluidics applications (part ii). *Analytical and Bioanalytical Chemistry, 385*(8), 1362–1369.

Powell, J. (1998). *CO_2 laser cutting* (2nd ed.). London: Springer.

Prakash, S., & Kumar, S. (2014). Fabrication of microchannels: A review. In: *Proceedings of the Institution of Mechanical Engineers, Part B: Journal of Engineering Manufacture*. doi:10.1177/0954405414535581.

Prakash, S., Acherjee, B., Kuar, A. S., & Mitra, S. (2013). An experimental investigation on ND: YAG laser microchanneling on polymethyl methacrylate submerged in water. *Proceedings of the Institution of Mechanical Engineers, Part B: Journal of Engineering Manufacture, 227*(4), 508–519.

Qi, H., Chen, T., Yao, L., & Zuo, T. (2009). Micromachining of microchannel on the polycarbonate substrate with CO_2 laser direct writing ablation. *Optics and Lasers in Engineering, 47*(5), 594–598.

Radice, J. J., Joyce, P., Tresansky, A., & Watkins, R. (2012). A comsol model of damage evolution due to high energy laser irradiation of partially absorptive materials. In*: Excerpts from the Proceedings of the 2012 COMSOL Conference in Boston.*

Russom, A., Gupta, A. K., Nagrath, S., Carlo, D. D., Edd, J. F., & Toner, M. (2009). Differential inertial focusing of particles in curved low-aspect-ratio microchannels. *New Journal of Physics, 11*(7), 075025.

Shulepov, M. A., Bulgakova, N. M., Zakharov, L. A., Panchenko, A. N., & Telminov, A. E. (2010). CO_2 laser ablation of poly (methyl methacrylate) and polyimide: Experiment and theory. In: *Proceedings of the 10th International Conference on Modification of Materials with Particle Beams and Plasma Flow* (pp. 132–135).

Snakenborg, D., Klank, H., & Kutter, J. P. (2004). Microstructure fabrication with a CO_2 laser system. *Journal of Micromechanics and Microengineering, 14*(2), 182.

Xiang, H., Fu, J., & Chen, Z. (2006). 3D finite element modeling of laser machining PMMA. In: *1st IEEE International Conference on Nano/Micro Engineered and Molecular Systems, 2006, NEMS'06* (pp.942–946).

Yuan, D., & Das, S. (2007). Experimental and theoretical analysis of direct-write laser micromachining of polymethyl methacrylate by CO_2 laser ablation. *Journal of Applied Physics, 101*(2), 024901.

Zhou, B. H., & Mahdavian, S. M. (2004). Experimental and theoretical analyses of cutting nonmetallic materials by low power CO_2-laser. *Journal of Materials Processing Technology, 146*(2), 188–192.

Fiber Laser Micro-machining of Ti-6Al-4V

A. Sen, B. Doloi and B. Bhattacharyya

Abstract Progress of laser micro-machining i.e. micro-cutting, micro-drilling, Micro-channeling, micro-grooving, micro-turning etc. in the field of aeronautic, automobile, semiconductor and biomedical industries such as turbine blades of aircraft engine, automotive fuel filters, combustion chambers, surgical needles and micro-fluidic devices have emerged extensively in the present era. The prime contributor to the success of laser micro-machining in the recent years is fiber laser technology that involves the combination of diode pumped solid state lasers and fiber technology. This is the most promising substitute to the high-power, bulk solid-state lasers and some gas lasers owing to its simplicity, ruggedness, cost effectiveness, low maintenance, higher efficiency, higher reliability and smaller spot size. Fiber lasers are mainly characterized by short pulse lengths which range from millisecond to picosecond and even femtosecond for precise micro-machining of different materials. Titanium alloys play crucial roles in the areas of advanced structures and technologies for aerospace and power industry, medicine, automatics and mechatronics and various measurement equipments, because of their high strength and stiffness at elevated temperatures, high corrosion resistance, fatigue resistance, high strength to weight ratio and ability to withstand moderately high temperatures without creeping. The conventional machining methods intended for cutting these alloys not only suffer due to poor thermal conductivity, low elastic modulus and high chemical affinity at elevated temperatures but also have to undergo higher cost associated with the machining of Ti-6Al-4V caused by lower cutting speeds and shorter tool life. Numerous research works have been conducted with laser beam micro-machining approaches on titanium alloys mainly on Ti-6Al-4V, in order to establish the optimal experimental conditions that are to be used in different applications. However, the physical mechanism which leads to the

A. Sen (✉) · B. Doloi · B. Bhattacharyya
Production Engineering Department, Jadavpur University, Kolkata 700032, India
e-mail: abhishek.sen1986@gmail.com

B. Doloi
e-mail: bdoloionline@rediffmail.com

B. Bhattacharyya
e-mail: bb13@rediffmail.com

S.N. Joshi and U.S. Dixit (eds.), *Lasers Based Manufacturing*,
Topics in Mining, Metallurgy and Materials Engineering,
DOI 10.1007/978-81-322-2352-8_15

observed geometry and surface roughness of the micro-machined surface is still behind the veil. The aim of the present chapter is to make an in depth study of the fiber laser micro-grooving of Ti-6Al-4V, discussing vividly the fiber laser machining system, the occurring physical processes and machining strategy and influence of various process parameters. Experimental results of present research work along with the work of various researchers are discussed so as to validate and run down hypotheses on the mechanisms involved.

Keywords Micro-machining · Fiber laser · Ti-6Al-4V alloy · Micro-grooving process

1 Introduction

Laser micro-machining process includes wide range of versatile processes which deals with the removal of stringent amount of molten material in the range of microns (Scientific Technical Committee of the Physical and Chemical Machining Processes of CIRP, 1–500 μm was adopted as the range for micro-machining work) (Masuzawa and Tonshoff 1997). In general other related manufacturing processes such as micro-joining and micro-adjustment (Meijer 2002) by the laser beam can also be termed as micro-machining process. Generally, the micro-machining term is broadly classified into the ablation, micro-cutting, micro-drilling, marking in the ranges of micron. This chapter aims to focus on the aspect of micro-grooving (broadly micro-machining i.e. micro-cutting) of titanium alloys of grade 5 (Ti-6Al-4V) by fiber laser machining system. The average power of various laser systems that are identified to be best suitable for micro-machining applications of wide range of engineering materials are 20–50 W. In brief, laser micro-machining is mainly characterized by a machining process in which small amount of material is removed by focusing highly intense laser beam on the work-piece surface as the generated heat from the laser beam melts and then vaporizes the work-piece throughout the thickness of the material (Meijer 2002). The molten material can be expelled from the cutting front by a pressurized assist gas jet that includes compressed air pressure, inert gaseous systems i.e. argon, helium, nitrogen and sometimes enhanced material removal can also be conducted through chemical reactions such as oxidation of the material (Sun and Brandt 2013).

Progress in engineering in the field of advanced structures and technologies for aerospace and power industry, biomedical, automatics and mechatronics, manufacturing of control and measurement equipment, is driven to a large extent by development and application of new or modified functional materials. It includes titanium and its alloys, although initially they were intended mainly for applications in structural material to carry loads and carrying airframe elements for aerospace industries. Titanium alloy of grade 5 enables implementation of new solutions in structures and technologies and orthopaedics for its unique properties such as low cytotoxicity and biocompatibility (Ratner et al.1996), corrosion resistance (Hayes and Mow 1997), wear resistance (Chesnutt et al. 1980) and fatigue resistance (Collings 1984; Williams 1984; Fasasi et al. 2009). In most applications, Ti-6Al-4V implants have relied on surface roughening and porous coatings to improve osseointegration (Fasasi et al. 2009). The potential for micro-parts as medical tools operating at cellular level is increasing which drives extensive research efforts to be concentrated on the evolution and control of the titanium alloys microstructure through the adjustment of the processing parameters in order to obtain desirable balance of properties for specific applications. The challenges in understanding the mechanisms involving machining of titanium alloys along with its low thermal conductivity (laser beam machining process) and the high tool wear associated with the reactivity of titanium with tool materials (traditional machining process) are still to be addressed.

Fiber lasers evolved as the most versatile and rapid growing laser systems during the last decade and successfully emerged into the various fields of manufacturing, medical, metrological and military applications that were previously dominated by conventional solid state lasers and gas lasers. Today, fiber lasers (average power of 10–50 W) hold a good market share in the domain of fine and precise micro-cutting that involves the combination of both continuous mode and pulse mode with the aid of fusion cutting (inert gas) to sublimation cutting (oxygen cutting). The initiation of combining optical fiber with laser system was started when the advantages of introducing a rare earth doped single-mode optical fiber in a laser cavity to provide a robust single spatial mode at the laser output was recognized in 1961 by Snitzer (1961). After a few years, high gain in neodymium-doped multimode silica optical fiber lasers pumped by flash lamps was demonstrated by Snitzer and Koester (1964). In the mid-1980s, practical work on single-mode optical fiber lasers commenced, after the initial advancement of rare-earth-doping methods that used modern optical fiber fabrication processes based on vapor-phase deposition technique (Hegarty et al. 1983; Nakazawa et al. 1989; Poole et al. 1985). For the majority of the 1990s, the highest-power single-mode optical Fiber lasers were usually pumped either by gas lasers or solid-state lasers, which turned out to be irrelevant in case of commercial, industrial applications owing to low average power drawn from most single-mode fiber lasers (Teodoro 2011). Finally, the booming of the telecommunications during 2001 caused the thrust needed for the development of high-power optical fiber lasers. Extensive amount of dedicated research works and development of multimode pump diodes brought about significantly more powerful and reliable multimode pump diodes at radically lower cost. It is worth noting that the contribution of the Military-funded programs, in a drive for directed energy weapon system and countermeasures which was another driving force for higher powers from fiber lasers and the development of related technologies (Hecht 2009).

1.1 Advantages of Fiber Lasers Over Other Solid State and Gas Lasers

Shorter wavelength in combination with its high focus ability, better system flexibility, high component yield, long uptime along with improved reliability, high repeatability, high aspect ratios with utmost precision, unlimited material coverage, low cost and fully automated processes seem be advantageous for fiber lasers in thin sheet metal cutting. Whereas, the CO_2 lasers are probably still capable of machining thicker materials more efficiently. Although, Olsen et al. (2009) study revealed efficiency of fiber laser cutting operation for thick materials is higher than of CO_2 laser. Their experimental results also revealed that burr free cuts can be achieved in 1 and 2 mm AISI 304 stainless steel, over a wide range of cutting rates using multi-beam fiber laser machining system. Further research studies in the domain of fiber laser machining process also brought about the observations that fiber lasers can produce better results as compare to CO_2 lasers when machining of copper and magnesium alloys. Fiber lasers have also been replacing Nd-YAG lasers in various micro-machining applications such as micro-cutting of stents, thin sheet of ferrous and non-metals in terms of cutting speed, cut edge quality and the length of micro cracks. Fiber laser cutting of thick polycrystalline silicon (silicon wafer) is another significant area where it produces better results compared to Nd-YAG laser. The fundamental wavelengths of fiber lasers lie between 1 and 2 μm (Schaeffer 2012) and can couple extremely well with wide range of metals. The long thin gain media of fiber lasers produce high quality beams which are focussed to a very small spot on the work piece surface, giving very high power density. The high efficiency of the laser diodes produce considerably high wall plug efficiency compared to other conventional lasers. Other important characteristics of fiber lasers have active regions of several kilometres long which can provide kilowatt levels of continuous output power. Fiber lasers are also benefited due to flexible coupling of the light with the fiber for which light can be delivered easily to a movable focusing element. Efficient cooling can be achieved because of long fiber length that leads to high surface to volume ratio. High vibrational stability, extended lifetime and maintenance free operation (Schaeffer 2012) appears to be advantageous and results into more compactness in the case of fiber laser system as compared to other conventional solid state and gas lasers. Therefore, in the present scenario, fiber lasers are outpacing market growths mainly in the areas of automotive, aerospace, wind turbines, solar processing and bio-medical devices.

1.2 Need of Laser Beam Micro-grooving Process

Micro-grooves are used as one of the key micro-features in different micro-products like micro-thermal devices, micro-heat exchangers, micro-reactors, micro-pumps and micro-mechanical systems. A high aspect ratio micro-groove not only provides a large surface area for heat dissipation and reaction systems but also provides high capillary force for micro-heat pipes and micro-pumps, as micro-fluidics, frictional force and capillarity vary according to the geometry of the micro-groove (Rathod et al. 2013). Micro-grooves can also be used as micro-channels in biomedical and biochemical applications such as super alloys, titanium alloys, aluminium alloys, copper alloys etc., which are hard and difficult to cut by conventional machining methods. The effectiveness of the laser micro-grooving process mechanism depends on the thermal properties and the optical properties to a certain extent. Mechanical properties of the materials that are to be machined do not play significant roles as compare to other mentioned properties. Energy transfer between the laser and the material occurs through irradiation for which no cutting forces are generated by the laser. As a consequence, mechanically induced material damage, tools wear and machine vibrations cannot be observed. Moreover, the material removal rate (MRR) for laser machining is not limited by constraints such as maximum tool force, built-up edge formation or tool chatter. In order to obtain fine and deep micro-grooves i.e. grooves with high aspect ratio (width/depth), importance of different properties such as reflectivity, thermal conductivity, specific heat and latent heats of melting and evaporation is crucial.

1.3 Importance of Ti-6Al-4V Micro-grooves in Research and Industrial Perspective

Ti-6Al-4V is the most used material out of the other titanium alloys and its total production is almost half of all titanium alloys. The present research work is dedicated towards the bio-medical features of this alloy, giving priority to both the geometrical aspects i.e. width and depth and surface topology i.e. surface roughness at the time of selecting the process parameters. Titanium alloys (grade 5) are often used in different bio-medical field of applications such as artificial hip joints, knee joint replacements and also dental implants. It is important that Ti-6Al-4V can bond firmly with the bones, for optimal function and durability. Therefore, enhanced bioactivity and improved implant-host interactions are important so as to reduce biological related implant failure. The process of osseointegration mainly depends upon the various surface properties such as surface chemistry, surface topography, surface roughness and the surface energy. Modified improved surfaces confer enhancement of cell-implant interactions and brings about more opportunities for focal attachment which are beneficial for orthopaedic applications (Brown and Arnold 2010). Therefore, selection of optimum machining parametric conditions holds the key behind obtaining precise micro-grooves with minimum surface roughness. Thereafter, these precise and fine micro-grooves can be optimized to integrate with the surrounding tissue.

1.4 Basic Mechanism of Laser Micro-grooving Process of Ti-6Al-4V

Most of the titanium alloys that are used in the industry contain α- and β- stabilizers which include Ti-6Al-4V, Ti-6Al-6V-2Sn and Ti-6Al-2Sn-4Zr-6Mo. The chemical compositions of Ti-6Al-4V are listed in table 1. These alloys are heat treatable and most are weldable especially with the lower β-stabilizer (Sieniawski et al. 2013). According to Ezugwu et al. (2003), machinability can be defined as the difficulty to machine a particular material under a given set of the machining parameters such as cutting speed, feed rate and depth of cut. Conventional machining process fails to meet stringent quality control at the time of machining of titanium alloys owing to excessive amount of heat, work hardening, low thermal conductivity, abrasiveness and high strength level. In contrast, laser beam machining is fast and repeatable process. The heat generated during the laser beam machining process, is used to remove material in a very small volume without mechanical engagement with work-piece material. The basic mechanism involved behind laser beam micro-grooving process can be identified as the laser beam micro-cutting process includes several crucial factors. During laser beam micro-machining of titanium alloy of grade 5, temperature zones can be divided into solidus temperature of 1877 K, liquidus temperature of 1923 K and evaporation temperature of 3533 K. Four different approaches which govern laser beam machining process are evaporative laser cutting, fusion cutting, reactive fusion cutting and controlled fracture technique. The selection of optimum laser cutting technique depends on the thermo-physical properties of the material, the thickness of the work piece and the type of laser employed. For a thick section cutting of titanium alloy, the most preferred process is reactive fusion cutting. In reactive fusion cutting, molten material is removed and the lost material is compensated by further melting of the work-piece at the solid-liquid interface below the cutting front (Mahrle and Beyer 2009). The present topic covers fiber laser micro-machining of Ti-6Al-4V having thickness of 1.1 mm. It is worth to take a note that all the research experiments were carried out in atmospheric conditions.

Fiber laser beam is focused on the surface of Ti-6Al-4V and some portion of the energy (optical energy is converted to heat energy) is absorbed. Theoretically, only 40–80 % of that energy may be absorbed in most cases of fiber lasers. Remaining part of the energy is reflected back and some portions of the energy may be scattered through the work-piece. Absorption occurs through a very thin surface layer and the absorbed energy diffuses into bulk of the material. As the pulse width in the present system is in nanosecond pulse regime, the heat flow can be assumed to be one dimensional (Meijer 2002). The absorbed laser beam penetrates on the

Table 1 Chemical composition of Ti-6Al-4V

Al (%)	Fe (%)	C (%)	N (%)	O (%)	V (%)	Ti
5.55–6.5	<0.25	<0.08	<0.05	<0.2	3.35–4.5	Balance

work-piece surface in the range of microns and sub microns. As the temperature increases, surface temperature will reach to the melting point at a certain period of time. The melting time reduces drastically when power density increases rapidly. Another important aspect that should be taken into consideration is that all the experiments were accomplished in pulse mode and thus the peak power is high at a considerable amount which can cause high vaporization rate. The high vaporization rate can create a shock wave and a high vapour pressure at the liquid surface considerably increases the boiling temperature (Meijer 2002). Finally the molten material is removed as vapour by the expulsion of melt, as the resultant of the high pressure in combination with super heated liquid after the end of the laser pulse. As all the research experiments were carried out in atmospheric conditions, heat affected zone (HAZ) and re-solidification of the material may occur in a form of rim (Meijer 2002). Overall laser beam micro-machining process of Ti-6Al-4V including representation of laser beam process of micro-groove generation is shown in Fig. 1 (a) & (b) respectively.

1.5 Literature Review on Laser Beam Micro-machining of Ti-6Al-4V

In the previous sub-chapter, various research aspects of Ti-6Al-4V were discussed and the difficulties of major issues of machining of these alloys were also addressed. Researchers studied laser beam micro-machining (LBMM) and laser assisted micro-machining (LAMM) approaches with different combinations of assist gas pressure system so as to optimize the various process parameters depending upon the applications. Ti-6Al-4V grooves can be machined in water and silicon oil conditions but formation of bubbles and cracks are obvious in such machining conditions. These cracks can be formed on the surface of the liquid-machined grooves because of the induced additional thermal stresses from the rapid quenching. This sub-chapter not only focuses on the various research works on laser beam micro-machining of Ti-6Al-4V that enhances bone healing and implant integration but also on various other research works related to laser beam micro-grooving process.

The mechanisms involved laser beam micro-grooving process that can be compared to micro-cutting process, except for the evacuation of scoria in the range of microns, which cannot flow out in the rear side of the micro-grooved surface as is realized in a micro-cut surface. Al-Hajri et al. (2005) studied the characterization of copper-finned micro-grooved surfaces for effective evaporation heat transfer, which can be applied to cooling of high flux electronics. Mauclair et al. (2013) reported cross sectional profile of micro-grooves machined in stainless steel by ultrafast laser machining system. They relied on spatial beam shaping technique in which the intensity distribution was utilized for precisely controlling of the pulse overlapping during the translation of the sample under femtosecond exposure. Experimental results revealed a good agreement with the priori calculations based on the ablation rate and pulse overlapping. Dhupal et al. (2008) investigated pulsed Nd:YAG

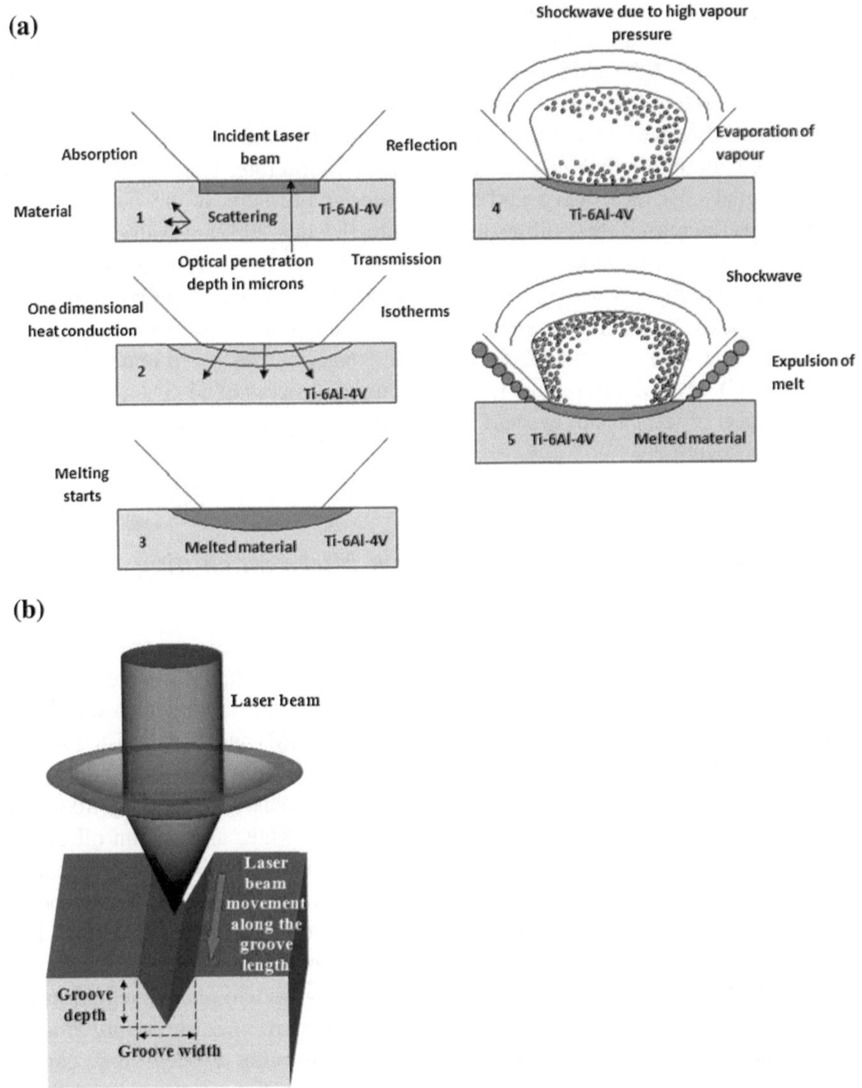

Fig. 1 **a** Representation of laser beam micro-machining of Ti-6Al-4V. **b** Representation of laser beam process of micro-groove generation in terms of width and depth

laser-turning operation to generate micro-groove on cylindrical work-piece of aluminum oxide ceramic material. They concluded that minimum depth deviation can be achieved by moderate settings of the lamp current with moderately high pulse frequency.

Various research efforts have been concentrated on improving the bone-implant interface, with the aid of either physical or chemical approaches. The physical approach is focused on the modification of the implant surface morphology and topography using mechanical methods such as machining, acid-etching, plasma spraying, grit-blasting and anodization in order to improve the micro-topography of the surface. Increase in surface roughness of the implant material would provide a higher level of surface energy which would improve bone anchorage, matrix protein adsorption, osteoblasts functions and ultimately osseointegration (Wang and Poh 2013). The most preferred assist gas system which involves the inert gasses nitrogen, argon and helium was found out. In contrast, when this alloy is cut with oxygen assist gas at even low pressure, the uncontrolled burning of cutting front starts, which results into wide kerf width and poor surface quality. At the time of air assisted laser cutting, the reaction of titanium with oxygen and nitrogen produces a thin layer of hard and brittle oxides and nitrides and also generates much thicker HAZ in comparison to that of argon (Zhang et al. (2007). Therefore, it was suggested that, in order to overcome these problems titanium alloys may be cut using inert gasses such as argon and helium. Chen et al. (2009) studied the effects of nano-second pulse UV laser multi scale laser texturing on Ti-6Al-4V substrate on the adhesion of osteoblast cells. The researchers found that groove width was not significantly affected either by the number of passes or the distance between pulses. The effect of pulse energy on groove width was critical. In contrast, groove depth was affected by the translation distance and the number of passes as well as laser pulse energy. They also reported that laser processing conditions also affect surface roughness and other sub-micron features created on the surface. Almeida et al. (2006) performed a factorial designed experimentation on Nd: YAG laser cutting of Ti and Ti alloy sheets. It was concluded that use of nitrogen assist gas increases surface hardness from 2 to 3 times due to the formation of titanium nitride (TiN) while a mixture of He and Ar gases reduce the irregular edges and also eliminate the nitride formation. It was also reported that micro-milling of titanium alloy for medical applications especially for implants can create free-form surfaces, which lead to improvement in biocompatibility. Therefore, in order to produce functional micro-products, not only part features and tolerances have to be concerned but also surface quality must be considered as well. Several approaches have been used in order to gain a control over the surface finish of micro-features; for example, process parameters optimization, surface generation modelling and simulation, effects of using lubrication etc. Sun et al. (2008) performed laser assisted machining trials on commercially pure (alpha) titanium alloy and observed significant reductions in the cutting forces due to laser pre-heating and smoother surface finish due to the diminished fluctuations in the cutting forces. Similarly, Dandekar et al. (2010) observed an increase in tool life up to a cutting speed of 107 m/min during laser assisted turning of Ti-6Al-4V alloy. Sun et al. (2011) also carried out laser assisted milling trials on Ti-6Al-4V alloy and reported a dramatic reduction in the feed force and the tool edge chipping which is the dominant tool failure mode in dry milling of this alloy. Shelton and Shin (2010b) conducted a study of LAMM slotting of Ti-6Al-4V stainless steels AISI 422 (422SS) and AISI 316 (316SS) with tungsten carbide (WC) micro end mills. Fasasi et al. (2009) showcased the effects of nano-second laser-processing parameters i.e. pulse repetition rate, scan speed and wavelength on micro-groove geometry. To fabricate micro-structure having approximately 11 μm width and depth, Ti-6Al-4V plate of 6 mm × 6 mm × 12 mm was utilized with Q-switched nanosecond laser. The researchers successfully demonstrated the laser ablated micro-grooves without cracks. Desired 8–12 μm micro-groove depths and widths could be achieved by controlling pulse frequency, scan speed and lens focal length that controls spot size. Shelton and Shin (2010a) again studied LAMM side cutting experiments of 316SS, 422SS, Ti-6Al-4V and Inconel 718 (IN718) of fin work-piece structure. Experimental results revealed that formation of burrs in Ti-6Al-4V and IN718 were drastically reduced with the aid of laser-assist machining (Rashid et 2012). Yang et al. (2010) developed a 3D finite element model so as to predict the formation of HAZ during laser-assisted milling of Ti-6Al-4V. They developed a proper correlation between the predicted HAZ and measurements. A transient, 3D finite volume prismatic thermal model was also developed by the researchers' group for laser-assisted milling (Tian et al. 2008). The thermal model was then utilized for the modeling of the temperature field during LAMM processes of different configurations. This prismatic thermal model provided the transient temperature distributions within the work-piece and validated through surface temperature measurements using an infrared camera and embedded thermocouples.

2 Working Principle of Fiber Laser Generation

The working principle of fiber laser can be identified with the production of laser light that requires material (a rare earth element) in the active medium which is to be raised from its ground state to an excited state (population inversion) in that medium. Normally atoms in the active medium reside in their ground states and some external form of energy must be added to raise these atoms to an excited state. Laser is generated when the laser diodes are pumped to excite lower energy molecules to excited states. The number of laser diodes depends on the application of the fiber laser and the output power required. Laser is generated by 'spontaneous emission' and then amplified by 'stimulated emission' along the axis of the resonator cavity. Inside the fiber, generated light propagates in the mode form. Mode of travel along the axis of the waveguide is constant with its characteristic propagation and group velocity. It is important that light which incidents on the boundary between fiber core and cladding, should be at angles greater than critical angle and must undergo a total internal reflection (Teodoro 2011). Light rays are guided through the core without refraction into cladding. The cavity mirrors consist of two primary optics, front and rear mirror; reflect the photons to and fro through the laser medium to increase amplification. Through the pumping process, only a fraction of energy passes through the front mirror. In conventional fibers doped single-mode core and a cladding, the laser power is restricted to about 1 W. To achieve higher output powers, fiber lasers use double-clad fibers as the amplifying medium (Schaeffer 2012). Most of fiber lasers employ Q switch mode of operation though

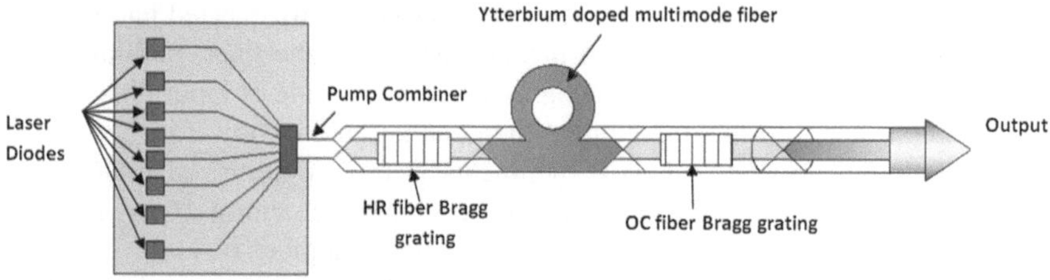

Fig. 2 Simple schematic diagram of Q switched ytterbium doped fiber laser

introduction of 'master oscillator, power amplifier' (MOPA) design enables independent control of pulse width, pulse repetition frequency and average power. End-to-end all-fusion-spliced-fiber MOPA designs, is a very common design for the use of fiber-coupled semiconductor laser. In Q switched fiber laser systems, pulse energy is independent of the fiber length. A simple schematic diagram of Q switched ytterbium doped fiber laser is shown in Fig. 2.

3 Experimental Studies on Fiber Laser Micro-machining of Ti-6Al-4V

Details of fiber laser micro-machine setup, along with the experimental plan and results have been thoroughly discussed in this chapter.

3.1 Fiber Laser Micro-machining Setup

The key elements of fiber laser are (a) optical fiber, (b) rare earth dopant ions, (c) mirrors, (d) pump sources and (e) fiber coupler. Laser interaction with the optical fiber material is dependent on several parameters i.e. laser source (wavelength and emission regime, mainly) and also on the characteristics of the material itself. Optical fiber (main component) is a cylindrical dielectric waveguide made of low-loss material such as a silica glass. The improvement of the optical properties is accomplished by doping with ytterbium (Yb^{3+}), neodium (Nd^{3+}) as rare earth elements. Yb^{3+} holds several advantages compared to other rare earth elements owing to a longer upper-state lifetime, a small quantum defect (results in lower thermal load per unit of pump power) and absence of the excited state absorption. To create laser cavity, various types of mirrors can be used in fiber lasers, i.e. bragg gratings, multilayer dielectric mirrors or semiconductor saturable absorber mirrors (SESAM) (Schaeffer 2012). In most of the cases, pumping is done predominantly by semiconductor diode lasers. Different semiconductor materials are used for different wavelengths such as aluminum gallium arsenide lasers, output ranging from 750 to 950 nm. The output of the diode laser is directly injected into the laser fiber. In order to increase the pump power reaching to the fiber, multiple diode lasers are used. Fiber couplers are being used either one side of input fibers or one side of output fibers so that the emitted light cannot go back towards the source.

In the present research, all the experiments were carried out on multi-diodes pumped Ytterbium (Yb^{3+}) doped fiber laser machining system of 50 W, made by M/S Sahajanand Laser Technology Limited. Total 8 number of laser diodes are used for pumping and each of the diodes is coupled with the multimode of optical fibers. The photographic view of the fiber laser machining system along with diode pointer and F-θ lens (shown separately) were used in the present research work as represented in Fig. 3a, b respectively. The laser head of the fiber laser consists of both electric and galvanometer unit from which the generated laser beam is focused on the work piece with the aid of F-θ lens of 100 mm. The mode of operation in which experiments were carried out was Q switched mode which can generate a high level of peak power in the regime of kilowatts. The present experimental setup is also boosted by nanosecond pulse regime for which material interaction time with the laser beam is less. The diode pointer assures that the work piece is on the focal plane. The work-piece remains stationary and the laser beam moves to and fro to the programmed directions (by software) and highly focused generated beam is directed to the work-piece. The detailed machining setup is listed in Table 2.

3.2 Experimental Planing

Due to the complexity of relationship between laser-cutting parameters and the cutting quality, four process parameters i.e. scan speed, pulse frequency, average power and number of pass were considered so as to find out the influence of fiber laser micro-grooving process parameters on various responses such as width, depth and surface roughness. The objective of the present research work was to study the minimization of width and surface roughness and maximization of the depth although the present research work exclusively deals with the basic experimental studies. Extensive pilot experiments were carried out previously so that all the desired results could be achieved. Another important objective was to fabricate micro-grooves with utmost precision and improved surface roughness in atmospheric conditions. The literature review revealed that only few works have been carried out in such conditions and the end results regarding surface quality were not satisfactory. The present research work was also important for narrowing the gap between the theoretical and experimental studies. After the micro-machining, the samples were cleaned with the ultrasonic cleaner for 15 min so as to remove dust particles. The dimensions of the depth and width were measured using Leica optical microscope using 5× lens. The measurements of the widths were taken inside four/five (depending upon the width dimensions' variation) different places of the micro-grooves and the final width was calculated on the basis of average of four/five values. The measurements of the depth were carried out from the surface edges of

(a)

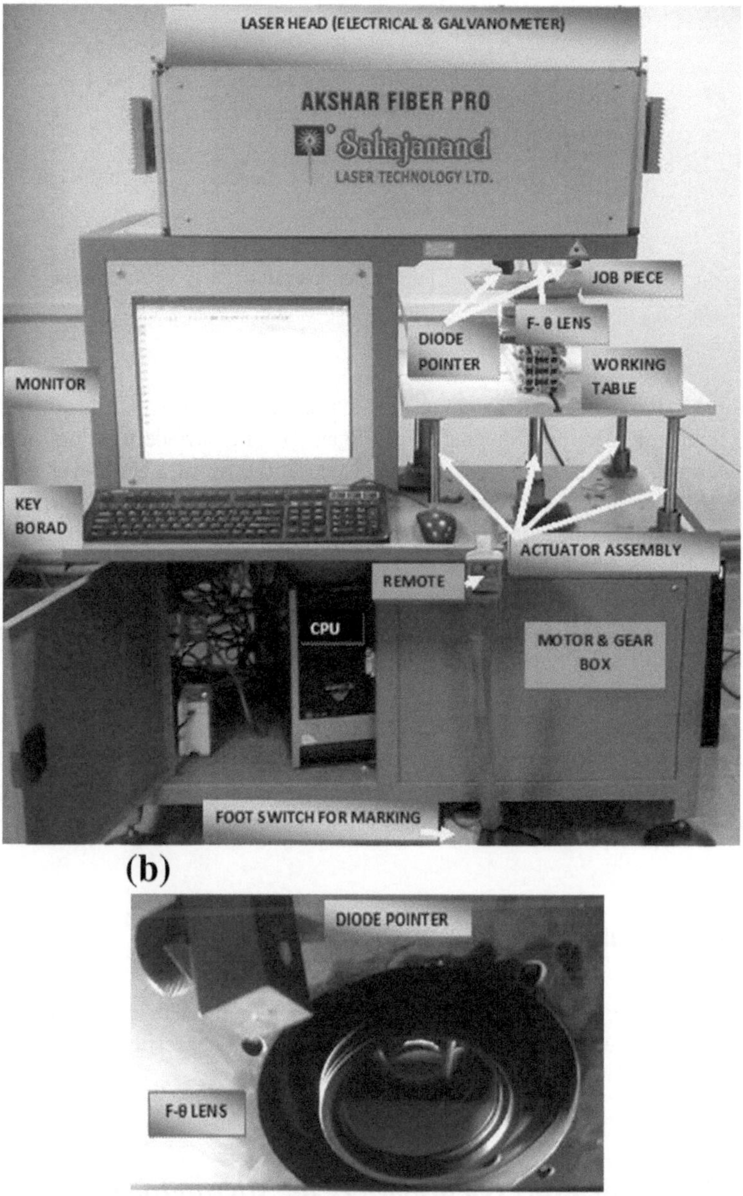

(b)

Fig. 3 **a** Photographic view of fiber laser machining system. **b** Photographic view of lens and diode pointer (Courtesy of Sahajanand Laser Technology Limited)

Table 2 Specification of the fiber laser

No.	Specification	Description
1	Laser type	Pulsed
2	Wavelength	1064 nm
3	Mode of operation	Pulsed
4	Mode of laser beam	TEM_{00}
5	Beam diameter	8.8–9 mm
6	Laser beam spot diameter	18–20 μm (after focus)
7	Average power	50 W
8	Pulse width	120 ns
9	Pulse repetition rate	50–120 kHz
10	Output fiber length	3 m

the sample material. Surface roughness (R_a) of the micro-grooves was computed by atomic force microscope (Nanosurf Easy Scan 2). The surface topology measurements were carried out by a small probe placed at the tip of a cantilever beam and the forward scanning were done from the left to right of the selected surface of the work-piece. All the measurements of the surface roughness using AFM are conducted by line fit which calculates the first order least squares fit (mean value and slope) for each line of data points and subtracts the fitted values from the raw measurement data for each data point of that line. The 3D view of the surface topology and measured surface roughness values of each groove in terms with R_a were utilized to analyze and validate the micro-grooving process. Origin 8 (data analysis and graphing software) was utilized for parametric plots with the measured values of width, depth and R_a so as to identify effects of the process parameters for the understanding of fiber laser micro-machining process for fabricating the micro-grooves on Ti-6Al-4V.

3.3 Experimental Results and Discussions

The results of the total 49 experiments consist of four set of experiments which include diferent level of micro-cutting parameters. The results of the total 49 experiments are listed in Table 3.

3.4 Influence of Process Parameters on Micro-groove Geometry and Surface Roughness

In this present sub-section, influence of four process parameters on the micro-groove geometry and surface roughness have been explained in detail.

Table 3 Results of the total 49 experiments

Ex.	Number of passes	Scan speed (mm/s)	Pulse frequency (kHz)	Average power (W)	Width (μm)	Depth (μm)	R_a (μm)
1	8	1000	80	30	145.852	113.701	0.419
2	8	900	80	30	144.451	198.033	0.450
3	8	800	80	30	124.645	112.285	0.210
4	8	700	80	30	155.512	144.958	0.145
5	8	600	80	30	149.855	136.230	0.130
6	8	500	80	30	171.548	164.612	0.103
7	8	400	80	30	174.580	175.964	0.180
8	8	300	80	30	169.541	349.098	0.209
9	8	200	80	30	224.72	367.372	0.109
10	8	100	80	30	197.638	515.670	0.044
11	8	90	80	30	220.689	387.885	0.205
12	8	80	80	30	207.658	415.315	0.505
13	8	70	80	30	205.693	415.258	0.511
14	8	60	80	30	213.668	401.594	0.245
15	8	50	80	30	232.772	481.512	0.236
16	8	40	80	30	232.772	454.673	0.115
17	8	30	80	30	236.893	481.868	0.179
18	8	20	80	30	188.611	613.832	0.212
19	8	10	80	30	186.274	750.829	0.100
20	8	05	80	30	321.385	748.923	0.169
21	8	40	50	30	319.007	821.558	0.087
22	8	40	60	30	351.115	880.702	0.263
23	8	40	70	30	349.805	839.754	0.108
24	8	40	80	30	375.595	791.742	0.213
25	8	40	90	30	350.871	803.632	0.119
26	8	40	100	30	340.456	746.961	0.118
27	1	40	50	30	279.094	759.799	0.130
28	2	40	50	30	279.094	812.343	0.140
29	3	40	50	30	340.464	812.019	0.169
30	4	40	50	30	324.245	824.667	0.237
31	5	40	50	30	246.853	862.59	0.228
32	6	40	50	30	247.779	867.667	0.140
33	7	40	50	30	262.206	905.615	0.137
34	8	40	50	30	261.834	779.146	0.151
35	9	40	50	30	249.980	747.445	0.167
36	10	40	50	30	238.465	755.312	0.262
37	11	40	50	30	235.065	735.082	0.276
38	12	40	50	30	234.118	738.054	0.277
42	5	100	50	2.5	320.017	475.500	0.198

(continued)

Table 3 (continued)

Ex.	Number of passes	Scan speed (mm/s)	Pulse frequency (kHz)	Average power (W)	Width (μm)	Depth (μm)	R_a (μm)
43	5	100	50	5	373.093	479.653	0.272
44	5	100	50	7.5	379.653	473.043	0.286
45	5	100	50	10	384.777	559.074	0.329
46	5	30	100	7.5	124.133	516.072	0.026
47	5	30	100	10	111.292	198.064	0.051
48	5	30	100	12.5	106.327	32.982	0.064
49	5	30	100	15	150.159	49.038	0.100

3.4.1 Influence of Scan Speed on Width, Depth and Surface Roughness on Ti-6Al-4V Micro-grooves

Previous research articles revealed that the fiber lasers are more effective in cutting operations than the conventional CO_2 lasers because of their linear energy per unit sheet thickness. Fiber lasers are capable of producing high energy intensity to cause the work-piece vaporization and create a keyhole owing to shorter wavelengths (Dahotre and Harimkar 2008). The cutting quality is mainly governed by the combination effects of pulse frequency and scan speed. In pulse mode of laser micro-machining operation, the extent of spot overlapping determines the surface roughness of the micro-groove geometries. During pulsed fiber laser micro-grooving operation of titanium alloys, minimization of surface roughness, width as well as increase in dimensions of depth of the micro-groove criterion is most important for various fields of engineering applications. Another important concerning factor is the formation of heat affected zone that can have a detrimental effect on the surface roughness of the fabricated micro-groove geometry. To overcome this phenomenon to a large extent, many of the experiments were carried out with higher scan speed in the moderate setting of pulse frequency with the combination of other process parameters that were kept as constants.

Figure 4 shows the influence of scan speed on width, depth and surface roughness on fiber laser machined micro-groove surfaces while other process parameters were kept constant such as 8 number of passes, pulse frequency of 80 kHz and average power of 30 W. Increase in scan speed causes the energy density to decrease and less amount of heat to be conducted to the work-piece resulting in reduction in thickness of HAZ layers. As mentioned previously, if the spot overlapping increases then the surface roughness will tend to decrease. It is worth mentioning that in this present 20 set of experiments, pulse frequency was set at moderately low value, but pulse width was relatively at high value. When scan speed was at lower value, it produced high level of spot overlapping and continuous power density per unit length. On the other hand, when scan speed was increased at moderate setting of pulse frequency, it produced low spot overlapping, discontinuous power density and less material interaction time. Due to continuous power

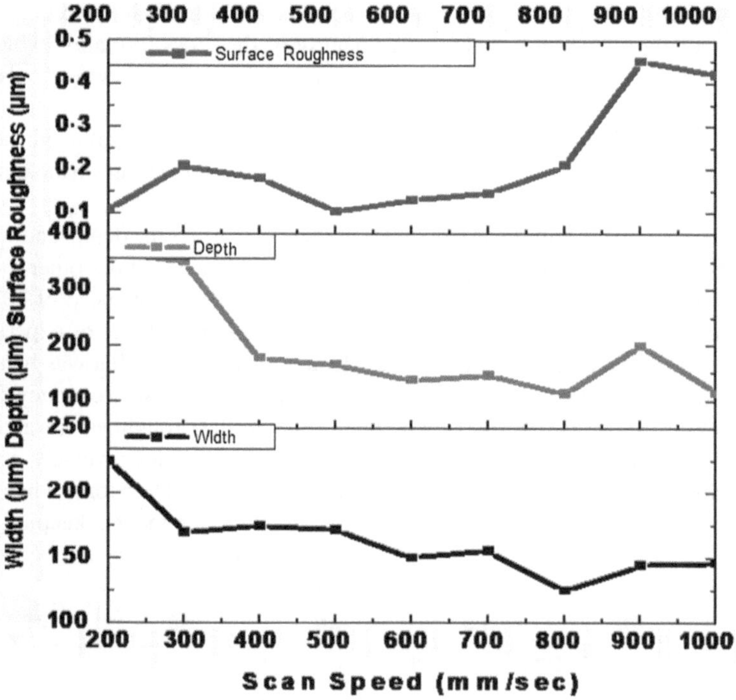

Fig. 4 Effect of scan speed on width, depth and surface roughness

density and more spot overlapping produced by lower scan speeds, dimensions of both width and depth were more as compared to the phenomenon observed at higher scan speed. In addition to that, reduction of penetration rate at each pass and sputtering of molten materials were observed at the micro-groove surfaces because of high number of passes. Each mentioned parameters had pivotal roles on Ti-6Al-4V micro-groove geometry and surface topology. Therefore, variations in both width and depth dimensions were observed at high scan speed and both had the tendency to reduce. In contrast, surface of Ti-6Al-4V micro-grooves showed reverse results for the same reasons. As a result, surface of the machined micro-grooves became rough at higher scan speed and vice versa.

3.4.2 Influence of Pulse Frequency on Width, Depth and Surface Roughness on Ti-6Al-4V Micro-grooves

Pulse mode operation of the fiber laser system provides a high instantaneous power for a period of pulse duration and followed with a period of power off. During pulse mode fiber laser micro-grooving operation, melt ejection occurs in each laser pulse during the time when melt front cools during pulse off-time. It is important for the micro-grooving operation to reduce the quantity of the dross formation which can be achieved by producing low heat input. The relationship between pulse frequency, laser beam energy and average power is shown by Eq. 1. Low heat input can also reduce the quantity of melt formed but also imparted sufficient cooling between two successive laser pulses which prevented overheating of the cut front (Steen and Mazumder 2010).

Figure 5 shows the influence of pulse frequency on width, depth and surface roughness of fiber laser machined micro-groove surfaces while other process parameters were kept constant such as 8 number of passes, scan speed of 40 mm/s and average power of 30 W. It can be observed in the present set of experiments that scan speed remained constant at a moderately high value. This was beneficial not only for understanding the complex phenomenon of pulse frequency on the micro-groove geometry but also for reducing the surface roughness in a considerable amount. Equation 1 revealed that pulse frequency varies inversely with the laser beam energy when laser average power is constant. Therefore, when pulse frequency was increased during the present set of experiments, keeping other

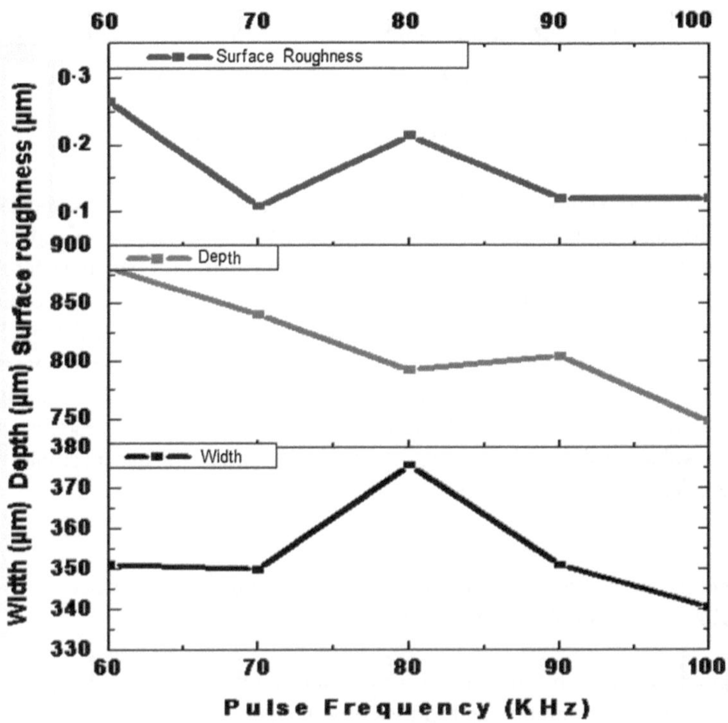

Fig. 5 Effect of pulse frequency on width, depth and surface roughness

process parameters constant, laser beam energy reduced inversely. This phenomenon holds true with the moderate high settings of scan speed. In contrast, this phenomenon reverts back with the combination of moderately low settings of scan speed with increased pulse frequency. Low settings of pulse frequency lead to the increase in laser beam energy. This causes more heat input to the micro-groove surface and thus melting of more amount of material was observed. Therefore, width and depth dimensions were relatively at high values in low pulse frequency. This phenomenon also holds true for producing rough micro-groove surfaces at low pulse frequency. In contrast, with the increase of pulse frequency with moderate high settings of number of passes, reduction of laser beam pulse energy as well as reduction of depth of penetration might occur. The effect of spot overlapping also has a pivotal role in micro-groove topology. In addition, the melted material may not be removed completely and the remaining melted material can be re-solidified at the micro-grooving edges at high settings of pulse frequency. Therefore, low heat generation at high settings of pulse frequency, led to moderate changes in the dimensions of both width and depth; even though a tendency to reduce in the dimensions of width and depth were observed. Along with this, reduction in surface roughness due to the low heat generation and more spot overlapping at these settings of pulse frequency was observed. Hence better micro-groove topology was produced with the increase of pulse frequency.

$$\text{Laser beam energy} = \frac{\text{Average power}}{\text{Pulse frequency}} \tag{1}$$

3.4.3 Influence of Number of Pass on Width, Depth and Surface Roughness on Ti-6Al-4V Micro-grooves

Figure 6 shows the influence of number of pass on width, depth and surface roughness of fiber laser machined micro-groove surfaces while other process parameters were kept constant as scan speed of 40 mm/s, pulse frequency of 50 kHz and average power of 30 W. The present parametric combinations of 12 set of experiments revealed that the experiments were carried out at high heat input on the micro-groove surfaces. It can be observed from Fig. 6 that the width increases with the number of passes, but reverts back after a certain increment of number of passes. This phenomenon happened due to the striking of laser beam on the work-piece surface in a repetitive manner during pulse on time. In addition, the length of focus was altered every time as the number of passes went on increasing. As a result of that spot size of the laser beam was increased significantly. Therefore, with the increased spot size, the penetration rate into the desired thickness of the work-piece

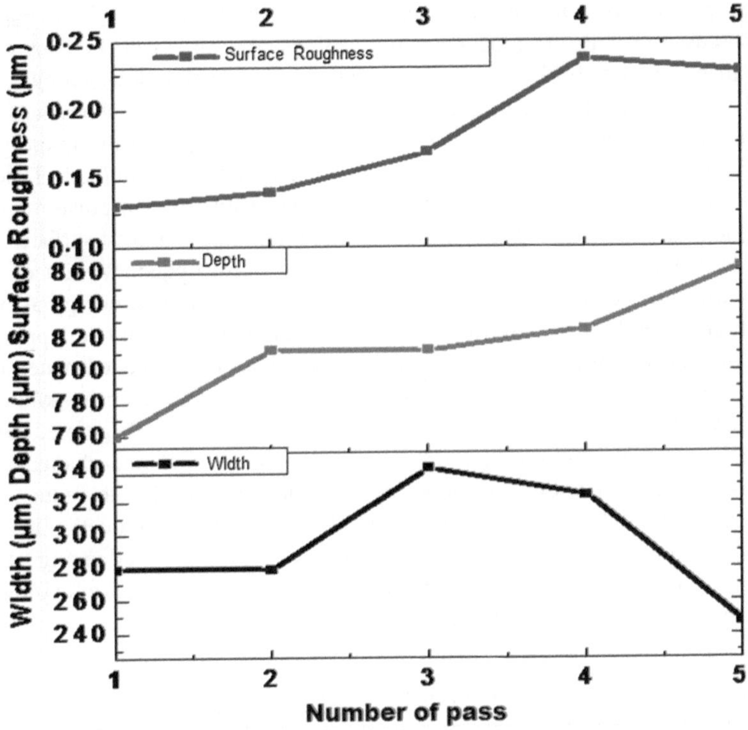

Fig. 6 Effect of number of pass on width, depth and surface roughness

could not be achieved. Due to repetitive striking of laser beam on the work-piece surface and pulse mode operation of fiber laser system, the depth increased. Due to the absence of inert gaseous systems in the present machining setup, some amount of molten material were re-solidified on the side wall of the micro-grooves. The moderate high setting of scan speed led to the phenomenon of sputtering at the edges of the micro-groove surfaces. Therefore, due to the combined effect of sputtering and re-solidification, the dimensions of width were reduced after certain increment of the number of passes as observed in Fig. 6. This phenomenon also holds true for micro-groove topology. The machined micro-grooves surfaces were rough as compared to those produced at low number of passes.

3.4.4 Influence of Average Power on Width, Depth and Surface Roughness

Laser power determines the direct energy input to the micro-cutting process. At lower laser powers, the energy supplied to the cutting front may be insufficient to cut through the desired depth of the material, whereas at higher powers, production

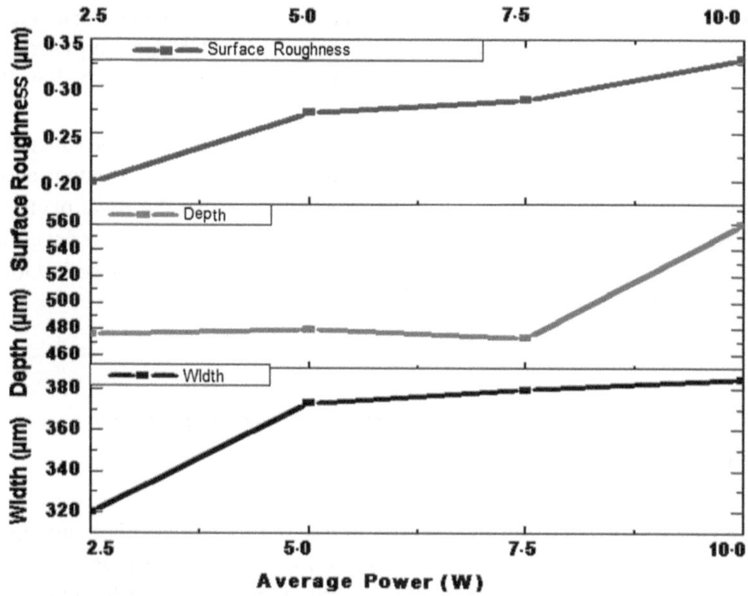

Fig. 7 Effect of average power on width, depth and surface roughness

of clean through cuts necessitates the reduction of cutting speed (Brown and Arnold 2010). Laser average power also determines the maximum cutting speed which is defined as the minimum speed at which through cut is produced. Both micro-cutting quality and performance is dependent on the laser average power.

Figure 7 represents the influence of the average power on width, depth and surface roughness of fiber laser machined micro-groove surfaces while other process parameters were kept constant such as 5 number of passes, scan speed of 100 mm/s and pulse frequency of 50 kHz. It can be observed from Fig. 7 that the amount of re-solidified material on the micro-groove surface increased with the increase in laser average power. Increase in laser power led to increase in the dimension of width up to a certain extent due to the increase in more power density and power input per unit area (refer to Eq. 1). In addition to this, low settings of pulse frequency produced more spot overlapping, continuos power density at the time of fiber laser micro-machining of Ti-6Al-4V micro-grooves. Hence, better micro-groove topologies were observed at low average power and vice versa. It was observed that the effect of laser energy in the geometry of the micro-grooves was not having the same functional dependence in both width and depth dimensions. The tendency of width dimensions was found to be increasing with the combine parametric effects of low laser average power, low pulse frequency and high scan speed. Large heat input led to the removal of more amount of molten material from the micro-groove surfaces, although the peak power was moderately low due to high setting of pulse width. Hence, dimensions of groove depths were changed insignificantly with the increment of average power. The same can be stated in case of width dimensions which increased moderately up to a certain extent and then remained almost constant after that. It was also obvious that the focal plane varied each time during pulse off/on time and the rate ofpenetration also reduced as the machining time increased. Therefore, changes in the dimensions of both width and depth were insignificant.

3.5 Photographic Exhibits of Ti-6Al-4V Micro-grooves and Its Surface Characteristics

Figure 8a shows the micro-groove profiles of two micro-grooves measured in Leica 5× lens when scan speed was at 1000 and 900 mm/s respectively. Figure 8b and c shows the two micro-groove profiles of the width dimensions at different parametric settings i.e. 30 W of average power, scan speed of 40 mm/s, 4 number of passes and pulse frequency of 50 kHz and 30W of average power, scan speed of 40 mm/s, 8 number of passes and pulse frequency of 50 kHz respectively. Better micro-groove profile and reduction in width dimensions were observed when the number of passes was 8 as compare to number of passes were 4. Figure 9a, b provides the 3D view of four graphical displays of the measured data and other elements at different parametric settings as shown in the figures. At the right side of the 3D graphs, there is a color scale that shows which measured signal level that is mapped by the respective color. Therefore, Z height data is encoded with that colour scale and displayed 2-dimensionally. From Fig. 9a, it can be understood that when scan speed is 700 mm/s at 80 kHz of pulse frequency as compared to the scan speed of 40 mm/s at 60 kHz of pulse frequency, better finished micro-groove surfaces can be obtained. It can also be also observed that from Fig. 9b that at 12.5 W of laser average power (other parameters were constant) better surface topology can be obtained compared to average power of 30 W (other parameters were constant).Thus, selecting a combination of optimum parameters can lead to a better surface quality for potential application in the field of bio-medical engineering.

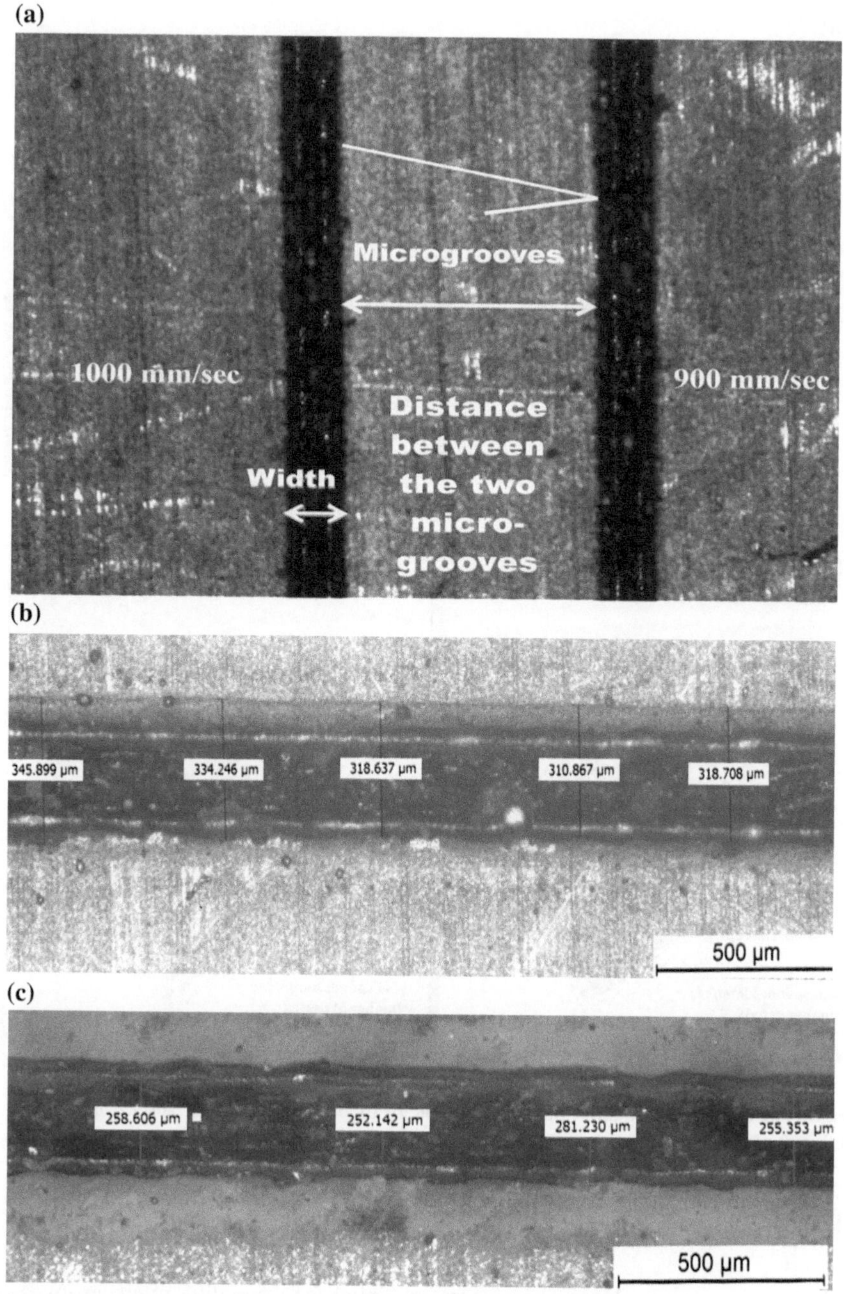

Fig. 8 **a** Microscopic view of different micro-groove profile at different scan speed. **b** and **c** Microscopic view of the micro-groove width at different number of passes

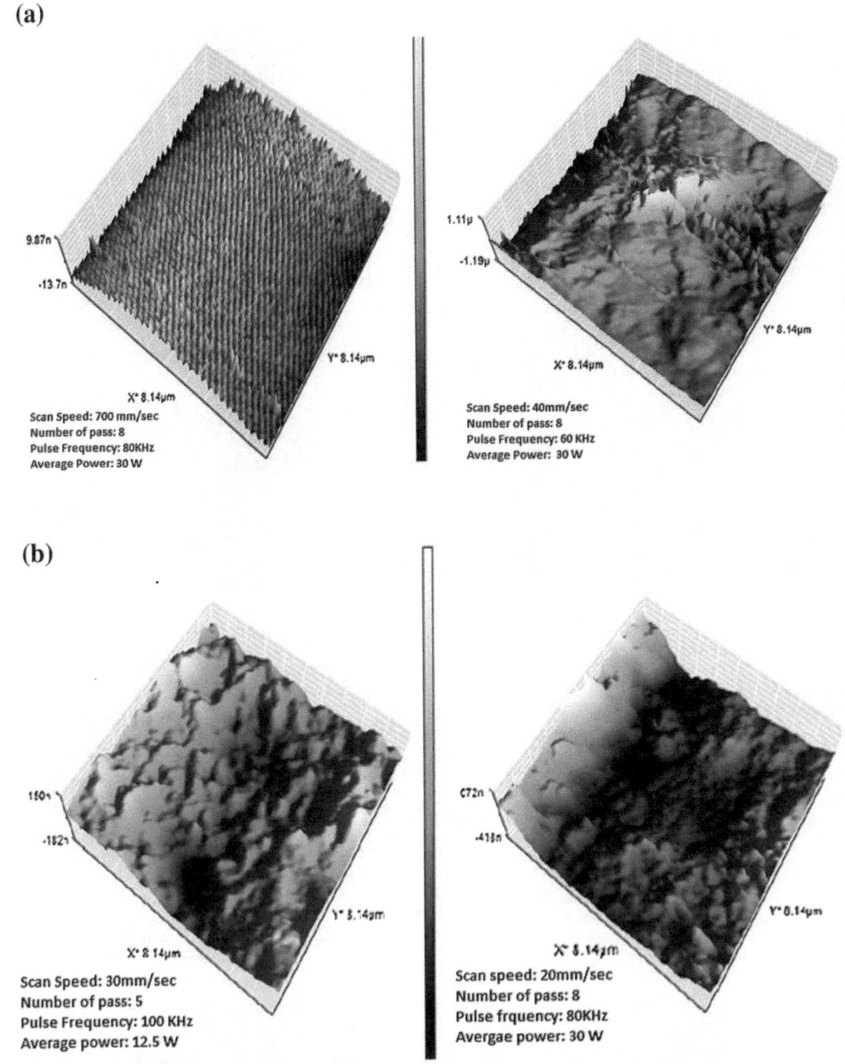

Fig. 9 **a**, **b** 3D representation of surface topology of the micro-grooved surface topologies at different parametric settings

4 Conclusions

Fiber laser micro-machining has a great potential for generating micro-grooves on Ti-6Al-4V. The present chapter mainly deals with the influence of four process parameters such as average power (2.5–30 W), number of passes (1–8), scan speed (40– 1000 mm/s), pulse frequency (50–100 kHz) during laser micro-grooving on

Ti-6Al-4V using 50 W diode pumped fiber laser. Different process parametric setting led to different width, depth and surface roughness. From the results of present experimentation, the following conclusions can be drawn.

(a) Depth and width dimensions tend to decrease with high scan speed with the combination of low pulse frequency, but surface roughness has a tendency to gradually increase owing to low spot overlapping and discontinuous power density at high scan speed.
(b) With the increase of pulse frequency combining moderate scan speed, reduction in the dimensions of width and depth can be observed owing to low heat generation. At higher values of pulse frequency, the more spot overlapping results continuous power density and smoother micro-groove surfaces.
(c) With the increment of number of pass, depth tends to increase while width decreases. After a certain number of pass, width will tend to decrease owing to re-solidification and sputtering of the molten material on the micro-grooved surface. Rough surface quality is observed due to the increase in the number of passes.
(d) As the average power increases, width and depth simultaneously increase due to more heat input and evaporate more amount of material from the micro-groove zone. Poor surface quality can also be observed with the increase of average power.

Fiber laser micro-machining is a unique process which can produce micro-grooves on titanium alloys with better accuracy. The growth of the fiber laser is commendable within a short span of time both in research and industrial perspectives. The present research work not only facilitates the use of fiber laser in the domain of micro-machining but also simplifies the complex phenomena of the micro-machining of the titanium alloys. Further research work can be carried out for determining parametric settings so as to achieve optimized values of better micro-machining quality along with minimum width and maximum depth.

Acknowledgments The authors acknowledge the financial support and assistance provided by CAS Ph-IV program of Production Engineering Department of Jadavpur University under University Grants Commission, New Delhi and TEQIP phase II program of Jadavpur University also M/S Sahajanand Laser Technology Limited for extending the machining facilities.

References

Al-Hajri, E., Ohadi, M., Dessiatoun, S. V., & Qi, J. (2005). Thermal performance of micro-structured evaporation surfaces: application to cooling of high flux microelectronics. *ASME Heat Transfer Division of Publication HTD, 376*, 799–805.

Almeida, I. A., Rossi, W. D., Lima, M. S. F., Berretta, J. R., Nogueira, G. E. C., Wetter, N. U., & Vieira, N. D, Jr. (2006). Optimization of titanium cutting by factorial analysis of the pulsed Nd: YAG laser parameters. *Journal of Materials Processing Technology, 179*, 105–110.

Brown, M. S., & Arnold, C. B. (2010). *Fundamentals of laser-material interaction and application to multiscale surface modification. Laser precision microfabrication*. Berlin: Springer.

Chen, J., Ulerich, J. P., Abelev, E., Fasasi, A., Arnold, C. B., & Soboyejo, W. O. (2009). An investigation of the initial attachment and orientation of osteoblast-like cells on laser grooved Ti-6Al-4V surfaces. *Materials Science and Engineering: C, 5*, 1442–1452.

Chesnutt, J., Thompson, A. W., & Williams, J. C. (1980). *Titanium 80. Science and technology* (pp. 1875). New York: AIME.

Collings, E. W. (1984). *The physical metallurgy of titanium alloys*. New York: American Society for Metals.

Dahotre, N. B., & Harimkar, S. P. (2008). *Laser fabrication and machining of materials*. New York: Springer.

Dandekar, C. R., Shin, Y. C., & Barnes, J. (2010). Machinability improvement of titanium alloy (Ti–6Al–4V) via LAM and hybrid machining. *International Journal of Machine Tools and Manufacture, 50*, 174–182.

Dhupal, D., Doloi, B., & Bhattacharyya, B. (2008). Pulsed Nd-YAG laser turning of micro-groove on aluminum oxide ceramic (Al_2O_3). *International Journal of Machine Tools and Manufacture, 48*, 236–248.

Ezugwu, E. O., Bonney, J., & Yamane, Y. (2003). An overview of the machinability of aeroengine alloys. *Journal of Materials Processing Technology, 134*, 233–253.

Fasasi, A. Y., Mwenifumbo, S., Rahbar, N., Chen, J., Li, M., Beye, A. C., et al. (2009). Nanosecond UV laser processed micro-grooves on Ti6Al4V for biomedical applications. *Materials Science and Engineering C, 29*, 5–13.

Hayes, W. C., & Mow, V. C. (1997). *Basic orthopaedic biomechanics*. Philadelphia: Lippincott-Raven.

Hecht, J. (2009). Half a century of laser weapons. *Optics Photonics News, 20*, 16–21.

Hegarty, J., Broer, M. M., Golding, B., Simpson, J. R., & MacChesney, J. B. (1983). Photon echoes below 1 K in a Nd^{3+}-doped glass fiber. *Physics Revised Letter, 51*, 2033–2035.

Koester, C. J., & Snitzer, E. (1964). Amplification in a fiber laser. *Applied Optics, 3*, 1182.

Mahrle, A., & Beyer, E. (2009). Theoretical aspects of fiber laser cutting. *Journal of Physics D: Applied Physics, 42*, 175507 (9 pp.).

Masuzawa, T., & Tonshoff, H. K. (1997). Three-dimensional micromachining by machine tools. *Annuals of the CIRP, 46*, 621–628.

Mauclair, C., Landon, S., Pietroy, D., Baubeau, E., Stoian, R., & Audouard, E. (2013). Ultrafast laser machining of micro grooves on stainless steel with spatially optimized intensity distribution. *JLMN-Journal of Laser Micro/Nano Engineering, 8*, 11.

Meijer, J. (2002). *Laser micromachining, micromachining of engineering materials*. USA: Marcel Dekker.

Nakazawa, M., Kimura, Y., & Suzuki, K. (1989). Efficient Er^{3+}-doped optical fiber amplifier pumped by a 1.48 μm InGasP laser diode. *Applied Physics Letter, 54*, 295−297.

Olsen, F. O., Hansen, K. S., & Nielsen, J. S. (2009). Multibeam fiber laser cutting. *Journal of Laser Applications, 21*, 133.

Poole, S. B., Payne, D. N., & Fermann, M. E. (1985). Fabrication of low loss optical fibers containing rare earth ions. *Electronics Letter, 21*, 737–773.

Rashid, R. A. R., Sun, S., Wang, G., & Dargusch, M. S. (2012). An investigation of cutting forces and cutting temperatures during laser-assisted machining of the Ti–6Cr–5Mo–5V–4Al beta titanium alloy. *International Journal of Machine Tools and Manufacture, 63*, 58–69.

Rathod, V., Doloi, B., & Bhattacharyya, B. (2013). Influence of electrochemical micromachining parameters during generation of micro-grooves. *The International Journal of Advanced Manufacturing Technology,*. doi:10.1007/s00170-013-5304-3.

Ratner, B. D., Hoffman, A. S., Schoen, F. J., & Lemons, J. E. (1996). *An introduction to materials in medicine, biomaterials science*. San Diego: Academic Press.

Schaeffer, R. D. (2012). *Fundamentals of laser micro-machining*. Boca Raton: CRC Press-Taylor & Francis Group.

Shelton, J. A., & Shin, Y. C. (2010a). Experimental evaluation of laser-assisted micromilling in a slotting configuration. *Journal of Manufacturing Science and Engineering, Transactions of the ASME, 132*, 0210081.

Shelton, J. A., & Shin, Y. C. (2010b). Comparative evaluation of laser-assisted micromilling for AISI 316, AISI 422, Ti–6Al–4V and inconel 718 in a side-cutting configuration. *Journal of Micromechanics and Micro engineering, 20*, 075012.

Sieniawski, J., Ziaja, W., Kubiak, K., & Motyka, M. (2013). *Microstructure and mechanical properties of high strength two-phase titanium alloys, titanium alloys—advances in properties control*. Croatia: In Tech.

Snitzer, E. (1961). Proposed fiber cavities for optical masers. *Journal of Applied Physics, 23*, 36–39.

Steen, W., & Mazumder, J. (2010). *Laser material processing*. London: Springer.

Sun, S., & Brandt, M. (2013). *Laser beam machining, nontraditional machining processes*. London: Springer.

Sun, S., Brandt, M., Barnes, J. E., & Dargusch, M. S. (2011). Experimental investigation of cutting forces and tool wear during laser-assisted milling of Ti–6Al–4V alloy. *Proceedings of the Institution of Mechanical Engineers, Part B: Journal of Engineering Manufacture, 225*, 1512–1527.

Sun, S. J., Harris, J., & Brandt, M. (2008). Parametric investigation of laser-assisted machining of commercially pure titanium. *Advanced Engineering Materials, 10*, 565–572.

Teodoro, F. D. (2011). *Pulsed fiber laser, high power laser hand book*. New York, USA: McGraw-Hill.

Tian, Y., Wu, B., Anderson, M., & Shin, Y. C. (2008). Laser-assisted milling of silicon nitride ceramics and Inconel 718. *Journal of Manufacturing Science and Engineering, Transactions of the ASME, 130*, 031013.

Wang, W., & Poh, C. K. (2013). *Titanium alloys in orthopaedics, titanium alloys—advances in properties control*. Croatia: In Tech.

Williams, D.F. (1984). *Titanium and titanium alloys, biocompatibility of clinical implant materials* (pp. 44–47). Boca Raton: CRC Press-Taylor & Fransis.

Yang, J., Sun, S., Brandt, M., & Yan, W. (2010). Experimental investigation and 3D finite element prediction of the heat affected zone during laser assisted machining ofTi6Al4V alloy. *Journal of Materials Processing Technology, 210*, 2215–2222.

Zhang, S. Y., Lin, X., Chen, J., & Huang, W. D. (2007). Influence of heat treatment on the microstructure and properties of Ti–6Al–4V titanium alloy by laser rapid forming. *Rare Metal Materials and Engineering, 36*, 1263–1266.

Nd:YAG Laser Marking on Zirconia Ceramic

Josephine Peter, B. Doloi and B. Bhattacharyya

Abstract Laser marking is one of the well-developed technologies of materials processing. Laser marking is the best and most applied permanent marking method. Ceramic is a difficult material to be processed by conventional marking techniques due to its high hardness and brittleness. This chapter deals with the artificial neural network (ANN) and the response surface methodology (RSM) based mathematical modelling and also an optimization analysis on marking characteristics i.e., mark width, mark depth and mark intensity on zirconia ceramic. The experiments have been planned and carried out based on Design of Experiment (DOE). The major influencing laser marking process parameter considered are pulse frequency, lamp current, pulse width, scanning speed and air pressure. The experiments have been planned and carried out based on RSM based modelling with 32 runs. ANN modelling is performed and the results are compared. The average percentage of prediction errors of the developed ANN model for mark width, mark depth and mark intensity are 2.52, 2.58 and 2.58 respectively and the overall percentage of prediction error is 2.6. The output of the RSM optimal data is validated through experimentation and ANN predictive model. A good agreement is observed between the results based on ANN predictive model of 82.8 μm, 46.3 μm and 0.605 for mark width, mark depth and mark intensity respectively and actual experimental observations.

Keywords Laser marking · Zirconia · RSM and ANN

J. Peter (✉) · B. Doloi · B. Bhattacharyya
Production Engineering Department, Jadavpur University, Kolkata 700032, India
e-mail: joshiriyapeter@gmail.com

B. Doloi
e-mail: bdoloionline@rediffmail.com

B. Bhattacharyya
e-mail: bb13@rediffmail.com

S.N. Joshi and U.S. Dixit (eds.), *Lasers Based Manufacturing*,
Topics in Mining, Metallurgy and Materials Engineering,
DOI 10.1007/978-81-322-2352-8_16

1 Introduction

Laser marking is flexible, programmable and environmentally clean method which employs a thermal process that utilises high intensity beam of focused laser light to create a contrasting mark on the material surface and is primarily used to mark or print bar codes, matrix codes, universal product codes (UPC) and serial numbers, etc. (Restrepo et al. 2006). The main application of laser marking is for the purpose of product identification and traceability (Chen and Darling 2005). Compared to traditional marking techniques, such as inkjet, hot stamping or mechanical scribing, laser marking offers many advantages such as flexibility, speed, indelibility, reliability, free of consumables and high cleanliness (Ng and Yeo 2001). Nd:YAG laser marking has great potential in diverse range of applications, i.e., electronics components, automotive products, jewellery, medical implants and surgical instruments, etc. The line width produced by Nd:YAG laser is narrower than that of CO_2 lasers, which is a major factor for higher resolution. In some cases, second and third harmonic Nd:YAG lasers at 532 and 355 nm, as well as ultraviolet excimer lasers are the optimum choice to mark highly transparent materials. The application of laser technology is pacing the demands from industry for smaller geometries and increasingly tighter specifications. At the same time, the requirements with regard to the quality, flexibility, speed and production costs of the marking process are becoming increasingly stringent. The diverse range of applications in the area of product recognition and identification are given as follows:

(a) Computer Equipment

Marking text and graphics on plastic computer peripherals. An entire keyboard can be marked in a single operation.

(b) Automotive Products

Marking text, graphics and part numbers on ball and roller bearing races, gears, shafts, instrument displays and audio bezels for identification. Marking layout, current ratings and protection information on fuse box covers.

(c) Electronic Components

Marking epoxy packaged ICs and other micro components with text and graphics for identification. Cleaning epoxy flash material from capacitor legs to ensure integrity of electrical connections.

(d) Electrical Switchgear

Production of text and graphics on plastic relays, circuit breakers and switch covers to meet specific requirements of approvals bodies in different markets.

(e) Communications Equipment

To create graphics on mobile phone keypads and production of text and logos on plastic bodies of telephone headsets.

(f) Agriculture Equipment

Replacement of hot foil marking to generate text, graphics and bar codes on livestock identification tags.

(g) Promotional Products and Giftware

Origination of text and graphics on pen barrels, key rings, knives, trophies, picture frames, plaques, etc.

(h) Light Shows

Removal of metallic coatings from glass discs to produce graphics on gobos used in laser light shows.

Technical demands from the electronic, medical, aerospace and other industries are pushing the boundaries into micron level for smaller components which need fine marking with more precise tolerances. Nd:YAG laser has some unique properties for using it in marking processes. The advantage of laser systems is the small beam width, allowing manufacturers to mark particularly small parts, an issue of increasing concern as the quest for miniaturization is being increasing very rapidly (Chen et al. 2008). The relative robustness and compactness of the laser and the possibility for the light it produces to be transmitted to the work piece via optical fibres are two features that contribute to its success. Laser beam can be focused and deliver sharp-edged energy directly to a tiny region without heating the rest of the material which allows for greater flexibility in the factory environment.

Advanced ceramic materials have great applications in the various fields due to their explicit properties. The typical properties exhibited by zirconia are high strength, high fracture toughness, excellent wear resistance and chemical resistance, high hardness and toughness which make its applications in wide fields like orthopaedic implants, manufacture of cutting tools, refractory applications and structural opacifiers etc. Zirconia has great potential as substitutes for traditional materials in many biomedical applications. These properties and features of zirconia add to value of the material, which makes laser marking on zirconia to have an important scope. On the other hand, the major difficulty in the process is the number of laser parameters involved.

Nd:YAG lasers are typically found in industrial applications as they are known for their ability to mark metal, either by ablation or by creating a colour change through annealing and therefore Nd:YAG lasers are able to provide more detailed graphics (Adelina and Dinu 2008). The absorption properties of Nd:YAG laser beam of wavelength of 1064 nm on various structural ceramics is also very high. However, this absorption also greatly depends on the optical properties, surface temperature and also on surface roughness of the ceramic work piece. The major concern in the marking process is the number of laser parameters involved. Hence there is a requirement to explore it in detail on laser marking of ceramics, which has different applications in various fields. In order to attain clear and excellent mark, it is most important that proper laser marking methods and the suitable process parameters are to be selected for the specific material, which leads to increase in the marking quality. Laser marking characteristics may be altered due to a small change in the marking parameters; therefore, monitoring and control of the laser marking parameters are necessary. This chapter deals with the modelling of laser process parameters like pulse frequency, lamp current, pulse width, scanning speed and air pressure on the marking characteristics i.e., mark width, mark depth and intensity using RSM. Moreover optimized process parameters for effective laser marking are determined. Artificial neural network (ANN) predictive model is also been developed to confirm the results and to train the data and predict the optimum response based on the empirical model obtained through RSM.

2 Basic Mechanism of Laser Marking

Lasers produce a significant amount of energy in a small defined area. This focused energy is far greater than that achieved by any other process as it heats, melts or vaporizes within selected areas. Lasers can achieve significant contrast by material melting, material annealing or material vaporization, all effects being uniquely permanent. The principle of laser marking is that a laser beam modifies the optical appearance of a surface that it hits. This can occur through a variety of mechanisms. Mechanisms for marking can broadly be divided into two categories:

(a) Marking by material removal from the surface.
(b) Marking by surface modification.

2.1 *Marking by Material Removal from the Surface*

In this mechanism, removing material with a laser creates a visible mark: either by melting, vaporization or photochemical decomposition or ablation and engraving. The resulting surface morphology creates a readable high contrast mark.

2.1.1 Etching

Laser etching is a process by which the heat applied to the surface is increased to a level that causes substrate surface melting. The laser beam heats the surface of a material and as it cools, the material takes on a different finish. The etching process actually removes small amounts of material from the metal piece to create a mark. The actual etching process is initiated by absorption of laser light by a solid in contact with a gas or liquid, a gas in contact with a solid, or by a both a solid and a gas simultaneously (Houle 1986). The excitation will lead to a reaction between the solid and gas or liquid. Depending on the wavelength of light used, absorption results in electronic excitation, which can subsequently melt, to heat, or in immediate direct heating. The former is referred to as photochemical, and the latter photo thermal. Laser marking process involves the application of high speed laser etching techniques on lead frames that were protected with polymer coating (Ho et al. 1997). The input energy required to initiate etching is relatively small, and so etching is chosen over ablation whenever it is desirable to minimize the thermal loading on a material.

Etching is often used for industrial purposes, such as for marking tools or parts with serial numbers, logos, and barcodes. It is only applicable to certain metals, metallic coated surfaces and some plastics. Excellent results can be routinely obtained at penetration depths of less than 25 μm (Dahotre and Harimkar 2008). This technique, however, should not be used on some metals where material integrity is critical to the part because cracks produced in the molten metal during cooling can propagate into the underlying surface material. These cracks can expand downward if the part is stressed and/or after repeated temperature cycling. These conditions have led to metal fatigue and part failures. Laser etching is not recommended for parts thinner than 1.5 mm.

2.1.2 Surface Melting

The material can be brought to a molten state as a result of melting. The laser beam removes a layer of coating such as black oxide, paint, anodization, chrome, etc., to reveal the base material. This method creates very sharp and detailed characters and graphics due to the high contrast between the coating and the base material. An analytical model for the laser melting of inhomogeneous work pieces with parabolic melt pool geometry were studied to find the melt pool depth during laser marking to determine the thickness of the resulting surface layer after solidification (i.e. the glaze thickness), which is related to the quality of the mark (Peligrad et al. 2001, 2002). It is frequently employed to induce a color change in plastics. A wide variety of commercial plastics yield excellent color contrast and high quality marking images. Excellent results are routinely obtained at penetration depths of less than 25 μm.

2.1.3 Ablation

Ablation refers to material being ejected by the interaction of the high intensity laser pulse with the material. Ablation is most often performed in vacuum or air. A part of the incident light energy is absorbed in the material and transformed into heat. Depending on material conductivity and thermal capacity the bulk material temperature rises. When temperature reaches a certain threshold value melting and vaporization appears. Parts of the heated material that reach such high temperatures are removed by means of vaporization and vapour caused ejection of melt. Solidified melt forms a recast layer on the crater wall (the sides and bottom of a laser mark) and condensed vapour is deposited as debris on the surface around the interaction region (Nedialkov et al. 2005). These phenomena (debris and recast layer) may adversely influence the laser mark visibility so they should be minimized by proper choice of processing parameters. Generally, shorter pulses are more efficient and produce less unwanted depositions near the processing area. In addition, at a given beam energy E the peak beam intensity (power/area) is higher at shorter beam pulse, so material removal threshold values are attainable.

Material is removed from the work piece, often resulting in with little or no change in coloration. This type of interaction is often found when carbon black is used for coloration due to the high laser energy absorption and heat generation. The technique may be used on homogeneous materials, in which case the mark arises as a result of the contrast between the substrate and the ablated regions. Marking in carbon black (CB)/polypropylene (PP) compounds using a 532 nm Nd:YAG laser were performed (Feng et al. 2001). Both thermo physical and photochemical reactions are found to co-operative during laser processing of conventional polymers. The polymer made the ablated region appears white, contrasting sharply with the black background of the original surface.

Alternatively, it may be applied to layered substrates; the ablated region reveals an underlying region that provides high contrast—a popular technique for marking polymers. The effects of laser ablation upon multiple-layered coloured inks were carried out upon four layered ink samples consisting of white/cyan/white/black layers showed that it was possible to selectively remove the inks to expose both the top white and the cyan layers, although charring occurred with deeper ablation (Stewart et al. 2000). Ablation is used in partial removal of coating layers from the base material that are marked with different colour.

2.1.4 Engraving

The laser beam penetrates the surface and removes material in the laser path through melting displacement and/or evaporation. Laser engraving involves more heat than laser etching and results in the removal of substrate material. Localized increase in temperatures above the melting point of the material causes melting. Once the material is resolidified, a modified surface structure occurs in the form of a mark. The high contrast obtained by laser colouring or etching cannot be obtained by laser engraving because the material is vaporized and ejected during the marking process. A Q-switched diode-pumped frequency-doubled Nd:YAG green laser to deep engrave different kinds of wood without carbonization of the surface was demonstrated (Leone et al. 2009). Although this method appears to be the most appropriate laser marking technique, it generally produces less damage to the substrate than laser etching; however, because most of the structural strength of materials is in the integrity of their surface and because it can produce micro cracking in some materials. Like laser etching, direct laser engraving can be easily determined by power microscope (10×) magnification. Laser engraving is not recommended for use on parts thinner than 2.5 mm. Nearly any material can be engraved. Laser engraved marks are exceptionally durable and can only be removed by heavy grinding. Therefore laser engraving is used for forgery-proof product marking.

2.2 *Marking by Surface Modification*

In this mechanism, the laser radiation affects the material composition to create a high contrast mark without material removal. It may locally melt the material, causing it to oxidize or chemically alter to form a visible mark.

2.2.1 Foaming

The foaming process occurs when the local temperature surrounding the absorption place is sufficiently high to generate gases via burning or evaporation. The hot gases are themselves surrounded by molten material and expand to form bubbles. If the energy of the laser is controlled, foaming can result in bubbles that scatter light in such a way that it results in white or light-on-dark marking contrast (Ion 2005). In order to obtain this type of result the laser needs to be set to high pulse frequencies, resulting in high heat transmission to the compound. White coloured markings on dark surfaces can be achieved through foaming which gives very high contrasts. Small gas bubbles form in the molten material which increases the volume creating a type of plastic foam. This makes the processed areas appear much brighter than the surrounding material. In some cases, the marking projects from the surface. This is referred to as a foamed marking. As a plastic material is heated by the laser beam, gas pockets form and the material starts to foam leaving a foamed light mark. The foam rises approximately 40 μm from the surface and extends to a depth of about 60 μm in the polymer surface. As the marking is partly raised on the surface, its wear resistance may be somewhat reduced.

2.2.2 Carbonisation

A dark or black marking can be achieved on a light surface by a laser heating which causes localised carbonization of the polymer matrix. The process occurs when the absorbed energy raises the local temperature of the material surrounding the adsorption site high enough to cause thermal degradation of the polymer. Laser oxidizes only the surface of material into black, resulting in a smooth and cut-less surface. This marking applies to lubricious parts. Also it is suited for metal of iron group, stainless steel, silicon wafer, etc., where visibility is an important issue. Certain particle substrates coated with certain metal oxides may be employed as laser sensitive pigments for use in laser marking of plastics using different types of lasers. Hence the components of the coating layer were chosen to optimize laser sensitivity. Sensitivity Nd:YAG laser is primarily attributable to the marking components of the metal oxide layer, namely CuO, MoO_3, WO_3, V_2O_5, Ag_2O, or PbO_2. The oxide layer may also become partially reduced upon laser irradiation, forming dark-colored sub-oxides which increase the marking contrast (Fu 2004). The effect is best seen on light colour materials. Extreme carbonisation may lead to excessive local heating and carbonisation which can affect the physical properties of the matrix. In some cases, this can produce dark brownish markings.

2.2.3 Annealing

The laser is used to heat metal to near melting points, which induces a colour change to the top layer of material (Sugioka et al. 2010). Annealing with the heat of the laser beam creates a dark, burr-free, permanent mark without noticeable surface penetration. Laser annealing creates an indelible mark induced by heat and oxygen without noticeable material ablation. Laser annealing is used with all metals which show a colour change when exposed to heat and oxygen, especially stainless steel. Annealing often gives a dark iridescent look, with a faint rainbow of greens, blues, and pinks that can sometimes be seen in the text or graphic. The colour depends strongly on the temperature to which it is heated. Annealing leaves no cuts or shallow marking and typically produces the darkest. Because no material is removed from the metal, this technique is often used for medical devices used within the human body. Annealing leaves no cuts or shallow engravings like those found in marking and etching and typically produces the darkest mark of these three methods. As a burr-free process, annealing offers significant advantages when marking already finished surfaces. Process speed is lower compared to laser engraving because annealing relies solely on thermal effects (Ready 2001). Comparatively low temperatures can be applied to metal to anneal the surface. The marking beam will produce a sharp, contrasting line to the surrounding material with very shallow material penetration. Marking by annealing has the advantage of not disrupting the surface that is important for some medical applications, specifically implantable devices.

2.2.4 Colouring

Laser colouring is a process used to discolour some metallic substrate materials without burning, melting, or vaporizing the material. This is done by passing a relatively low power laser beam across the surface at slow speed to discolour the area of the mark. This laser marking method produces a high-quality; high-contrast marking that does not disrupt the surface (Ready 2001). The process, however, does not work with all materials and can have an adverse affect on materials that have been previously heat-treated and can reduce the corrosion resistant properties of some stainless steel alloys. These mixed colorant systems are dependent on specific colorant stabilities and not all colour changes are possible. Laser formulations cannot be toxic or adversely affect the product's appearance, physical properties, or functional properties. Additives can help improving absorption properties. Applicable laser sources are Nd:YAG, Nd:Vanadate or fiber lasers. Furthermore, harmonic wavelengths of 532 nm (green) and 355 nm (UV) can be used for bleaching and photo reduction processes. Nd:YAG laser beam which is operated in continuous wave mode as a tool to print patterns of coloured enamels on tile substrates was performed (Fernandez-Pradas et al. 2007). The laser marking technology for textile material to remove indigo-dye from denim is also applied. A Nd:YAG laser used with 1064 nm and its second harmonic 532 nm with energy per pulse as high as 40 mJ and a repetition rate of 60 Hz. The denim-fading process occurs when the fluency of the beam is higher than the threshold of interaction (Dascalu et al. 2000). UV-induced photo reduction is often called cold marking, as there is no noticeable heating of the material. This is a most common method for marking plastics such as polycarbonate and ABS.

3 Nd:YAG Laser Marking Process Parameters

The key parameters of the laser marking process that define a given laser's marking ability are the pulse frequency, lamp current, scanning speed, spot size, the peak and average power of the pulse. Therefore, monitoring and control of the laser marking parameters are necessary so that marking is done with high efficiency.

3.1 Pulse Frequency

The pulse repetition rate determines the amount of time between lasers 'bursts' allowed to strike a surface. The high peak power pulses at low frequencies will increase the surface temperature rapidly resulting in material vaporization and minimal heat conduction into the part. At higher repetition rates, the lower peak power will produce much less vaporization but will result in significantly more heat conduction. The greater number of pulses in a given timeframe will also increase the heat conducted into the work surface. The Q-switch pulse rate is probably the most important variable for control of the thermal process. The ability to Q switch the laser provides short pulses with very high peak powers that enable the laser to produce high quality contrasting marks on a wide range of materials. The laser parameters such as pulse duration, peak power and average power can be varied as per requirement according to the material and type of the mark. Typically plastics and fine marking are achieved at lower Q-switch frequencies. Increasing Q-switch frequency reduces peak power, increases pulse width and average power and is more suited to higher threshold materials such as metals. For graphics and large fill areas, high fill rates can be achieved with frequency which provides the thickest line width of marking.

3.2 Scanning Speed

In laser marking process the scanning speed of laser beam with respect to the work piece is another important variable in the thermal process and must be set to achieve the desired process results. For deep marking (typically >0.05 mm) each point on the marked line will require exposure to several pulses to achieve depth. The beam speed must be reduced until the desired depth is achieved. For shallow marking, the speed may be increased to the system's maximum velocity or until the separation between pulses is aesthetically unacceptable at the selected pulse rate setting. The result of pulse frequency and beam speed is the mark point overlap. As a general rule, pulses should overlap at least 50 % to give the appearance of a continuous mark line and around 70–90 % to ensure a good marking. Hence adjustment should be made to compromise between process speed and the quality of a mark.

3.3 Focused Spot Size

Focal spot size determines the maximum energy density that can be achieved when the laser beam power is set, so the focal spot size is very important for material processing. The diameter of the focused laser beam on the work surface determines both the marking line width and marking efficiency. The spot diameter is a function of focal length of the lens and the divergence of the laser beam. This implies that when the focal length is increased the focusing spot size is correspondingly enlarged. The setting of focal length has effects two properties i.e., first, the incident spot size and the energy density, second, the marking field area. The spot diameter (d_0) of laser beam at focal point is determined as:

$$d_0 = (2f\lambda)/D \quad (1)$$

where, d_0 is the diameter of the focused spot, f is the focal length of the focus lens, D is the entrance beam diameter, and λ is the wavelength (Klimt 1988). Figure 1 shows the relation between spot size and percentage of overlap.

3.4 Laser Power

The most fundamental method of checking the performance of a laser is to measure its power or energy output. Laser output directly affects a laser's ability to perform a process. Lasers operate in either continuous wave mode or pulsed mode. Figure 2 shows the difference between continuous and pulsed mode laser power. Both operation states have lot of applications. For pulsed laser, an important parameter is the peak power. In general, CO_2 lasers have relatively high continuous wave power,

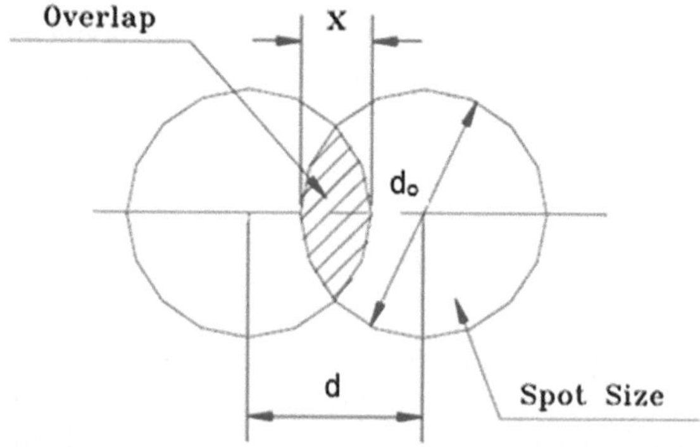

Fig. 1 The relation between spot size and percentage of overlap

while Nd:YAG lasers can provide relatively high peak power for pulsed operation. Output power is closely related with processing time and operation expense. If the selected laser power is lower, the processing time will be increased, if the selected laser has too high power than necessary, the operation expense will be higher than necessary. So the proper choosing of laser power is very important. Laser beam is pulsed at medium to high frequencies so as to get the maximum energy in a short period of time. The lower the frequency with which the energy is released the higher will be the pulse power and the consequent effect. High peak power and power density are required for reflective materials, such as aluminium, brass and copper which can be achieved by lasers with high peak pulse energy. It represents the power averaged over the period of the pulse, and is the product of the pulse energy and the pulse frequency.

For a pulse laser with high pulse frequency, the peak power is usually expressed in terms of average power as (Peter et al. 2013),

$$Peak\ Power\ (P_{peak}) = \frac{Average\ Power\ (P_{avg})}{Pulse\ Frequency\ (f) \times Pulse\ Duration\ (\mu)} \tag{2}$$

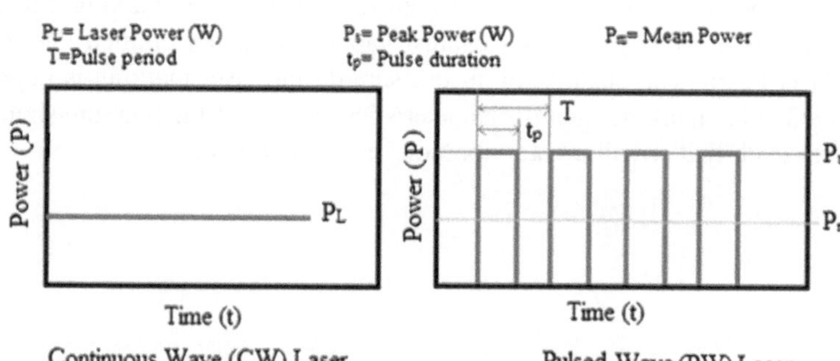

Fig. 2 Difference between continuous and pulsed laser

It is observed that a high peak power is often required for marking. Shorter pulses have higher peak power. The thermal interaction time is also shorter, which lead to smaller heat-affected-zones, and thus better mark quality.

3.5 Lamp Current

The lamp current determines the power or energy being used to stimulate a material into photoemission, i.e., increasing the lamp power increases the laser amplification. When a laser heats up the targeted material, the energy absorbed is conducted into surrounding material; at higher absorption rate the surface region of the material will melt and began to vaporize. Increasing or decreasing the electric current to the krypton arc lamp adjusts the output power of the laser. As the current is changed, the light output from the lamp and the rate of laser amplification increases or decreases accordingly.

3.6 Pulse Width

Pulse width refers to the temporal length of laser pulse; that is, the time during which the laser actually emits energy. Pulse width is expressed as the % of duty cycle. Required pulse width depends on the time needed to vaporize the material. High quality beam can be produced by accurately controlling this parameter in order to achieve quality marking.

3.7 Air Pressure

Air pressure plays an important role in laser marking. It mainly assists to remove the material and to minimize the heat affected zone during laser marking. Moreover, the air protects the focusing optics against vapour or spatter emitted due to high intense laser beam. The quantity of air pressure during laser marking is important, as it can affect the marking quality characteristics. However, the pressure should not be so high such that it will create vibration of machine parts.

4 Evaluation of Nd:YAG Laser Marking Quality Characteristics

Evaluation of laser marking quality characteristics is done by estimating mark width, mark depth and mark intensity. The quality of a mark is assessed by its legibility characteristics. The quality characteristics are usually evaluated using complementary techniques such as optical microscopy, surface roughness measurement, and contrast evaluation. The acceptance of level of each of these quality characteristics generally depends on the customer's requirements.

4.1 Mark Width

The width of the line segment that formed by laser marking is referred to mark width. The marking width is controlled by the focused beam spot size. Other parameters such as beam scanning speed, laser power density and material absorptivity also affect the line width. Generally, mark width is measured using an optical microscope. Sometimes, surface texture measuring equipment is also used for determining the width measurement.

4.2 Mark Depth

The marking depth is mostly affected by the laser energy density, the type of material and the interaction time between the material and laser beam. Marking depth is measured using a surface profile instrument. The depth of penetration in the material that can be achieved varies from a few microns to several tens of millimetres. The assist compressed gas enhances the evaporation of material from marking zone which further results in more marking depth. A further enhancement of the effect on the material can be realized by the supply of gases such as oxygen or compressed air, which assist material removal.

4.3 Mark Intensity

The visual difference between the visible brightness of a particular mark and unmarked portion is referred as mark intensity. The measurement of marking contrast is done with image analysis software. The sharpness or resolution of the marked edges affects the marking contrast or intensity. This parameter is particularly important in marking "bar code", as poor edge sharpness may cause failure of bar code reader. Better edge resolution can be produced due to high peak power.

The marking intensity or the mark contrast (c) can be determined by,

$$c = \frac{g_{background} - g_{mark}}{g_{white} - g_{black}} \quad (3)$$

where $g_{background}$ is the grey level value of unmarked area and g_{mark} is grey level value of marked area. g_{black} is the grey level of black colour and g_{white} is the grey level of white colour. After taking the optical photographs using optical microscope, the image is analyzed in a PC, which computes the grey level of mark of different portion.

5 Laser Marking Technique and Procedure Used for Experimentation

Laser marking can be done by either moving the laser in accordance with the profile which is needed to be generated or by moving the work piece by keeping the laser stationary. The latter technique is followed to mark the work piece in this present work. The schematic representation of the marking technique is shown in Fig. 3. The system consists of various subsystems such as laser generation and beam delivery unit, cooling unit, assist air supply unit and computer numerical control (CNC) for X-Y-Z axis movement. Laser from the laser head is transmitted through the beam bender that bends the laser beam at 90° so that it could be focussed on to the work piece using a lens. Pulsed Nd:YAG laser with a wavelength of 1064 nm with a TEM00 mode of operation is used to conduct experiments. The laser beam is focused with a lens of 50 mm focal length having a spot size of 0.1 mm. The work piece is mounted on the CNC controlled X-Y table. The work piece movement

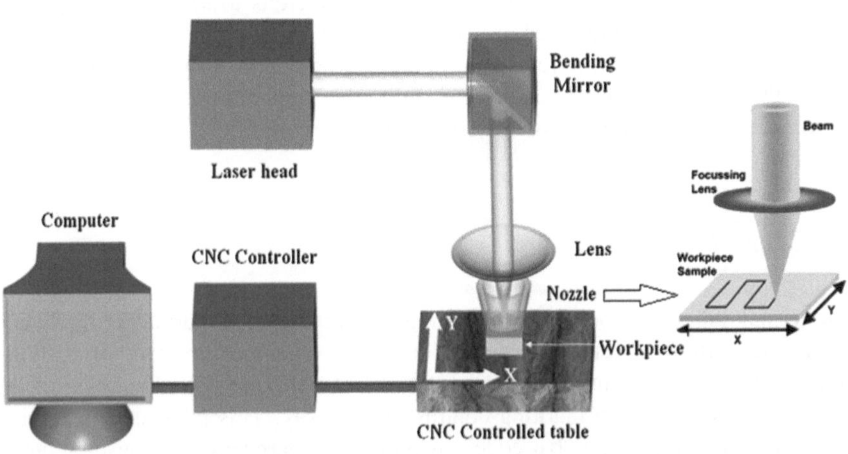

Fig. 3 Schematic representation of CNC controlled pulsed Nd:YAG laser marking system

along the X-Y axis and the focusing lens movement along Z axis are CNC controlled and can be operated via a computer for marking after proper focusing of laser beam by means of focusing lens. The surface of the flat work piece is set as the focal plane. The feed rate is given by X-Y table movement so as the marking is done appropriately. Laser beam is directed to the work piece through a nozzle by which assist gas mainly air is supplied to the surface of the work piece so as to aid in material removal and to prevent resolidification. With the use of Q-switched Nd: YAG laser, laser beam with high peak power is achieved. This enables to mark different materials with good quality. Moreover, the alignment of focusing lens is very important because if the beam centre is not co-axial with the centre of the lens, then the beam after the lens will not be straight and therefore the marking efficiency drastically decreases. Hence special care has to done to check if focussing lens is well aligned. The work piece is held in a fixture and is monitored with the help of charge couple device (CCD) camera. A closed-circuit television (CCTV) is used to view the work piece for focusing the beam accurately for proper marking operation.

Experiments are performed with the aid of compressed air with different air pressures. Figure 4 shows the focussing arrangement of laser marking system. The drawing file is utilised through the software MULTISAW that is interfaced with the CNC controller for the axis movement of the work piece. Any special figures, logos and characters can be plotted by using any plotting software namely AutoCAD or CorelDraw and can be exported to the laser marking software for generating CNC program. Marking with multiple fonts and desired sizes, orientations and rotations can be done by controlling the motion of work piece as per CNC program with the aid of CNC drive systems.

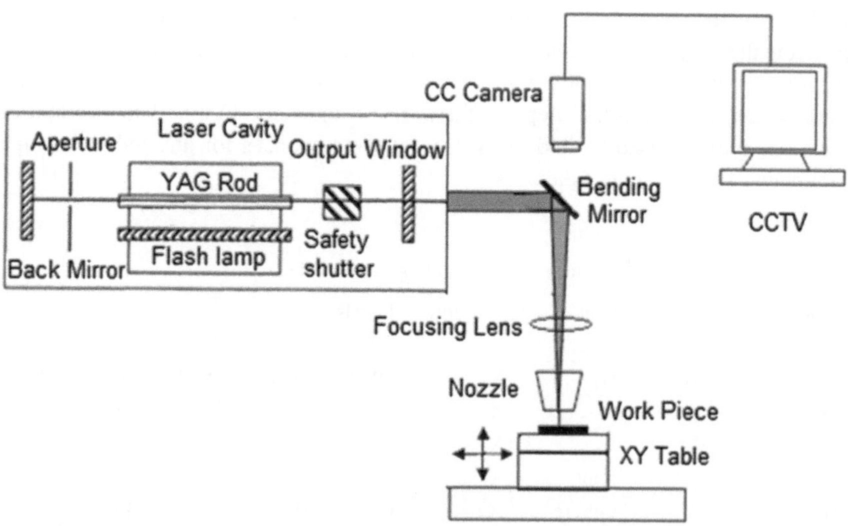

Fig. 4 Focusing arrangement of laser marking system

6 Optimization of Nd:YAG Laser Marking Based on RSM and ANN Model

Modelling of the laser marking operation on zirconia ceramic by using experimental design and response surface methodology (RSM) are performed so as to determine optimal parametric combinations for quality marking. Artificial neural network is also utilized to model laser marking process because of its versatile capability and reliability.

6.1 Experimental Planning and Development of Empirical Model Based on RSM

Response surface methodology based analysis is an effective method for establishing empirical models for optimal analysis on marking quality characteristics of the pulsed Nd:YAG laser marking operations. The design of experiments (DOE) based on RSM are planned for conducting experiments. In order to find the quality characteristics of the laser marking, geometric shapes marked on a flat zirconia work piece of 20 × 30 mm dimension. Central composite rotatable half design (CCD) with 32 runs is selected with an alpha value of 2.0. CCRD is a very efficient method for fitting a second-order model. Experiments were carried out based on the design with six centre points (Douglas 2001). Five factors with five levels were considered for experimental purpose. The actual and the coded values of Nd:YAG laser marking process parameters are listed in Table 1.

RSM based empirical models for responses i.e., mark width, mark depth, and mark intensity have been established to show the mathematical relationship between the response, Y and the laser marking process parameters. The corresponding empirical equation for all the response characteristic can be represented as mark width (Ywz), mark depth (Ydz), mark intensity (Yiz). These empirical equations may be used to find the values of the responses for any values of input parameters for the range taken.

Table 1 Actual and coded values of laser marking of zirconia

Factors	Unit	Symbol	Levels				
			−2	−1	0	+1	+2
Pulse frequency (PF)	Hz	X_1	4000	5000	6000	7000	8000
Lamp current (LC)	A	X_2	25	26	27	28	29
Pulse width (PW)	%	X_3	2	4	6	8	10
Scanning speed (SS)	mm/s	X_4	1	2	3	4	5
Air pressure (AP)	kgf/cm^2	X_5	0.3	0.8	1.3	1.8	2.3

$$
\begin{aligned}
Y_{wz} = {} & 3764.70 - 0.14X_1 - 250.36X_2 + 18.87X_3 + 17.59X_4 - 33.02X_5 \\
& + 4.94X_2^2 + 0.96X_3^2 - 3.03X_4^2 + 18.25\,X_5^2 - 1.41X_2X_3 - 1.36X_2X_4 \\
& - 4.03X_2X_5 - 0.46X_3X_4 + 1.17X_3X_5 + 13.66X_4X_5
\end{aligned} \quad (4)
$$

$$
\begin{aligned}
Y_{dz} = {} & 3158.28 + 0.02X_1 - 223.49X_2 + 14.90X_3 - 71.52X_4 - 124.87X_5 \\
& + 4.06X_2^2 - 0.07X_3^2 + 1.12X_4^2 - 0.27X_5^2 + 0.01X_1X_5 - 0.19X_2X_3 \\
& + 1.58X_2X_4 + 2.90X_2X_5 - 1.0X_3X_4 - 2.\,31X_3X_5 + 7.0X_4X_5
\end{aligned} \quad (5)
$$

$$
\begin{aligned}
Y_{iz} = {} & 12.1248 - 0.0004X_1 - 0.8544X_2 + 0.3100X_3 - 0.0719X_4 - 0.4462X_5 \\
& + 0.0165X_2^2 + 0.0010X_3^2 + 0.0127X_4^2 + 0.2059X_5^2 - 0.0075X_2X_3 \\
& - 0.0062X_2X_4 + 0.0069X_3X_4 - 0.0225X_3X_5 - 0.00250X_4X_5
\end{aligned} \quad (6)
$$

6.1.1 Analysis of Process Parameters on Marking Quality Characteristics on Zirconia

The parametric effects of different process parameters like the pulse frequency, lamp current, pulse width, scanning speed and air pressure on marking quality characteristics like the mark width, mark depth, and mark intensity during Nd:YAG laser marking on zirconia ceramic are analysed based on response surface plots.

Parametric Influences on Mark Width

The surface plot of mark width with pulse frequency and lamp current by keeping all other process parameters constant at pulse width as 6 %, scanning speed as 3 mm/s and air pressure as 1.3 kgf/cm^2 is shown in Fig. 5. It is seen that mark width decreases more steeply with the increase in the pulse frequency for both low and high values of lamp current. As the pulse frequency increases there is a decrease in the mark width. It is because of the fact that the energy of laser beam mainly depends on lamp current and pulse frequency. With respect to lamp current there is an increase in the mark width with increase in lamp current. As the lamp current increases and pulse frequency decreases, the laser beam energy increases and the top surface melted at a faster rate, thus increasing the mark width.

The variation of mark width with the change in lamp current and pulse width is shown in Fig. 6 keeping all other parameters constant at pulse frequency of 6000 Hz, scanning speed of 3 mm/s and air pressure 1.3 kgf/cm^2 as constant. As the lamp current increases the power intensity increases, which aids in more material removal, which is indicated, by the increase in the mark width. At low values of percentage of pulse width, there is a steep increase in the mark width with the

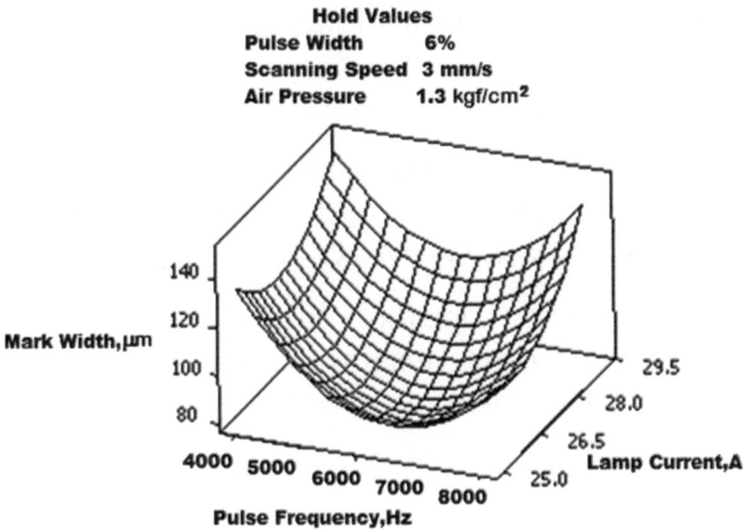

Fig. 5 Variation of mark width with pulse frequency and lamp current

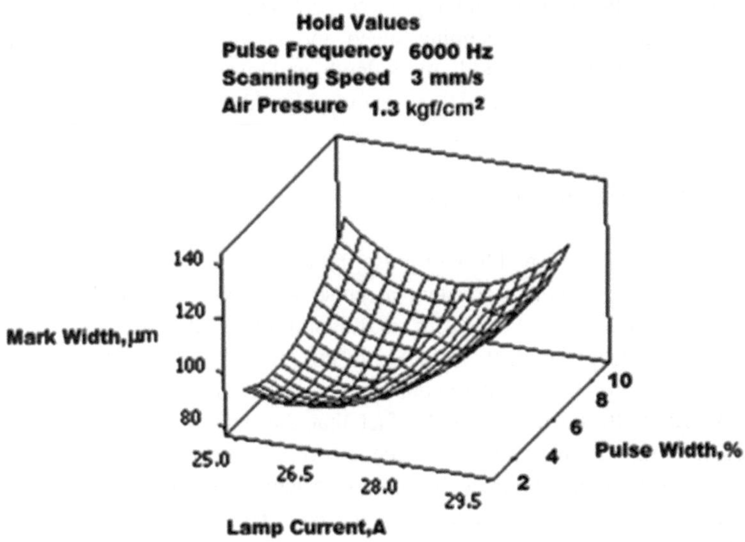

Fig. 6 Variation of mark width with lamp current and pulse width

increase in the lamp current. When the pulse width is increased, there is a decrease in the mark width. At low pulse width Nd:YAG laser produces high quality beam which makes the laser marking more accurate and hence there is an increase in mark width.

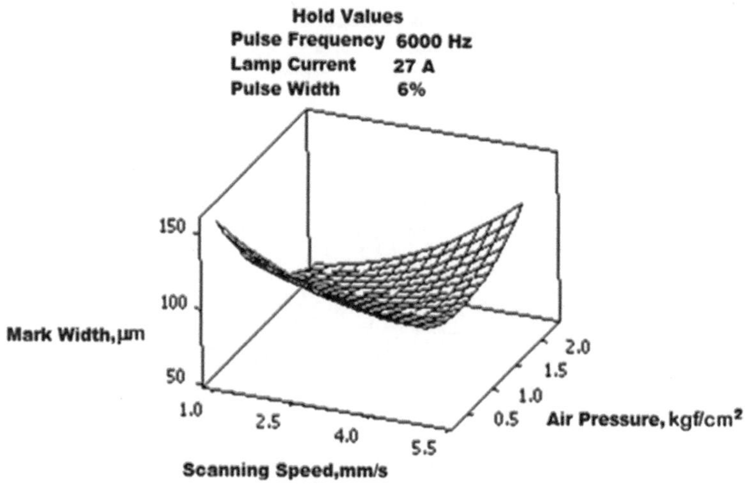

Fig. 7 Variation of mark width with scanning speed and air pressure

The change in the mark width with respect to change in scanning speed and air pressure is shown in Fig. 7 keeping all other parameters constant at pulse frequency of 6000 Hz, lamp current of 27 A and pulse width of 6 % as constant. From the surface plot it is seen that there is a gradual decrease in the mark width with rise in the scanning speed for low air pressure. At high scanning speed, the time available for the laser at the marking surface is very less hence there is a decrease in mark width. On the contrary, there is an increase in the mark width at high air pressure with the variation in the scanning speed.

Parametric Influences on Mark Depth

The variation of mark depth with the change in pulse frequency and lamp current by keeping all other process parameters as constant at pulse width as 6 %, scanning speed as 3 mm/s and air pressure as 1.3 kgf/cm^2 is shown in Fig. 8. From the surface plot it is observed that mark depth decreases gradually when the lamp current is increased up to the mid value after which it shows a slight increase in the mark depth for low and high values of pulse frequency. It is inferred that a change in pulse frequency and lamp current has a major influence to attain the mark depth. However with the rise in the lamp current there is an increase in the mark depth as the laser beam intensity increases the amount of melted and vaporized material so as to form a deeper marking. Similar phenomenon is observed at a low pulse frequency and with the change in the lamp current.

The effect of change in lamp current and pulse width on mark depth is seen from Fig. 9 by keeping other process parameters at pulse frequency of 6000 Hz, scanning speed of 3 mm/s and air pressure 1.3 kgf/cm^2 as constant. It is observed that the

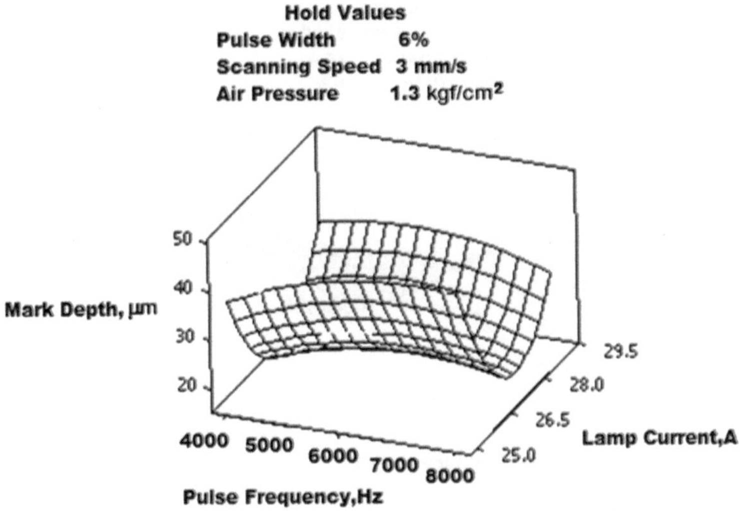

Fig. 8 Variation of mark depth with pulse frequency and lamp current

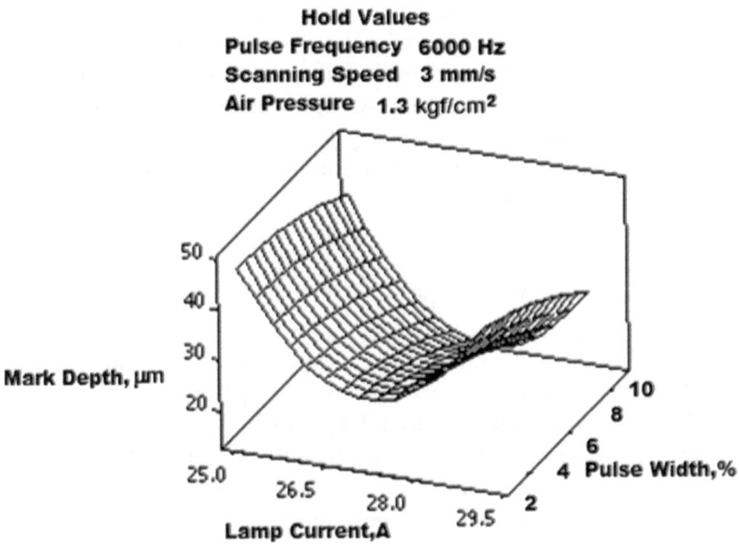

Fig. 9 Variation of mark depth with lamp current and pulse width

mark depth decreases with the change in lamp current up to mid level after which it shows a slight increase in the mark depth at low pulse width and high pulse width. But at low lamp current with the variation in the pulse width, there is a small rise in the mark depth. At high lamp current with the change in pulse width, there is steep increase in the mark depth.

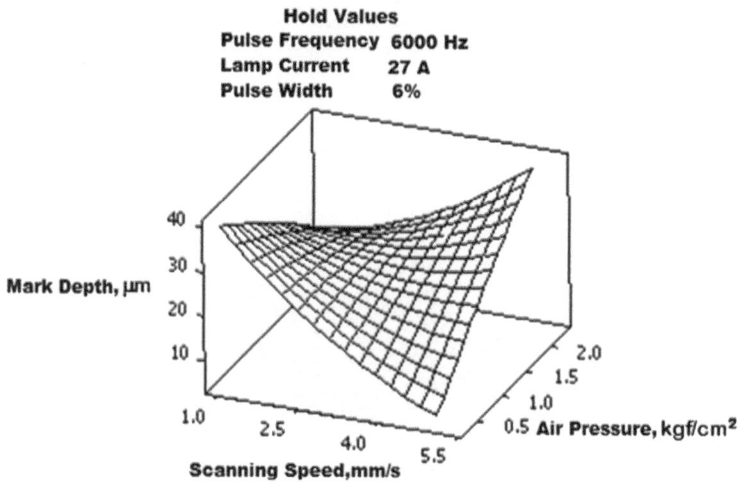

Fig. 10 Variation of mark depth with scanning speed and air pressure

The influence of scanning speed and air pressure on mark depth is studied from Fig. 10 by maintaining other process parameters like pulse frequency of 6000 Hz, lamp current of 27 A and pulse width of 6 % as constant. With the change in the scanning speed there is a steep fall in the mark depth for low range of air pressure compared to high range of air pressure. It is observed that at low scanning speed with the change in the air pressure there is small decrease in the mark depth. When the air pressure is high, the material is ejected out thoroughly which aids to the increase in the mark depth. In addition to it, at low scanning speed the material of the work piece absorbs sufficient amount of heat energy for longer time, as result, the mark depth is increased.

Parametric Influences on Mark Intensity

The variation of mark intensity with the change in pulse frequency and lamp current is studied from Fig. 11 with pulse width of 6 %, scanning speed of 3 mm/s and air pressure of 1.3 kgf/cm^2. With the change in the pulse frequency there is a decrease in the mark intensity for low and high lamp current up to 6000 Hz after which it shows a gradual increase with the change in lamp current. But at low and high pulse frequency, there is an increase in mark intensity with the change in lamp current. This is due to the fact that as the lamp current increases the intensity of laser beam reaching the work piece also increases which ultimately increases the intensity of the mark.

The surface plot of mark intensity with the change in lamp current and pulse width keeping all other laser marking process parameters like pulse frequency of 6000 Hz, scanning speed of 3 mm/s and air pressure 1.3 kgf/cm^2 as constant is

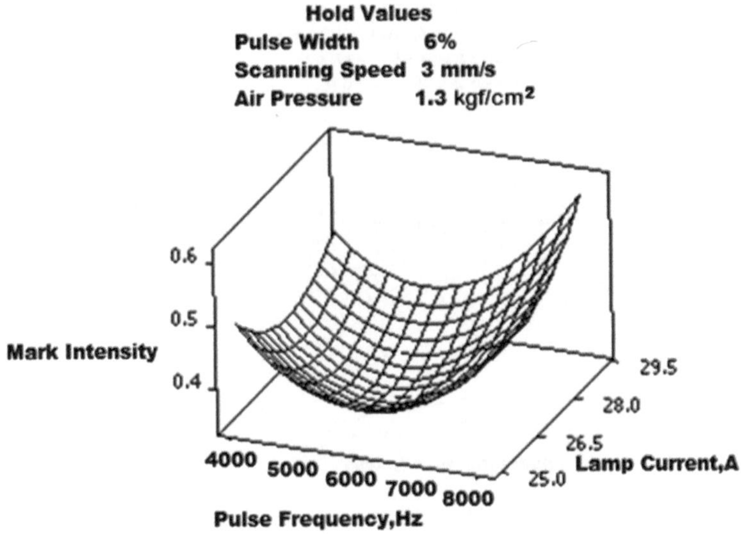

Fig. 11 Variation of mark intensity with pulse frequency and lamp current

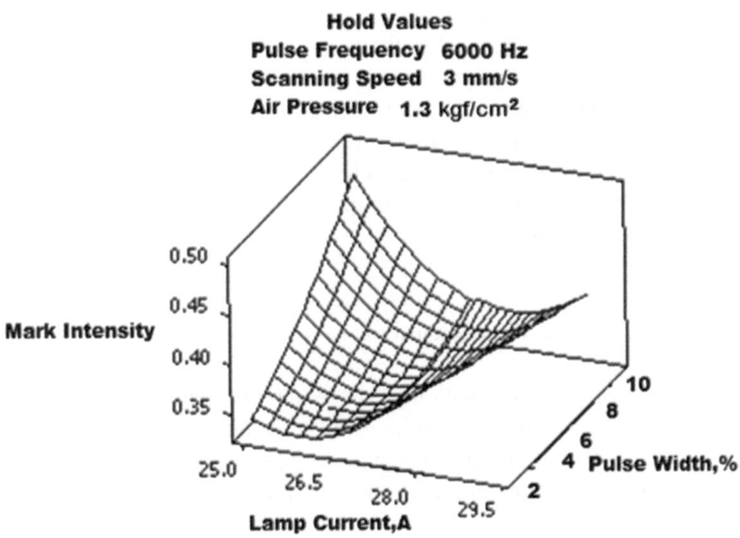

Fig. 12 Variation of mark intensity with lamp current and pulse width

shown in Fig. 12. As the lamp current increases there is a steady change in the mark intensity at low pulse width. Because at low pulse width high concentrated laser beam energy causes faster rate of penetration compared to high pulse width, as a result there is decrease in the mark intensity. It is seen that there is a fall in the mark

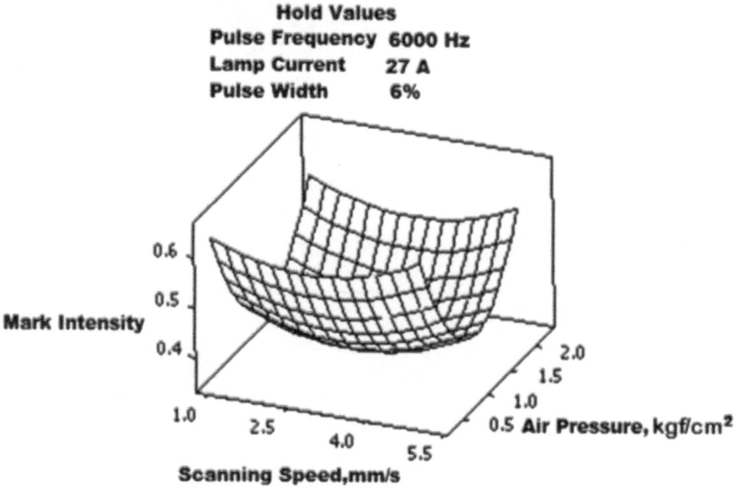

Fig. 13 Variation of mark intensity with scanning speed and air pressure

intensity at high pulse width with the change in lamp current. Similarly, the mark intensity increases at low values of lamp current with the increase in pulse width.

The variation in mark intensity with change in scanning speed and air pressure is shown in Fig. 13 keeping all other parameters like pulse frequency of 6000 Hz, lamp current of 27 A and pulse width of 6 % as constant. When the scanning speed is increased there is a slight increase in mark intensity at low and high range of air pressure. But there is a decrease in mark intensity at mid level of air pressure. At low scanning speed with the variation of air pressure, there is a slow decrease up to mid level after which there is a slow increase in the mark intensity.

6.1.2 Multi-objective Optimization of Nd:YAG Laser Marking Quality Characteristics

The multi-objective optimization analysis is performed to achieve the minimum mark width, maximum mark depth and maximum mark intensity of the laser marking on zirconia based on the developed empirical equations. Figure 14 shows the result of multi objective optimization. The optimal combinations of laser marking process parametric setting is obtained as pulse frequency of 4.484 kHz, lamp current of 25.68 A, pulse width of 5.17 %, scanning speed of 1.06 mm/s, air pressure of 2.3 kgf/cm^2. The minimum mark width is obtained as 78.34 μm. The maximum mark depth is found as 49.99 μm. The maximum intensity value is 0.60. The value of composite desirability is taken as 0.999. Figure 15 shows the SEM micrographs of the laser marking quality characteristics like mark width, mark depth and mark intensity at optimal setting.

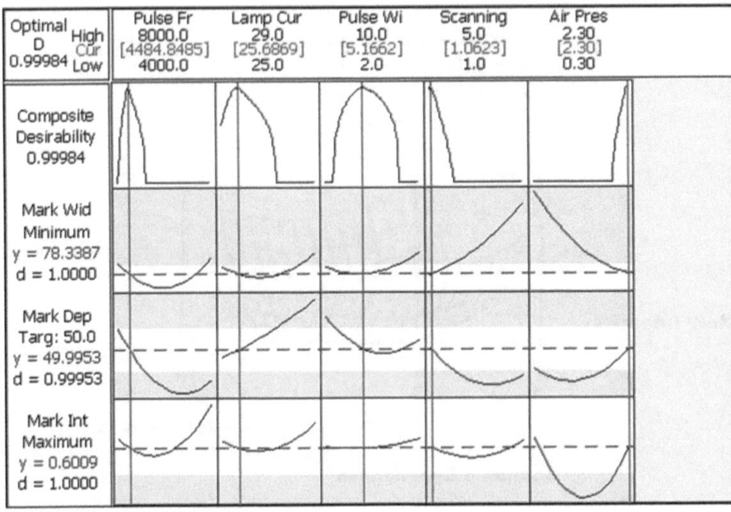

Fig. 14 Multi objective optimization of zirconia ceramic

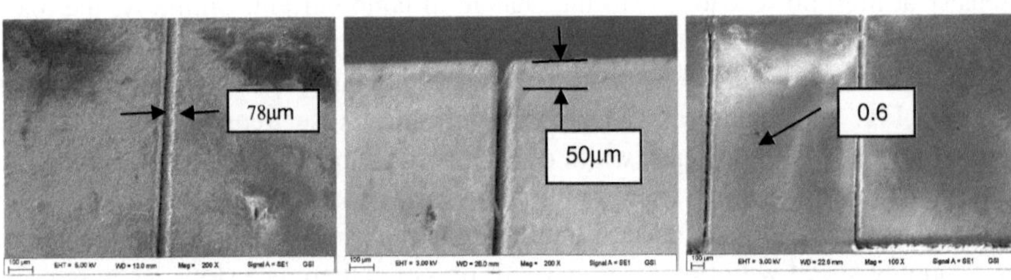

Fig. 15 SEM views of the mark at optimal setting

6.2 Development of Artificial Neural Network

The application of artificial intelligence is remarkable in virtually all fields of engineering. Modeling and optimization are necessary for the understanding and control of any process. One of the main applications of artificial neural networks is that it can be used to model complex relationships between inputs and outputs or to find patterns in data. Artificial neural network is a massively parallel distributed information processing structure consisting of processing elements interconnected via unidirectional signal elements called connection weights (Hassoun 1995). The basic model of the neuron is founded upon the functionality of a biological neuron. The basic processing elements of neural networks are called artificial neurons, or simply neurons or nodes. In a simplified mathematical model of the neuron, the effects of the synapses are represented by connection weights that modulate the effect of the associated input signals, and the nonlinear characteristic exhibited by neurons is represented by a transfer function.

The neuron impulse is then computed as the weighted sum of the input signals, transformed by the transfer function. The learning capability of an artificial neuron is achieved by adjusting the weights in accordance to the chosen learning algorithm. Depending on the weights, the computation of the neuron will be different. By adjusting the weights of an artificial neuron, the desired output can be obtained (Muthukrishnan and Davim 2009). A typical artificial neuron and the modelling of a multilayered neural network are illustrated in Fig. 16.

From Fig. 16, the signal flow from inputs x_1, x_2, …, x_n is considered to be unidirectional, which are indicated by arrows, as is a neuron's output signal flow (O). The neuron output signal O is given

$$O = f(net) = f\left(\sum_{j=1}^{n} w_j x_j\right) \tag{7}$$

where w_j is the weight vector and the function f(net) is referred to as an activation (transfer) function. The concept for the development of ANN model is implemented right from development of ANN model to validation, and optimization of laser marking parameters have been carried out and the methodological approach for the development of ANN model is depicted in Fig. 17. At first, the input and the output data are normalized. The number of neurons for the input layer and the output layer is decided according to the number of inputs variables and the number of outputs required. Different networks are selected by varying the number of neurons in the hidden and trained to search out the optimal ANN model for prediction of laser marking. The error available at the neurons of the hidden layer is calculated by iterations and back-propagated. The number of neurons in the hidden layer is selected with precisely in order to reach the goal in lesser number of iterations with minimum percentage of prediction errors to obtain the desired result in the shortest possible time. Further network model is trained and the predicted values are

Fig. 16 Artificial neuron

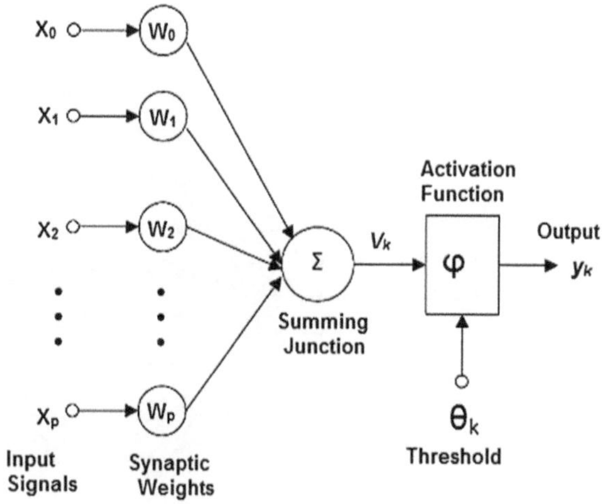

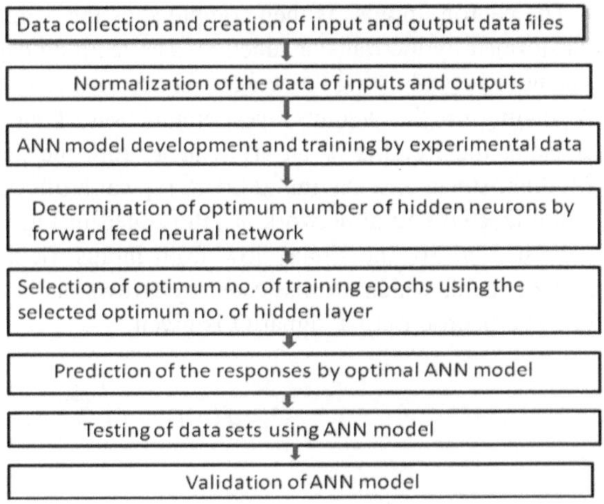

Fig. 17 Flow diagram of methodological approach of ANN model development

validated with the experimental data. The percentage of prediction error is calculated as,

$$\text{Prediction error } (\%) = \frac{[(\text{Experimental result} - \text{Predicted result})]}{\text{Experimental result}} \times 100 \quad (8)$$

The number of neurons for the input layer and the output layer is decided according to the number of inputs variables and the number of outputs required. Finally, a neural network model with a structure of 5-8-3 is found to be the most efficient model for zirconia ceramic and the architecture of the developed ANN model is shown in Fig. 18 with a hidden layer. The ANN model is developed based on multi-layer feed-forward backpropagation algorithm using 'trainlm'. With this training program, the model is trained with 32 sets of experimental data and the results are shown in Table 2.

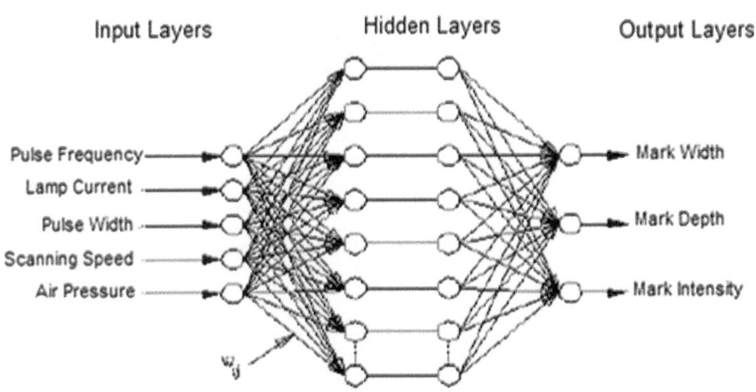

Fig. 18 Architecture of developed neural network

Table 2 Design of experimental plan with responses

No.	Actual values laser parameters					Responses			ANN predicted results		
	PF (Hz)	LC (A)	PW (%)	SS (mm/s)	AP (kgf/cm^2)	Mark width (μm)	Mark depth (μm)	Mark intensity	Mark width (μm)	Mark depth (μm)	Mark intensity
1	5000	26	4	2	1.8	78	27	0.39	78.5	27.5	0.37
2	7000	26	4	2	0.8	101.25	32	0.46	102	32.3	0.45
3	5000	28	4	2	0.8	137.75	27	0.43	138.6	27.9	0.44
4	7000	28	4	2	1.8	91.5	26.5	0.58	92.8	25.7	0.57
5	5000	26	8	2	0.8	118.5	44	0.5	117.4	44.8	0.55
6	7000	26	8	2	1.8	90.25	26	0.39	91	26.9	0.35
7	5000	28	8	2	1.8	87.5	16.4	0.44	86.1	16	0.43
8	7000	28	8	2	0.8	130	16.4	0.48	131.9	15.7	0.47
9	5000	26	4	4	0.8	115	25.5	0.37	116.4	25	0.36
10	7000	26	4	4	1.8	90.75	48	0.48	91.8	47.4	0.47
11	5000	28	4	4	1.8	112.5	29	0.43	114	28.6	0.44
12	7000	28	4	4	0.8	113.75	23	0.52	112.1	23.8	0.5
13	5000	26	8	4	1.8	107.75	16.4	0.49	106	16	0.47
14	7000	26	8	4	0.8	101.5	23	0.5	100.4	22.4	0.51
15	5000	28	8	4	0.8	118.75	16.4	0.54	117.5	15.7	0.53
16	7000	28	8	4	1.8	107	27	0.44	108.2	27.9	0.43
17	4000	27	6	3	1.3	127.5	18	0.41	127	18.7	0.42
18	8000	27	6	3	1.3	104.5	16.4	0.5	103.7	16	0.52
19	6000	25	6	3	1.3	89.25	31	0.42	89.9	30.3	0.44
20	6000	29	6	3	1.3	118.5	42	0.39	117.6	41.2	0.4
21	6000	27	2	3	1.3	101.25	22	0.36	100.4	21.4	0.37
22	6000	27	10	3	1.3	97.75	16.4	0.35	97.1	16	0.36
23	6000	27	6	1	1.3	92.5	33	0.4	93.4	32.3	0.42
24	6000	27	6	5	1.3	100	16.5	0.38	101.8	16.9	0.36
25	6000	27	6	3	0.3	115	14	0.62	116.9	14.9	0.6
26	6000	27	6	3	2.3	89.75	26	0.47	88.7	26.9	0.46
27	6000	27	6	3	1.3	82.25	30	0.35	83.4	31.7	0.34
28	6000	27	6	3	1.3	81.7	25.5	0.36	81	25	0.35
29	6000	27	6	3	1.3	83	22	0.34	82.6	22.8	0.33
30	6000	27	6	3	1.3	82.95	25	0.33	81.4	25.7	0.34
31	6000	27	6	3	1.3	82.5	16	0.35	83.9	15.5	0.36
32	6000	27	6	3	1.3	81.9	14.5	0.36	82	14	0.37

6.2.1 ANN Results and Analysis

Training of the neural network model is performed using 32 experimental data. To validate the results, the trained network model has been tested using ten experimental data (verification data), which are not used in the training process is given in Table 3. The process parametric data are taken from the range as mentioned in Table 1 and their results are compared with the RSM model. Table 4 shows the

Table 3 Verification experiment (test data) for zirconia ceramic

No.	Process parameters					Responses		
	PF (Hz)	LC (A)	PW (%)	SS (mm/s)	AP (kgf/cm^2)	Mark width (μm)	Mark depth (μm)	Mark intensity
1	6500	26	8	1	0.8	90.5	30	0.38
2	4000	28	6	3	2.3	127.5	18	0.41
3	5500	25.5	7	4	1	116.5	36	0.54
4	7000	27.5	4	5	1.8	90	25.3	0.57
5	4500	25	3	2	1.3	115.6	16	0.59
6	5000	28.5	5	4	2	112.6	29	0.49
7	5500	29	9	3	0.5	130.4	21.6	0.4
8	6000	26.5	7	1	0.7	89.25	31	0.42
9	7500	27	10	5	1.5	98.2	26	0.36
10	8000	25	2	2	0.3	104.5	16.4	0.5

Table 4 Comparison between RSM based results and ANN predicted results of test data for zirconia

Expt. No.	RSM estimated results			ANN predicted results		
	Mark width (μm)	Mark depth (μm)	Mark intensity	Mark width (μm)	Mark depth (μm)	Mark intensity
1	93.1	30.8	0.395	92.8	30.7	0.389
2	123.5	18.9	0.425	123.2	18.4	0.42
3	112.7	37.5	0.56	113.5	35.9	0.553
4	92.3	24.4	0.593	92.5	26.1	0.588
5	113.1	16.6	0.605	113	16.5	0.608
6	109	30.4	0.509	114.6	29.7	0.499
7	134	20.6	0.386	133.2	22.3	0.391
8	85.2	29.9	0.435	91.2	30.3	0.429
9	102.5	26.6	0.341	100.8	26.7	0.351
10	107.8	15.8	0.48	101.7	16.8	0.486

comparison of experimental results and ANN predicted results along with the percentage of prediction error. The average percentage of prediction errors of the developed ANN model for mark width, mark depth and mark intensity are 2.52, 2.59 and 2.58 respectively and the overall percentage of prediction error is 2.6.

Results of the responses for the ten sets of data calculated based on RSM and ANN predicted results are given in Table 4. The percentage of estimation errors for two models is shown in Table 5. For RSM predicted results, the average percentage of estimation error for mark width, mark depth and mark intensity are 3.34, 3.83 and 3.93 respectively. The overall average percentage of prediction error is found as

Table 5 Comparison of percentage of estimation and prediction errors of RSM model estimated and ANN model predicted results of test data for zirconia

Expt. No.	% of estimation error of RSM model			% of prediction error of ANN model		
	Mark width (μm)	Mark depth (μm)	Mark intensity	Mark width (μm)	Mark depth (μm)	Mark intensity
1	2.76	2.66	3.94	2.54	2.33	2.37
2	3.52	5	3.66	3.37	2.22	2.44
3	3.26	4.12	3.7	2.57	2.25	2.41
4	2.55	3.56	4.06	2.78	3.16	3.15
5	2.42	3.75	3.05	2.24	3.12	2.71
6	3.2	4.8	3.88	2.1	2.41	3.05
7	3.5	4.17	4.21	2.14	3.2	2.25
8	4.5	3.54	3.57	2.4	2.25	2.14
9	4.35	3.07	5.2	2.64	2.69	2.5
10	3.34	3.65	4	2.4	2.24	2.8
Average percentage of prediction error	3.34	3.832	3.927	2.518	2.587	2.582
Overall percentage of prediction error	3.99			2.6		

3.99. It can be seen from these two prediction errors that the overall prediction error for RSM model is nearly 1.53 times the overall prediction error of ANN model. Thus, by evaluating the percentage of prediction error of the two models, it is observed that the developed ANN model for laser marking process predicts the responses quite satisfactorily.

The comparison between experimental and ANN model predicted values for mark width, mark depth and mark intensity graphically for all 32 experiments for all the responses are shown in Figs. 19, 20 and 21. It is seen from this figure that the developed ANN model predicted all responses at different process parametric setting effectively. It is also observed that the data points of experimental and ANN

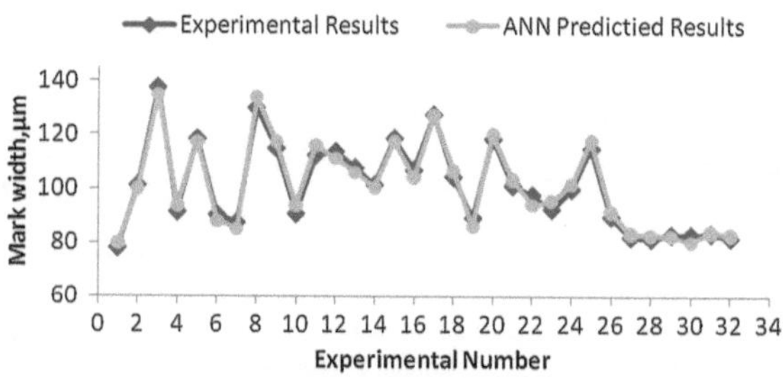

Fig. 19 Comparison of experimental and ANN predicted results of mark width

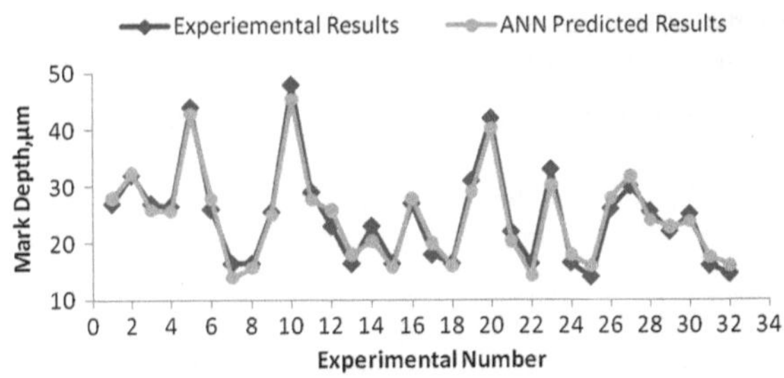

Fig. 20 Comparison of experimental and ANN predicted results of mark depth

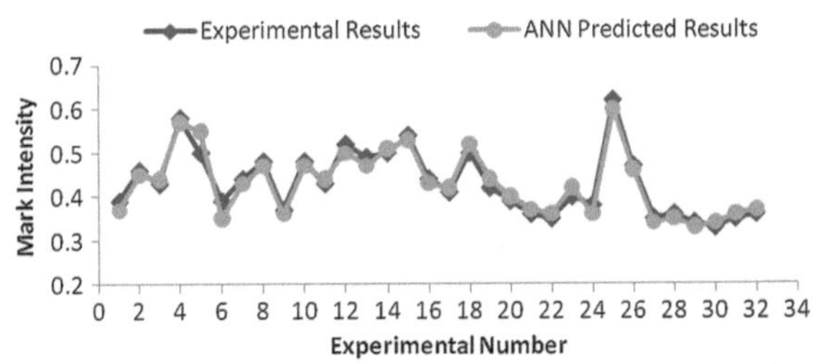

Fig. 21 Comparison of experimental and ANN predicted results of mark intensity

predicted results are very close to one another. Also comparison of experimental and ANN predicted results for all the responses for test data are shown in Figs. 22, 23 and 24 for the responses like mark width, mark depth and mark intensity respectively. It is also observed that the data points of experimental and ANN predicted results are very close to one another. It is seen from this figure that the developed ANN model predicted all responses at different process parametric setting effectively. It is concluded that ANN models developed for prediction of the laser marking quality characteristics such as mark width, mark depth and mark intensity at different process parametric setting are adequate.

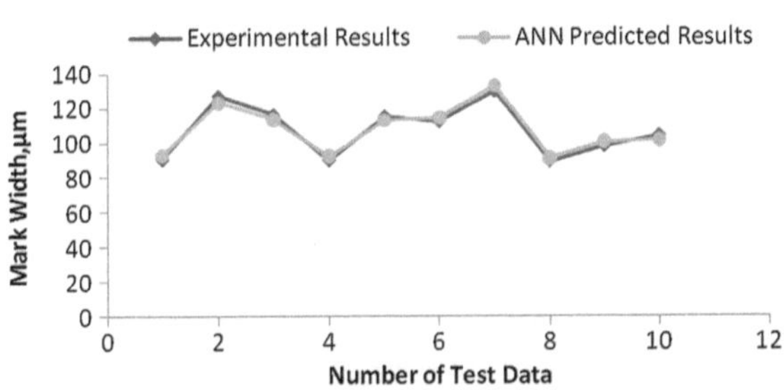

Fig. 22 Comparison of experimental and ANN predicted results of test data for mark width

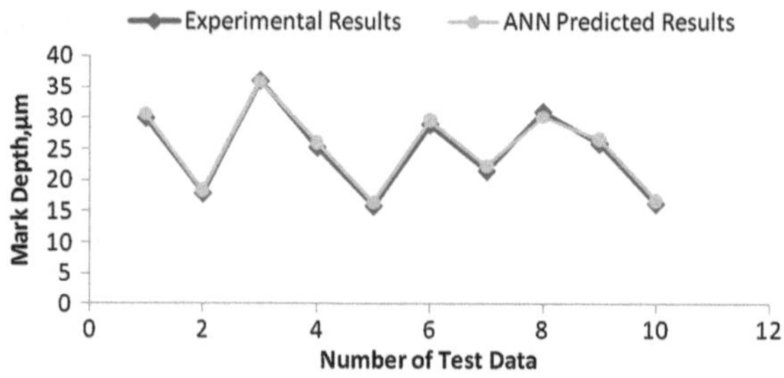

Fig. 23 Comparison of experimental and ANN predicted results of test data for mark depth

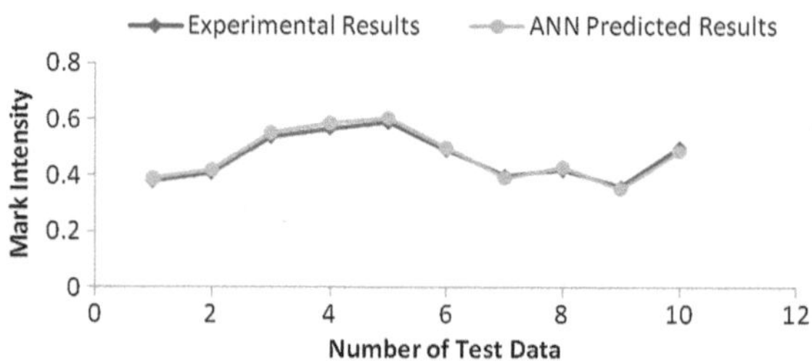

Fig. 24 Comparison of experimental and ANN predicted results of test data for mark intensity

6.3 Optimization and Prediction Through ANN

The developed ANN model has been used to predict the response parameters, i.e. mark width, mark depth and mark intensity for all combinations of input factors. Hence all the possible combinations of five process parameters with each at five levels are generated. From the developed ANN model, the marking characteristics like mark width, mark depth and mark intensity are also predicted for all the 3125 combinations of laser marking process parameters. RSM based multi objective optimization is carried out for laser marking on zirconia ceramic and the optimal process parameters obtained are pulse frequency of 4.484 kHz, lamp current of 25.68 A, pulse width of 5.17 %, scanning speed of 1 mm/s and air pressure of 2.3 kgf/cm^2. The optimized values are 78.34 μm for mark width, 49.99 μm for mark depth and 0.60 for mark intensity. Experimentally obtained results of the responses are compared with the RSM and ANN model based predicted results and the percentage of error in the responses of both models are calculated and the results obtained is shown in Table 6. The results of the response are also predicted using the developed ANN model from the RSM predicted optimal parametric combination. The results obtained by ANN prediction are 82.8 μm, 46.3 μm and 0.605 for mark width, mark depth and mark intensity respectively. It is noticed that the ANN model which is developed has predicted the response quite acceptably as the percentage of prediction errors are 2.5, 3.13 and 1.62 respectively which are much lower than that of results predicted through RSM model. Therefore it is confirmed

Table 6 Comparison of RSM predicted, ANN predicted and experimental results at optimal RSM based parametric setting

Terms	Experimental	RSM predicted	ANN predicted
Parametric combination	Near feasible setting of RSM predicted values	Pulse frequency of 4.484 kHz, lamp current of 25.68 A	
	Pulse frequency of 4.4 kHz, Lamp current of 25.7 A	Pulse width of 5.17 %	
		Scanning speed of 1.06 mm/s	
	Pulse width of 5 %	Air pressure of 2.3 kgf/cm^2	
	Scanning speed of 1 mm/s		
	Air pressure of 2.3 kgf/cm^2		
Mark width (μm)	80.78	78.34	82.8
Mark depth (μm)	47.8	49.99	46.3
Mark intensity	0.615	0.60	0.605
Percentage of prediction error in RSM approach			
Mark width = 3.02 %			
Mark depth = 4.5 %			
Mark intensity = 2.43 %			
Percentage of prediction error in ANN approach			
Mark width = 2.5 %			
Mark depth = 3.13 %			
Mark intensity = 1.62 %			

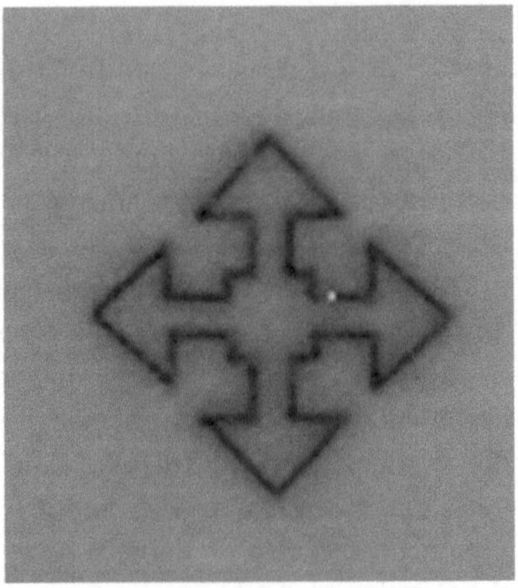

Fig. 25 Marking on zirconia with optimal combination of PF—4.4 kHz, LC—25.7 A, PW—5 %, SS—1 mm/s and AP—2.3 kg/cm^2

that the developed ANN model predicts better results than the developed RSM model. Figure 25 shows the marking of a logo with the predicted optimal parametric setting.

7 Conclusions

Laser marking is found to be an effective and permanent marking, which offers greater flexibility. Artificial neural network (ANN) model is developed, which has been tested and validated using test data sets for the prediction of various responses. The overall percentage of prediction error is 2.6 for the developed ANN model and the overall average percentage of estimation error is 3.99 for the RSM model. Comparing RSM and ANN models it is found that the overall estimation error for RSM model is nearly 1.53 times the overall prediction error of ANN model and this clearly indicates that the ANN model is superior to RSM model. Also the ANN predicted optimal values of mark width of 82.8 μm, mark depth of 46.3 μm and mark intensity 0.605 are obtained from the multiple parametric setting and the percentage of prediction errors for mark width, mark depth and mark intensity are obtained as 2.5, 3.13 and 1.62 respectively. The small value of errors of various responses of the laser marking process indicates that the ANN model performs the precise prediction of optimal parametric setting for Nd:YAG laser marking on zirconia ceramic.

Acknowledgments The authors acknowledge the financial support and also assistance provided by CAS Ph-IV program of Production Engineering Department of Jadavpur University under University Grants Commission (UGC), New Delhi.

References

Adelina, H., & Dinu, G. (2008). Analysis of the laser marking technologies. *Nonconventional Technologies Review* (4), 17–22.

Chen, T. C., & Darling, R. B. (2005). Parametric studies on pulsed near ultraviolet frequency tripled Nd:YAG laser micromachining of sapphire and silicon. *Journal of Materials Processing Technology, 169*, 214–218.

Chen, M.-F., Chen, Y.-P., Hsiao, W.-T., Wu, S.-Y., Hu, C.-W., & Gu, Z.-P. (2008). A scribing laser marking system using DSP controller. *Optics and Lasers in Engineering, 46*, 410–418.

Dahotre, N. B., & Harimkar, S. (2008). *Laser fabrication and machining of materials*. Boston: Springer Science & Business Media.

Dascalu, T., Acosta-Ortiz, S. E., Ortiz-Morales, M., & Compean, I. (2000). Removal of the indigo color by laser beam denim interaction. *Optics and Lasers in Engineering, 34*, 179–189.

Douglas, C. M. (2001). *Design and analysis of experiments* (5th ed.). New York: Wiley.

Feng, Y., Liu, Z. Q., & Yi, X. S. (2001). Marking carbon black/polypropylene compounds using a Nd:YAG laser. *Journal of Materials Science Letters, 20*, 517–519.

Fernandez-Pradas, J. M., Restrepo, J. W., Gomez, M. A., Serra, P., & Morenza, J. L. (2007). Laser printing of enamels on tiles. *Applied Surface Science, 253*, 7733–7737.

Fu, G. (2004). Methods and compositions related to laser sensitive pigments for laser marking of plastics. *United States Patent* 6706785.

Hassoun, M. H. (1995). *Fundamentals of artificial neural networks*. Cambridge: MIT Press.

Ho, C. K, Man, H. C, Yue, T. M., & Yuen, C. W. (1997). Laser etching of polymer masked leadframes. *Applied Surface Science, 109*, 236–241.

Houle, F. A. (1986). Basic mechanisms in laser etching and deposition. *Applied Physics A, 41*, 315–330.

Ion, J. C. (2005). *Laser processing of engineering materials principles, procedure and industrial application*. Oxford: Elsevier's Publication.

Josephine, P., Doloi, B., & Bhattacharyya, B. (2013). *Parametric analysis of Nd:YAG laser marking on alumina ceramics* (Vol. 94)., The Institution of Engineers (India): Series C Berlin: Springer.

Klimt, B. H. (1988). Review of laser marking and engraving. *Lasers and Optronics, 7*, 61–67.

Leone, C., Lopresto, V., & DeIorio, I. (2009). Wood engraving by Q-switched diode-pumped frequency-doubled Nd:YAG green laser. *Optics and Lasers in Engineering, 47*, 161–168.

Muthukrishnan, N., & Davim, J. P. (2009). Optimization of machining parameters of Al/SiC-MMC with ANOVA and ANN analysis. *Journal of Materials Processing Technology, 209*, 225–232.

Nedialkov, N. N., Imamavao, S. E., Berger, P. A., & Dasinger, E. (2005). Mechanism of ultrashort laser ablation of metals: Molecular dynamics simulation. *Applied Surface Science, 247*, 243–248.

Ng, T. W., & Yeo, S. C. (2001). Aesthetic laser marking assessment using luminance ratios. *Optics and Lasers in Engineering, 35*, 177–186.

Peligrad, A. A., Zhou, E., Morton, D., & Li, L. (2001). A melt depth prediction model for quality control of laser surface glazing of inhomogeneous materials. *Optics and Laser Technology, 33*, 7–13.

Peligrad, A. A., Zhou, E., Morton, D., & Li, L. (2002). Transient temperature behaviour and dynamic process models in laser surface melting of clay tiles. *Surface and Coatings Technology, 150*, 15–23.

Ready, J. F. (2001). *Industrial applications of lasers* (2nd ed.). San Diego: Academic Press.

Restrepo, J. W., Fernandez-Pradas, J. M., Gomez, M. A., Serra, P., & Morenza, J. L. (2006). Influence of preheating and hematite content of clay brick pavers on the characteristics of lines marked with a Nd:YAG laser. *Applied Surface Science, 253*, 2272–2277.

Stewart, R., Li, L., & Thomas, D. (2000). Laser ablation of multi layers of ink from a paper substrate for tactile printing. *Optics and Laser Technology, 32*, 301–305.

Sugioka, K., Meunier, M., & Piqué, A. (2010). *Laser precision microfabrication*., Springer Series in Materials Science Berlin: Springer.

Nd:YAG Laser Microdrilling of SiC-30BN Nanocomposite: Experimental Study and Process Optimization

N. Roy, A.S. Kuar, S. Mitra and B. Acherjee

Abstract Nd:YAG laser microdrilling of SiC30BN nanocomposite material is studied here. Taguchi based grey relational analysis is used to simultaneously determine the optimum setting for minimum hole taper and HAZ width. Grey relation analysis is adopted for combining multiple quality characteristics into one integrated numerical value called Grey relational grade. A L27 orthogonal array has been used for conducting experiments. Lamp current, pulse frequency, pulse width, assist gas pressure and focal distance are considered as input process parameters whereas hole taper and HAZ width are considered as machining responses. It is observed that the quality characteristics of drilled micro holes are improved markedly at the optimized parameter settings as compared to quality levels achieved for initial machine parameter settings.

Keywords Nanocomposite · Hole taper · HAZ width · Grey relational analysis

N. Roy (✉) · A.S. Kuar · S. Mitra
Department of Production Engineering, Jadavpur University,
Kolkata 700032, West Bengal, India
e-mail: nilanjan_83@yahoo.co.in

A.S. Kuar
e-mail: askuar@rediffmail.com

S. Mitra
e-mail: Mail.souren.mitra@gmail.com

B. Acherjee
Department of Production Engineering, BIT Mesra, Extension Centre,
Deoghar 815142, Jharkhand, India
e-mail: a.bappa@yahoo.com

S.N. Joshi and U.S. Dixit (eds.), *Lasers Based Manufacturing*,
Topics in Mining, Metallurgy and Materials Engineering,
DOI 10.1007/978-81-322-2352-8_17

1 Introduction

From the beginning of civilization, the materials have been processed into useful components or tools, by modifying their physical properties like size, shape and texture. This has been done through different processes, namely cutting, welding, bending, machining, punching, marking, cladding, alloying etc. Machining is one of the important areas in the engineering discipline. The advancement of technology during the last few decades causes to the development of many hard-to-machine materials, such as titanium, stainless steel, high-strength temperature—resistant alloys, ceramics, refractories, fibre reinforced composites, superalloys, nanomaterials etc. These materials are not suitable to be machined by the conventional machining processes because of their high hardness, strength, brittleness, toughness and low machinability properties. Sometimes, the machined components require high surface finish and dimensional accuracy which cannot be achieved by the conventional machining processes. Moreover, the rise in temperature and the residual stress generated in the workpiece by those machining processes are not at all desirable or acceptable. These requirements have led to the development of newer processes in machining, known as non-traditional machining (NTM) processes. These are non-traditional in the sense that the conventional cutting tools are not employed in these machining processes for metal removal instead energy in its direct form is utilized, e.g. mechanical energy in abrasive jet machining, electro chemical energy in electrochemical machining, chemical energy in chemical machining, thermoelectric energy in electric discharge machining processes, etc. These NTM processes are not limited by the hardness, toughness, and brittleness and machinability criteria of the work materials. Moreover, using these NTM processes, any kind of intricate shape can be generated with high accuracy and precision with minimum residual stress generation (Pandey and Shan 2004). During the last few decades, the varieties of these NTM processes have increased significantly. All these NTM processes have their own process (machining/operating) parameter sand performance measures (responses). For effective utilization of the NTM processes, it is very much important to find out the optimal NTM process to get machining performance with high dimensional accuracy.

1.1 Laser Beam Machining

Lasers can be considered as one of non-conventional machining process because of their precision of operation, low cost, localized processing, wide range of material processing, flexibility and high speed of operation. Laser beam machining is the machining processes involving a laser beam as a heat source. Laser, Light Amplification by Stimulated Emission of Radiation, is a coherent, monochromatic, collimated, low divergence, high-energy beam of electromagnetic radiation (Meijer 2004). It is a thermal process where laser beam is used as a heat source to remove materials without mechanical engagement with workpiece material. In laser beam machining process, material removal isn't dependent on mechanical, physical or electrical properties of workpiece but on thermoptical properties of the material (Davim 2013). During laser beam machining process laser beam interact with workpiece, the beam energy is partially absorbed by the workpiece material first heats up the surface to boiling point followed by melting or sublimation, evaporation, plasma formation and ablation. Based on the kinematics of the machining zone where material removal takes place, laser beam machining is classified into one, two, and three dimensional machining. Laser drilling is a one-dimensional laser machining process where the laser beam is fixed relative to the workpiece. Laser cutting operation is a two dimensional laser machining process where, laser beam have a relative motion with a respect to workpiece. During three dimensional laser machining process two or more laser beams are used and each beam forms a surface with relative motion respect to the workpiece (Samant and Dahotre 2009). Different types of lasers such as CO_2, Nd:YAG and Excimer lasers are used for machining with each type of laser having its own wavelength of absorption and machining applications. CO_2 laser is a gas laser that use gas molecules (combination of carbon dioxide, nitrogen and helium) as the lasing medium. The actual excitation takes place by an AC or DC electrical discharge and this laser emits light at a wavelength of 10.6 μm in the far infrared region of the electromagnetic spectrum. CO_2 lasers are widely used in industry for applications in laser machining, heat treatment and welding. On the other hand, Nd:YAG laser is a solid state laser that use dopants (Neodymium (Nd3+)) dispersed in a crystalline matrix (complex crystal of Yttrium–Aluminum–Garnet (YAG) with chemical composition $Y_3A_{15}O_{12}$) to generate laser light. Krypton or xenon flash lamp is used as an exciter and an output wavelength of 1.06 μm in the near-infrared region of the spectrum can be obtained. Nd:YAG fibre lasers are used in applications requiring low pulse repetition rate and high pulse energies (up to 100 J/pulse) such as hole piercing and deep keyhole welding applications. Excimer lasers are an increasingly popular type of gas lasers made up of a compound of two identical species that exist only in an excited state. Commonly used excimer complexes include argon fluoride (ArF), krypton fluoride (KrF), xenon fluoride (XeF) and xenon chloride (XeCl) with the output wavelengths varying from 0.193 to 0.351 μm in the ultraviolet to near-ultraviolet spectra. These compounds can be formed by inducing the noble gas (Ar, Kr, or Xe) of the compound into an excited state with an electron beam, an electrical discharge or a combination of the two. Excimer lasers are used for machining solid polymer workpieces, removing metal films from polymer substrates, micromachining ceramics and semi-conductors, and marking thermally sensitive materials (Chryssolouris 1991). These different types of lasers can be operated either in the continuous wave (CW) or in the pulsed mode. In continuous mode lasers, continuous pumping of the laser emits incessant light. While in pulsed mode laser, there is a laser power-off period between two successive pulses which varies from nano, pico to femto seconds.

1.1.1 Laser Beam Drilling

In different types of laser machining processes laser drilling is the simplest one. Laser drilling is most extensively used in the aerospace, aircraft, biomedical, MEMS, renewable energy source instrument, microelectronic and automotive industries (Tam et al. 1994). One of the most important applications of laser drilling is in the aerospace industry for drilling of a large number of closely spaced effusion holes with high aspect ratio and high quality to improve the cooling capacity of turbine engine components. In general, there are four approaches to laser drilling, namely, single pulse, percussion, trepanning, and helical drilling. Figure 1 shows the different types of laser drilling operation.

Single Pulse drilling: only single shot is used to make the hole for which high pulse energy is supplied to vaporize the material in single pulse (Fig. 1a). The hole diameter can be smaller or greater than laser beam diameter which primarily depends on temporal and spatial intensity profiles of the focused laser beam, and material thickness (Ready and Farson 2001). Examples of its application are drilling of the sieves and filters.

Percussion Drilling: multiple shot is used to make the hole with no relative movement between laser beam and workpiece (Fig. 1b). Percussion drilling is mainly applied for drilling of cooling holes in aerospace engine components, such

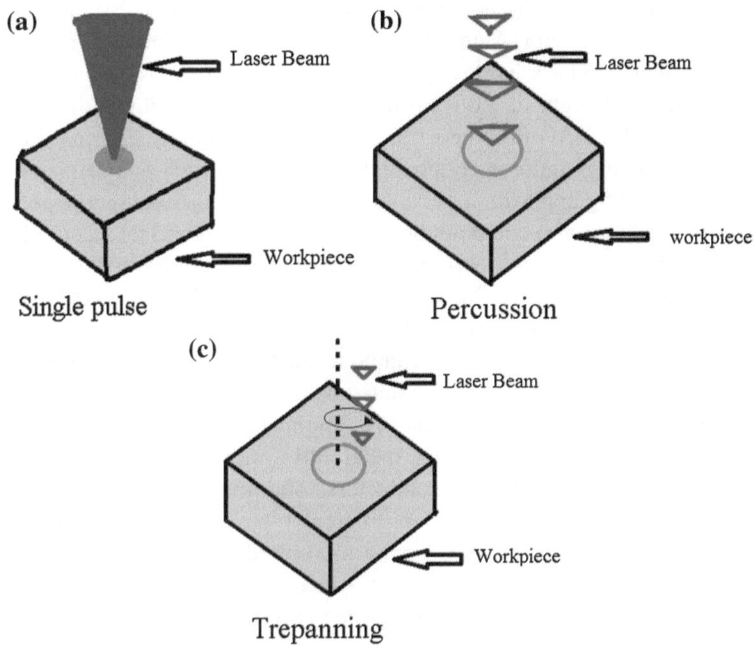

Fig. 1 Types of laser drilling. **a** Single pulse laser drilling. **b** Percussion laser drilling. **c** Trepanning laser drilling

as turbine blades, combustion chamber, guide vanes, blind hole in surgical needle and most kind of the lubrication holes (Sanikommu et al. 2007).

Trepanning Drilling: trepanning drilling process begins by percussion drilling of a central hole followed by contour cutting to achieve the desired hole diameter (Fig. 1c). To produce good hole circularity with less metallurgical defects trepanning drilling technique is used (Corcoran et al. 2002).

Helical Drilling: spiral ablation of the laser beam into the workpiece at several small steps produces helical drilling. The hole drilled by this technique shows significant reduction in the recast layer and good hole circularity, however, the processing time is comparatively slow (Sanikommu et al. 2007).

Laser Beam Microdrilling

The world of laser machining production is divided into micro and macro machining. Laser micromachining has been widely used in microelectronics, aerospace, biomedical and other higher-precision applications. Generally pulsed mode is used for laser micromachining with high resolution in depth and lateral dimensions. To minimize the thermal effects such as melting and burr formation, extremely short pulse durations (nano or pico or even femtosecond), wide range of wavelengths and repetition rates (from single pulse to Megahertz) are used during laser micromachining thus eliminating the need for any post processing measures (Lasagni and Lasagni 2011). Nd:YAG lasers has expanded high beam quality laser machining to micron-scale precision. The high peak powers obtainable with these lasers enable material removal through thermal ablation (Russo 1995). The process of micro drilled printed circuit board (PCBs) having diameter as small as 10 μm is developed by Bachmann (1989). Parametric study of pulsed Nd:YAG laser microdrilling of gamma-titanium aluminide has been done by Biswas et al. (2010a) for effective utilization of this material in modern manufacturing industry. Pulsed Nd:YAG laser microdrilling on titanium nitride alumina composite have been carried out to study the effect of process parameter e.g., lamp current, pulse frequency, pulse width, assist air pressure, focal length on hole circularity at entry and exit and hole taper (Biswas et al. 2010b). Modelling and analysis of pulsed Nd:YAG laser machining characteristics during microdrilling of zirconia (ZrO_2) has been carried out by Kuar et al. (2006).

From the manufacturing point of view, the geometrical as well as metallurgical features of the laser drilled section are very important. Formations of taper, heat-affected zone (HAZ), recast layer are most undesired machining responses during laser microdrilling. Yilbas has studied the effects of focus setting on resulting mean hole diameter (Yilbas 1988). Drilling characteristics are determined by exit hole diameter as a function of material thickness and pulse energy for the single pulse drilling of the material (Rodden et al. 2002). Spatter-free laser percussion drilling closely spaced array holes are performed by Low et al. (2001). Non-uniform melts ejection, erosion to the hole walls and laser power reduction as the beam propagated to the hole are the most significant reasons for taper formation (Yilbas 1997). Okamoto et al. (2010) have experimentally investigated the effects of wavelengths on machining characteristics of silicon nitride and AlN in Nd:YAG laser microdrilling. The effect of drilling angle on recast formation, HAZ, oxide layer has studied by Sezer et al. during laser beam machining of thermal barrier coated (TBC) Nimonic 263 superalloy by varying the beam angles to workpiece surface from 30 to 90° (Low et al. 2000). Rega et al. (2011) have experimentally studied the influence of processing gas on laser microdrilling and ablation of aluminum and stainless steel. During which it is observed that compressed air is better than argon or nitrogen as a processing gas for laser microdrilling of aluminum and stainless steel. Some prior researchers attempted to find out the optimal process parameter settings for laser microdrilling process with less amount of HAZ (Heat affected zone), taper and recast layer. The Taguchi method (TM) is a statistical approach for the purpose of designing and improving product quality. Taguchi method is designed to optimize only a single quality characteristic (Ross 1996). Optimization of multiple quality characteristics, however, is much more complicated than the former. Improving one particular quality characteristic would possibly lead to serious degradation of other critical quality characteristics (Anawa and Olabi 2008). As a result, uncertainty will be increased during the decision making process. Grey theory, first established by Dr. Deng in 1982, can provide a solution to a system in which the model is uncertain or the information is incomplete (Pan et al. 2007). Taguchi based Grey analysis can effectively recommend a method of optimization technique for determination of optimal parameter settings and effectively solve the complicated interrelationships among multiple quality characteristics, which are applied in different manufacturing, processes (Acherjee et al. 2010).

Laser drilling process can be applied on a wide range of materials including off ceramics, composites, superalloys and advanced materials like nanocomposite. Recent times in industries like automobile, electronics, aerospace, biomedical, use of new advanced composite materials are increased rapidly due to their enhanced properties. Properties like higher toughness, very high elastic modulus, ductility, high temperature stability, low thermal conductivity, high strength and wear resistance make nanocomposite material highly competitive against other conventional materials. Composite materials are inhomogeneous and highly abrasive by nature which makes it difficult to machine material. In which nanocomposite offers a remarkable potential for new properties and technologies (Dandekar and Shin in press).

A nanocomposite is a multiphase solid material where one of the phases has one, two or three dimensions of less than 100 nm, or structures having nano-scale repeat distances between the different phases that make up the material. The mechanical, electrical, thermal, optical, electrochemical, catalytic properties of the nanocomposite differ vastly from that of the parent materials. Nanocomposite materials, mechanical properties are different from conventional composite material due to its very high aspect ratio (Ajayan et al. 2003). Pulsed Nd:YAG laser has great ability for micromachining because of higher laser beam intensity at low mean power due to small pulse duration and good focusing characteristics. Due to formation of a thermal gradient at that laser beam and work piece interface, the bond strength of nanocomposite material is reduced to a great extent which helps for easy material removal. That makes application of laser beam machining more advantageous for micromachining of nanocomposites over other conventional and non conventional manufacturing technique (Li et al. 2011). Das et al. (2008) studied laser micromachining of barium titanate ($BaTiO_3$)-polymer nanocomposite to produce novel micromachined 3D capacitors. A laser micromachining technique is also developed by Das et al. (2008) to control the surface morphology of sol-gel thin films. Laser beam microchanelling of Ni–YSZ cermets has been experimentally studied by J. Gurauskis et al. to improve gas permeation (Gurauskis et al. 2008). Benga et al. (2010) experimentally studied laser micromachining of HAP based biocomposites for hard tissue grafting. The laser surface modification of electrodeposited nanostructured binary Ni-Co alloy has been studied by Aqeeli (2011) to study the corrosion behaviour. However, no such comprehensive research work or any technology guidance is available for Nd:YAG laser beam microdrilling of Silicon Carbide 30 % Boron Nitride (SiC-30BN) nanomaterial.

Pulsed Nd:YAG laser microdrilling of Silicon Carbide 30 % Boron Nitride (SiC-30BN) nanocomposite material has been performed for determination of optimum laser machining Process parameters to get desired machining responses. From review of past research works, it is observed that the application of air as assist gas may causes some undesired characteristics during laser microdrilling. In the present research work, argon is used as assist gas during laser microdrilling of silicon carbide-30 boron nitride (SiC-30BN) nanocomposite material. Lamp current, pulse frequency, pulse width, gas pressure and focal distance with argon as assist gas have been chosen as independently controllable process variables. The settings of machining parameters are determined by using Taguchi L_{27} orthogonal array. Hole taper and heat affected zone (HAZ) width are considered as process responses. Optical measuring microscope at 10× magnification has been used to capture the image of drilled microhole from where the top and bottom micro-drilled hole diameter are measured by image analyzing software to calculate hole taper. Taguchi based Grey Relational Analysis method is used in this study to provide an efficient solution to complicated interrelationships among multiple response parameters. The application of this optimization technique converts the multiple quality characteristics to a single performance characteristic called grey relational grade from which finally the optimum machining parameters setting are obtained that simplifies the optimization procedure (Pan et al. 2007).

2 Mechanism of Material Removal During Laser Beam Drilling

Reflection, absorption, scattering and transmission are the physical phenomena that take place when the laser beam is incident on the workpiece surface. Depending on the focused beam diameter at the surface, laser power intensity, reflectivity,

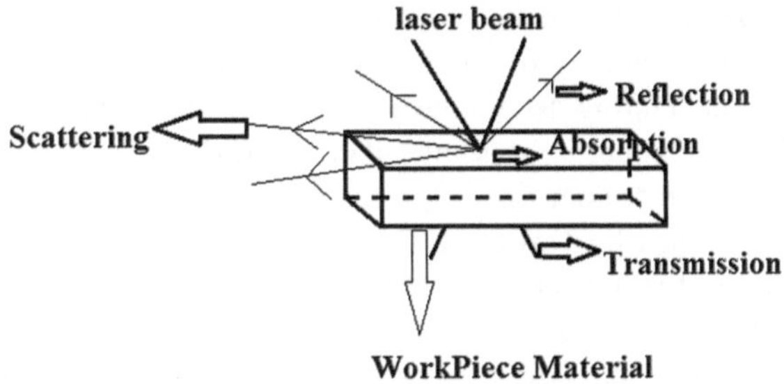

Fig. 2 Interaction of laser beam with workpiece material

absorptivity and physical properties of workpiece material, the laser radiation penetrates on different depth beneath the surface due to instant solid heating, melting and evaporation. Figure 2 shows a schematic of interaction of laser beam with work material.

Material removal during laser drilling is governed by melting, sublimation, vaporization, dissociation, plasma formation and ablation (Samant and Dahotre 2009). The surface temperature increases with increasing irradiation time and the surface temperature of the workpiece may reach the melting point and material removal takes place by melting (Salonitis et al. 2007). As the heating progresses further, the surface temperature of workpiece reaches the boiling point; the material removal takes place by evaporation instead of melting. After vaporization starts at the material surface, the liquid–vapour interface moves further beneath the surface with supply of laser energy and material is removed by evaporation from the surface above the liquid–vapour interface (Dahotre and Harimkar 2008). The pulse length and power intensities along with surface temperatures affect the recoil pressure which plays a vital role in material removal in molten state. When the laser energy density surpasses a certain threshold limit, the material immediately vaporizes, gets ionized and forms plasma having temperatures as high as 50,000 °C and pressures up to 500 MPa (Tönshoff and Kappel 1998). When the surface temperature exceeds a certain threshold value, plasma plume forms a shield over the machining area and reduces the energy available to the workpiece which has a harmful effect on machining operation. When laser energy exceeds the characteristic threshold laser energy, the temperature of the surface exceeds the boiling point of the material causing rapid vaporization and subsequent material removal by the process known as thermal ablation (Bäuerle 2000). For the long pulses liquid ejection from the cavity takes place, depending on the pulse length and power intensities whereas incase of short pulses, liquid ejection takes place without surface ablation. The liquid ejection improves the material removal rates from the cavity (Yilbas 2013). The physical phenomena of material removal mechanism are shown in Fig. 3.

Fig. 3 Various governing physical phenomena of material removal mechanism

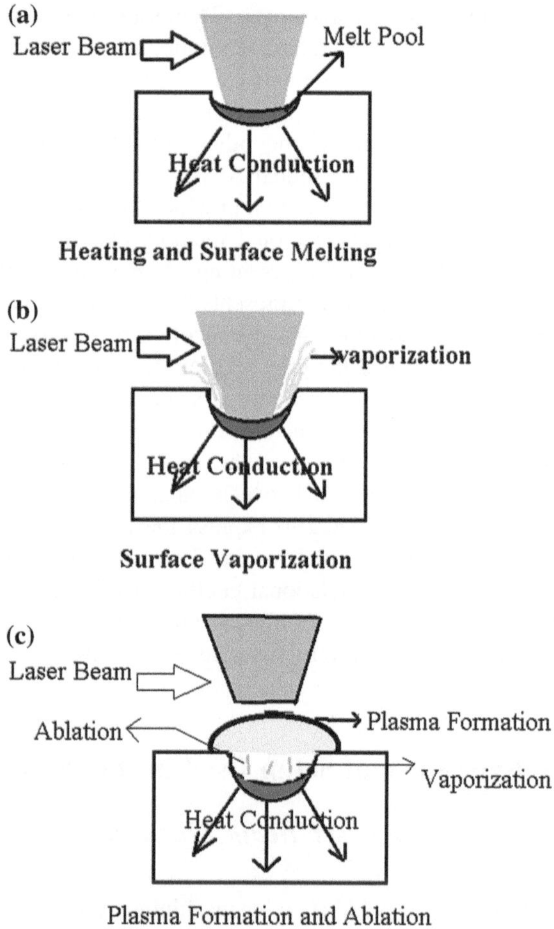

3 Overview of Proposed Optimization Methodology

Taguchi method of robust parameter design has been extensively adopted in manufacturing for nearly three decades. Taguchi method uses a special design of orthogonal arrays to study the entire process parameter space with a small number of experiments only (Phadke 1989). In Taguchi method, a loss function is defined to calculate the deviation between the experimental value and the desired value. The value of the loss function is further transformed into a signal-to-noise (S/N) ratio where 'signal' represents the desirable value and the 'noise' represents the undesirable value and S/N ratio expresses the scatter around the desired value. Usually, there are three categories of the quality characteristic in the analysis of the S/N ratio, i.e., the-larger-the-better, the-smaller-the-better, and the-nominal-the-best. Regardless of the category of the quality characteristic, the larger S/N ratio corresponds to the better quality characteristic. Therefore, optimal level of the process parameters is the level with the highest S/N ratio. Optimization of multiple quality characteristics is much more complicated than that of a single quality characteristic (Ross 1996). Improving one particular quality characteristic would possibly lead to serious degradation of the other critical quality characteristics. To solve this problem, the grey Relational Analysis first established by Dr. Deng in 1982, can provide a solution to a system in which the model is uncertain or the information is incomplete. In grey relational method, experimental information is classified as white, black, or grey. Representing white as full of information and black as lack of information, The grey represents the primitive data with poor, incomplete and uncertain information, and the incomplete relationship of the information among these data is called the grey relation. Grey relation analysis is used to measure an approximate correlation between sequences (Fung 2003). In grey relational analysis, the dimensions of factors considered are usually different and their magnitude difference is large. Therefore, the original data is normalized to make the magnitude of the original data in between range of zero and one. Next, a grey relational coefficient is calculated to express the relationship between the ideal and actual normalized experimental results. Then, the grey relational grade is computed by averaging the grey relational coefficient corresponding to each quality characteristic. The level with the highest grey relational grade is considered as optimum process parameter level (Acherjee et al. 2010).

4 Experimental Setup Used in the Present Research

4.1 Nd:YAG Laser Beam Machining Set up

A CNC pulsed Nd:YAG laser machining system manufactured by M/s Sahajanand Laser Technology, India, is used for the experimental study. The CNC Nd:YAG laser machining system consists of various subsystems such as; Laser source and beam delivery unit, power supply unit, radio frequency (RF) Q-switch driver unit, cooling unit and CNC controller for X,Y and Z axes movement.

Laser Source: Laser source consists of laser head and intra-cavity aperture.

Laser Head: In Nd:YAG Lasers, Neodymium (Nd) atoms (lasing media) are embedded in an Yttrium Aluminum Garnet (YAG) crystal host. The pump source is a krypton arc lamp positioned parallel to the Nd:YAG crystal. Both the Nd:YAG rod and the krypton arc lamp are located inside a gold plated elliptical cavity. The elliptical shaped cavity is used because of the special property of the ellipse that it has two focuses, and the light emitted from one focus is totally concentrated on another focus. Therefore in the cavity the krypton arc lamp is kept at one focus and the Nd:YAG rod is kept on another focus. The gold plated cavity is used for its high reflectivity to the wavelength of the pump light.

Intra cavity aperture: intra cavity aperture restricts the amplification to occur along the off axis of the resonator. An intracavity safety shutter is used to block the path of reflection between two mirrors.

Beam delivery unit: it consists rear mirror of 100 % reflectivity, front mirror of 80 % reflectivity, expander, bending mirror, focusing lens and protective lens. The output from the Q-switched Nd:YAG laser is directed to the workpiece using a beam delivery system that first bends the laser beam at 90°, and then focuses it on the workspot through the focusing lens and protective lens.

Power Supply unit: This is the main power supply unit that controls the laser output. It mainly ignites and controls the intensity of light emitted by the Krypton arc lamp. The intensity of light produced by the lamp is used for pumping the Nd atoms in Nd:YAG rod. Photographic view of RF-Q switch driver and Lamp current controller unit are given below.

Radio frequency (RF) Q-switch driver unit: The Q-switching is an excellent method to change the continuous mode operation into pulsed mode operation. RF Q-switch driver unit supplies RF signal to Q-switch for its operation. As it is being switched with a high frequency, it is also cooled by chilled water for repetitive operations and therefore the unit is provided with over temperature interlock. Figure 4 shows the pictorial view of RF-Q Switch driver and the power supply.

Cooling unit: This unit consists of two substances such as 3-phase chiller system, which is used for providing the chilled water to the heat exchanger. The specifications of chiller unit are given in Table 1. Heat exchanger and pump systems, which are mainly used for circulating the de-ionized water through resin filter

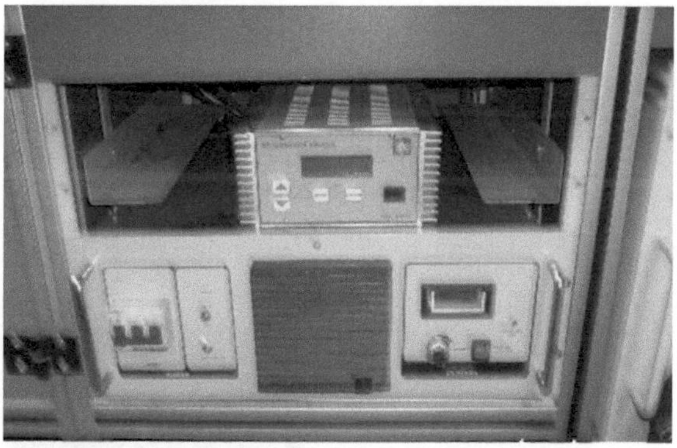

Fig. 4 The RF-Q Switch driver and the power supply (Lamp current controller) unit

Table 1 Specification of the chiller unit

Property	Description
Electrical requirement	415 V AC supply
De-ionized water tank capacity	5 L
De-ionized water resistance	≥200 kΩ/cm
De-ionized water flow rate	20 l/min
De-ionized water operating temperature	24 ± 1 °C

via water to water heat exchanger to the laser head & Q-switch. The de-ionized water is used because it has high optical transparency and low electrical conductivity. The heat from de-ionized water subsequently removed by water-to water heat exchanges connected to an outside water source. The de-ionized temperature is regulated means of a solenoid, which turns the outside flow 'ON' and 'OFF' as required.

CNC controller for X, Y and Z axes movement: The CNC controller consists of X-Y-Z axis translation stage and a control unit. The specifications of CNC table are given in Table 2. A stepper motor is attached to each of the axes and this is connected to the controlling unit.

The CNC Z axis controller unit controls the movement of focusing lens. The workpiece is held on the CNC controlled X-Y translation stage table. A specially developed fixture is placed on the X-Y translation stage table. CCD camera together with CCTV monitor is used for viewing the location of workpiece and also for checking the proper focusing condition on the workpiece surface. Photographic view of CCD camera and CNC controller unit are shown in Figs. 5 and 6 respectively.

Table 2 Specification of CNC table unit

Properties	Description
Axis of travel (X-Y axis)	150 mm × 150 mm
Focusing vertical travel	50 mm
Position accuracy	00.02 mm over travel 250 mm
Repeatability	0.01 mm over travel of 250 mm
Resolution	1 μm
Table working area	150 mm × 150 mm
Clamping	Using newly developed fixture
Feed rate (X, Y, and Z axis)	25 mm/s
Control system	CNC open loop system

Fig. 5 The CCD camera

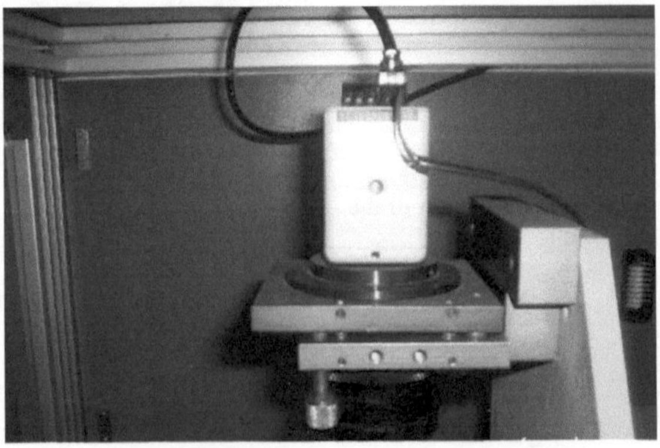

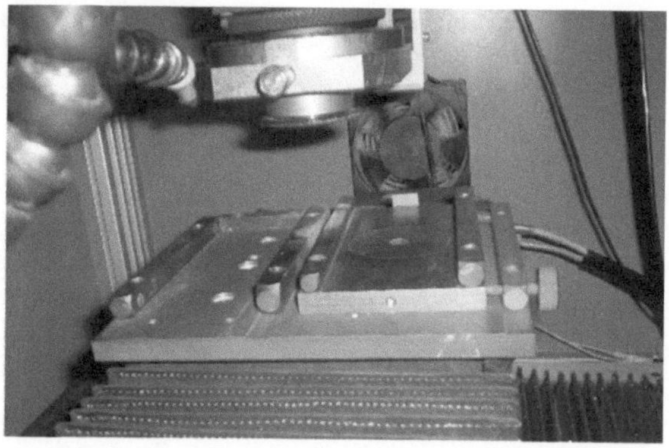

Fig. 6 The CNC controller unit (work table)

Assist Gas supply Unit: An assist gas supply unit has attached with LBM system which is mainly used to remove the machined particle from the machining zone as well as for cooling of machining zone. Assist gas supply unit consists of air compressor or specific gas container, oil moisture separator, pressure valve. Oil and moisture separator is used for purification of compressed air or gas. Pressure regulator is used to supply compressed air or gas at various pressure to the machining zone. Photographic view of Nd:YAG Laser beam machine is given in Fig. 7. The specifications of Nd:YAG Laser beam machine are given in Table 3.

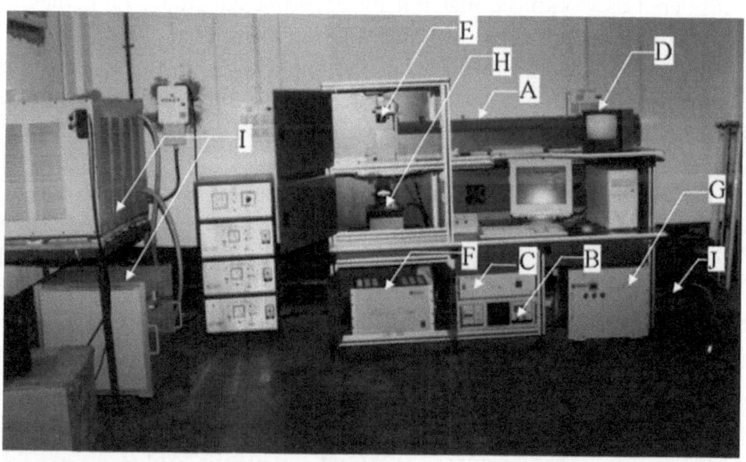

Fig. 7 Photographic view of laser beam machining. *A* = laser head, *B* = power supply unit, *C* = RF Q-switch driver unit, CNC X-Y-Z work table, *D* = CCTV, *E* = CCD camera, *F* = RF Q-switch driver unit, *G* = chiller unit, *H* = CNC controller unit, *I* = heat exchanger unit, *J* = air compressor

Table 3 Specification details of Nd:YAG laser machining setup

Specification	Description
Laser type	Nd:YAG laser
Wave length	1064 nm (Fixed)
Mode of operation	Pulsed (by RF-Q switch)
Type of Q switch	Acousto-optic Q switch
Mode of laser beam	Fundamental mode TEM00 (Fixed)
Mirror reflectivities	Rear mirror 100 %, front mirror 80 %
Beam diameter(1/e2)	1 mm (by default)
Laser beam spot diameter	75 μm
Average power	75 W
Pulse width	120–150 ns

4.2 Selection of Process Parameters and Workpiece Material

Lamp current, pulse frequency, assist gas pressure, pulse width and focal distance are considered as the controllable process parameters to carry out the experiment. Schematic illustration of focal distance change is shown in Fig. 8.

The geometrical aspects i.e. hole taper and Heat affected zone (HAZ) width are considered as process responses. The ranges of these parameters are selected on the basis of trial experiments conducted by using one factor at a time approach. 1.57 mm thick Silicon Carbide 30 % Boron Nitride (SiC-30BN) nanocomposite has been considered as workpiece material for experimental study. Properties of SiC-30BN nanocomposite are listed in Table 4 (Xiangdong et al. 2004).

Previous research works reveals that the application of air as assist gas may cause some undesired hole characteristics during laser microdrilling. Argon is used as assist gas during laser micro drilling of silicon carbide-30 boron nitride (SiC-30BN) nanocomposite material. Taguchi methodology for five factors at three levels (given in Table 5) is used for the implementation of the plan of orthogonal array experiments. An L27 orthogonal array with five columns and 27 rows is employed in this work. Experiments are performed accordingly with the

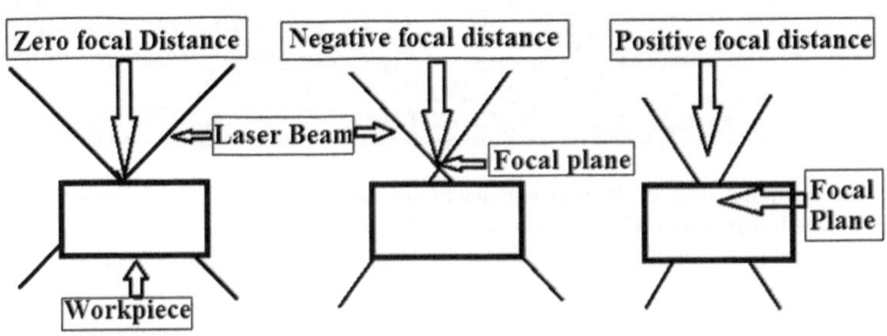

Fig. 8 Schematic diagram illustrating different focal distance

Table 4 Properties of SiC-30BN nanocomposite

Property	Unit	Values
Yield stress	MPa	263.2 ± 6
Young's modulus	GPa	230 ± 5
Fracture strength	MPa	380 ± 10
Density	g/cc	2.906

Table 5 Machining parameters and their levels

Process parameters	Level		
	1	2	3
Lamp current (*L*) (A)	23.5	24.5	25.5
Pulse frequency (γ) (kHz)	1	2	3
Pulse width (*w*) (%)	2	4	6
Argon gas pr. (*p*) (kg/cm^2)	0.4	0.7	1.0
Focal distance (*F*) (μm)	−200	0	200

arrangement of the orthogonal array. The interaction between the process parameters are not considered for this study. Therefore, the five controllable process parameters yield 10° of freedom which greater than number of controllable process parameters.

Optical measuring microscope has been used to capture the image of drilled microhole at 10× magnification and to determine the top and bottom diameters of the microholes. Thickness of the workpiece is measured by a digital micrometer. From where HAZ width width and hole taper are measured. Microscopic views of machined microholes are also taken for further analysis. HAZ width and taper of the drilled microhole has been calculated as follows:

$$\text{HAZ width (mm)} = \frac{\text{HAZ\_dia}_{\text{Top}} - \text{Hole\_dia}_{\text{Top}}}{2} \tag{1}$$

$$\text{Taper (rad)} = \frac{\text{Hole\_dia}_{\text{Top}} - \text{Hole\_dia}_{\text{Bot}}}{2 \times \text{Thickness}} \tag{2}$$

5 Experimental Observation

The experimental layout for the laser microdrilling process parameters using the L27 orthogonal array and the experimental results are presented in Table 6.

The effects of the process parameters such as the lamp current, pulse frequency, pulse width, assist gas pressure and focal distance on the hole taper and HAZ width, during laser microdrilling are presented in Fig. 9a, b respectively. Taper has been found to be a decreasing tendency with increase of lamp current up to 24.5 A, and then it increases sharply with increase of lamps current. HAZ width increases

Table 6 Experimental lay out with multi-performance results

Lamp current	Pulse frequency	Pulse width	Assist gas pressure	Focal distance	Taper	HAZ
1	1	1	1	1	0.0821	0.2038
1	1	1	1	2	0.0605	0.2179
1	1	1	1	3	0.0435	0.2084
1	2	2	2	1	0.1005	0.1966
1	2	2	2	2	0.0835	0.1943
1	2	2	2	3	0.0793	0.1784
1	3	3	3	1	0.0901	0.1825
1	3	3	3	2	0.0865	0.1798
1	3	3	3	3	0.0649	0.1854
2	1	2	3	1	0.0482	0.2504
2	1	2	3	2	0.0350	0.2465
2	1	2	3	3	0.0274	0.2436
2	2	3	1	1	0.0989	0.2037
2	2	3	1	2	0.0903	0.2074
2	2	3	1	3	0.0582	0.2123
2	3	1	2	1	0.1121	0.1976
2	3	1	2	2	0.0995	0.2020
2	3	1	2	3	0.0649	0.1916
3	1	3	2	1	0.1250	0.2350
3	1	3	2	2	0.1200	0.2263
3	1	3	2	3	0.0787	0.2259
3	2	1	3	1	0.1203	0.2503
3	2	1	3	2	0.0905	0.2501
3	2	1	3	3	0.0834	0.2371
3	3	2	1	1	0.1470	0.2376
3	3	2	1	2	0.1117	0.2409
3	3	2	1	3	0.0870	0.2203

monotonically over the entire range. Generally, higher laser beam power generates higher thermal energy which produces large taper and greater HAZ width. From these figures it is observed that increase in the pulse frequency results in increase in taper due to instant melting and vaporization from the top surface at high pulse frequency and vice versa. But HAZ width decreases due to less energy beam generation at higher pulse frequency. Increase in pulse width results in uniform material removal from machining zone due to high peak power. For which taper shows a decreasing tendency up to 2nd level of pulse width but further increase in pulse width causes higher heating value in top surface than bottom surface results in increase in taper. At lower pulse widths less HAZ is generated, than that of higher pulse widths due to less interaction time and heat diffusion between laser beam and workpiece material. It is found that, Taper increases with gas pressure up to 2nd

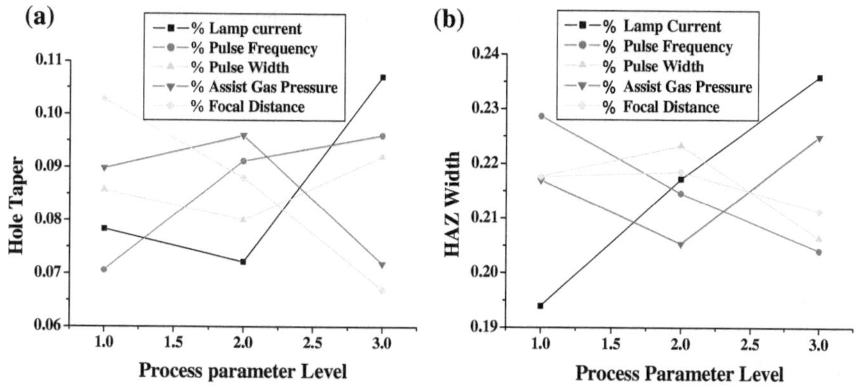

Fig. 9 **a** Effects of Nd:YAG laser microdrilling parameters on taper. **b** Effects of Nd:YAG laser microdrilling parameters on HAZ width

level and thereafter starts decreasing but HAZ width shows the reverse effect. Increase in assist gas pressure causes rapid removal of molten material as well as excess heat from machining zone results in increase in taper but decrease in HAZ width. Further increase in assist gas pressure cause uneven heating value on the top surface of the workpiece results in formation of low taper and high HAZ width. It has been shown from the figure that taper decreases with increase in focal distance. Difference in the material removal from the top and bottom surface decreases gradually with increase in focal distance, generates low hole taper. But the HAZ width marginally increases up to level 2 then decreases. The increased beam spot area with increased focal distance causes high HAZ width. With further increases in focal distance laser beam propagates through the hole for which less HAZ width generates in top surface.

6 Determination of Optimal Parameter Settings

The steps of the grey Relational Analysis are as follows:

Step 1: Grey relational generating
Step 2: Calculating the grey relational coefficient
Step 3: Calculating the grey relational grade

6.1 Grey Relational Generating

In Grey Relation analysis, the orders of magnitude of the factors are different. Because of this the input data must be pre processed to transform the original sequence to a comparable sequence. For this purpose, the experimental results are normalized in the range of zero and one, the process is called grey relational generating. Microhole taper and HAZ width have lower-the-better quality characteristic. The normalized results, x_{ij}, for lower-the-better quality characteristic can be expressed (Pan et al. 2007) as

$$x_{ij} = \frac{\max_j y_{ij} - y_{ij}}{\max_j y_{ij} - \min_j y_{ij}} \tag{3}$$

where y_{ij} is the ith quality characteristic in the jth experiment. Larger normalized results correspond to the better quality and the best-normalized result should be equal to 1.

6.2 Determination of Grey Relational Coefficients

A grey relational coefficient is calculated to display the relationship between the optimal and actual normalized experimental results. The grey relational coefficient can be expressed (Pan et al. 2007) as

$$\xi_{ij} = \frac{\min_i \min_j \left|x_i^0 - x_{ij}\right| + \zeta \max_i \max_j \left|x_i^0 - x_{ij}\right|}{\left|x_i^0 - x_{ij}\right| + \zeta \max_i \max_j \left|x_i^0 - x_{ij}\right|} \tag{4}$$

where x_i^0 is the ideal normalized result (i.e. best normalized result = 1) for the ith quality characteristics and ζ is a distinguishing coefficient, which is defined in the range of $0 \leq \zeta \leq 1$. In this study, the value of ζ is taken as 0.5 (which is generally adopted in most previous studies) for calculating the grey relational coefficients.

6.3 Grey Relational Grades Determination

The grey relation grade can be calculated from Eq. 5 (Pan et al. 2007) by taking the average of the grey relational coefficients and applying the same weight on both quality characteristics under investigation.

$$\gamma_j = \frac{1}{m}\sum_{i-1}^{m} w_i \xi_{ij} \tag{5}$$

where γ_j is the grey relational grade for the jth experiment, w_i is the weighting factor for the ith quality characteristic and m is the number of quality characteristics. The calculated value of grey relation grade is given in Table 7.

Table 7 Results of grey relation analysis

Exp. no.	Taper (rad)	HAZ width (mm)	Grey relation generating		Grey relation coefficient		Grey relation grade	Order
			Taper	HAZ width	Taper	HAZ width		
1	0.0821	0.2038	0.5426	0.6472	0.5223	0.5863	0.5543	13
2	0.0605	0.2179	0.7232	0.4514	0.6437	0.4768	0.5603	11
3	0.0435	0.2084	0.8654	0.5833	0.7879	0.5455	0.6667	7
4	0.1005	0.1966	0.3888	0.7472	0.4500	0.6642	0.5571	12
5	0.0835	0.1943	0.5309	0.7792	0.5160	0.6936	0.6048	9
6	0.0793	0.1784	0.5661	1.0000	0.5354	1.0000	0.7677	1
7	0.0901	0.1825	0.4758	0.9431	0.4882	0.8978	0.6930	4
8	0.0865	0.1798	0.5059	0.9806	0.5029	0.9626	0.7328	2
9	0.0649	0.1854	0.6865	0.9028	0.6146	0.8372	0.7259	3
10	0.0482	0.2504	0.8261	0.0000	0.7419	0.3333	0.5376	14
11	0.0350	0.2465	0.9365	0.0542	0.8872	0.3458	0.6165	8
12	0.0274	0.2436	1.0000	0.0944	1.0000	0.3557	0.6779	5
13	0.0989	0.2037	0.4022	0.6486	0.4554	0.5873	0.5214	17
14	0.0903	0.2074	0.4741	0.5972	0.4874	0.5538	0.5206	18
15	0.0582	0.2123	0.7425	0.5292	0.6600	0.5150	0.5875	10
16	0.1121	0.1976	0.2918	0.7333	0.4138	0.6522	0.5330	15
17	0.0995	0.2020	0.3972	0.6722	0.4534	0.6040	0.5287	16
18	0.0649	0.1916	0.6865	0.8167	0.6146	0.7317	0.6732	6
19	0.1250	0.2350	0.1839	0.2139	0.3799	0.3888	0.3843	25
20	0.1200	0.2263	0.2258	0.3347	0.3924	0.4291	0.4107	22
21	0.0787	0.2259	0.5711	0.3403	0.5383	0.4311	0.4847	19
22	0.1203	0.2503	0.2232	0.0014	0.3916	0.3336	0.3626	26
23	0.0905	0.2501	0.4724	0.0042	0.4866	0.3343	0.4104	23
24	0.0834	0.2371	0.5318	0.1847	0.5164	0.3801	0.4483	21
25	0.1470	0.2376	0.0000	0.1778	0.3333	0.3782	0.3557	27
26	0.1117	0.2409	0.2952	0.1319	0.4150	0.3655	0.3902	24
27	0.0870	0.2203	0.5017	0.4181	0.5008	0.4621	0.4815	20

The optimal parameter setting determined by using grey based Taguchi method is $L1\gamma3w3p3f3$. The grey relational grade of each factor level is calculated as the average for the same level in each column. The grey relational grade graph (Fig. 10) shows the trend of each factor with respect to different levels with grey relation grade.

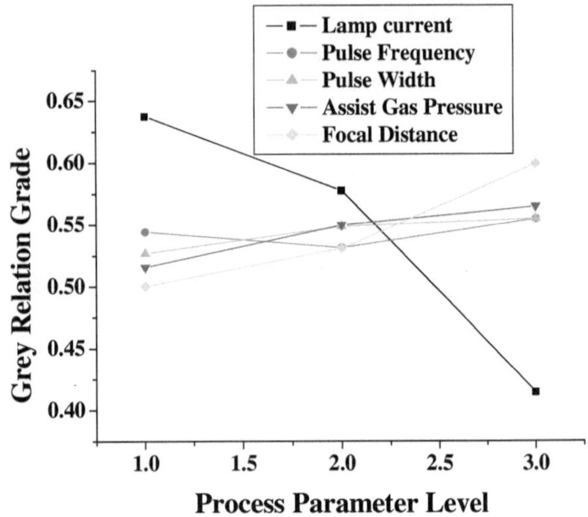

Fig. 10 Grey relation grade graph

7 ANOVA Analysis

ANOVA (Analysis of Variance) is a computational technique to estimate the effect of each process parameter on microhole quality characteristics quantitatively. This is accomplished by separating the total variability of the grey relational grades. In ANOVA, the ratio between the variance of the process parameter and the error variance is called Fisher's ratio and it is used to determine the significance of process parameter on machining response by comparing the F-test value of the parameter with the standard F table value ($F_{0.05}$) at the 5 % significance (95 % confidence) level. The ANOVA results for HAZ width and hole taper are listed in Tables 8 and 9 respectively.

If the value of P is less than 0.05, the process parameter is considered significant. From the Table 8 it is found that all the process parameters other than focal distance are statistically significant for HAZ width. Percentage contribution column in ANOVA result table indicates at the degree of influence of all the process parameters on the microhole characteristics. According to the table it is found that lamp current (56.62 %) has the most dominant factor followed by pulse frequency (19.46 %), assist gas pressure (12.75 %), pulse width (9.34 %) and focal distance (1.83 %). From the table it has been observed that S-values of the responses are smaller and R-Sq, R-Sq (adj) and R-Sq (pred) values of responses are moderately high, from which it can be concluded that the data for each response are well fitted in the developed models.

Whereas from the Table 9 it is found that all the process parameters other than pulse width are statistically significant for hole taper. Percentage contribution column in ANOVA result table indicates at the degree of influence of all the process parameters on the hole taper. According to the table it is found that lamp current

Table 8 Results of ANOVA for HAZ width

Source	Degrees of freedom	Sum of square	Mean of square	F	P	% contribution
Lamp current	2	0.007900	0.003950	107.63	0.000	56.62
Pulse frequency	2	0.002714	0.001357	36.98	0.000	19.46
Pulse width	2	0.001303	0.000651	17.75	0.000	9.34
Assist gas pressure	2	0.001778	0.000889	24.23	0.000	12.75
Focal distance	2	0.000255	0.000128	3.48	0.056	1.83
Error	16	0.000587	0.000037			
Total	26	0.014538				
S	R-sq		R-sq(adj)		R-sq(pred)	
0.0088199	94.13 %		90.46 %		83.28 %	

Table 9 Results of ANOVA for hole taper

Source	Degrees of freedom	Sum of square	Mean of square	F	P	% contribution
Lamp current	2	0.006883	0.003442	44.24	0.000	34.49
Pulse frequency	2	0.003581	0.001791	23.02	0.000	17.95
Pulse width	2	0.000487	0.000243	3.13	0.071	2.44
Assist gas pressure	2	0.002665	0.001332	17.13	0.000	13.35
Focal distance	2	0.006341	0.003170	40.75	0.00	31.77
Error	16	0.001245	0.000078			
Total	26	0.021202				
S	R-sq		R-sq(adj)		R-sq(pred)	
0.0060581	95.96 %		93.44 %		88.50 %	

(34.49 %) has the most dominant factor followed by focal distance (31.77 %), pulse frequency (17.95 %), assist gas pressure (13.35 %) and pulse width (2.44 %). It can be concluded that the data for each response are well fitted in the developed models from the results of regression analysis for hole taper given in Table 9.

8 Confirmation Test

After getting the optimal parameter setting for desired microhole characteristics, the next step is to verify the feasibility of the proposed Taguchi based grey method. The optimum grey relational grade, γ_{opt}, is calculated as:

$$\gamma_{opt} = \gamma_m + \sum_{i-1}^{q} (\overline{\gamma}_i - \gamma_m) \tag{6}$$

where, γ_m is the total mean of the grey relational grade, γ_i, is the mean of the grey relational grade of ith parameter at the optimal level, and q is the number most influencing of microhole drilling parameters.

Three additional experiments are performed with level at optimal parameter setting and average of those three results is taken for confirmation test. Table 10 furnishes the predicted value of grey relation grade and result of confirmation test. From the table it is found that hole quality characteristics viz. Hole taper is improved from 0.0649 rad to 0.055 rad (15.25 %) and HAZ width is improved from 0.1854 to 0.1828 mm (1.40 %). The microscopic view of drilled hole at optimum parameter condition is given in Fig. 11.

This result shows that the use of Taguchi based grey relation analysis is effective for improvement of microhole quality characteristics during laser microdrilling of SiC-30BN. The overall quality characteristic is improved from 0.7259 to 0.7877

Table 10 Results of confirmation test

Level	Initial parameter setting	Optimal parameters	
		Prediction	Experiment
	$L1\ \gamma\ 3w3\ p3f3$	$L1\gamma3w3\ p3f3$	$L1\gamma3w3p3\ f3$
Taper (rad)	0.0649	–	0.055
HAZ width (mm)	0.1854	–	0.1828
Grey relation grade	0.7259	0.7822	0.7877

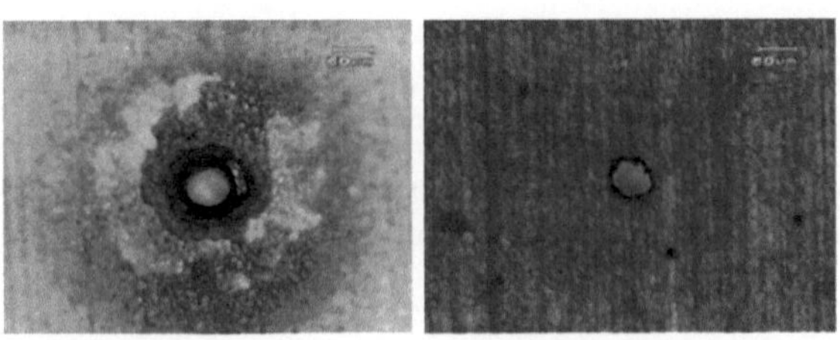

Top Surface **Bottom Surface**

Fig. 11 Microscopic view of the hole drilled at optimum parametric condition

(8.51%). The result of confirmation test justify the application of the proposed method and leads to continuous improvement of microhole geometry in laser microdrilling of nanocomposite materials.

9 Conclusion

From the study it is concluded that laser beam microdrilling technique can be used successfully for microdrilling of SiC-30BN nanocomposite. Taguchi based grey relational analysis is used in this paper to optimize the laser microdrilling of SiC-30BN nanocomposite with multiple performance characteristics. Multiobjective optimization is simplified by converting multiple machining responses into a single performance characteristic called grey relational grade. It is shown from the study that the performance characteristics of the laser microdrilling process such as HAZ width and hole taper are improved simultaneously by using the proposed method. From the experimental observation it has been found that pulse frequency, pulse width, applied gas pressure, lamp current and focal distance have influence on both the machining responses hole taper and HAZ width. The mechanisms for the phenomenon observed during the experimental observation are also discussed. The outcome of the study helps to obtain optimum parameter settings for desired microhole quality characteristics and indicates the feasibility of the proposed multi-objective optimization technique for laser microdrilling of SiC-30BN nanocomposite material.

References

Acherjee, B., Kuar, A. S., Mitra, S., & Misra, D. (2010). Selection of process parameters for optimizing the weld strength in laser transmission welding of acrylic. *Proceedings of the IMechE Part B: Journal of Engineering Manufacture, 224*(B10), 1529–1536.

Ajayan, P. M., Schadler, L. S., & Braun, P. V. (2003) *Nanocomposite science and technology*. New York: Wiley. ISBN 3527303596.

Anawa, E. M., & Olabi, A. G. (2008). Using Taguchi method to optimize welding pool of dissimilar laser-welded component. *Optics and Laser Technology, 40*, 379–388.

Aqeeli, N. A. (2011). The corrosion behavior of electrodeposited and laser irradiated ni-co nanostructured alloy. *Reviews Advanced Material Science, 27*, 75–82.

Bachman, F. (1989). Excimer lasers in a fabrication line for a highly integrated printed circuit board. *Chemtronics, 4*, 149.

Bäuerle, D. (2000). *Laser Processing and Chemistry*. New York: Springer. ISBN 3540668918.

Benga, G., Gingu, O., Ciupitu, I., Gruionu L., Pascu, I., & Moreno, J. C. (2010). Processing and laser micromachining of HAP based biocomposites. *Engineering the Future*. ISBN 978-953-307-210-4.

Biswas, R., Kuar, A. S., Sarkar, S., & Mitra, S. (2010a). A parametric study of pulsed Nd:YAG laser micro-drilling of gamma-titanium aluminide. *Optics and Laser Technology, 42*, 23–31.

Biswas, R., Kuar, A. S., Biswas, S. K., & Mitra, S. (2010b). Effects of process parameters on hole circularity and taper in pulsed Nd:YAG laser microdrilling of Tin-Al_2O_3 composites. *Materials and Manufacturing Processes, 25*, 503–514.

Chryssolouris, G. (1991). *Laser machining theory and practice*. New York: Springer.

Corcoran, A., Sexton, L., Seaman, B., Ryan, P., & Byrne, G. (2002). The laser drilling of multi-layer aerospace material systems. *Journal of Materials Processing Technology, 123*, 100–106.

Dahotre, N. B., & Harimkar, S. P. (2008). *Laser fabrication and machining of materials*. New York, NY: Springer.

Dandekar, C. R., & Shin, Y. C. (in press). Modelling of machining of composite materials: A review. *International Journal of Machine Tools & Manufacture*.

Das, R. N., Egitto, F. D., Lauffer, J. M., & Markovich, V. R. (2008). Laser Micromachining of barium titanate ($BaTiO_3$)-epoxy nanocomposite-based flexible/rollable capacitors: New approach for making library of capacitors. *Electronics Packaging Manufacturing, IEEE Transactions, 31*, 97–103.

Davim, P. J. (2013). *Nontraditional machining processes, research advances*. London: Springer. doi:10.1007/978-1-4471-5179-1. ISBN 978-1-4471-5178-4.

Fung, C. P. (2003). Manufacturing process optimization for wear property of fiber-reinforced polybutylene terephthalate composites with grey relational analysis. *Wear, 254*, 298–306.

Gurauskis, J., Sola, D., Pena, J. I., & Orera, V. M. (2008). Laser drilling of Ni–YSZ cermets. *Journal of the European Ceramic Society, 28*, 2673–2680.

Kuar, A. S., Doloi, B., & Bhattacharyya, B. (2006). Modelling and analysis of pulsed Nd:YAG laser machining characteristics during micro-drilling of zirconia (ZrO_2). *International Journal of Machine Tools and Manufacturing, 46*(12–13), 1301–1310.

Lasagni, F. A., & Lasagni, A. F. (2011). Fabrication and characterization in the micro-nano range. *Advanced Structured Materials, 10*. doi:10.1007/978-3-642-17782-8_2.

Low, D. K. Y., Li, L., & Corfe, A. G. (2001). Spatter-free laser percussion drilling of closely spaced array holes. *International Journal of Machine Tools and Manufacture, 41*, 361–377.

Low, D. K. Y, Li, L., & Byrd, P. J. (2000). Taper formation and control during laser drilling in nimonic 263 alloy. In *Proceedings of 33rd International MATADOR Conference*, pp. 461–66.

Li, L., Hong, M., Schmidt, M., Zhong, M., Malshe, A., Huis. B. H. & Kovalenko, V. (2011). Laser nano-manufacturing—State of the art and challenges. *CIRP Annals—Manufacturing Technology, 60*, 735–755.

Meijer, J. (2004). Laser beam machining (LBM), state of the art and new opportunities. *Journal of Material Process Technology, 149*, 2–17.

Okamoto, Y., Sakagawa, T., Nakamura, H., & Uno, Y. (2010). Micro-machining characteristics of ceramics by harmonics of Nd:YAG laser. *Journal of Advanced Mechanical Design, Systems, and Manufacturing, 4*(2), 661–667.

Pan, L. K., Wang, C. C., Wei, S. L., & Sher, H. F. (2007). Optimizing multiple quality characteristics via Taguchi method-based Grey analysis. *Journal of Materials Processing Technology, 182*, 107–116.

Pandey, P. C. & Shan, H. S. (2004). *Modern machining processes*. India: Tata McGraw Hill BookCom. 0-07096-553-6.

Phadke, M. S. (1989). *Quality engineering using robust design*. Englewood Cliffs, New Jersey: Prentice Hall.

Ready, J. F., & Farson, D. F. (2001). *LIA handbook of laser materials processing*. Orlando: Laser Institute of America.

Rega, Y., Leitzb, B. & Schmidta, M. (2011). Influence of processing gas on the ablation quality at ns-laser beam ablatio. *Physics Procedia, 12*, 182–187.

Rodden, W. S. O., Kudesia, S. S., Hand, D. P., & Jones, J. D. C. (2002). A comprehensive study of the long pulse Nd:YAG laser drilling of multi-layer carbon fibre composites. *Optics Communication, 210*, 319–328.

Ross, P. J. (1996). *Taguchi techniques for quality engineering* (2nd ed.). New Delhi, India: Tata McGraw-Hill Publishing Company Ltd.

Russo, R. (1995). Laser ablation. *Applied Spectroscopy, 49*(9), 14A–28A.

Salonitis, K., Stournaras, A., Tsoukantas, G., Stavropoulos, P., & Chryssolouris, G. (2007). A theoretical and experimental investigation on limitations of pulsed laser drilling. *Journal of Material Processing and Technology, 183*(1), 96–103.

Samant, A. N., & Dahotre, N. B. (2009). Laser machining of structural ceramics—A review. *Journal of the European Ceramic Society, 29*, 969–993.

Sanikommu, N., Bathe, R., & Joshi, A. S. (2007). Detection of breakthrough in laser percussion drilling. *Lasers in Engineering, 17*, 361–369.

Tam, S. C., Yeo, C. Y., Jana, S., & Lau, M. W. (1994). A technical review of the laser drilling of aerospace materials. *Journal of Materials Processing Technology, 42*, 15–49.

Tönshoff, H. K., & Kappel, H. (1998). Surface modification of ceramics by laser machining. *CIRP Annal—Manufacturing Technology*, *47*(1), 471–474.

Xiangdong, W., Guanjun, Q., & Zhihao, J. (2004). Fabrication of machinable silicon carbide-boron nitride ceramic nanocomposites. *Journal of the American Ceramic Society, 87*, 565–570.

Yilbas, B. S. (1988). The examination of optimal focus setting with material thickness at atmospheric pressures of air and oxygen in laser drilling of metals. *Proceedings of Institution of Mechanical Engineers Part B, Journal of Engineering Manufacture, 202*, 123–127.

Yilbas, B. S. (1997). Parametric study to improve laser hole drilling process. *Journal of Materials Processing Technology, 70*, 264–273.

Yilbas, B. S. (2013). Laser *drilling, springer briefs in manufacturing and surface engineering*. doi:10.1007/978-3-642-34982-9_1, ISBN 978-3-642-34981-2.

Pulsed Nd:YAG Laser Micro-turning Process of Alumina Ceramics

G. Kibria, B. Doloi and B. Bhattacharyya

Abstract Laser micro-turning process is one of the new and emerging technologies in the area of laser material processing (LMP) of engineering materials. It is employed for generation of micro-turning surface of particular surface profile and dimensional accuracy on cylindrical workpiece with specific length and depth of turn within tight tolerance. As the process is recently developed micro manufacturing technique, a well planned research study and experimental investigation should be conducted considering various laser micro-turning process parameters. Therefore, various experimental schemes are adapted to study and analysis of significant process parameters on response criteria such as surface roughness and machining depth. A servo controller based fixture is designed and developed indigenously to hold and rotate the cylindrical shaped work samples at various *workpiece rotating speed*. Overlap between two successive spots (i.e. *spot overlap*) and overlap between two successive micro-groove widths (i.e. *circumferential overlap*) play major role for generating quality surface features during laser micro-turning process. Therefore, mathematical formulations of *spot overlap* and *circumferential overlap* are developed for better understanding of the laser micro-turning process and also to study the effects of these overlap factors on performance characteristics. Moreover, attempt has been made to carry out experimental investigation to micro-turn cylindrical shaped engineering ceramics at laser defocus conditions of laser beam. Moreover, comparative study and analyse is performed to explore the effect of focused and defocused conditions of laser beam on surface roughness criteria. SEM micrographs of the laser turned surface captured at various parametric combinations have also been studied for qualitative analysis of the process.

G. Kibria (✉)
Department of Mechanical Engineering, Aliah University, Kolkata 700156, India
e-mail: prince_me16@rediffmail.com

B. Doloi · B. Bhattacharyya
Department of Production Engineering, Jadavpur University, Kolkata 700032, India
e-mail: bdoloionline@rediffmail.com

B. Bhattacharyya
e-mail: bb13@rediffmail.com

S.N. Joshi and U.S. Dixit (eds.), *Lasers Based Manufacturing*,
Topics in Mining, Metallurgy and Materials Engineering,
DOI 10.1007/978-81-322-2352-8_18

Keywords Laser micro-turning process · Alumina (Al_2O_3) · Surface roughness · Defocusing conditions · Pulsed Nd:YAG laser · Spot overlap · Circumferential overlap

1 Introduction

Increased interest in micromachining technology has captured the imagination of every manufacturing and industry segment such as aerospace, medical appliance and automotive as well as military applications. The potential for product miniaturization continues to grow and while posing numerous technical challenges. In response to this continued miniaturization, researchers are developing new technologies to meet the high demands of miniature products and unique challenges posed by micro manufacturing and must develop appropriate micromachining systems or processes to support this growth. Several micro-machining technologies are being employed in the manufacture of a wide variety of micro-products and devices. Manufacturing techniques, both traditional and non-traditional, have been developed significantly for the realization of diverse range of microstructures used for various micro-system applications. Demand for various micro-features in macro as well as micro-components has necessitated the urgent need to develop and employ a number of alternative manufacturing techniques, other than traditional processes, so that a wide range of materials such as ceramics, polymer, composites, high strength temperature resistant (HSTR) alloys can be used as parent material for micro-components or parts. Non-traditional micro-machining processes can be efficiently and effectively applied to machine these types of hard-to-machine materials irrespective of their hardness, toughness and brittleness. Laser beam machining (LBM) is one of the advanced non-traditional machining processes, which can be applied to manufacture wide variety of geometries and complex shapes on difficult-to-machine materials with greater accuracy and faster rate.

Laser beam machining is classified based on type of source, mode of operations and wavelength. Among a wide variety of lasers, Nd:YAG lasers, Excimer lasers and Diode lasers are widely used for micro-machining applications. In pulsed mode, the vaporization of material occurs in a very short time duration so that it does not allow the heat to be conducted into the bulk material. The pulsing of laser beam is done in various ways, such as normal pulsing, Q-switching and mode locking. Nd:YAG laser compiled with Q-switch pulsing offers outstanding features and quality of machining during laser micro-machining operations such as micro-cutting, micro-drilling, micro-grooving, micro-turning and micro-milling.

Among the various micro-machining techniques of laser beam machining processes, laser micro-turning process is a newly developed machining process in the modern cutting-edge technology era. In this innovative technique of micro-machining process, cylindrical surface of engineering ceramics and composites can be micro-turned with high accuracy and good quality surface. Moreover, laser micro-turning process using pulsed Nd:YAG laser has machining features including melting and vaporization of material in very short time that avoids formation of recast/re-solidified layer on machined surface and high repetition rate machining that avoids cracks in machined surfaces and sub-surfaces. By taking into account all these significant advantages, this new micro-machining technique i.e. laser micro-turning process can be employed to remove very thin layer (depth) from cylindrical work-piece of ceramics, which are difficult-to-turn in conventional turning processes. By this process, one can produce desired dimensional micro-grooves or slots for a specific depth and length on cylindrical surface. This kind of micro-turned surface with specific surface texture is required for assembly purpose of various micro-parts in the micro-engineering applications. In ceramic whistle stud assembly purpose, high-alumina ceramic spikes are often used. In lining applications, the alumina ceramic bearing rings are also being used for assembling several precision components. Further, on the shop floor, the reduction of the number of micro cutting edges of ceramic grinding wheel due to worn out of abrasive particles from the wheel surface is a major problem. The generation of micro-cutting edges on grinding wheel is a major challenge for the manufacturing technocrats of high precision industries. Laser micro-turning process can be a novel technology to generate new cutting edges of desired surface topography on grinding wheel surface.

2 Laser Micro-turning Process

In 1988, three dimensional laser beam machining technology was introduced for removal of material in bulk mode while using two intersecting laser beams (Chrysolouris et al. 1988). However, to produce a laser turned surface of a specific length and micron sized depth or to machine micro-groove for a specific length, the use of two intersecting beam fails. Most of the past research works are related to laser micro-drilling and micro-grooving operations of structural ceramics such as Si_3N_4, Al_2O_3, MgO and SiC (Dahotre and Harimkar 2008; Kibria et al. 2010a). In the area of laser micro-turning process, few research investigation has been reported in the literature. On the other hand, Aluminium Oxide (Al_2O_3) ceramic is one of the most important technical ceramics, which are being utilized in several applications ranging from automotive (bearing, valve, etc.), biomedical (orthopedic implants particularly in hip replacement surgery) to aerospace (seals for gas turbine engines, fuel line assembly, and thermocouples) and military (modern weapons, ceramic armor, etc.) due to its several properties like high mechanical strength, the ability to retain dimensional stability and excellent chemical resistance in aggressive environments (Basu and Balani 2010; Gitzen 1970; Kim et al. 2009). In many applications, micro-turning process is carried out using diamond tool for processing hard-to-machine cylindrical shaped materials. Yet, the use of diamond tool calls upon many limitations while machining of advanced ceramics and composites. Firstly, the diamond tool may undergo irregular and excessive wear as it is a contact-type machining process and it further results in uneconomic production and also cracks formation on the machined surface and sub-surface due to brittle nature of ceramic materials (Buttery et al. 1979; Shih 2000). In addition, wire electrical discharge turning (WEDT) process is a very recently developed novel micro-turning technology used for fabricating micro-structures like small-pins, micro-tool electrodes etc. (Haddad et al. 2008, 2010). However, in this process, non-conductive materials like ceramics cannot be used as workpiece materials. Laser micro-turning process, which removes materials layer-by-layer from cylindrical shaped work sample, can be effectively utilized to generate such micro-turning surface or micro-groove on difficult-to-machine materials. The basic concept of laser micro-turning process is of producing multiple side-by-side micro-grooves, where each groove width makes certain amount of overlap with previous one (Kruth et al. 2007). The amount of groove width overlap is controlled by two important process parameters of laser micro-turning process, namely axial feed motion (*Y feed rate*) and rotational speed of workpiece. To achieve a certain micro-turned depth and desired surface texture on the laser scanned surface, continuous scanning process of laser beam is carried out by focusing the beam on the preceding micro-turning surface. The amount of overlap between groove widths is measured in terms of *circumferential overlapping percentage* (CO_p), while the overlap between two consecutive spots is measured in terms of *spot overlapping percentage* (SO_p). The values of these two parameters are chosen so carefully that the percentage values lay more than 50 % to achieve quality surface during laser micro-turning operation. The laser spot diameter, *spot overlap*, *pulse frequency* and *linear scanning speed* are related as follow (Thawari et al. 2005)

$$(Spot\ diameter) \times (1 - spot\ overlap) \times (pulse\ frequency) = linear\ speed \quad (1)$$

Or,

$$Spot\ overlapping\ percentage\ (SO_p) = \left(1 - \frac{v}{D \times F_p}\right) \times 100\,\% \quad (2)$$

Here, v, D and F_p are *linear scanning speed* (mm/s), laser spot diameter (mm) and *pulse frequency* (Hz), respectively. By converting the *linear scanning speed*, v (mm/s) with *workpiece rotating speed*, N (rpm) and taking the diameter of workpiece as 10 mm, Eq. 2 can be written as

$$Spot\ overlapping\ percentage\ (SO_p) = (1 - \frac{0.5238 \times N}{D \times F_p}) \quad (3)$$

During laser beam scanning over the work surface, as shown in schematic view in Fig. 1, two successive spots 1 and 2 produce an overlap length of x. Due to Gaussian energy distribution of laser beam, the respective spots generate a small crater through melting and/or evaporation from irradiated spots. The micro-turned profile produced can be controlled by increasing the spot overlap area, i.e., increasing the *pulse frequency* or/and decreasing *workpiece rotating speed*, as per

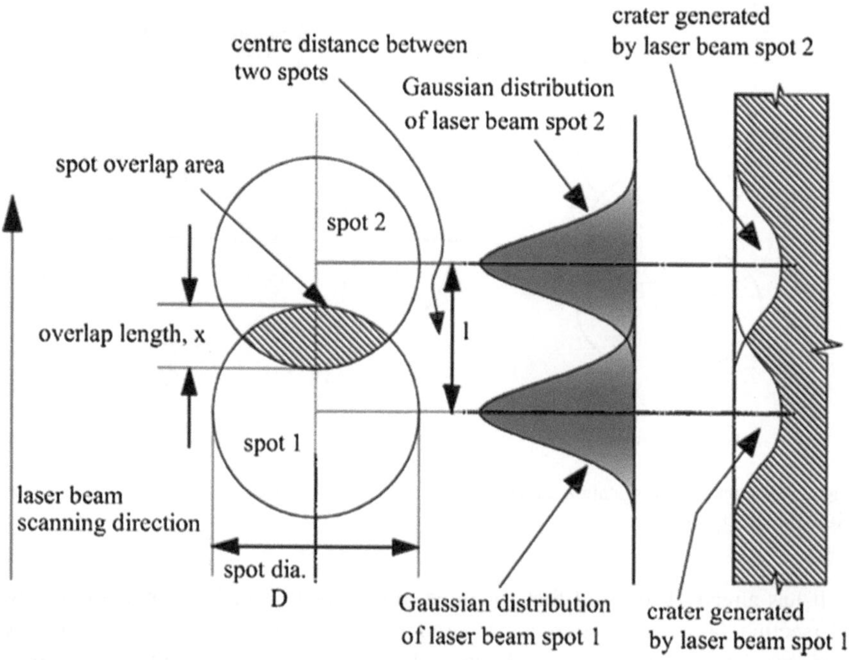

Fig. 1 Schematic view of overlap between consecutive spots showing Gaussian energy distribution (Kibria et al. 2012)

Eq. 3. *Circumferential overlap percentage (CO_p)* is another important criterion during laser micro-turning process. The relationship of this factor with laser process parameters can be formulated as follows.

The schematic view of *circumferential overlap* during laser beam scanning for two successive rotations is shown in Fig. 2. Let, N and D_s are the *rotating speed* in rpm and diameter of sample in mm, respectively. Now, the linear *scanning speed* of laser beam is, $v = \frac{\pi \times D_s \times N}{60}$ in mm/s. The circumferential length of sample is, $L = \pi \times D_s$. Now, the time taken by the laser beam to scan one complete rotation is, $t_s = \frac{L}{v} = \frac{60}{N}$ in seconds. If, the sample is moved through an *axial feed rate*, f (mm/s), then the movement of sample in t_s seconds is, $l_c = f \times t_s = \frac{60 \times f}{N}$ in mm. Therefore, from Fig. 1, the *circumferential overlap percentage* can be calculated from,

$$CO_p = \frac{overlap\ length}{spot\ size} = \frac{x_c}{D} = (1 - \frac{l_c}{D}) = (1 - \frac{60 \times f}{D \times N}) \times 100\,\% \quad (4)$$

In the above equation, x_c and D are laser scan overlap length (in mm) and laser spot diameter (in mm), respectively. In Fig. 2, it is observed that the two successive beam scans i.e. (n − 1)th and nth produce an overlap length of x_c during rotation of workpiece and the workpiece moves l_c in its axial direction.

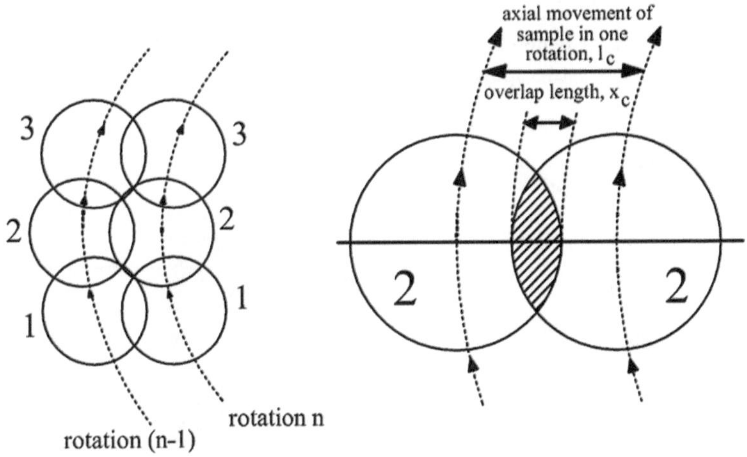

Fig. 2 Schematic representation of circumferential overlap between two consecutive rotations (Kibria et al. 2012)

It has already been said that for three dimensional laser beam machining, two intersecting laser beam sources are used (Chrysolouris et al. 1988). Each laser produces the corresponding blind cutting kerfs which converges and results in the solid stock removal from the machining zone. However, it is very difficult to remove thin layer from cylindrical shaped work surface. Experimental as well as theoretical research investigation into single and two dimensional laser beam machining of advanced ceramics like aluminum oxide, silicon nitride, silicon carbide, magnesia etc. have been conducted successfully by researchers across the globe (Samant and Dahotre 2008a, b; Dhupal et al. 2008). However, possibilities of micro-turning of these ceramics are in developing stage. Hence, extensive research is needed for developing laser materials processing technology for machining these advanced engineering materials with required geometrical shape and higher dimensional accuracy. To study and investigate the effect of various process parameters during laser micro-turning process, which removes very thin layer from work surface using single-laser beam and also to achieve desired surface profile on machined components, in the present investigation, laser micro-turning of cylindrical shaped alumina ceramic material by pulsed Nd:YAG laser have been performed based on various experimental schemes.

Figures 3a and b depict the schematic views for lower and higher values of *circumferential overlap* at high and low value of workpiece feed rate, respectively. From Fig. 3a, it is obvious that at low *circumferential overlap percentage*, the laser micro-turning surface is rendered with lots of micro-peaks, which cause higher surface roughness values. However, this generation of micro-peaks can be avoided at higher *circumferential overlap* values (as shown in Fig. 3b).

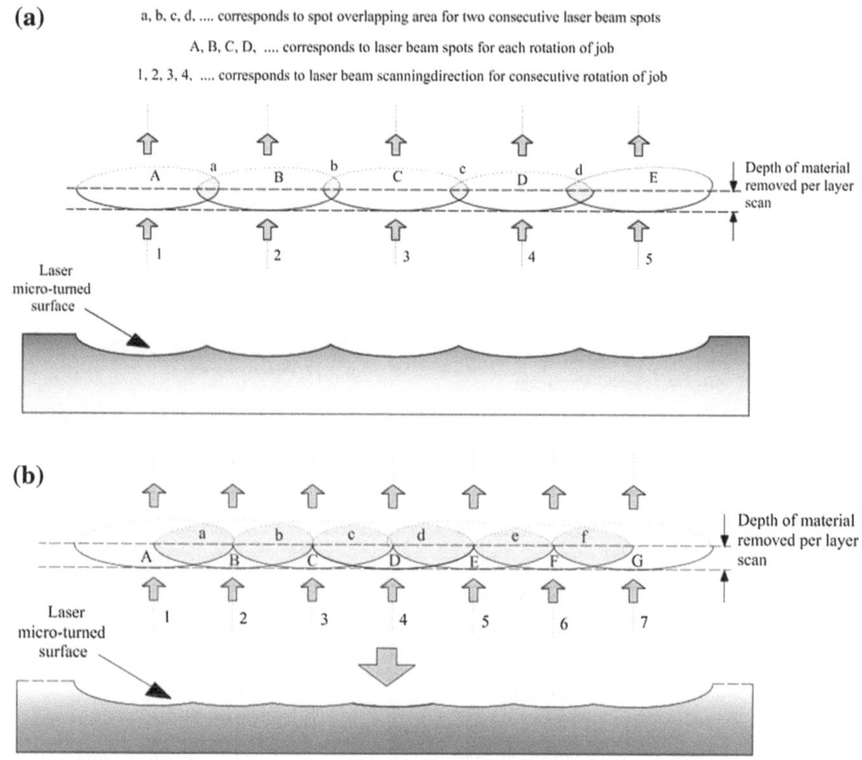

Fig. 3 Schematic view of laser micro-turned surface for (**a**) low and (**b**) high percentage of circumferential overlap for consecutive rotations (Kibria et al. 2013a)

3 Development of Laser Micro-turning System

To carryout laser micro-turning operation on cylindrical workpiece using the pulsed Nd:YAG laser set-up, a separate work holding and rotating unit has been developed indigenously. The complete work holding and rotating system consists of three components, (i) servomotor and developed workpiece holding fixture, (ii) servomotor amplifier, and (iii) power supply and interfacing unit. Figure 4 shows the complete schematic diagram of the developed workpiece rotating system. In Table 1, the specifications of the servomotor used for developing the workpiece rotating system are enlisted. A collet, which can grip cylindrical workpiece of diameter ranging from 2 to 16 mm, is mounted to the shaft of the servomotor very rigidly and accurately so that when the servomotor rotates in any desirable speed, it produces minimum amount of eccentricity due to rotation. The eccentricity of rotation was measured by using a dial gauge and the error was about 3 μm. The workpiece gripped with the collet can be rotated by the signals coming from servo amplifier at any rotational speed within the range of rotation of the servomotor.

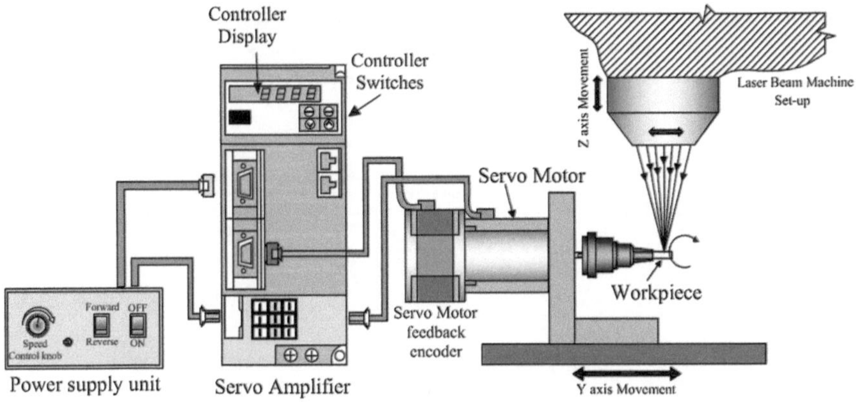

Fig. 4 Schematic representation of workpiece holding and rotating system (Kibria et al. 2012)

Table 1 Detailed specifications of servomotor used

Specifications	Description
Servomotor type	GYS101DC2-T2A
Input voltage	AC 200 V
Rated output	100 W
Rated speed	3000 rpm
Encoder	17-bit INC
Winding insulation class	IP67
Oil seal/shaft	Without an oil seal

The base of the fixture was made in such a dimension so that the whole fixture can be fixed by the slot provided on the X-Y worktable of the pulsed Nd:YAG laser system.

The speed of the servomotor was controlled using a servo amplifier. In Table 2, the detailed specifications of the servo amplifier are enlisted. The connections between the servomotor and servo amplifier are already shown in Fig. 4. The servo amplifier has four keys, (i) Mode change key (ii) Shift/Enter key (iii) Sub-mode selection key (UP) and (iv) Sub-mode selection key (DOWN). The amplifier has four-digit display. By systematic method and using the four keys, one can provide the input speed to the servo amplifier to get a particular rotational speed of the servomotor. Apart from the servomotor and servo amplifier, a power supply and

Table 2 Detailed specifications of servo amplifier

Specifications	Description
Amplifier type	RYC101D3-VVT2
Input voltage	Single-phase 200–230 V
Rated output	100 W
Circuit breaker	EA33AC/5

interfacing unit has been developed indigenously. The unit provides power to both the servomotor and servo amplifier. Moreover, there are some additional switches and knobs by which one can change the *rotating speed* of servomotor manually. In addition, the user can alter the rotational direction of servomotor by various switches. The power supply and interfacing unit gets power from the two-phase AC power supply line. The servomotor and its fixture is placed on the CNC X-Y worktable and by using CNC servo controller and personal computer system, feed is given along the axis to the rotating workpiece during laser micro-turning operation. After one pass of laser scan along the desired length of turning, the focusing lens was moved down by Z axis feed rate with the help of CNC controlled Z axis motion so as to focus the laser beam to the laser micro-turned surface generated by previous laser beam scanning. Thus, the desired depth and length of micro-turning surface is generated on cylindrical workpiece of a particular diameter.

4 Experimental Methodology of Laser Micro-turning

The entire experimentation was performed in a CNC controlled pulsed Nd:YAG laser system (manufactured by Sahajanand Laser Technology Ltd, India) as shown in Fig. 5. The beam of the system operates in Gaussian mode (TEM_{00}) and the wavelength of the beam is 1064 nm. The pulse duration of the laser beam is in the range of 120–150 ns. In the present pulsed Nd:YAG laser system, the polarization of the laser is random. The beam was focused to spot size of 100 μm (measurement of single spot on alumina workpiece surface) using a 50 mm focal length lens. The indigenously developed work holding and rotating fixture, which was manufactured

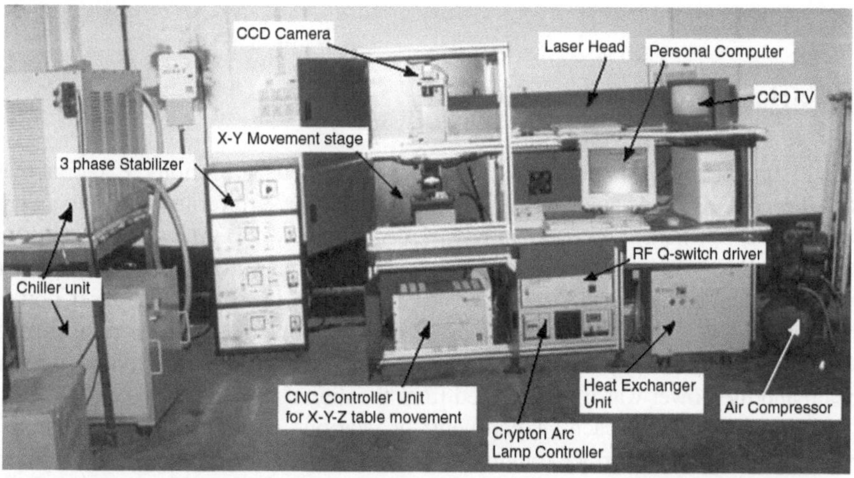

Fig. 5 Photographic view of the pulsed Nd:YAG laser machining system (Kibria et al. 2012)

Table 3 Major thermo physical properties of aluminium oxide ceramics

Properties	Values and units
Melting temperature	2050 °C
Creep rate	10^{-9} s^{-1} at 150 MPa
Density	3.984 g/cm^3
Specific heat	755 J/kg K
Activation energy	325(25) kJ/mol
Thermal conductivity	33 W/mK
Tensile strength	267 MPa
Thermal diffusivity	0.111 cm^2/s
Bulk modulus	257 GPa
Shear modulus	167 GPa
Grain size	5 μm

to facilitate the laser micro-turning operation, is used to hold and rotate the workpiece. A collet attached to the axis of the motor was used to grip the cylindrical work sample and the collet with the job was rotated through different speeds. Axial feed to the sample up to the desirable length of turn was provided by the Y directional feed rate of the CNC X-Y table operated by a personal computer attached to the Nd:YAG laser system. To encompass layer-by-layer removal of material from the cylindrical surface, after each scan of entire micro-turning surface, the focusing lens was moved down to provide the focal point to the previously machined micro-turning surface. Using the Multisawing software installed in the computer attached to pulsed Nd:YAG laser system, the length of micro-turning was selected as 5 mm. Moreover, the *Y feed rate* and *Z feed rate* was controlled by this software. Cylindrical shaped aluminium oxide ceramic (99 % pure) of size 10 mm in diameter and 40 mm in length was used as workpiece material. The major thermo and physical properties of aluminum oxide are enlisted in Table 3.

5 Various Measurement Schemes

Laser micro-turning experimentation was conducted at various experimental schemes. The process parameters involved during laser micro-turning operation were *laser average power, pulse frequency, workpiece rotational speed, assist air pressure* and *Y feed rate*. The *average power* of laser beam is measured using a laser power measuring device (make: Ophir Optronics Solutions Ltd, Israel). The laser power measuring device consists of power meter and the power sensor. To measure the *average power* of the laser beam, the sensor was kept at defocusing plane of laser beam and the power was directly noted from the digital display of the power meter.

After conducting each experiment at various parametric combinations, machining depth and surface roughness (Ra and Rt) was measured. The machining depth of the laser micro-turned surface was measured with the help of optical

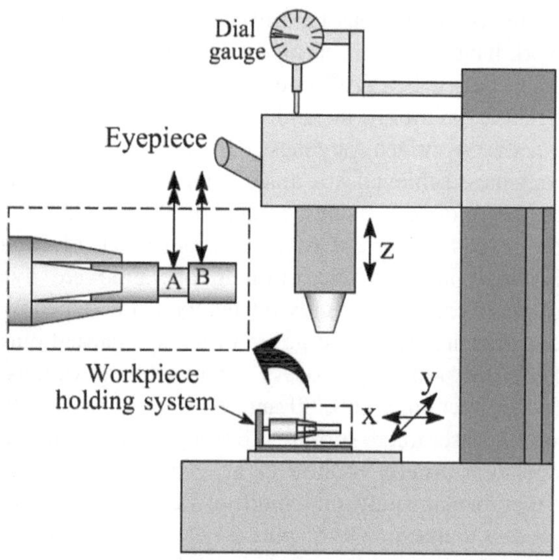

Fig. 6 Scheme of micro-turning depth measurement using optical microscope

measuring microscope (Olympus STM6, minimum measurable dimension = 0.5 μm) using a 10 × magnification lens. The scheme of measurement of depth of cut using the optical measuring microscope is shown in Fig. 6. As shown in figure, at first, the lens of the microscope is focused onto the non-turned workpiece surface such as point B and then focused onto the micro-turned surface such as at point A. The values of these heights are measured by using a high precision dial gauge kept on the Z-axis stage of optical microscope. The result of these height difference measures the machining depth for the experiment. In experimental scheme, where depth deviation was selected as response criterion, the depth deviation was measured by simply differentiating the machining depth and target depth (taken as 100 μm). The surface roughness (Ra and Rt) was measured along the axis of the cylindrical work sample by using SURFCOM 120A-TSK roughness measuring instrument. During roughness measurement, the cutoff length (λ_c) has been carefully chosen so that variations due to form errors are excluded and surface finish related to the closely spaced irregularities can be measured correctly. During roughness measurement, the cut-off length and total length of measurement were set as 0.25 and 2.5 mm, respectively.

6 Results and Discussion

Very few researches have been carried out in laser micro-turning process by the authors (Kibria et al. 2010b, 2013a, 2014a). Kibria et al. (2010b) have introduced the laser micro-turning operation as an advanced machining technology to machine or micro-turn various hard-to-machine engineering ceramics. A very few research work have been carried out on micro-turning process of alumina (Al_2O_3) ceramic using pulsed Nd:YAG laser system. Authors have done basic experimental investigation describing the influence of *lamp current, scanning speed* and *pulse frequency* on surface roughness (Ra) and depth of cut. In the experimentation, surface roughness achieved was much higher i.e. in the range of 6.45–11.48 μm. Moreover, the depth of cut was in the range of 38 and 223 μm. This amount of fluctuations in the results of surface roughness and depth of cut reveals that process variables involved in laser micro-turning operation are dynamic in nature and process responses are significantly related with simultaneous variation of parameters during experiments. The same group of authors carried out experimental investigation on laser micro-turning process for machining various grades of Aluminum Oxide (Al_2O_3) ceramics (60, 80 and 99 % pure Al_2O_3 ceramics) to study the effects of various laser micro-turning process parameters on surface roughness (Ra) and depth deviation criteria (Kibria et al. 2014a). The same group of authors performed experimental study and multi-objective optimization during laser micro-turning process to machine 99 % pure Al_2O_3 ceramics using combined approach of Taguchi Methodology and Grey Relational Analysis (Kibria et al. 2013a). Experimentation and analysis have also been performed varying the significant process parameters revealed from Taguchi based experiments. All these experimentations deal with basic study of laser micro-turning process to search out the significant process parameters and their suitable working ranges. However for successful study and analysis of laser micro-turning process, *spot overlapping percentage* (SO_p) and *circumferential overlap percentage* (CO_p) play major roles. Moreover, for successful implementation of this novel process in various micro-engineering industries, surface roughness achieved as in (Kibria et al. 2010b, 2014a, 2013a) should be reduced by considering some novel machining strategies during laser micro-turning. The experimental results and analysis considering some novel machining strategies are shown and described hereafter.

6.1 Study the Influences of Overlap Factors on Surface Roughness Criterion

To study the effects of various percentages of *spot* as well as *circumferential overlap*, experiments have been conducted at various parametric combinations of process parameters, which are related to the overlap factors (Kibria et al. 2014b). A well-designed set of experiments has been performed to vary one factor at a time (OFAT) and surface roughness is measured. The levels and ranges of various parameters like *pulse frequency, workpiece rotating speed* and *Y feed rate* for achieving the *spot overlapping percentage* (SO_p) and *circumferential overlap percentage* (CO_p) is shown in Table 4. For each of the experiments, *the number of laser scan passes* was 100. In each of the workpiece, three experiments were

Table 4 Experimental conditions and ranges of process parameters (Kibria et al. 2014b)

Condition	Description
Average power, W	7, 8
Pulse frequency, Hz	3000, 4000, 5000, 6000, 7000
Workpiece rotating speed, rpm	200, 300, 400, 500, 600
Y feed rate, mm/s	0.1, 0.2, 0.3, 0.4, 0.5
Pulse width, % of duty cycle	variable
Air pressure, kgf/cm^2	1.3
Z feed rate, mm/s	0.01

conducted. During the experiments, surface roughness (Ra) was measured as the process criterion. A length of 5 mm was micro-turned on the cylindrical workpiece. In the present experimental study, the levels of two process parameters i.e. *pulse frequency* and *workpiece rotating speed* were chosen carefully so that the *spot overlapping percentages* (SO_p) of the successive laser spots have definite and positive values. In Fig. 7 shows the variation of *spot overlapping percentage* while varying *workpiece rotating speed* and *pulse frequency* values. Moreover, the *axial feed rate* or *Y feed rate* of work sample was chosen carefully so that its values compiled with *workpiece rotating speed* give positive values of *circumferential overlap percentages*. The variation of *circumferential overlapping percentage* at various parametric combinations of *Y feed rate* and *workpiece rotating speed* is shown in Fig. 8. These two plots have been drawn based on the considered range of all the related process parameters of *spot overlap* as well as *Circumferential overlap*. It is shown in both plots of *spot overlap* as well as *circumferential overlap percentage* that some points have been considered for negative overlap percentage to show the effect of the conditions of no overlap on surface roughness criterion.

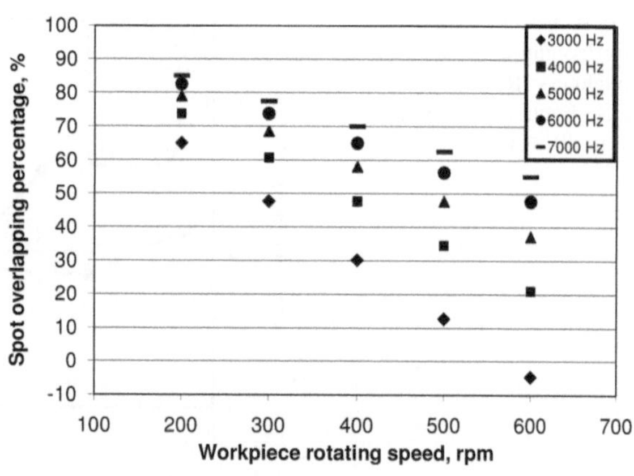

Fig. 7 Variation of spot overlapping percentage at various workpiece rotating speed and pulse frequency

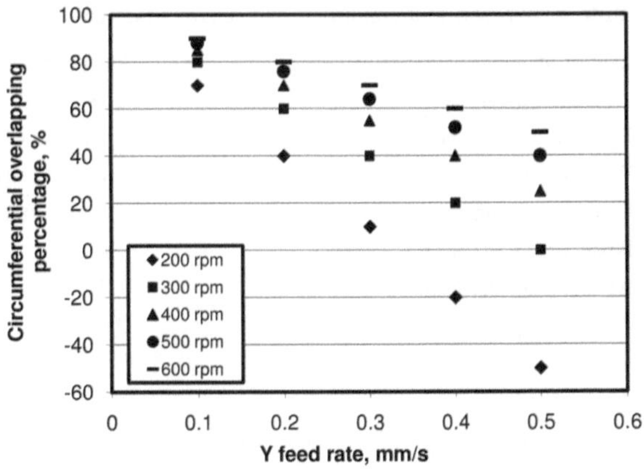

Fig. 8 Variation of circumferential overlapping percentage at various Y feed rate and workpiece rotating speed

6.1.1 Influence of Spot Overlap on Surface Roughness (Ra)

Each experiment was repeated three times and the average of three surface roughness values was taken to plot various graphs. Figures 9 and 10 show the variation of surface roughness (Ra) for varying *spot overlapping percentage* at different *pulse frequency* and *workpiece rotating speed*, respectively. It is already described that the *spot overlap percentage* can be varied by two methods, (i) by varying *workpiece rotating speed* keeping the *pulse frequency* at a constant value, and (ii) by varying *pulse frequency* keeping the *workpiece rotating speed* at constant value. For the results shown in Fig. 9, the *average power* and *Y feed rate* were kept constant as 7 W and 0.3 mm/s, respectively. From this figure, it is observed that roughness of the micro-turned surface reduces with the increase of *spot overlap* for each of the *pulse frequency* settings. From Eq. 3, it is obvious that the reduction in *workpiece rotating speed* results in high overlap value between successive laser spots. This high amount of overlap results in even machined surface and reduces the

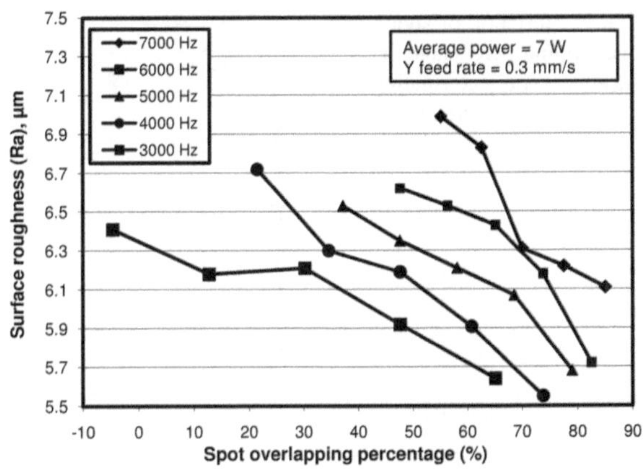

Fig. 9 Effect of spot overlapping on surface roughness (Ra) at different pulse frequency (Kibria et al. 2014b)

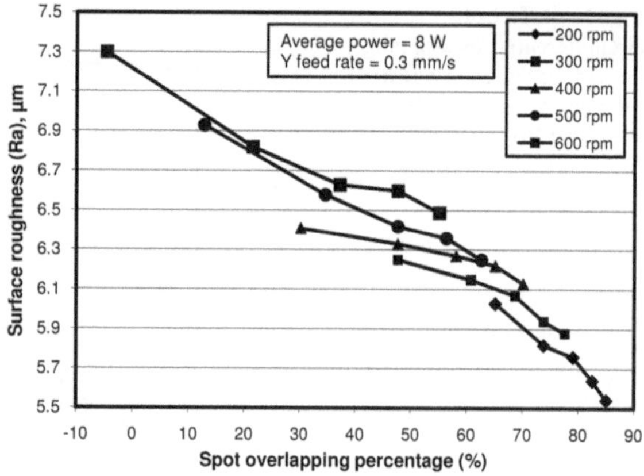

Fig. 10 Effect of spot overlapping on surface roughness (Ra) at different rotating speed (Kibria et al. 2014b)

surface roughness. It is also seen from the same plot that the surface roughness value increases with *pulse frequency* at any particular *spot overlap percentage* value. With the increase of *pulse frequency*, according to Eq. 3, the *spot overlap percentage* increases. However, at high *pulse frequency* condition, the time duration between two successive pulses is very less. Therefore, the material at the focused zone on the sample gets adequate thermal energy for melting and vaporization instantly. Thus, ultimately, the total amount of irradiated laser beam energy is high enough and it results in more depth of crater formed on the surface, which results higher value of surface roughness.

For the experimental results shown in Fig. 10, the *average power* and *Y feed rate* were kept constant at 8 W and 0.3 mm/s, respectively. The figure clearly indicates that with the increase of *spot overlap percentage* for any constant value of *workpiece rotating speed*, surface roughness (Ra) decreases. Due to high *percentage of spot overlap*, the material removal from the workpiece surface occurs uniformly and therefore, high quality machining surface is achieved. Furthermore, for same *spot overlap percentage* value, the micro-turned roughness is more for higher value of *workpiece rotating speed*. At higher setting of *workpiece rotating speed*, consecutive *spot overlap* is less, resulting uneven micro-turning surface.

6.1.2 Influence of Circumferential Overlap on Surface Roughness (Ra)

In Figs. 11 and 12, the effect of varying *circumferential overlap* values on surface roughness (Ra) is shown at different *Y feed rate* and *workpiece rotating speed*, respectively. In Fig. 11, the surface roughness has been plotted keeping *average power* at 8 W and laser *pulse frequency* at 5000 Hz. From Eq. 4, it is obvious that the *circumferential overlap percentage* mainly depends onto two process parameters i.e. *Y feed rate* and *workpiece rotational speed*. It is clear from this plot that the increase in *circumferential overlap percentages* i.e. increase in *workpiece rotating speed* with constant *Y feed rate* setting results in slight decrease in surface

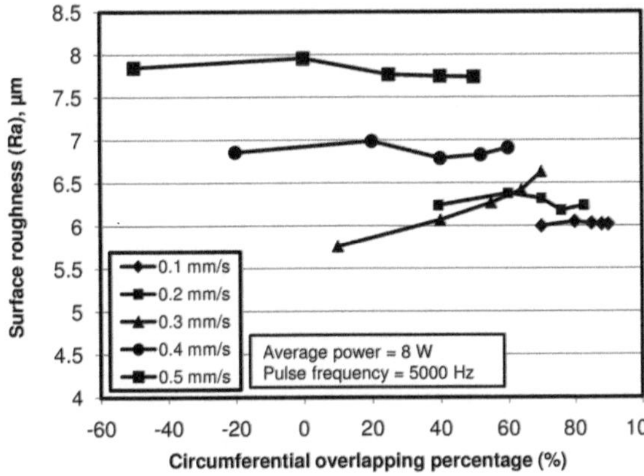

Fig. 11 Effect of circumferential overlapping on surface roughness (Ra) at different Y feed rate (Kibria et al. 2014b)

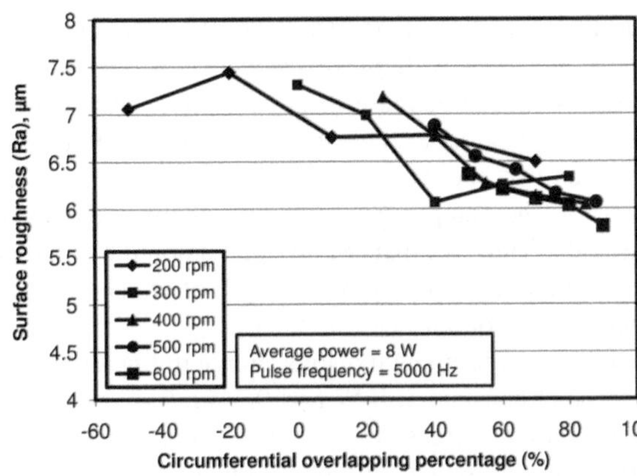

Fig. 12 Effect of circumferential overlapping on surface roughness (Ra) at different rotating speed (Kibria et al. 2014b)

roughness values. From Eq. 4, it is clear that *circumferential overlap percentage* increases with increasing *workpiece rotating speed* and this phenomenon has directly resulted for reduction of roughness of laser micro-turned surface. High *circumferential overlap percentage* results in uniform removal of material from laser irradiated surface and high finish surface is achieved. Moreover, it is also clearly seen from Fig. 11 that for a constant value of *circumferential overlap*, the increase in *Y feed rate* results in rougher micro-turned surface. Increase in *Y feed rate* of sample is resulted more null and void spacing between two successive scan widths and this phenomenon consequently results uneven and rough micro-turning surface.

In Fig. 12, the influence of *circumferential overlap* on surface roughness is shown at *average power* of 8 W and *pulse frequency* of 5000 Hz, respectively. It is observed from this plot that with the increase in *circumferential overlap*, roughness of the machined surface decreases for each *workpiece rotating speed* setting. Due to high *circumferential overlap*, the laser scan widths for successive rotation of workpiece come closer to each other and in turn, this phenomenon results in even and regular micro-turning surface It shows that at higher *workpiece rotating speed* settings, the amount of *circumferential overlap* is high and further it results lower value of surface roughness of the surface profile produced during laser micro-turning operation.

6.1.3 Microscopic Analysis of Laser Micro-turned Surface

Studies of scanning electron microscopy (SEM) micrographs of the laser micro-turned surface have also been conducted to get a clear picture of the effect for varying the laser micro-turning process parameters. Figure 13a–d show and compare the micro-turned surface characteristics for different *spot overlap* and *circumferential overlap* values, accounted for varying the related process parameters i.e. *workpiece rotating speed*, laser *pulse frequency* and *Y feed rate* of workpiece. Comparing the experimental parametric settings and SEM micrographs in Fig. 13a and b, it is seen that with increased *spot overlap percentage* from 21.43 to 73.81 %, due to increase in *pulse frequency* from 4000 to 6000 Hz and decrease in *workpiece rotating speed* from 600 to 300 rpm, the Ra value of micro-turned surface is reduced from 6.72 to 6.18 μm. From Fig. 13c and d, it is observed that when the *circumferential overlap percentage* is increased from 25 to 90 % by reducing the *Y*

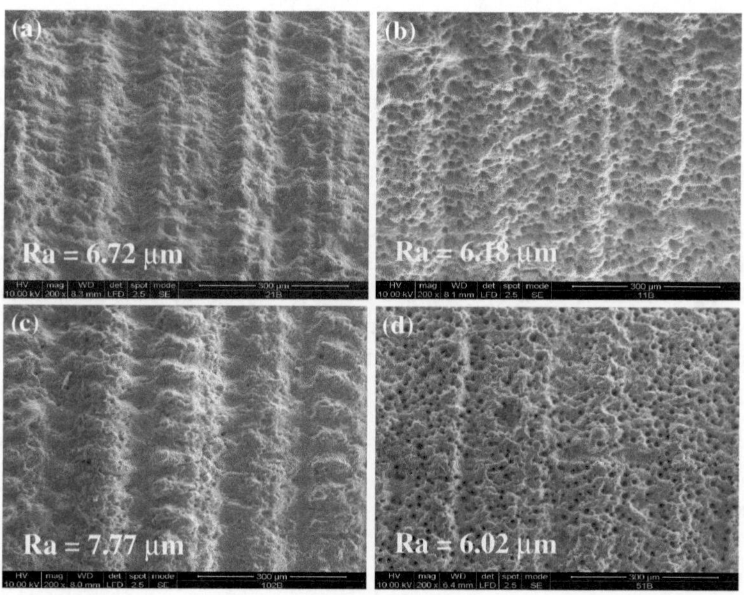

Fig. 13 SEM micrographs of the laser micro-turned surface of work samples machined at parametric combinations of average power, pulse frequency, workpiece rotating speed and Y feed rate of (**a**) 7 W/4000 Hz/600 rpm/0.3 mm/s, (**b**) 7 W/6000 Hz/300 rpm/0.3 mm/s, (**c**) 8 W/5000 Hz/400 rpm/0.5 mm/s and (**d**) 8 W/5000 Hz/600 rpm/0.1 mm/s (Kibria et al. 2014b)

feed rate from 0.5 to 0.1 mm/s and increasing the *workpiece rotating speed* from 400 to 600 rpm simultaneously, quite uniform micro-turning surface was realized. Ra value is reduced from 7.77 to 6.02 μm.

6.2 Study of Defocusing Conditions at Multi-objective Optimization to Achieve Better Quality Surface

Response surface methodology (RSM) based approach has been considered to carry out micro-turning operation of cylindrical shaped alumina ceramic (Kibria et al. 2012, 2013b). A lot of pilot experiments have been conducted before selecting the actual process parametric ranges. Five levels of each process parameters i.e. *average power, pulse frequency, workpiece rotational speed, air pressure, Y feed rate* were considered as shown in Table 5. The experimentation has been carried out based on central composite rotatable second-order design (CCRD). After each experiment, surface roughness (Ra) and micro-turning depth deviation have been measured. The measurement scheme of surface roughness and machining depth are already discussed in previous sections. The target machining depth was taken as 100 μm. The deviation in depth was calculated according to the following equation (Kibria et al. 2013b).

$$Y_{depth\ dev.} = Y_{td} - Y_{md} \quad (5)$$

where, $Y_{depth\ dev.}$ is the amount of depth deviation, Y_{td} is the targeted depth and Y_{md} is the machined depth accounted during experiments. MINITAB™ statistical software was used to develop the design of matrix of experimentation and further analysis of obtained values of responses from measurement schemes. Multi-objective optimal parametric combination of process variables has been obtained for achieving least values of responses i.e. surface roughness and depth deviation. To achieve further high quality machined surface, machining was done at various defocusing conditions of laser beam and also by varying the *number of laser scan passes*. Other process parameters were kept constant at optimal parametric combination obtained in multi-objective optimization.

Table 5 Factors and various levels of the process parameters for RSM experimentation (Kibria et al. 2013b)

Parameters	Unit	Levels				
		−2	−1	0	1	2
Average power	W	6	7	8	9	10
Pulse frequency	Hz	3000	4000	5000	6000	7000
Rotational speed	rpm	200	300	400	500	600
Air pressure	kgf/cm^2	0.3	0.8	1.3	1.8	2.3
Y feed rate	mm/s	0.1	0.2	0.3	0.4	0.5

6.2.1 Multi-objective Optimization of Surface Roughness and Depth Deviation

To achieve the targeted shape as well as machined surface during laser micro-turning operation considering the performance criteria, multi objective optimization has been performed (Kibria et al. 2013b). Figure 14 shows the multi response optimization result during laser micro-turning operation of alumina ceramic with pulsed Nd:YAG laser system. In the mentioned figure, columns of the plot represents each parameter with their ranges considered and each row of the plot represents the performance criteria measured in this experimentation. Moreover, each cell shows how the process criteria changes as function of one of the process parameters while keeping the other parameters fixed. From this figure, it can be seen that the parametric combination for multi-objective optimization i.e. minimization of all the process responses was achieved as *average power* of 7.81 W, *pulse frequency* of 5601.59 Hz, *workpiece rotational speed* of 435.60 rpm, *air pressure* of 0.30 kgf/cm^2 and *Y feed rate* of 0.4434 mm/s. During multi-objective optimization, all the responses were optimized with composite desirability (D) value of 1. Due to the constraints of the existing machine settings, the parameters values were taken nearer to the value obtained during optimization. Experimentation has been conducted nearer the feasible setting of process parameters achieved during multi-objective optimization to evaluate the actual process criteria values. The percentage error of experimental results for the optimized responses were compared with RSM based estimated results as obtained from developed empirical models and the results are tabulated in Table 6. The percentage errors calculated are within acceptable range (<5 %). This shows a good agreement of the optimized responses at the optimal parameters setting during laser micro-turning of alumina ceramic.

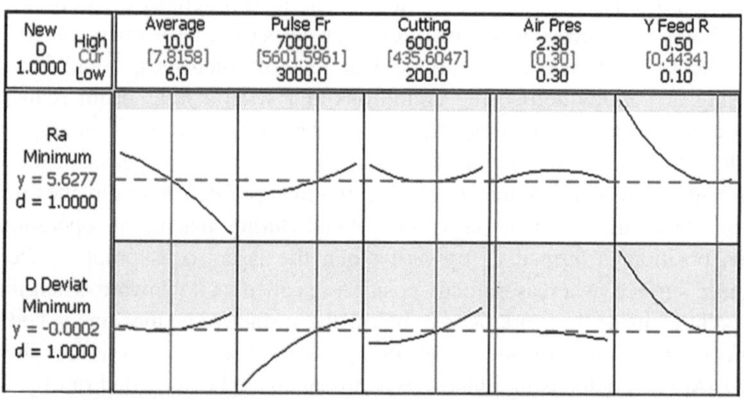

Fig. 14 Multi objective optimization results for surface roughness (Ra) and depth deviation (Kibria et al. 2014b)

Table 6 Comparison of experimental and RSM estimated results at optimal parametric combination (Kibria et al. 2013b)

Optimization	Experimental parameters settings			Responses	Experimental results	RSM estimated results	Percentage of error (%)
	Parameters	Unit	Value				
Multi-objective optimization	Average power	W	7.81	Surface roughness, Ra	5.91 μm	5.63 μm	4.78
	Pulse frequency	Hz	5600	depth deviation	−0.00021 mm	−0.00020 mm	4.76
	Rotating speed	rpm	436				
	Air pressure	kgf/cm^2	0.30				
	Y feed rate	mm/s	0.443				

6.2.2 Study of Laser Defocusing Conditions During Laser Micro-turning

Although, surface roughness has been considered as minimum is the best type response, the value of optimal surface roughness is quite high (i.e. 5.63 μm in RSM and 5.91 μm in validation). This amount of roughness may be due to generation of micro-peaks on the machined surface. However, in actual practice, to produce an acceptable machined surface and to manufacture durable and reliable products, which can provide high-end performances, the roughness of machined surface should be lowered down further. Therefore, it is felt that further experiments should be carried out to minimize the surface roughness values without impeding the depth deviation value. Therefore, to further improve the quality of machined surface and to reduce the laser micro-turned surface roughness, experiments have been conducted at RSM based multi-objective optimization setting as shown in Table 6 (Kibria et al. 2015). Two process parameters i.e. *defocusing position* and *number of laser scan pass* were varied during defocused laser micro-turn machining. Figure 15 shows the process mechanism of laser defocused laser micro-turning operation. The optimal machined surface has lots of micro-peaks and by further machining at various defocusing conditions and several laser beam scan passes, these micro-peaks can be removed by lowering the laser beam energy density through melting and vaporization of material from these micro-peaks. A lot of pilot experiments have been conducted to select the range of defocusing planes in which the workpiece surface to be kept and rotated during machining operation. The defocus position is termed as upward when the laser focus point is above the workpiece surface, whereas defocus position is called as downward when the laser focus point is inside the workpiece or below the surface to be machined. The values of defocus positions considered in the present experimentation are shown in Table 7. At each defocusing planes, experiments have been carried out by varying the number of laser scans. Here, surface roughness, Rt has been considered as additional response to study the laser micro-turning surface features in more details.

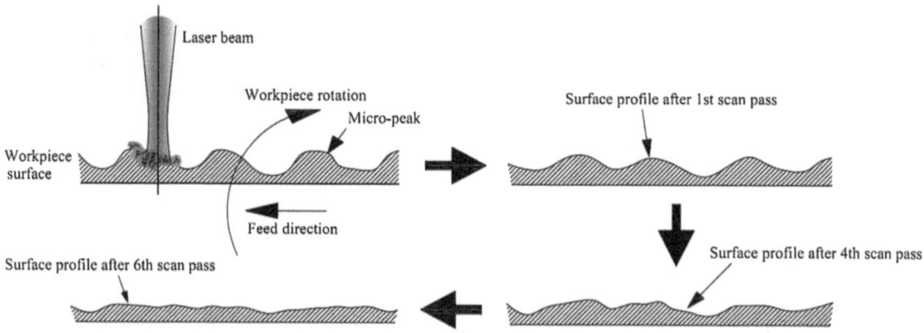

Fig. 15 Process of removal of micro-peaks during defocused laser micro-turning process (Kibria et al. 2013c)

Table 7 Factors and various levels for further experimentation (Kibria et al. 2013c)

Parameters	Unit	Levels	
Defocus position	mm	Upward	Downward
		−0.2, −0.4, −0.6, −0.8	+0.2, +0.4, +0.6, +0.8
Number of passes	–	1, 2, 3, 4, 5, 6, 7, 8, 9, 10	

This response of surface roughness is much responsive to the high peaks or deep scratches on the surface of assessment. It is defined as the vertical measurement of distance between the highest peak and the lowest valley within the assessment length of surface profile. It is the summation of maximum height of peak from mean line (Rp) and maximum depth of valleys from mean line (Rv) of the surface profile (Gadelmawla et al. 2002). Although, the present set of experiments have been conducted to further reduce the surface roughness (Ra and Rt) criteria through various experiments of varying the focal positions and number of laser beam passes, as it is basically a laser micro-turning operation, therefore, the dimensional accuracy features of the machined parts or components would be worsen, mainly the depth deviation of micro-turning surface. The depth deviation of the machined parts has been measured after conducting each two consecutive laser beam passes. After each experiment, the surface roughness (Ra and Rt) and depth deviation of the machined surface were measured by same technique as discussed in measurement schemes. The depth deviation is measured using the following equation (Kibria et al. 2015).

$$Depth\,deviation = Machined\,depth - Machined\,depth\,achieved\,at\,optimal\,setting \quad (6)$$

Based on the obtained results of responses i.e. surface roughness (Ra and Rt), various plots have been drawn and shown in Figs. 16, 17, 18 and 19. In Fig. 16, the effect of *number of laser scan passes* on surface roughness (Ra) is shown at various *upward defocusing positions*. The other process parameters were kept constant at

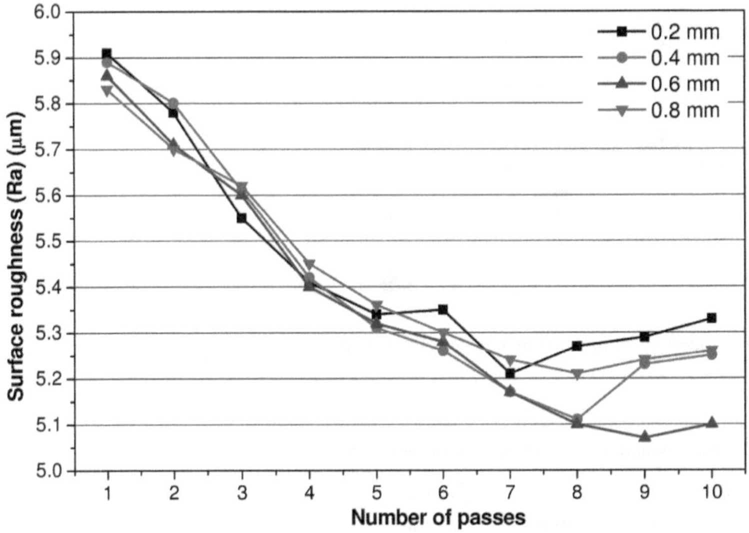

Fig. 16 Variation of surface roughness (Ra) with number of laser scan passes at various upward defocus positions (Kibria et al. 2013c)

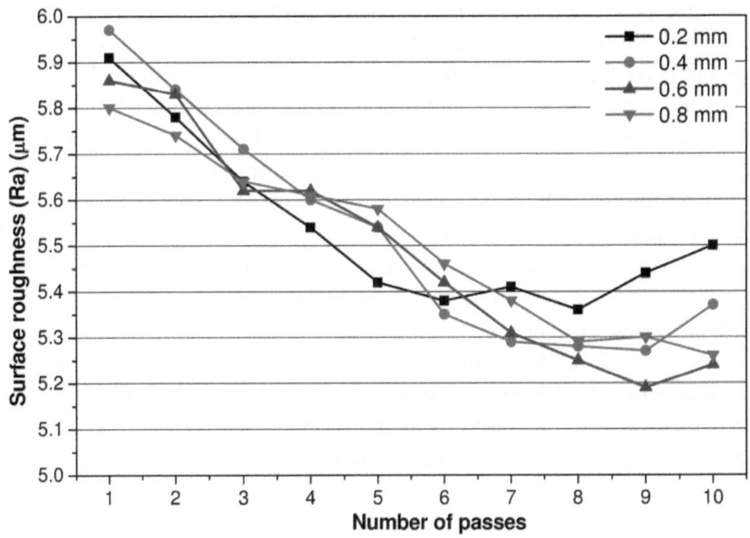

Fig. 17 Variation of surface roughness (Ra) with number of laser scan passes at various downward defocus positions (Kibria et al. 2013c)

7.81 W/5600 Hz/436 rpm/0.30 kgf-cm^{-2}/0.443 mm/s of *average power*, *pulse frequency*, *workpiece rotating speed*, *assist air pressure* and *Y feed rate*, respectively. In Fig. 16, it is observed that surface roughness of laser micro-turning

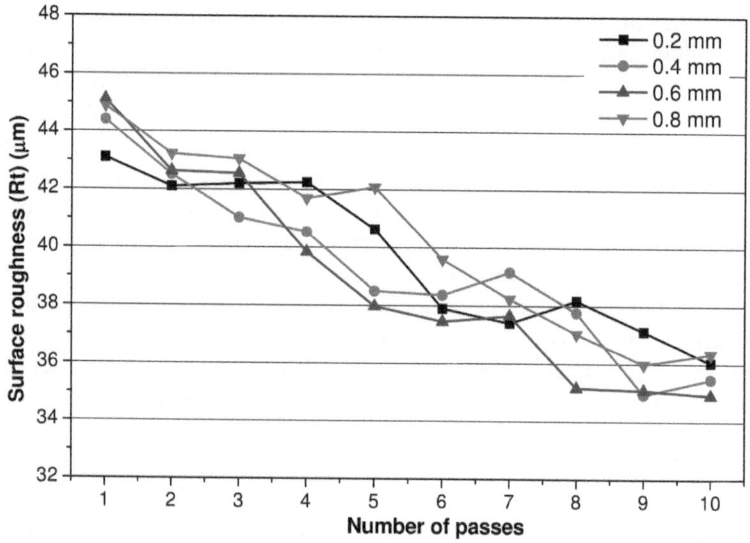

Fig. 18 Variation of surface roughness (Rt) with number of laser scan passes at various upward defocus positions (Kibria et al. 2013c)

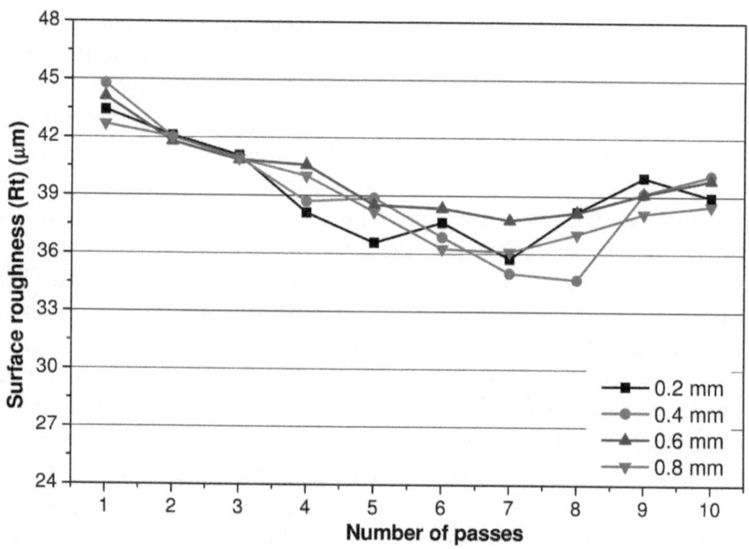

Fig. 19 Variation of surface roughness (Rt) with number of laser scan passes at various downward defocus positions (Kibria et al. 2013c)

surface is decreasing with the increase in *number of laser beam passes* at each setting of *downward defocusing positions*, i.e. 0.2, 0.4, 0.6 and 0.8 mm. During machining in defocusing conditions, with the increase of *number of laser scan passes*, the height of various micro-sized peaks on the machining surface reduces due to melting and evaporation of material from micro-peaks and therefore, the quality of the laser micro-turning surface is improved. The figure also depicts that at various defocus positions, different values of surface roughness are achieved for any particular *number of laser passes*. However, with the increase of defocus position, surface roughness of the laser turning surface is decreasing due to adequate laser beam energy, which removes the rough and irregular peaks from the workpiece surface and makes the surface even. Conversely, it is also observed that at higher value of defocus condition, i.e. at 0.8 mm, the roughness is much higher than 0.4 and 0.6 mm defocus positions. It is so because at 0.8 mm position, the energy density of the laser beam on the workpiece surface is very low and this amount of energy is unable to remove the micro-peaks from the laser micro-turned surface. These phenomena result in higher amount of surface roughness of the machined surface.

The effects of variation of *number of laser scan passes* on surface roughness (Ra) criterion is shown in Fig. 17 at various *downward defocusing positions*. During these experiments, the other process parameters were kept constant at 7.81 W/5600 Hz/436 rpm/0.30 kgf-cm^{-2}/0.443 mm/s of *average power*, *pulse frequency*, *workpiece rotating speed*, *assist air pressure* and *Y feed rate*, respectively. From this plot, it is obvious that the *number of laser scan passes* has significant effect on surface roughness (Ra) criteria. The surface roughness value decreases with the increase in *number of laser beam passes* at each of the defocus positions. It is due to the removal of sharp micro-sized peaks from the workpiece surface by the irradiated laser beam. However, after a certain *number of laser beam passes*, the surface roughness increases to a little extent for each of the focusing conditions. This is because, after certain *number of passes*, the removal of material from workpiece surface is not uniform and due to the high intense laser beam, the machined surface achieved is irregular and uneven. At higher setting of *downward defocusing position* i.e. at 0.8 mm, the laser beam spot size is much high and correspondingly the energy density on the workpiece surface is very low. This low energy beam is rather incapable to remove the micro-sized peaks from the rough workpiece surface. Therefore, at higher value of *defocusing position*, the surface roughness is low compared to other defocusing conditions i.e. 0.4 and 0.6 mm.

Figure 18 shows the effect of varying the number of laser scan passes on surface roughness (Rt) at various *upward defocusing positions*. The other process parameters were kept constant at 7.81 W/5600 Hz/436 rpm/0.30 kgf-cm^{-2}/0.443 mm/s of *average power*, *pulse frequency*, *workpiece rotating speed*, *assist air pressure* and *Y feed rate*, respectively. It is observed from this plot that the surface roughness (Rt) decreases with the increase in *number of laser scan passes* for all the settings of *defocusing positions*. It is due to the removal of micro-sized peaks from the rough laser micro-turning surface by the irradiated laser beam. From this figure, it is also observed that at same *number of laser beam passes*, lower setting of *defocusing position* results in higher value of surface roughness (Rt) compared to higher setting of *defocus positions*. This is mainly due to differentiation of energy densities at various focal positions and requirement of amount of laser beam energy to reduce the height of micro-peaks. However, at 0.8 mm defocusing condition, the surface roughness (Rt) is high compared to 0.6 and 0.4 mm defocus conditions. This is because at 0.8 mm *defocusing position*, the energy density of laser beam is very low, which is unable to remove or reduce the micron-sized peaks from the workpiece surface.

The effect of variation of *number of laser scan passes* on surface roughness (Rt) value is illustrated in Fig. 19, while machining at various *downward defocusing positions*. The other process parameters were kept constant at 7.81 W/5600 Hz/436 rpm/0.30 kgf-cm^{-2}/0.443 mm/s of *average power*, *pulse frequency*, *workpiece rotating speed*, *assist air pressure* and *Y feed rate*, respectively. The plot shows that with the increase in *number of laser scan passes* results in decrease of surface roughness (Rt) values up to a certain *number of passes*. With the increase of *number of passes* of laser beam, the overall heights of the uneven peaks on the laser micro-turned surface decreases. However, after certain *number of passes*, the roughness of micro-turning surface does not significantly vary. In some defocusing conditions such as 0.4 and 0.8 mm, the surface roughness increases after 7th *number of passes*. It is due to generation of small sized uneven peaks again on the laser micro-turned surface. The plot also shows that effect of *defocusing positions* on surface roughness criterion is very random. The minimum surface roughness is achieved at 0.4 mm defocus position and at 8th *number of laser beam passes*. At high defocusing condition, the laser beam energy density is very low and this amount of energy density is incapable of removing further micro-peaks from the rough and uneven surface of the workpiece.

It has been observed from various plots of surface roughness (Ra and Rt) that with the increase of *number of laser beam passes*, the surface roughness (Ra and Rt) decreases at all settings of upward defocus positions. However, at the same settings of *downward defocus positions*, the surface roughness (Ra) is having a trend of decreasing, whereas Rt is moderately varying with *number of laser beam passes* during the machining process. From various test results of the present set of experiments, the individual parametric combinations have been identified for achieving minimum values of individual responses and enlisted in Table 8. At 0.6 mm *upward defocus position* and 9th *number of laser beam passes*, the surface

Table 8 Parametric combinations to achieve minimum surface roughness (Ra and Rt) (Kibria et al. 2013c)

Responses	Parametric combination of average power, pulse frequency, workpiece rotating speed, air pressure and Y feed rate	Parametric setting of additional parameters		Minimum value of responses	
		Number of passes	Defocusing position	Surface roughness, Ra	Surface roughness, Rt
Surface roughness, (Ra, Rt)	7.81 W/5600 Hz/436 rpm/ 0.30 kgf-cm^{-2}/0.443 mm/s	9	0.6 mm (upward)	5.07 μm	35.08 μm
		8	0.4 mm (downward)	5.28 μm	34.65 μm

roughness (Ra and Rt) values are 5.07 and 35.08 μm, which is the smallest value. However, surface roughness, Rt is achieved as minimum as 34.65 μm (Ra = 5.28 μm) at defocus position of 0.4 mm downward and at 8th *number of laser beam passes*. Moreover, it is found from various test results that depth deviation of the laser micro-turned surface increases with the increase in *number of laser beam passes* for all the settings of *defocusing positions*. It is due to reduction of the overall height of the small-sized peaks on the laser turning surface. With lower setting of defocus value, higher value of depth deviation is achieved. With the increase in defocus length, the beam energy density decreases due to increase in laser beam spot size.

6.2.3 Analysis Based on SEM Micrographs of Laser Micro-turned Surface

To study the laser micro-turned machined surface comparatively, the microscopic analysis of the micro-turned surfaces machined at various parametric combinations also been analysed to get a clear picture of the effect of laser beam *defocusing positions* and *number of laser passes* during experiments. Figure 20a–f show SEM images (different magnifications) of the laser micro-turned surface machined at 0.2 mm *downward defocusing* after 10th passes (Ra = 5.50 μm, Rt = 38.96 μm), 0.8 mm *upward defocusing* after 10th passes (Ra = 5.26 μm, Rt = 36.34 μm) and 0.6 mm *upward defocusing* after 10th passes (Ra = 5.10 μm, Rt = 34.91 μm) of laser beam. Comparing these SEM images as well as surface roughness values, it is obvious that the quality of micro-turned surface machined at 0.6 mm *upward defocus position* and 10th *number of laser beam passes* is high compared to the machined surface shown in Fig. 20d and f.

The high magnification SEM images of the laser micro-turned surfaces machined at above-mentioned parametric combinations are shown in Fig. 21a–c, respectively. In Fig. 21a, it is seen that still the machined surface has some micro-peaks after machining at *defocusing position* of 0.2 mm and 10th *number of laser beam passes*. However, it is seen in Fig. 21b, there is no such micro-peaks on the machined surface. Because of irradiation of defocused laser beam on the surface, the micro-peaks melted as well as evaporated and some melted materials resolidified on the machined surface. On the other hand, the SEM image of the machined surface shown in Fig. 21c has no such micro-peaks as well as resolidified materials because of adequate laser beam energy irradiated on the material surface to completely evaporate and remove the micro-peaks from alumina surface.

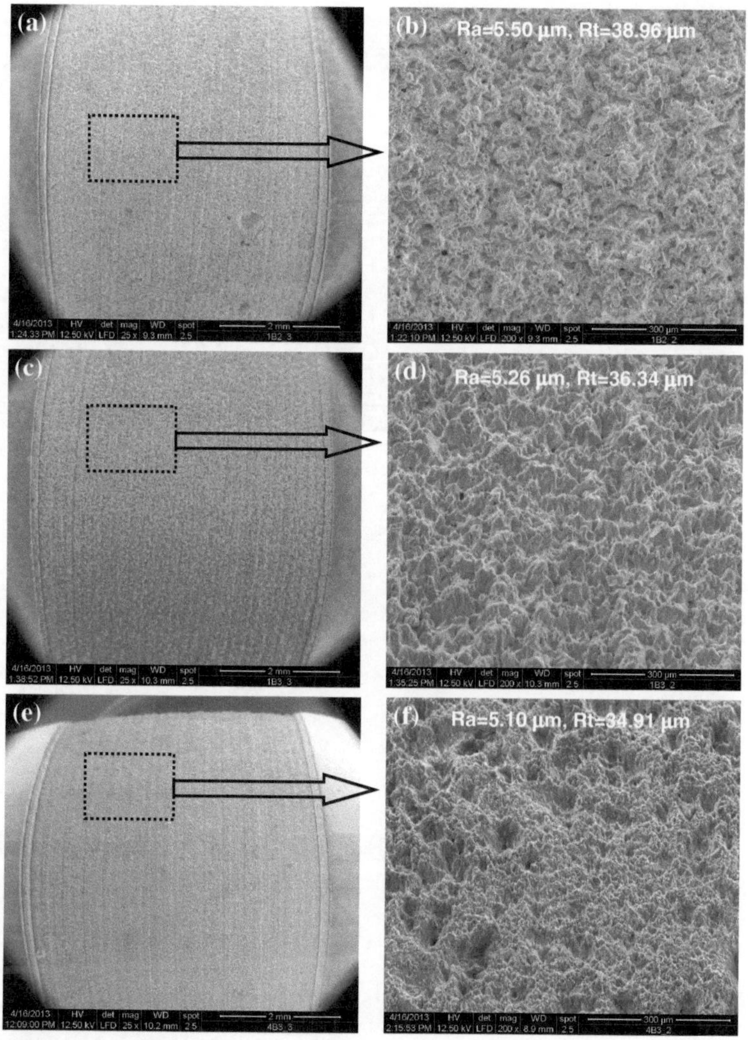

Fig. 20 SEM micrographs of laser micro-turned surface of alumina ceramic machined at 0.2 mm downward/10th pass (**a** and **b**), 0.8 mm upward/10th pass (**c** and **d**), and 0.6 mm upward/10th pass (**e** and **f**) of defocusing position and number of passes of laser beam

6.3 Comparative Study of Surface Roughness Criteria at Laser Focused and Defocused Conditions During Laser Micro-turning

In this experimental scheme, comparative investigation and analysis is carried out at focused and defocused conditions of laser beam during laser micro-turning process of aluminium oxide (Al_2O_3) ceramics to explore the effects of various process

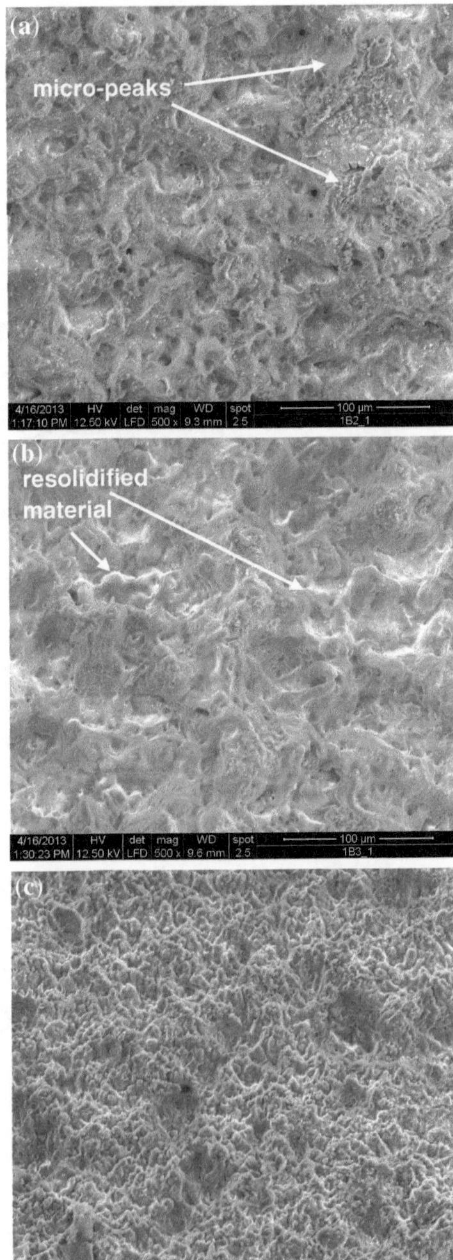

Fig. 21 High magnification SEM micrographs of laser micro-turned surface machined at (**a**) 0.2 mm downward/10th pass, (**b**) 0.8 mm upward/10th pass and (**c**) 0.6 mm upward/10th pass of defocusing position and number of passes of laser beam

Table 9 Process parameters levels and other machining conditions

Variable parameters	
Process parameters	Levels
Average power	5, 6, 7, 8, 9, 10 W
Pulse frequency	4000, 4500, 5000, 5500, 6000 Hz
Rotational speed	200, 300, 400, 500, 600 rpm
Y feed rate	0.2, 0.3, 0.4, 0.5, 0.6 mm/s
Condition of lens	Focused, defocused
Defocused	0.2, 0.4, 0.6 mm upward
Constant parameters	
Air pressure	0.3 kgf/cm^2
Mode of operation	Q-switched (pulsed)
Type of Q-switch	Acousto optic
Mirror reflectivity	Rear mirror–100 % Front mirror–80 %
Beam diameter 1/e^2	1 mm
Laser beam spot diameter	100 μm

parameters such as *average power*, *pulse frequency*, *workpiece rotating speed* and *Y feed rate*. The experiments have been conducted utilizing one factor at a time (OFAT) experimental scheme. The target depth was set at 100 μm. Table 9 shows the levels of process parameters considered. The table also shows the other process parameters which were kept constant during experimentation. In the previous experimentation, it is revealed that least surface roughness (Ra) is achieved as 5.07 μm at 0.6 mm *upward defocusing condition*. Therefore, all the experiments in the present experimental scheme are carried out at various *upward defocused* and as well as *focused conditions*. After each experiment, surface roughness (Ra and Rt) has been measured.

Based on the results of surface roughness (Ra and Rt) values measured both the machining conditions i.e. focused and defocused, various plots are drawn and analysed in details hereunder. In Fig. 22, the effect of *average power* on surface roughness (Ra) is shown at focused and various upward defocusing conditions. The other process parameters were kept constant at *pulse frequency* of 5000 Hz, *workpiece rotating speed* of 400 rpm and *Y feed rate* of 0.4 mm/s. From this plot, it is observed that low value setting of *average power* results in high surface finish due to a smaller amount of melting and/or ablation of materials from work sample. As *average power* increases, the roughness of micro-turning surface also increases. It is also seen from the same plot that surface roughness values are less when laser micro-turning is carried out in *defocused conditions* compared to focus condition machining. In *defocus machining condition*, the energy density of laser beam is low, therefore, material from top surface micro-peaks is removed through melting and evaporation and surface conditions of machined surface is improved. The figure

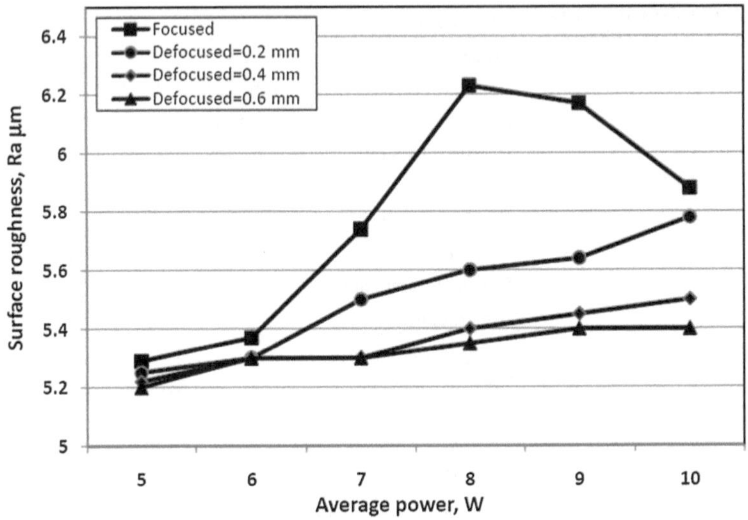

Fig. 22 Variation of surface roughness (Ra) with average power at focused and various upward defocus conditions

also depicts that at various *defocus positions*, different values of surface roughness are achieved for any particular *average power* value. With the increase of defocus position, surface roughness of the laser turning surface is decreasing due to adequate laser beam energy, which removes the rough and irregular peaks from the workpiece surface and makes the surface even.

Figure 23 shows the effect of *pulse frequency* on surface roughness (Ra) at *focused* and various *upward* defocusing conditions. The other process parameters were kept constant at *average power* of 8 W, *workpiece rotating speed* of 400 rpm and *Y feed rate* of 0.4 mm/s. It is evident from this plot that surface roughness (Ra) slightly increases at focused and other *defocused conditions.* According to the Eq. 7, it is clear that with the increase of *pulse frequency*, *peak power* of laser beam decreases (Kibria et al. 2013b). However, the time interval between two consecutive pulses of laser beam is very small, therefore the top surface of the workpiece gets adequate energy to melt and vaporize rapidly and it results in more peaks and valleys on the micro-turned surface.

$$Peak\,power\,(P_P) = \frac{Average\,power\,(P_A)}{Pulse\,frequency\,(F_P) \times Pulse\,duration\,(\mu)} \tag{7}$$

It is also seen in the same plot that the values of surface roughness (Ra) is less at all parametric combinations during machining at all *defocus conditions* compared to machining at *focused condition.* It is due to adequate thermal energy consumed by the workpiece surface and as a result of this, the height of micro-peaks is less. The surface roughness values are less at higher *defocused machining conditions* i.e.

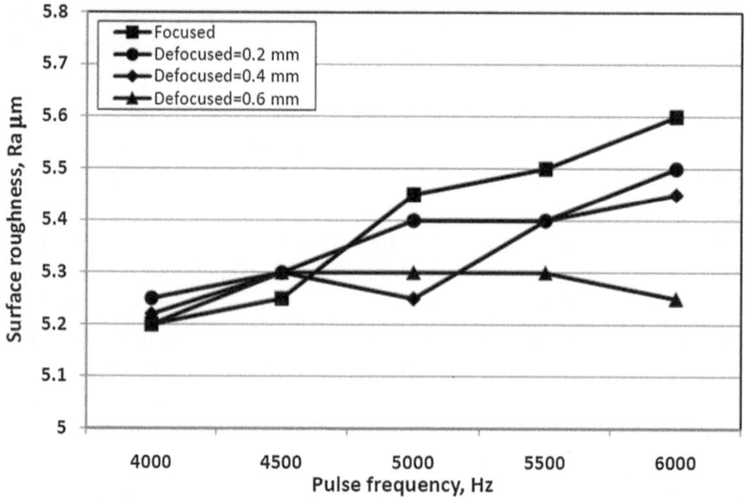

Fig. 23 Variation of surface roughness (Ra) with pulse frequency at focused and various upward defocus conditions

more defocused, more quality surface and it is due to adequate laser power density at laser material interaction zone.

In Fig. 24, the effect of *workpiece rotating speed* on surface roughness (Ra) is shown at *focused* and *various upward defocusing conditions*. The other process parameters were kept constant at *average power* at 8 W, *pulse frequency* of

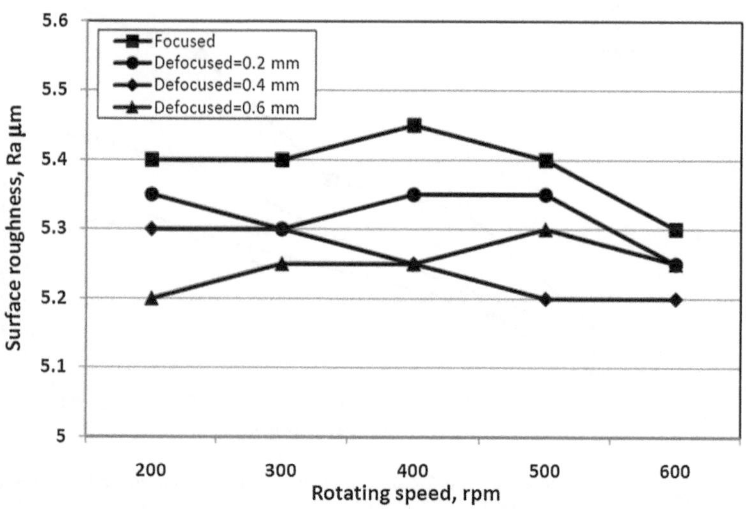

Fig. 24 Variation of surface roughness (Ra) with rotating speed at focused and various upward defocus conditions

5000 Hz and *Y feed rate* of 0.4 mm/s. From this plot, it is observed that surface roughness values are less for the micro-turned surface machined at *defocused positions* compared to *focused conditions*. It is known that at various *defocused conditions*, the irradiated laser beam energy density is less; therefore, the heights of micro-peaks generated on the machined surface are less. It is also revealed that increase of *rotating speed* results in lower values of surface roughness. It is due to the fact that increase of *rotating speed* results in increase of *circumferential overlap* value according to Eq. 4. Therefore, more even surface is achieved. Moreover, higher *defocused conditions* result in higher surface finish and it is mainly due to melting and instant solidification of workpiece material on the machined surface.

Figure 25 shows the effect of *Y feed rate* on surface roughness (Ra) at focused and various *upward defocusing conditions*. The other process parameters were kept constant at *average power* of 8 W, *workpiece rotating speed* of 400 rpm and *pulse frequency* of 5000 Hz. It is evident from this plot that the increase of *Y feed rate* results in deterioration of surface finish of work sample. As increase of *Y feed rate* results in reduction in *circumferential overlap* over two successive laser scan tracks and this phenomenon creates uneven surface profile over the micro-turned surface and therefore, the roughness of machined surface deteriorated. It is also observed that the surface roughness (Ra) is high at *focused condition* laser micro-turning operation compared to machining at all *defocused positions*. With the increase of defocusing of the laser beam, the laser beam energy density decreases, and as a result, the material removal from the workpiece surface is less.

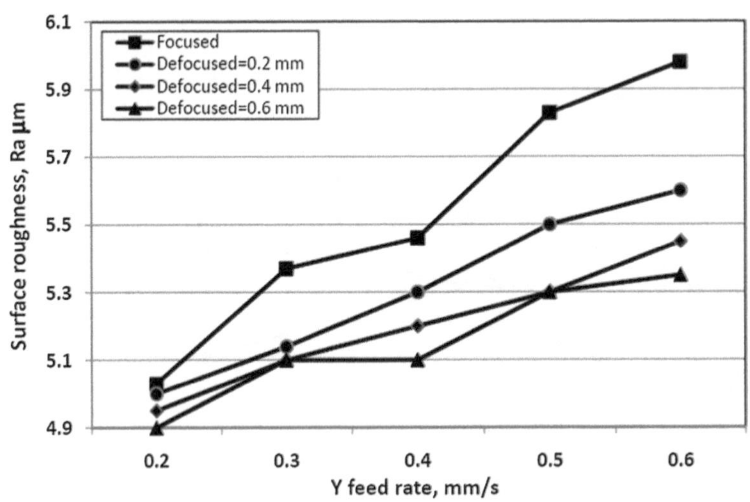

Fig. 25 Variation of surface roughness (Ra) with Y feed rate at focused and various upward defocus conditions

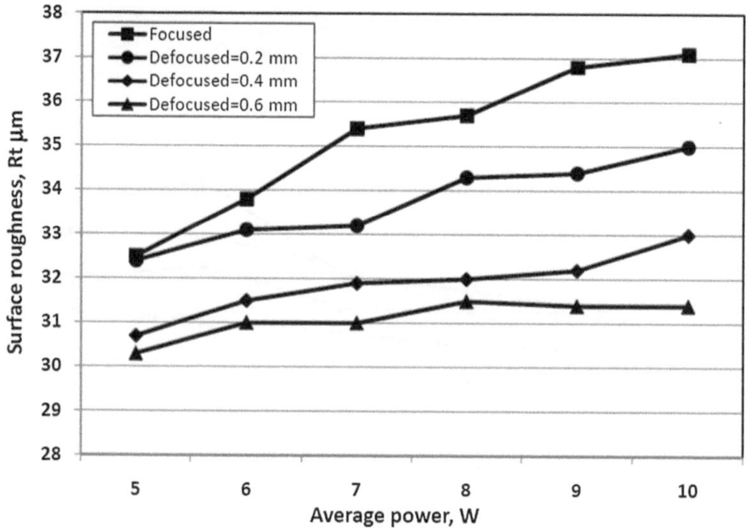

Fig. 26 Variation of surface roughness (Rt) with average power at focused and various upward defocus conditions

In Fig. 26, the effect of *average power* on surface roughness (Rt) is shown at *focused* and various *upward defocusing conditions*. The other process parameters were kept constant as same as Fig. 22. From this plot, it is revealed that with the increase of *average power*, the surface roughness (Rt) is slightly increases and it is due to generation of micro-peaks and valley on the machined surface. It is also observed from the same plot that *defocused condition* machining results in lower value of surface roughness compared to surface roughness (Rt) achieved during machining at *focused condition*. However, at higher value of defocused machining i.e. at 0.4 and 0.8 mm defocusing, the *average power* has no such effect on Rt values.

Figure 27 shows the effect of *pulse frequency* on surface roughness (Rt) at *focused* and various *upward defocusing conditions*. The other process parameters were kept constant as same as Fig. 23. It is evident from this plot that surface roughness (Rt) slightly increases at *focused condition* machining. However, there is no such increment of Rt value of the machined surface at *defocused conditions*. Rt value lie in the range of 30–33 µm. Moreover, the values of Rt is less at *defocusing conditions* compared to machining at *focused position*.

In Fig. 28, the effect of *workpiece rotating speed* on surface roughness (Rt) is shown at *focused* and various *upward defocusing conditions*. The other process parameters were kept constant as same as Fig. 24. It is observed from Fig. 28 that with the increase of *workpiece rotating speed*, surface roughness value decreases.

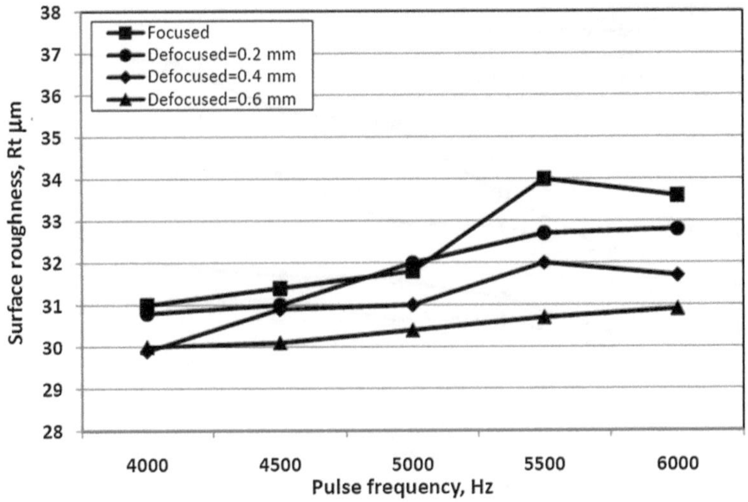

Fig. 27 Variation of surface roughness (Rt) with pulse frequency at focused and various upward defocus conditions

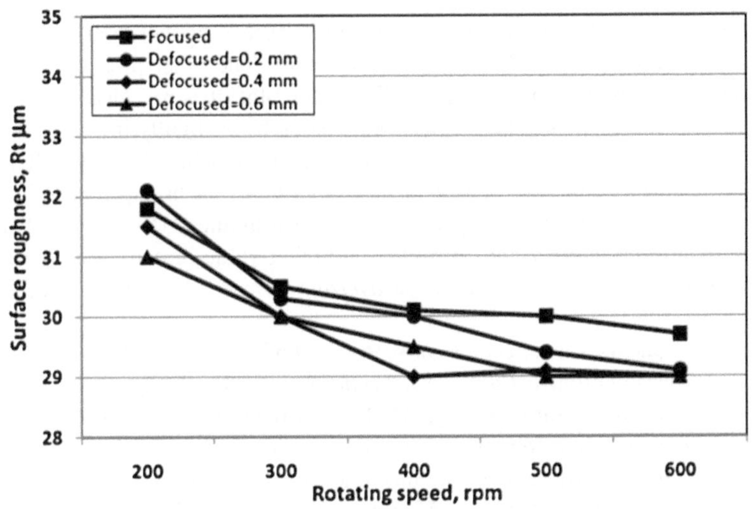

Fig. 28 Variation of surface roughness (Rt) with rotating speed at focused and various upward defocus conditions

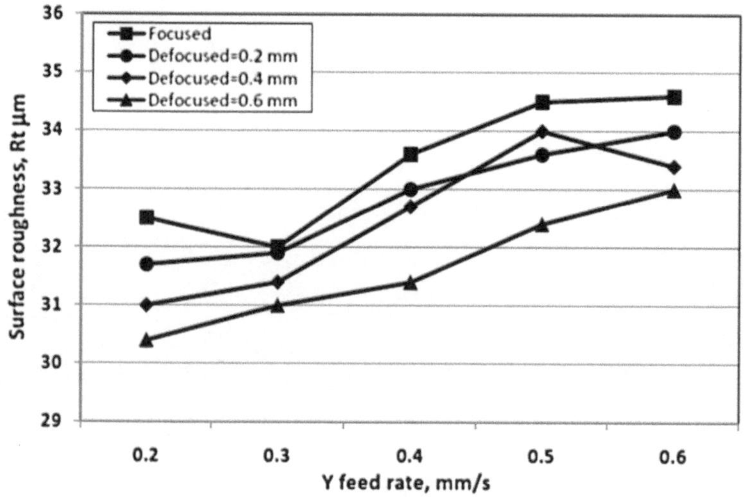

Fig. 29 Variation of surface roughness (Rt) with Y feed rate at focused and various upward defocus conditions

It is mainly due to the increase of *circumferential overlap percentage* during machining. From this plot, it is also revealed that there is no such effect of amount of *defocusing conditions*. However, the value of Rt is higher at *focused condition* compared to machining at *defocused conditions*. Figure 29 shows the effect of *Y feed rate* on surface roughness (Rt) at *focused* and various *upward defocusing conditions*. The other process parameters were kept constant as same as Fig. 25. The figure depicts that the increase of *Y feed rate* results in increase of Rt values at focused as well as all *defocused machining conditions*. It is due to decrease of *circumferential overlap percentage* according to Eq. 4. The values of Rt are less at *defocused machining conditions* compared to *focused machining*.

To analyse the laser micro-turned surface qualitatively, Scanning Electron Microscopic (SEM) images were captured for the surfaces machined at various parametric combinations. In Fig. 30, the exhibited photographs show the surfaces obtained through laser micro-turning on alumina at parametric setting of *average power* of 8 W, *workpiece rotating speed* of 400 rpm and *pulse frequency* of 5000 Hz, *Y feed rate* of 0.4 mm/s both at *focused* and *defocused machining condition*. The surface roughness (Ra) values are 5.46 and 5.1 µm, respectively for *focused* and *defocused* (=0.6 mm) *machining conditions*. The corresponding Rt values are 33.6 and 31.4 µm, respectively. From these figures and obtained results of Ra and Rt, it is obvious that laser micro-turned surface obtained at *defocused machining conditions* are qualitatively high compared to *focused machining condition*. There are lots of micro-peaks observed on the laser micro-turned surface obtained during *focused condition machining*.

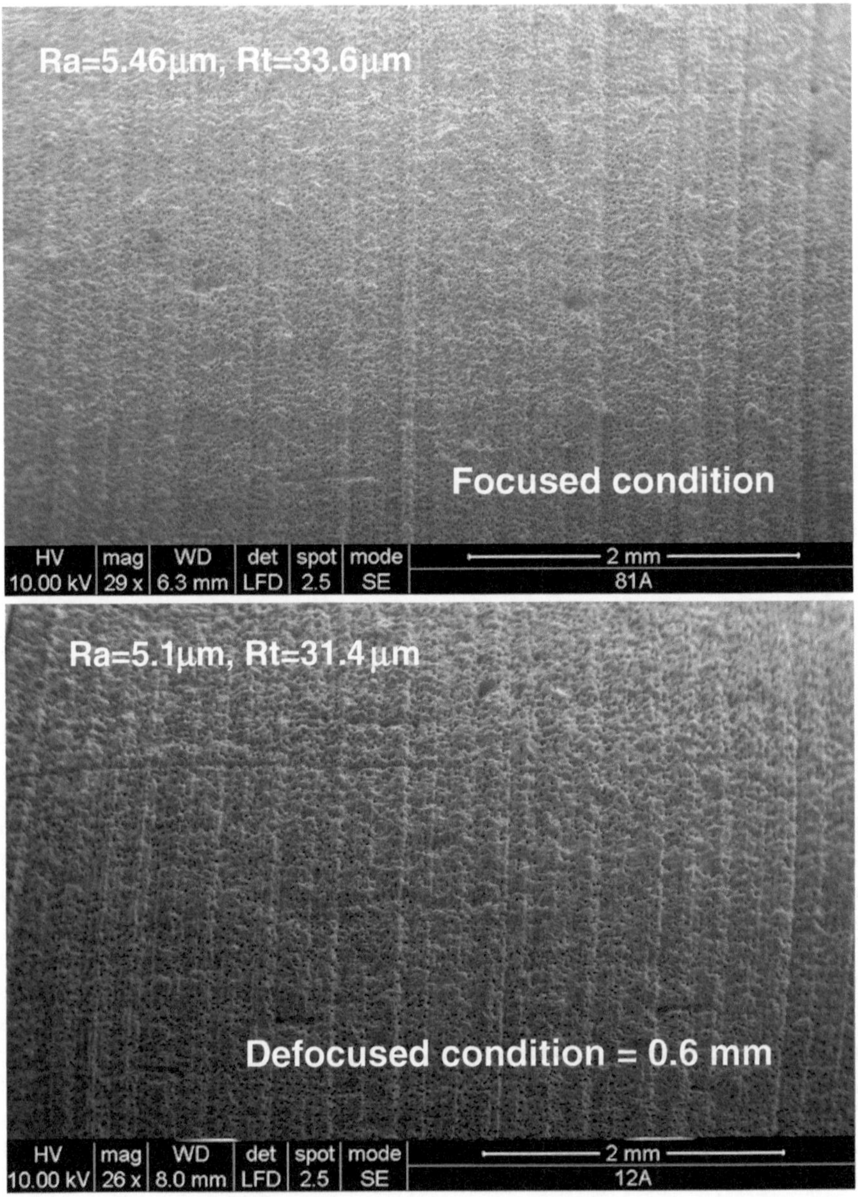

Fig. 30 SEM views of laser micro-turning surfaces machined at average power of 8 W, workpiece rotating speed of 400 rpm and pulse frequency of 5000 Hz, Y feed rate of 0.4 mm/s

7 Conclusions

The pulsed Nd:YAG laser machining system has the competence to perform micro machining process like laser micro-turning operations on difficult-to-machine ceramic materials such as aluminum oxide (Al_2O_3) to achieve a particular depth and surface finish of the cylindrical workpiece. In this chapter, Nd:YAG laser micro-turning operation has been successfully performed to produce laser micro-turning surface up to desired depth on cylindrical workpiece of aluminium oxide (Al_2O_3) ceramic. By proper controlling of the various laser beam parameters like *average power* and *pulse frequency* and various process parameters like *workpiece rotating speed*, *assist air pressure* and *Y feed rate*, the desired machined surface features like accurate dimensional aspect in terms of micro-turning depth and surface roughness were achieved. The present set of research findings will open up new insights and directions to the fundamental and applied research in the area of laser micro-turning to achieve and improve the accuracy, repeatability and finally the reliability of this novel material machining technique in the context of cylindrical micro part fabrication in modern shop floor. The present chapter opens up further future scope of research work related to analysis of surface structure, surface morphology and micro-hardness of the machined surface during laser micro-turning process of various engineering ceramic materials such as zirconia (ZrO_2), aluminium titanate (Al_2TiO_5), magnesia (MgO_2) and silicon nitride (Si_3N_4) under various process parametric conditions. Laser micro-turning can be carried out for examining the phase transformation as well as crack formation on laser micro-tuned surface at various machining conditions.

Acknowledgments The authors acknowledge the financial support and assistance provided by CAS Ph-IV programme of Production Engineering Department of Jadavpur University under University Grants Commission (UGC), New Delhi, India.

References

Basu, B., & Balani, K. (2010). *Sintering of Ceramics, in Advanced Structural Ceramics, Wiley-American Ceramic Society* (1st ed.). Hoboken: Wiley.

Buttery, T. C., Statham, A., Percival, J. B., & Hamed, M. S. (1979). Some effects of dressing on grinding performance. *Wear, 55*, 195–219.

Chrysolouris, G., Bredt, J., Kordas, S., & Wilson, E. (1988). Theoretical aspects of a laser machine tool. *Journal of Engineering for Industry, 110*, 65–70.

Dahotre, N. B., & Harimkar, S. P. (2008). *Laser Fabrication and Machining of Materials* (1st ed.). Berlin: Springer.

Dhupal, D., Doloi, B., & Bhattacharyya, B. (2008). Pulsed Nd:YAG laser turning of micro-groove on aluminum oxide ceramic (Al_2O_3). *International Journal of Machine Tools and Manufacture, 48*, 236–248.

Gadelmawla, E. S., Koura, M. M., Maksoud, T. M. A., Elewa, I. M., & Soliman, H. H. (2002). Roughness parameters. *Journal of Materials Processing Technology, 123*, 133–145.

Gitzen, W. H. (1970). *Alumina as a Ceramic Material* (1st ed.). USA: Wiley-American Ceramic Society.

Haddad, M. J., & Tehrani, A. F. (2008). Investigation of cylindrical wire electrical discharge turning (CWEDT) of AISI D3 tool steel based on statistical analysis. *Journal of Materials Processing Technology, 198*, 77–85.

Haddad, M. J., Alihoseini, F., Hadi, M., Hadad, M., Tehrani, A. F., & Mohammadi, A. (2010). An experimental investigation of cylindrical wire electrical discharge turning process. *International Journal of Advanced Manufacturing Technology, 46*, 1119–1132.

Kibria, G., Doloi, B, & Bhattacharyya, B. (2010a). Optimization of Nd:YAG laser micro-turning process using response surface methodology. *Proceedings of the 3rd International and 24th All India Manufacturing Technology, Design and Research Conference* (pp. 333–338), December 13–15, 2010, Visakhapatnam.

Kibria, G., Doloi, B., & Bhattacharyya, B. (2010b). Experimental analysis on Nd:YAG laser micro-turning of alumina ceramic. *International Journal of Advanced Manufacturing Technology, 50*, 643–650.

Kibria, G., Doloi, B., & Bhattacharyya, B. (2012). Optimisation of Nd:YAG laser micro-turning process using response surface methodology. *International Journal of Precision Technology, 3*, 14–36.

Kibria, G., Doloi, B., & Bhattacharyya, B. (2013a). Experimental investigation and multi-objective optimization of Nd:YAG laser micro-turning process of alumina ceramic using orthogonal array and grey relational analysis. *Journal of Optics and Laser Technology, 48*, 16–27.

Kibria, G., Doloi, B., & Bhattacharyya, B. (2013b). Predictive model and process parameters optimization of Nd:YAG laser micro-turning of ceramics. *International Journal of Advanced Manufacturing Technology, 65*, 213–229.

Kibria, G., Doloi, B., & Bhattacharyya, B. (2014a). Parametric study and optimization of Nd:YAG laser micro-turning process of different grade of alumina ceramics based on Taguchi methodology. *International Journal of Surface Engineering and Interdisciplinary Materials Science, 2*, 15–46.

Kibria, G., Doloi, B., & Bhattacharyya, B. (2014b). Investigation into the effect of overlap factors and process parameters on surface roughness and machined depth during micro-turning process with Nd:YAG laser. *Journal of Optics and Laser Technology, 60*, 90–98.

Kibria, G., Doloi, B., & Bhattacharyya, B. (2015). Investigation and analysis on pulsed Nd:YAG laser micro-turning process of aluminium oxide (Al_2O_3) ceramic at various laser defocusing conditions. *International Journal of Advanced Manufacturing Technology*, *76*(1–4), 17–27.

Kim, S. H., Sohn, I. B., & Jeong, S. (2009). Ablation characteristics of aluminum oxide and nitride ceramics during femtosecond laser micromachining. *Applied Surface Science, 255*, 9717–9720.

Kruth, J. P., Yasa, E., Vanparys, M., & Vaerenbergh J. V. (2007). The enhancement of micro-machining ability of selective laser melting by selective laser erosion. *Proceedings of the International Conference on Polymers and Moulds Innovations*, PMI Gent, Belgium.

Samant, A. N., & Dahotre, N. B. (2008a). Ab initio physical analysis of single dimensional laser machining of silicon nitride. *Advanced Engineering Materials, 10*, 978–981.

Samant, A. N., & Dahotre, N. B. (2008b). Computational predictions in single-dimensional laser machining of alumina. *International Journal of Machine Tools and Manufacture, 48*, 1345–1353.

Shih, A. J. (2000). An experimental investigation of rotary diamond truing and dressing of vitreous bond wheels for ceramic grinding. *International Journal of Machine Tools and Manufacture, 40*, 1755–1774.

Thawari, G., Sarin Sundar, J. K., Sundararajan, G., & Joshi, S. V. (2005). Influence of process parameters during pulsed Nd:YAG laser cutting of nickel-base superalloys. *Journal of Materials Processing Technology, 170*, 229–239.

A Literature Review on CO_2 Laser Welding

Rakesh Bhadra, Pankaj Biswas and M. Ravi Sankar

Abstract Laser welding is a sophisticated, high accuracy and high speed welding process. Laser welding is a process of joining components where laser beam used as a heat source. In this present study a literature review on welding by laser as a heat source has been addressed. In the present review, emphasis has been given especially on the laser welding numerical and experimental temperature field analysis, thermo-mechanical analysis. The time frame of the review is 1992–2013.

Keywords Laser welding · Numerical and experiment temperature distribution · Deformation · Microstructure

1 Introduction

The continuous development and demands of various parts for products assemble operation such as automotive components, aerospace structures and various other machine component and production of those parts at high rate is possible by automation. Accuracy is an important demand for manufacturing process (Kim et al. 2011; Geoffray et al. 2011). Any assemble operation joining is one of the most significant manufacturing requirements. Welding is an improved conventional joining process so welding can be applied to joining of various dimensional machine components. There are several welding method already been established (Sakagawa et al. 2011). The welding offers some benefits over mechanical fasteners

R. Bhadra (✉)
Department of Mechanical Engineering, Tezpur University, Tezpur 784028, India
e-mail: r.bhadr@tezu.ernet.in

P. Biswas · M. Ravi Sankar
Department of Mechanical Engineering, IIT Guwhati, Guwhati 781039, India
e-mail: pankaj.biswas@iitg.ernet.in

M. Ravi Sankar
e-mail: evmrs@iitg.ernet.in

S.N. Joshi and U.S. Dixit (eds.), *Lasers Based Manufacturing*,
Topics in Mining, Metallurgy and Materials Engineering,
DOI 10.1007/978-81-322-2352-8_19

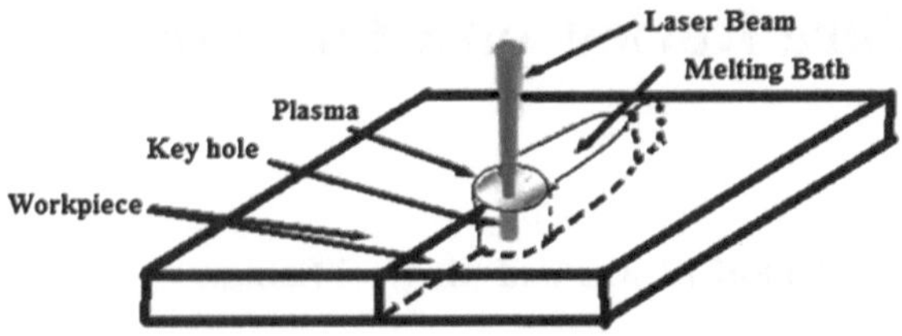

Fig. 1 Keyhole and plasma formation during laser welding (Belhadj et al. 2010)

such as weight reduction and absence of notches induced by machining operations. Several welding process have establish for joining of several parts. The laser welding is one of the important welding process because laser has provide high energy density in ambient air, the laser can be used to perform high-precision, high speed, deep penetration welding under low-strain conditions and easy to welding process automation. Figure 1 shows a schematic of keyhole and plasma formation in laser welding.

Applications of high power density laser have been increasingly utilized in industrial cutting and joining process. Laser is use due to lower width and shape distortion of laser beam as well as greater control and high velocity of laser beam movement at the time of laser welding. The CO_2 laser is one of the earliest gas lasers to be developed by CO_2 gas. At the earliest development of CO_2 laser has been used for cutting but after continuous development of laser quality and applications, it has been used for welding of various thickness plates. It was seen that the development of high-power laser systems 20–50 kW class CO_2 lasers, expected to the establishment of deep-penetration and high-speed welding method. Welding of various parts by using laser, it quality can have a little satisfying, but it has many problems for controlling the laser source's stabilization, laser beams quality, focus dimension, pulse energy, pulse width, waveform, defocusing amount power supply's stabilization and trigger frequency, etc. in the aspect of precisions (Hess et al. 2011).

The present book chapter is focused on analysis of current achievement and gaps in the CO_2 laser joining process applications.

2 Laser Welding

2.1 CO_2 Laser Welding

This is a non-contact process that requires access to the weld zone from one side of the parts being welded. The weld is formed as the intense laser light rapidly heats the material typically calculated in milliseconds. Laser welding is a joining process where lasers are used as a heating source for joining the two metal parts. The advantages of laser welding technology such as high speed welding process and non-contact welding process make the use of the technology more attractive in the automotive industry. A number of researchers work have been studied on CO_2 laser welding. Tu et al. (2003) have used 20 kW CO_2 laser for welding of mild steel to determine the plasma absorption in the keyhole, based on photodiode measurements and spectroscopic measurements.

Kurosaki et al. (2009) have developed a CO_2 laser welding method for overlapped welding of PFA polymer film sheet. The procedure feasibility was confirmed by joining of thinner 0.3 mm PFA film sheet.

Liu et al. (2006) studied on the welding of low carbon steel by using the 4 kW CO_2 laser. The welding parameters were laser power 4 kW, welding speed 1.2 m/min, shielding gas Helium and its flow rate 11 l/min. Authors observed that the isogrey lines of the image behaviour on the basis of the high-speed photography which was taken by NAC-10 made in Japan. Shanmugarajan and Padmanabham (2012) studied on CO_2 laser welding of dissimilar Ti–SS (304) combination. They have used continuous wave 3.5 kW CO_2 laser for the autogenously as well as using vanadium (V) and Tantalum (Ta) as interlayer for welding studies. The experimental results was exhibited that the welding with vanadium as interlayer provide less cracking compared with autogenously welding. Zhu et al. (2005) investigated welding of AZ31 alloys by using 1.5 kW diode laser and 2.0 kW CO_2 laser. Observation was made on keyhole welding and conduction welding mode for both CO_2 and diode lasers. It was proposed that the CO_2 laser can produce a keyhole weld with good quality but diode laser was limited to conduction welding, due to its spot size limitation. In both cases Mg vaporization was inevitable and oxygen content increases in the welding zone and protecting from weld zone oxidation glove box are more effective.

Sibillano et al. (2006) studied the effect of helium shielding gas on AA5083 aluminium alloys welding by CO_2 laser under different operating conditions. It was noted the effect of helium gas based on the microstructures of welded specimen and variation of intensity of emission lines and found that welding nozzle can affect the welding quality as well as the shielding gas flow rate.

2.2 *Hybrid Laser Welding*

Welding by using CO_2 laser with other welding heat sources also being used for welding of several components, this type of welding process is called hybrid laser welding process. Casalino (2007) has developed the MIG-CO_2 laser beam welding of 5005 aluminium alloy (0.6 % magnesium). Author has used Al 5356 as a filler material, which contains up to 5 % Mg. The bead appearances were evaluated and the weld penetrations were measured. The hybrid welding processes wear investigated on the basis of the regression model and factorial experiment output. The set of data was interpolated and the blanks in a four levels factorial design were filled and full factorial statistical analysis have been performed. It was observed the weld penetration is deeper when higher laser power and the position of the workpiece at focus height was used, as well as smaller inclinations for the MIG torch and interactions between other parameters have no effect on weld penetration.

Zhu et al. (2005) have made a comparative study on diode laser (1.5 kW) and CO_2 laser (2.0 kW) welding of studied for the welding of AZ31 alloys. It was observed that the effect on keyhole and conduction mode on both welding process and characterizes in both welding modes. It was investigated the welding process characteristics and mechanisms involved in CO_2 laser and diode laser welding of AZ31 alloy. Authors proposed that the CO_2 laser can produce a keyhole weld with good quality but diode laser was limited to conduction welding, due to the spot size limitation. The diode laser welded specimens has less porosity relative to CO_2 laser welded specimens, less under-cut also one of the positive side of diode laser over CO_2 laser. In both cases Mg vaporization was inevitable and oxygen content increases in the welding zone and protecting from weld zone oxidation glove box are more effective.

3 Studies on Mechanical Properties of Welded Joints

The by using CO_2 laser with other welding heat sources also being used for welding of several components, this type of welding process is called hybrid laser welding process. Zhu et al. (2005) have made a comparative study on diode laser (1.5 kW) and CO_2 laser (2.0 kW) welding of studied for the welding of AZ31 alloys. It was observed that the effect on keyhole and conduction mode on both welding process and characterizes in both welding modes. It was investigated the welding process characteristics and mechanisms involved in CO_2 laser and diode laser welding of AZ31 alloy. Authors proposed that the CO_2 laser can produce a keyhole weld with good quality but diode laser was limited to conduction welding, due to the spot size limitation. The diode laser welded specimens has less porosity relative to CO_2 laser welded specimens, less under-cut also one of the positive side of diode laser over CO_2 laser. In both cases Mg vaporization was inevitable and oxygen content increases in the welding zone and protecting from weld zone oxidation glove box are more effective.

Gao et al. (2007) have investigated the effect of shielding gas parameter on CO_2 laser-tungsten inert gas (TIG) hybrid welding method for SS-316L stainless plate welding. They observed that the shielding gas parameter can be effect on the process stability as well as weld penetration. The welding penetration is depending on the shape of the plasma which is mostly depending on plasma interacting height. The experimental results was shown that increasing of the interacting height of plasma provide shallower the weld penetration. The effect of shielding gas parameters on the plasma shape is achieved by two ways: laser-arc plasma interaction and gas flowing direction and velocity. For full penetrated welding, the hybrid welding protecting method combining the torch and paraxial nozzle was very narrow, but that of the hybrid protecting method coupling the torch and coaxial nozzle was considerably wide.

4 Microstructural Studies of Welded Joints

The quality of weld is depending on the microstructures of various weld zones. The physical properties of welded portion have change due to change of various phases as well as change in the dimension of grin size of the materials. Perez-Medina et al. (2013) observed the metallurgical development of CO_2 laser welding of TRIP-800 steel, where they found that the laser promoted the development of up to 23 % martensite in the fusion zone (FZ) and up to 30 % of the heat-affected zone (HAZ). Figure 2 shows the micrographs of the weld zones of TRIP-800 steel.

Author observed the microstructure using by optical, scanning, and transmission electron microscopy (TEM) and did the tensile test of the welded specimen, results indicated that the strength and ductility decreases in the welded region. The relative amounts of RA in the BM, FZ and HAZ were also determined using XRD and Cu–Ka radiation by considering the expression as,

$$V_\alpha = \frac{1.4 I_\gamma}{I_\alpha + I_\gamma} \tag{1}$$

where, I_α the average intensity of the austenite is peaks, and I_γ is the highest intensity of the ferrite peak.

Liu et al. (2013) compared the microstructure and mechanical properties of the CO_2 laser-metal active gas (MAG) hybrid welding and MAG welding of iron-based powder metallurgical steels. The micro structure of hybrid weld zone was columnar dendrite and fine acicular dendrite grain between the columnar dendrites grains well as narrow compare with MAG weld zone but for MAG there were only one zone that was columnar dendrite grain. Authors also observed that the hardness of hybrid CO_2 laser weld samples was better than the MAG weld sample. Kim et al. (2011) investigated on lab joint of galvanized steel plate (i.e. with 54 gm/m^2 zinc coating of approximately 50 μm thickness) which was welded by CO_2 laser, plasma arc, laser-TIG, laser-micro plasma hybrid welding. Authors studied the effect of process

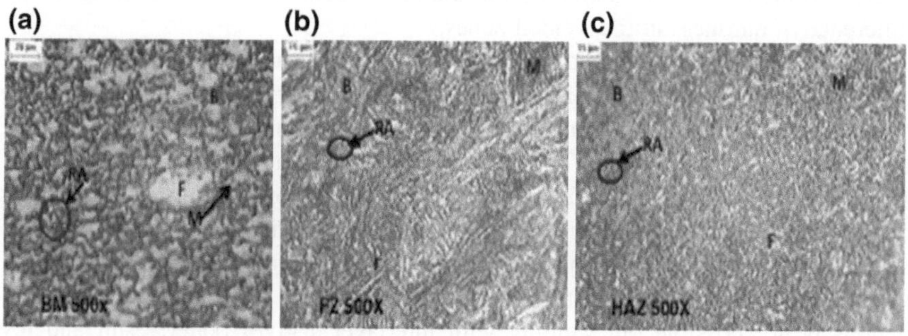

Fig. 2 Optical micrographs of the AHSS TRIP800 in **a** BM, **b** FZ, and **c** HAZ. Etchants: Nital, Picral, and KlemmÕs I, *B* bainite, *F* ferrite, *M* martensite, *RA* retained austenite (Perez-Medina et al. 2013)

parameters and the laser-arc interspacing distance on welding quality. It was also found that the laser-micro plasma hybrid welding with zero gap gives good weld bead shape and obtained 40 % smaller arc heat input than that found in laser-TIG hybrid welding.

Honga et al. (2008) studied that the CO_2 laser welding characteristics of Inconel 718 having two different grain sizes ASTM #4 and #10 under optimum conditions of welding power, welding speed and focused position of laser and uses pre-heat and post-heat treatment by 955 °C solution treatment, and 955 °C solution treatment + aging (955STA) for improving weld quality. Authors have analyzed microstructures of HAZ and fused zone using optical microscope and scanning electron microscope (SEM). Authors suggested that the modified cyclic solution heat treatment showed reduction of micro-fissures. Chen et al. (2009) studied CO_2 laser welding of hot-dip galvanized steel sheets, pre-drilled have used by pulsed Nd:YAG laser to form vent holes along the weld line. The liquid metal flows into the vent holes freely, thus creating rivet-shaped welds along the weld seam. The weld quality was evaluated through optical and scanning electron microscopy and tensile/hardness tests. It was also established that the using this process with no gap produced welding without defect such as porosity, spatter, and loss of penetration and possible to neglect the difficult to produce gap between welding plates. Boutarek et al. (2009) studied on CO_2 laser welding of zirconium alloy i.e. Zircaloy-4. Authors investigated the optimum process parameter to minimize the defect. The material characterization was done by optical microscopy, scanning electron microscopy, X-ray diffraction and micro hardness test. It was found that the micro-structure in different zone is Widemansttäten structure. Wang et al. (2011b) studied on effect of side assisting gas on 2 kW CO_2 laser welding of zinc coated steel sheets. Authors investigated the effect of side assisting gas based on the two set of experiments i.e. (1) Absence and presence of side assisting gas and (2) With different flow rates of side assisting gas. The optical emission and spectrum signals from the laser-induced plasma were obtained and compared. The metallurgical characterizations of various phase present in the various zones investigated on the basis of microstructure by using scanning electron microscopy and X-ray diffraction technique. The hardness determined by the Vickers micro hardness test showed that the variation of grin size did not exhibit significant difference of hardness in the welded zones.

Shanmugarajan and Padmanabham (2012) investigated on continuous wave 3.5 kW CO_2 laser welding of dissimilar Ti–SS (304) combination. The welding was done autogenously and with interlayer's such as Vanadium (V) and Tantalum (Ta) in the form of laser cladding. The welding wear conducted to recognize and optimized laser welding parameters using depth of penetration as criteria. The welding quality investigated on the basis of macrostructure analysis, SEM/EDAX and XRD and as well as tensile test. From the experimental results and microstructure analysis, it was found that the autogenous laser welding of Ti–SS dissimilar combinations is not possible but welding with interlayer's gives better result and less defect. Yu et al. (2011) studied the CO_2 laser lap welding of dissimilar metals of 5A05 aluminum alloy and ST04Z galvanized steel sheet (both 1 mm thick). The welding quality was tested and analyzed on the basis of microscopic structure, mechanical performance of weld joints. Authors found that the time of welding 5A05 aluminium alloy part was melted but ST04Z sheet part was not melted and fractures occurred in heat-affected zone (HAZ) of 5A05 aluminum alloy. VedantAkgun et al. (1995) studied on effect of heat treatment on corrosion behaviour of laser surface melted 304L stainless steel at the time of welding. The improved properties due to laser surface melting were attributed to removal and/or redistribution of sulphur based inclusions as well as micro-structural changes that come about as a result of rapid cooling. Micro-structural and compositional changes were investigated by the electrochemical behaviors. It was observed that chemical composition rather than micro-structural changes that took place during laser surface melting had a greater effect on corrosion behavior. Kawaguch et al. (2006) developed a high power (20 kW) CO_2 laser welding technology for preventing defects in welds at a penetration depth of more than 20 mm. They used nitrogen as a shielding gas for preventing porosity. Authors studied the effects of various welding parameters on plasma generation, penetration depth, porosity and as well as the effect of nitrogen shielding gas on the depth of penetration. It was found that the penetration depth inversely proportional to the welding speed and directly proportional to the formation of porosity. It also found that the penetration depth sharply decreases under the effect of periodic nitrogen plasma generation on the basis of micro structure, beam characteristics and power fluctuation response. Lakshminarayanan and Balasubramaniam (2012) studied on welding of 409 M stainless steel by using 3.5 kW CO_2 laser and evaluated the micro-structural and mechanical properties of weld zone. Figure 3 shows the SEM fractographs of tensile and impact tested specimens. Base material was observed to have coarse

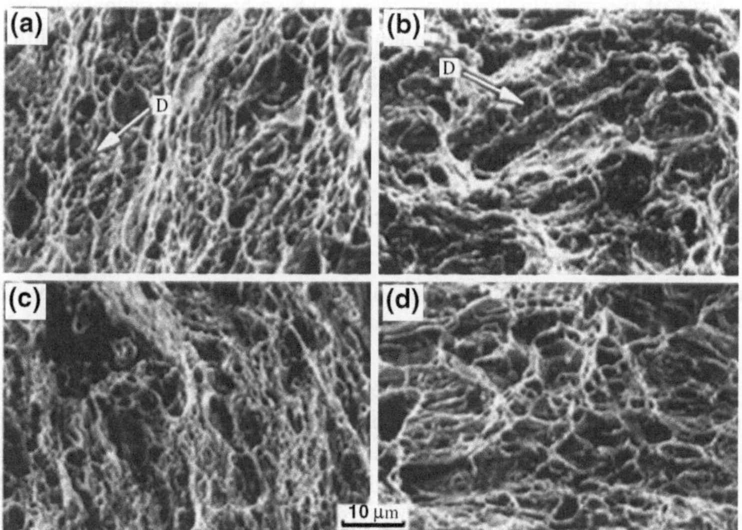

Fig. 3 SEM fractographs of tensile and impact tested specimens (Lakshminarayanan and Balasubramaniam 2012). **a** Base metal tensile; **b** base metal impact; **c** LBW tensile; **d** LBW impact

grain structure, whereas welded reason was seen to have fine dendritic structure and equiaxed axial grains of ferrite. Impact toughness of laser welded joint was found to be higher than base metal due to existence of equiaxed and dendritic grain perpendicular to the crack path.

El-Batahgy (1997) observed the effect of welding parameters on CO_2 laser (maximum 5 kW) welding of 304L stainless steel on fusion zone shape and solidification structure. It was found that the Laser power have more pronounced effect on penetration depth (directly proportional to the laser power) in comparison to its effect on weld profile and HAZ. With increase of welding speed, depth/width ratio was found to increase and the fusion zone therefore was found to be decreased. In terms of defocusing distance, maximum penetration and acceptable weld profile was obtained using defocusing distance of 0 to + and −1. Authors observed that the shielding gas is most effective due to its high ionization energy. It reduced plasma effect to a great extent and therefore helped laser power to reach work-piece better. Study of micro structure showed that high welding speed and low laser power resulted in finer structure, though effect of welding speed was more significant. High cooling rate supported directional nature of microstructure around the laser beam axis.

5 Modelling of Laser Welding Processes

The modelling can be possible by analytical methods as well as numerical methods. Several numerical methods are widely used for either modelling or optimizing the performance of the manufacturing processes, which has been advanced due to the large diffusion of the personal computer and the numerical algorithms. The knowledge of these methods and the ability in integrating their functions can make both the manufacturing engineer and the researcher ace their duties.

5.1 Analytical Modelling for Laser Welding

The mechanism of the laser plasma and characteristics of laser is providing a mathematical relation, which is used in weld quality estimation. Abderrazak et al. (2008) investigated the weldability of magnesium alloy (WE43) by using an industrial (CO_2) laser source. Authors provide an analytical thermal model which is the function of the penetration depth and the bead width, aiming at predicting the joint shape. Where, the weld bed width is the function of the incident laser power and welding speed. It was also considered the convection, conduction and radiation phenomena and a protecting gas from avoiding joint oxidation in the analytical thermal model of welding area. The thermal model based on the Davis thermal approach where largely considered for the characterization of the liquid zone average radius. Studying the effect of the laser beam power and welding speed.

The capillary size and the bead width in the work piece were among the priorities of this analysis. The consideration of convection and radiation phenomena is an important factor because this amount is about 20 % of laser power. It was also compared the result of thermal model with the experiment results and a satisfactory correlation was observed, indicating the reliability of the model. Park and Rhee (1999) observed that the effect of welding speed, laser power and flow rate of assist gas on light intensity and spatter formation in the CO_2 laser welding. The light emitted from the laser and spatter generation has been detected by photo-diodes. Authors have made a suitable relationship among the plasma, spatter and the bead shape along with the analysis of mechanism of plasma and spatter. It was also found that the volume of plasma in the keyhole and spatter increase with the increase of heat input and signals abruptly decrease at full penetration. Plasma plume signal decreases with the increase of heat input because of the effect of assist gas. The relationship between the bead shape and the plasma and spatter signal have certain patterns that can be used for estimation of bead shape.

Piekarska and Kubiak (2011) developed a analytical and finite element model for computational analysis process which can estimate the temperature field and velocity field of melted material for the laser–arc hybrid welding. The results from the numerical and analytical models were validated with the experimental results. Authors proposed that the developed analytical and numerical model can be used for analyzing welding parameter such as welding speed and distance between arc and laser beam.

5.2 Numerical Modelling for Laser Welding

Cho et al. (2010) have developed a numerical model by considering the physical assumptions for predicting the molten metal flow and allowing element distribution at the time of CO_2 laser-GMA hybrid welding. The three-dimensional transient simulations was conducted on the basis of combining arc welding and laser welding model without interaction between the two heat sources and modified it to introduce the optical geometry of the laser system and conservation equation. It was found that theoretical ε (strain) value of 0.08, rather than numerically compensated ε value of 0.2. The bead shape can be predicted very accurately by numerical model and flow of alloying element can be predicted along with the excess amount of alloying element flow at the rear part of the keyhole. Casalino and Ghorbel (2008) simulated thermo-morphological and mechanical behaviour by means of the finite element method (FEM) of moulded thermoplastic polymers during CO_2 laser welding. Keyhole and conduction welding for butt and overlap configurations were analyzed. It was collected data from the available literatures for validating the result from simulated FEM model. Under the variation of welding speed and power, recrystallized zones and penetration depth were calculated. Authors investigated that the thermoplastic polymers can be reasonable modelled by FEM. The mechanical, thermal, optical and solid-state characteristic was observed in this model.

Fortunato et al. (2012) studied nanosecond pulsed laser welding using two eutectoid steel components of 1 mm thickness of 1070 normalized sheet metal plates, which were used and about 1 mm long bead was maintained on plate weldments. It was performed with different welding process parameters. The Pulse duration (100 ns) and maximum energy per pulse (1 mJ) were kept constant with the variation of welding speed, frequency, laser power. It was observed that lower welding speed results in higher penetration depth. The pulse frequency and scanning velocity determined the number of pulses in the weld bead for a particular penetration depth. The specific energy per unit volume of heat source was required for achieving micro-welding with pulse nano-second, which was found to be in between 0.3 and 10 MJ/cm^3. Above specific energy of 10 MJ/cm^3, high value of specific energy per pulse resulted in pure ablation whereas low specific energy per pulse resulted in depression of the top of weld bead. For a given specific energy the higher penetration was obtained at the minimum specific energy per pulse (high pulse number), while for a given laser power high pulse frequency and high pulse duration lead to higher penetration depth. Also it has been seen that with respect to the amount of time, the material remained above the eutectoid temperature was less than that of required for the transformation. Finally no cracks were observed in the test specimens. Arif et al. (2011) studied on laser welding of mild steel tube under the nitrogen assisting gas ambient. For predicting the temperature and stress field they have used finite element method (FEM) along with incorporation of temperature-dependent thermal and mechanical properties with 3D volumetric moving heat source. Sharp increase in Von-mises stress was observed in the region of the laser heat source due to attainment of high temperature gradient. The residual stress developed in the welding region was measured using the XRD technique and results were compared with the predictions. Optical microscopy and SEM were used for metallurgical examination of the welding site. Increase of welding speed with constant laser power resulted in reduction of the width of high stress zone. Deng and Kiyoshima (2010) have studied the effect of existing residual stresses before welding on residual stresses produced by laser welding of SUS316 stainless steel pipe. Two cases were considered one without prior heat treatment and one with prior heat treatment to induce residual stresses. At the time of heat treatment the heat exchange was simulated using conduction, convection, and radiation laws. A volumetric heat source with uniform density distribution was used. The numerical simulation was done based on thermal elastic plastic FEM. The results showed that both the hoop and axial stress distribution were independent of the central angle around the pipe. For both inside and outside surfaces of pipe it was seen that within the critical length of plastic strain, only laser welding governed the hoop residual stresses and thus residual stress in both the cases are almost similar. But beyond plastic strain critical length, the difference between residual stresses of both the cases increases. Also beyond this point residual stresses induced by laser beam were observed up to a point after which the stresses became equal to that induced by heat treatment only. This point was referred to as critical length of hoop residual stress induced by laser beam welding and axial stress has same results.

GuoMing et al. (2007) studied the dynamic temperature field of laser welding on stainless steel. The thermo-mechanical properties of the material changed rapidly with the change in temperature. It was dynamically simulated by the FEA software —ANSYS using transient heat conduction equation. Surface thermal source with gauss distribution was adopted. Position of thermal source was based on time. The radiant heat exchange and convection contact with the difference of temperature on the surface of object were considered with the total exchange of heat coefficient in the calculation for convenience. The moving load was implemented by circulation of APDL language. With the help of calculation it was shown that the simulation results of weld shape were in concurrence with the experimental results. Duggan et al. (2011) developed a 2D numerical model for laser spot welding of an aluminium alloy at meso-scale by using FORTRAN 90/95. This model deals with temperature distribution during melting and solidification phenomenon at the time of welding process. An enthalpy method and Gaussian distribution was used for predicting melting and heat distribution phenomena. Authors were observed the formation of metallurgical structure after welding on the basis of heat distribution and it was incorporated in the numerical model. A mechanical blocking criterion was used to define dendrite coherency and the columnar-to-equiaxed transition within the weld pool was predicted. Smith et al. (2000) developed a model for determining residual stress after welding. Authors have used thermo-mechanical FE analysis, using the ABAQUS code for predicting residual stresses and compared with deep hole residual stresses measuring method in thick weld joints. Experimental tests were also conducted and were compared with the analysis results. Both the analytical results were observed to be in good agreement with each other which are slightly higher than experimental ones.

Belhadj et al. (2010) worked on finite element simulation of laser beam welding on magnesium alloys. A three dimensional finite element model was developed for simulating thermal history. Moving heat source was applied as load, which is dependent on process parameters such as power density, laser beam dimensions and welding speed. Experimental studies were conducted for validating numerical model simulation. The temperature measurement results and metallurgical investigations were in good agreement with thermal model. Wang et al. (2011a) numerically simulated the transient temperature field during laser welding of 304 stainless steels by using FLUENT software. Figure 4 shows the temperature profile obtained during numerical simulation. A volumetric heat source with Gaussian distribution was assumed. Phenomena such as phase transition, heat and mass transfer, fluid flow etc. were modeled analytically. According to the simulated results, recoil pressure played a vital role in the keyhole formation and the large temperature gradients were observed in front vicinity of this keyhole. Good agreement was made between the numerical simulation and experimental data.

Mert (2009) worked with analysis software ANSYS for the study of the effect of weld toe radius and root gap in T-fillet welded joint on fatigue life. In all cases, maximum von-Misses equivalent stresses were observed at weld toe. It was observed that small toe radius; small and big root gap promoted stress concentration and therefore decreases in fatigue life. Big toe radius and medium root gap was seen

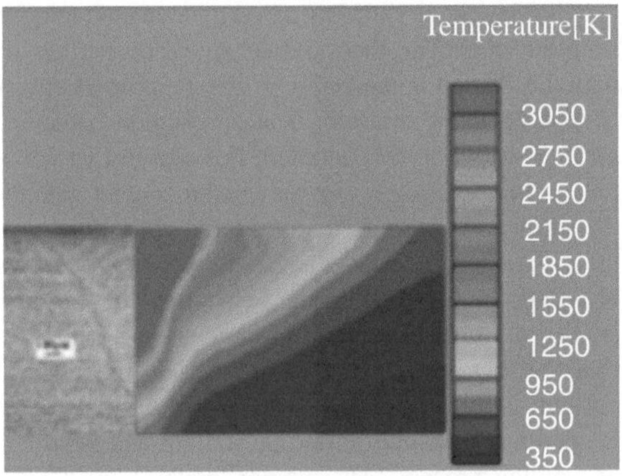

Fig. 4 Comparison of weld shape by experiment with that by calculation (Wang et al. 2011a)

to give best results in terms of fatigue life. These results were verified experimentally well.

6 Studies on the Laser Welding Process Parameters Optimization

Welding process parameter optimization is widely depending on its application. Ruggiero et al. (2011) investigated the butt welding on the dissimilar metal of low carbon steel and austenitic steel AISI316 with full depth of penetration by CO_2 laser. It was also investigated the effect of laser power (1.1–1.43 kW), welding speed (25–75 cm/min) and focal point position (−0.8 to −0.2 mm) on the shape of weld bead geometry using response surface methodology (RSM). Authors proposed that the optimal welding combination such as laser power of 1.1 kW, welding speed 72.6 and 75 cm/min and focused position being around −0.44 mm that would minimize the middle width, the weld-bead area and the cost. Welding speed is the parameter that most significantly influences the weld bead dimensions. Hamatani et al. (2005) studied the CO_2 laser welding of mild steel SS45 with a thickness of 0.8, 2.0, or 3.2 mm with TIG used for applying the voltage potential to increase the bead root width. It was also investigated enhancement of the welding speed and the butt joint gap tolerance and the influences of the experimental conditions such as: supplied voltage between plates and backside electrode, welding speed, plasma operates gaseous species, and the butt joint gap on the bead root. Authors found that the welding root bead width enhanced at optimal welding condition and applied suitable current is about 30 A. Olabi et al. (2006) have used neural network and the Taguchi approach for the optimization of the welding speed, the laser power and the focal position for CO_2 laser keyhole welding of medium carbon steel butt joining. The penetration-to-fuse-zone-width (p/w) and the penetration-to-HAZ-width (p/w) ratios were the process metrics used for the optimization. The training of the neural networks, welding parameters is being used from literature. They have found that the optimized solution is valid up to used parameters. Analysis of data from this approach is reasonably correct. They did not found any drawbacks to the use of this approach.

Olabi et al. (2013) optimized welding parameters such as welding speed, laser power and focal length of 1.5 kW CO_2 laser joining process, when dissimilar materials [316 stainless steel and low carbon steel (160 × 80 × 3 mm)] are involved. Charpy impact test and tensile strength tests were performed. Increase in parameters i.e. welding speed, laser power and focal length, all resulted in decrease of tensile strength. It has been observed that welding speed being the most significant parameter and focal length being the least. While in case of impact strength, laser power and welding speed were found to have opposite effect i.e. with increase in laser power impact strength increased proportionally, whereas with increase in welding speed it was found to decrease. Mathematical model involving important parameters was determined by design expert software. A laser power of 1.1 kW and welding speed of 72.66 cm/min was found optimum under given conditions.

7 Study the Effect of External Process Parameters on Laser Welding

Apart from the above important studies on laser welding, the other important parameters which affect the weld quality in case of laser welding were also studied. They are:

- Study the effect of assistance gases.
- Study the effect of electric, magnetic field and electrical potentia.

7.1 Study the Effect of Assistance Gases

Several researchers have been observed the effect of shielding gas that is protect the welding zone from oxidation and other un-desirable effects. Wang et al. (2011b) studied the role of side assisting gas on plasma and energy transmission during CO_2 laser welding of galvanized steel sheet. The plasma volume was found to decrease sharply with initial increase of gas flow rate, after which stable plasma was observed. This resulted in inference that the plasma consists of two parts one was diffusive plasma, which could be easily blown away and another was stable residual plasma which could not be blown away by strong gas flow rate. This assumption was supported by the photographs taken by high speed camera. Sathiya and AbdulJaleel (2010) studied on micro-structural characterization of CO_2 laser welded on AISI904L super austenitic stainless steel. The laser welding employing two shielding gases namely Argon and Helium was carried out. The Helium shielded weld metal aspect ratio was slightly higher than Argon shielded weld metal aspect ratio. Also hardness value for He shielded weld metal was observed to be higher. The ratio of Cr/Ni was less than 1.35 for both the cases and hence all the weld metal solidifies in austenitic mode. The Helium shielded weld metal comprised of finer and more equiaxed grains. All this was attributed to the higher ionization potential of He gas.

7.2 Study the Effect of Electric, Magnetic Field and Electrical Potential

Tse et al. (1999a, b) have investigated the effect of magnetic field on the shielding gas helium, at the time of CO_2 laser welding on AISI 304 stainless steel. It was investigated the effect of magnetic field strength, laser power, welding speed, field direction and shielding gas helium and argon on depth of penetration and weld bead width. Authors proposed that the effect magnetic field on helium shielding gas is increased the penetration depth about 7 % but no significant effect on weld bed width. It was also observed that the penetration depth was significantly increases at low magnetic field strength with argon shielding gas. Tse et al. (1999a, b) investigated on feasibility of using electric and magnetic fields as a plasma control during CO_2 laser welding of AISI 304 stainless steel sheet. It was found that the penetration depth was increases for both field and it can be increased by more than 13 %, week field strength penetration depth can be improve with decreasing the bead width. Authors make the interrelation between the field direction, penetration depth and the width of bead. It was found that the optimum field strength was directly proportion to the laser power. Authors also observed that the penetration depth was decreases with increases field strength above the optimum value.

Tse et al. (2000) observed that the effect of magnetic field on the CO_2 laser welding of AISI 304 stainless-steel (8 mm thick) plate. They have also observed that the effect of the magnetic field strength and direction with Helium shielding gas on the weld bead profile and laser power. They also observed that the penetration depth for shielding gas Helium, Argon and Helium with magnetic field. They wear found that the penetration depth 16 and 8 % increases for Argon and Helium with magnetic field compared with only use of Helium. The penetration depth maximum increases at optimum field strength but it was decrease at lower and higher then that the optimal field strength. They also observed that the negative electric field was increases the penetration depth but the improvement was smaller with compared to positive electric field.

8 Image Processing and Feature Extraction

Characterization of welding quality image processing and feature extraction is playing a vital role in welding process. Li et al. (2009) have studied on the effect of surface oil Metaflux 70-03 and blowing gas Helium flow rate on CO_2 laser welding of E-grade ship-building steel. Authors have developed a method to acquire the stability information from plasma images taken by high-speed photography. It was found that the presence of surface oil and less or more than the optimum gas flow rate reduces the stability of the welding process. Cheng et al. (2012) observed that the key-hole characteristic by a high-speed camera during deep penetrated CO_2 laser welding on aluminum alloy 6016. The spectrograph, plasma spectra was used for calculating the electron temperature. Authors also compute electron density by using Lorenz nonlinear fitting, the FWHM of Stark broadening lines. It was found that the key hole plasma electron temperature gradually decreases towards edge and bottom of the key hole and electron density increases at the keyhole depth but approximately same along radius.

Seto et al. (2002) observed on the porosity formation in CO_2 laser welding of SUS304 stainless steel. The micro-focusing X-ray transmission imaging device was used to perform real time observations. The results showed that the porosity formation readily in partially penetrated welds but did not form in fully penetrated welds. The larger pores were seen with high laser power and less welding speed. The Nitrogen was found successful in suppression of porosity by increasing fluidity of molten metal. The metal flow observation suggested that bubbles not only migrate under their own buoyancy, but are also transported along hot metal flow.

9 Summary

Although CO_2 laser welding have been successfully used to join various materials, it is still at an early stage. So far, the development of the CO_2 laser welding process for each new application has remained largely empirical. The numerical and experimental studies are significant help in understanding the CO_2 laser welding process.

In this book chapter, the research activities and progress to date in the development of CO_2 laser welding are reviewed. In the present review, emphasis has been given, especially on the laser welding numerical and analytical modelling, soft computing approaches and welding characterization of the present technique.

There are several gaps remaining in the experiment and numerical analysis of CO_2 laser welding. Accurate and reliable experimental and numerical analysis of the CO_2 laser welding is still having some difficulties. The main objective of the book chapter is to review recent progress in CO_2 laser welding and to provide a foundation for further research.

References

Abderrazaka, K., Salem, W. B., Mhiri, H., Lepalec, G., & Autric, M. (2008). Modelling of CO_2 laser welding of magnesium alloys. *Optics and Laser Technology, 40*, 581–588.

Arif, A. F. M., Al-Omari, A. S., Yilbas, B. S., & Al-Nassar, Y. N. (2011). Thermal stress analysis of spiral laser welded tube. *Journal of Materials Processing Technology, 211*(4), 675–687.

Belhadj, A., Bessrour, J., Masse, J.-E., Bouhafs, M., & Barallier, L. (2010). Finite element simulation of Mg alloys laser beam welding. *Journal of Material Processing Technology, 210*, 1131–1137.

Boutarek, N., Azzougui, B., Saidi, D., & Neggache, M. (2009). Microstructure change in the interface of CO_2 laser welded zirconium alloys. *Physics Procedia, 2*, 1159–1165.

Casalino, G. (2007). Statistical analysis of MIG-laser CO_2 hybrid welding of Al–Mg alloy. *Materials Processing Technology, 191*, 106–110.

Casalino, G., & Ghorbel, E. (2008). Numerical model of CO_2 laser welding of thermoplastic polymers. *Journal of Materials Processing Technology, 207*, 63–71.

Chen, W., Ackerson, P., & Molian, P. (2009). CO_2 laser welding of galvanized steel sheets using vent holes. *Materials and Design, 3*, 245–251.

Cheng, Y., Jin, X., Li, S., & Zeng, L. (2012). Fresnel absorption and inverse bremsstrahlung absorption in an actual 3D keyhole during deep penetration CO_2 laser welding of aluminum 6016. *Optics and Laser Technology, 44*, 1426–1436.

Cho, W.-I., Na, S.-J., Cho, M.-H., & Lee, J.-S. (2010). Numerical study of alloying element distribution in CO_2 laser—GMA hybrid welding. *Computational Materials Science, 49*, 792–800.

Deng, D., & Kiyoshima, S. (2010). Numerical simulation of residual stresses induced by laser beam welding in SUS316 stainless steel pipe with considering initial residual stress influences. *Nuclear Energy and Design, 240*(4), 688–696.

Duggan, G., Tong, M., & Browne, D. J. (2011). An integrated meso-scale numerical model of melting and solidification in laser welding. *Materials Science and Engineering, 27*, 012–077.

El-Batahgy, A.-M. (1997). Effect of laser welding parameters on fusion welding zone and shape and solidification structure of austenitic stainless steels. *Materials Letters, 32*, 155–163.

Fortunato, A., Ascari, A., Orazi, L., Cuccolini, G., Campana, G., & Tani, G. (2012). Numerical simulation of nano-second pulsed laser welding of eutectoid steel components. *Optics and Laser Technology, 44*, 1999–2003.

Gao, M., Zeng, X., & Hu, Q. (2007). Effects of gas shielding parameters on weld penetration of CO_2 laser-TIG hybrid welding. *Journal of Materials Processing Technology, 184*, 177–183.

Geoffray, I., Bourdenet, R., & Theobald, M. (2011). Laser micro-welding applied to target manufacturing. *Physics Procedia, 12*, 363–369.

GuoMing, H., Jian, Z., & JianQuang, L. (2007). Dynamic simulation of the temperature field of stainless steel laser welding. *Materials and Design, 28*, 240–245.

Hamatani, H., Miyazaki, Y., & Ohara, M. (2005). Effect of applying electrical potential to a CO_2 laser welding of different thickness plates. *Science and Technology of Advanced Materials, 6*, 712–719.

Hess, A., Schuster, R., Heider, A., Weber, R., & Graf, T. (2011). Continuous wave laser welding of copper with combined beams at wavelengths of 1030 nm and of 515 nm. *Physics Procedia, 12*, 88–94.

Honga, J. K., Parkb, J. H., Parka, N. K., Eomc, I. S., Kimc, M. B., & Kangc, C. Y. (2008). Microstructures and mechanical properties of Inconel 718 welds by CO_2 laser welding. *Materials Processing Technology, 201*, 515–520.

Kawaguchi, I., Tsukamoto, S., Arakane, G., & Honda, H. (2006). Characteristics of high-power CO_2 laser welding and porosity suppression mechanism with nitrogen shielding. Study of high-power laser welding phenomena. *Welding International, 20*(2), 100–105.

Kim, C. H., Ahn, Y. N., & Kim, J. H. (2011). CO_2 laser-micro plasma arc hybrid welding for galvanized steel sheets. *Transactions of the Nonferrous Metals Society of China, 21*, 47–53.

Kurosaki, Y., Satoh, K., Koyanagi, H., & Miyahara, H. (2009). CO_2 laser welding of the PFA disk top on a circular tube assisted by a transparent solid heat sink. *67th Annual Technical Conference of the Society of Plastics Engineers, 3*, 1283–1288.

Lakshminarayanan, A. K., & Balasubramaniam, V. (2012). Evaluation of micro-structure and mechanical properties of laser beam welded 409M grade ferritic stainless steel. *Journal of Iron and Steel Research, 19*(1), 72–78.

Li, G., Cai, Y., & Wu, Y. (2009). Stability information in plasma image of high-power CO_2 laser welding. *Optics and Lasers in Engineering, 47*, 990–994.

Liu, J., Ma, L., Xie, Y., & Zhang, Y. (2006). Primary research on image of plasma in CO_2 laser welding with high-speed photography, *SPIE, 61*, 5016-1-5.

Liu, S., Zhang, H., Hu, J., & Shi, Y. (2013). Microstructure of laser-MAG hybrid welds of sintered P/M steel. *JMEPEG, 22*, 251–257.

Mert, T. (2009). Finite element analysis of effect of weld toe radius and root gap on fatigue life of T-fillet welded joint. *Conference of the International Journal of Arts and Sciences, 1*, 119–127.

Olabi, A. G., Alsinani, F. O., Alabdulkarim, A., Ruggiero, A., Tricarico, L., & Benyounis, K. Y. (2013). Optimizing the CO_2 laser welding process for dissimilar materials. *Optics and Lasers in Engineering, 51*, 832–839.

Olabi, A. G., Casalino, G., Benyounis, K. Y., & Hashmi, M. S. J. (2006). An ANN and Taguchi algorithms integrated approach to the optimization of CO_2 laser welding. *Advances in Engineering Software, 37*, 643–648.

Park, H., & Rhee, S. (1999). Analysis of mechanism of plasma and spatter in CO_2 laser welding of galvanized steel. *Optics and Laser Technology, 31*, 119–126.

Perez-Medina, G. Y., López, H. F., Zambrano, P., & Reyes-Valdés, F. A. (2013). Microstructural effects on the mechanical integrity of a TRIP-800 steel welded by laser-CO_2 process. *ASM International JMEPEG, 22*, 607–612.

Piekarska, W., & Kubiak, M. (2011). Three-dimensional model for numerical analysis of thermal phenomena in laser–arc hybrid welding process. *International Journal of Heat and Mass Transfer, 54*, 4966–4974.

Ruggiero, A., Tricarico, L., Olabi, A. G., & Benyounis, K. Y. (2011). Weld-bead profile and costs optimization of the CO_2 dissimilar laser welding process of low carbon steel and austenitic steel AISI316. *Optics and Laser Technology, 43*, 82–90.

Sakagawa, T., Nakashiba, S-i, & Hiejima, H. (2011). Laser micro welding system and its application to seam welding of rechargeable battery. *Physics Procedia, 12*, 6–10.

Sathiya, P., & AbdulJaleel, M. Y. (2010). Measurement of the bead profile and micro-structural characterization of a CO_2 laser welded AISI 904L super-austenitic stainless steel. *Optics and Laser Technology, 42*(6), 960–968.

Seto, N., Katayama, S., & Matsunawa, A. (2002). Porosity formation mechanism and reduction method in CO_2 laser welding of stainless steel. *Welding International, 16*(6), 451–460.

Shanmugarajann, B., & Padmanabham, G. (2012). Fusion welding studies using laser on Ti–SS dissimilar combination. *Optics and Lasers in Engineering, 50*, 1621–1627.

Sibillano, T., Ancona, A., Berardi, V., Schingaro, E., Basilea, G., & Lugara, P. M. (2006). A study of the shielding gas influence on the laser beam welding of AA5083 aluminium alloys by in-process spectroscopic investigation. *Optics and Lasers in Engineering, 44*, 1039–1051.

Smith, D. J., Bouchard, P. J., & George, D. (2000). Measurement and prediction of residual stresses in thick-sectioned welds. *Journal of Strain Analysis, 35*, 287–305.

Tse, H. C., Man, H. C., & Yue, T. M. (1999a). Effect of magnetic field on plasma control during CO_2 laser welding. *Optics and Laser Technology, 31*, 363–368.

Tse, H. C., Man, H. C., & Yue, T. M. (1999b). Effect of electric and magnetic fields on plasma control during CO_2 laser welding. *Optics and Lasers in Engineering, 32*, 55–63.

Tse, H. C., Man, H. C., & Yue, T. M. (2000). Effect of electric field on plasma control during CO_2 laser welding. *Optics and Lasers in Engineering, 33*, 181–189.

Tu, J. F., Inoue, T., & Miyamoto, I. (2003). Quantitative characterization of keyhole absorption mechanisms in 20 kW-class CO_2 laser welding processes. *Journal of Physics. D: Applied Physics, 36*, 192–203.

VedantAkgun, O., Urgen, Mustafa, & FuatCakir, Ali. (1995). The effect of heat treatment on corrosion behavior of laser surface melted 304L stainless steel. *Materials Science and Engineering, A203*, 324–331.

Wang, R., Lei, Y., & Shi, Y. (2011a). Numerical simulation of transient temperature field during laser keyhole welding of 304 stainless steel sheet. *Optics and Laser Technology, 43*, 870–873.

Wang, C.-M., Meng, X.-X., Huang, W., Hu, X.-Y., & Duan, A.-Q. (2011b). Role of side assisting gas on plasma and energy transmission during CO_2 laser welding. *Journal of Materials Processing Technology, 211*, 668–674.

Yu, S., Fan, D., & Dong, B. (2011). Laser beam welding of dissimilar metals of aluminum alloy and galvanized steel sheet. In *Second International Conference on Mechanic Automation and Control Engineering* (pp. 5030–5032).

Zhu, J., Li, L., & Liu, Z. (2005). CO_2 and diode laser welding of AZ31 magnesium alloy. *Applied Surface Science, 247*, 300–306.

Fiber Laser Welding in a Controlled Inert Gas Atmosphere: An Experimental and Numerical Investigation

Yadaiah Nirsanametla, Swarup Bag, C.P. Paul and L.M. Kukreja

Abstract In fusion welding, thermo-chemical reactions may take place among surrounding atmosphere particles and molten weld pool at high temperature gradients. The atmosphere particles such as oxygen, hydrogen and nitrogen may become part of final weld joint that severely affects the weld joint quality and weld metal properties. Therefore, the welding atmosphere and protection of weld pool plays a noticeable role on the quality of the final weld joint. Henceforth, in this chapter, fiber laser welding of austenitic stainless steel plates have been examined in two different ambient atmospheres. Firstly, the experiments are conducted in open atmosphere and in argon ambient atmosphere to study the characteristic difference between them. The experimental investigation specifies that the weld bead dimensions and aspect ratio are higher in case of argon atmosphere as compared to open atmosphere. The microstructures of heat affected zone (HAZ) and fusion zone (FZ) at both atmospheric conditions are analyzed. It is obvious from the experimental results that the top surface profile is smoother and very clear in case of welds at argon atmosphere. Moreover, in this work, the authors also reported an efficient conduction mode finite element based heat transfer model of linear fiber laser welding process using a volumetric heat source. The calculated weld bead dimensions using finite element model are compared with the experimentally measured results at similar process variables. Relatively fair agreement of the experimental results with model results entitles the robustness of the modeling approach followed here and reported in this work.

Keywords Fiber laser · Fusion welding · Inert gas atmosphere · Open atmosphere · Metallographic analysis · Volumetric heat source · Heat transfer · Finite element method

---

Y. Nirsanametla (✉) · S. Bag
Department of Mechanical Engineering, Indian Institute of Technology Guwahati, Guwahati, Assam 781039, India
e-mail: yadaiah@iitg.ernet.in

C.P. Paul · L.M. Kukreja
Laser Material Processing Division, Raja Ramanna Centre for Advanced Technology, Indore, Madhya Pradesh 452013, India

S.N. Joshi and U.S. Dixit (eds.), *Lasers Based Manufacturing*,
Topics in Mining, Metallurgy and Materials Engineering,
DOI 10.1007/978-81-322-2352-8_20

1 Introduction

In recent past, high energy beam welding processes, like laser welding have been progressively employing in industrial manufacturing with respect to traditional welding processes due to its advantages such as lower dimension and shape distortion of pieces, greater processing speed, high energy density, deep penetration, focalization, and high efficiency. Currently, gas, solid-state and semi-conductor lasers are used for joining of different materials. Recently, fiber lasers have been receiving greater attention and emerging because of its advantages like high beam quality which can produce an ultra-high peak power density of MW/mm^2 levels corresponding to a focused electron beam, and is favorable to become one of the desirable heat sources for deep-penetration and high-speed welding. Moreover, the advantage of fiber laser is that the beam transmitted from the fiber is usually of very high quality and can be focused to a small spot size. Furthermore, in a fiber laser the active gain media is an optical fiber doped with rare earth elements, such as ytterbium, erbium, dysprosium, praseodymium, thulium and neodymium. They are related to doped fiber amplifiers, which deliver light amplification without lasing. A fiber laser creates the beam inside the fiber; therefore, the delivery of the beam does not necessitate any sensitive or complicated optics. This makes it very suitable and easy to use. In fiber laser, the beam is generated and confined inside the small core of the fiber. During high power density fusion welding processes there is the possibility of thermo-chemical reaction among welding atmosphere particles and molten weld pool. These inclusions become part of final weld joint and initiate the weld defects such as pinholes. Therefore, most of the fusion welding processes involve protection of weld molten pool from the atmosphere by several means such as shielding gas or flux to avoid the reaction of molten pool with welding atmosphere.

Various molten pool protection techniques offer different degree of protection from atmosphere. Several shielding methods such as slag, gas, gas and slag, self-protection and vacuum offer different degree of weld molten pool protection from the surrounding atmosphere during various welding processes (Kim et al. 1998; Kou 2002). Sahoo et al. (1988) have studied the influence of the surface active elements such as sulphur and oxygen in blocking vaporization sites on weld pool surface. The presence of oxygen and sulphur in weld metal leads to increase in metal vaporization rate. Ramirez et al. (1994) have investigated the influence of welding process parameters on solidification substructure to cause weld metal porosity. In their investigations, the amount of porosity increases with increase of nitrogen gas introduced in argon shielding gas atmosphere. Dong et al. (2003) and Dong et al. (2005) have studied the nitrogen absorption and desorption during CO_2 laser welding of stainless steel in controlled atmosphere of $Ar\text{-}N_2$ mixed gas. Ramazan and Koray (2005) have studied the influence of the welding atmosphere during gas metal arc welding (GMAW) of low carbon steel. In their study, the authors realized that the toughness of the weld metal is higher and mechanical properties of the weldment are better due to less porosity in controlled atmosphere of argon when compared with the classical GMAW. Bayram et al. (2008) have studied the influence of the welding atmosphere and heat input during resistance spot welding of 316L austenitic stainless steel. In their study, the authors observed that the tensile shear load bearing capacity increases when heat input increases due to the enlargement of the nugget size. This occurrence is slightly higher in nitrogen atmospheric conditions as compared to welding in open atmosphere conditions. Dursun (2008) have investigated the influence of welding atmosphere and welding current during resistance spot welding of 304L austenitic stainless steel. In their study the author determined that in a nitrogen atmosphere welding, an optimum weld quality was achieved by using 9 kA peak weld current. Kang et al. (2009) have studied the GMA welding process with various shielding gases and its combinations such as helium, argon, mixture of argon and helium, and supplying pure argon and pure helium alternatively. In their study, the authors observed that the lowest degree of weld induced distortion is obtained if an alternative supply of shielding gas is used for molten weld pool protection when compared to individual supply and mixture of them with similar welding conditions. Ostsemin (2009) have studied the temperature measurement of the electrode-metal drop during carbon dioxide controlled atmosphere of arc welding. A critical review of literature survey suggests that the way of protection of molten weld pool and welding atmosphere plays a significant role in the weld metal properties and weld joint quality. Furthermore, the degree of molten weld pool protection can be enriched by avoiding poor methods of shielding media. Argon gas ambient atmosphere or self-protective atmosphere of argon welding process avoids chemical reaction among open atmosphere particles and molten weld pool. This type of self-created inert gas atmosphere leads to less degree of weld defects.

Laser welding process is an expensive process where the trial-and-error method of investigation is not appropriate. Therefore, prior to welding manufactured goods a comprehensive understanding of the operating process conditions and variables are necessary, which can only be realized by using modeling and simulation. The modeling and simulation provides an enhanced understanding of the manufacturing process and the interaction between the process variables. The utmost widespread analytical methods for calculating the thermal cycles in fusion welded plates was started in late 1940s by Rosenthal (1946, 1947) and Rykalin (1974) based on fundamental theory of heat flow for linear welding process. The heat source models, like point and line used by Rosenthal and Rykalin are subject to serious errors for predicting temperatures in and around the fusion and heat affected zones. In reality, the fusion welding heat sources either arc or beam should not be either a point or a line in nature. Moreover, the heat sources should be spread over some area. Pavelic et al. (1969) introduced a Gaussian distributed 'disc' heat source model on the surface of the work piece. Even though Pavelic's 'disc' model is undoubtedly an important step forward in modeling of fusion welding heat source models, the digging action in plate thickness direction was not reflected in this model. Goldak et al. (1984) proposed a three dimensional non-axisymmetric heat source model of 'double ellipsoid' configuration with Gaussian distribution. This type of heat source model is most appropriate for linear welding process modeling and which include the digging action of the welding heat source. Wu et al. (2006) developed a new heat source model based on conical heat source model for deep penetration plasma arc welding. However, the conical shape is more appropriate for high aspect ratio and deep penetration welds.

Several investigators studied the numerical modeling of fusion welding process for measuring the weld thermal cycles and macrographs using an appropriate heat source models (De et al. 2003; Frewin and Scott 1999). Bag et al. (2009) investigated the conduction mode laser spot welding process using an adaptive volumetric heat source model with finite element method which is one of the forward step in simulation of conduction based heat transfer model that does not need the weld pool dimensions *a priori*. Prior definition of heat source model parameters is one of the major hindrances in widely used double ellipsoidal heat source. The model parameters of Gaussian distributed double ellipsoidal heat source model can be realized from the experimentally measured weld bead dimensions. Moreover, if the experimental weld bead dimensions are not available in advance; the model parameters can also be calculated using the non-dimensional model presented by Christensen et al. (1965). However, the rear and front lengths of double ellipsoidal heat source model changes with welding speed which is independent of measured weld dimensions. Yadaiah and Bag (2012) studied the influence of the rear and front length ratio with respect to welding speed using a kernel of genetic algorithm based optimization module and recommended a cubic functional form of this ratio. To reduce the model parameters such as front and rear length selection in double-ellipsoidal heat source model, Yadaiah and Bag (2014) developed a new heat source model for simulation of linear fusion welding process. The flow of molten material in weld pool plays a significant role in the assessment of heat transport within weld pool and subsequently the weld joint quality. However, in the present work the thermal conductivity of molten material is increased artificially for several folds to account the enhanced heat transfer due to high convective flow of liquid molten metal within the weld pool. Few researchers also investigated the heat transfer and fluid flow in fusion welding process using finite element method along with optimization of uncertain model parameters (Bag and De 2010; De and DebRoy 2005, 2006; Yadaiah and Bag 2013). Transport phenomena based heat transfer and fluid flow models in fusion welding process are always more realistic than conduction heat transfer analysis. However, conduction heat transfer based heat transfer analysis is often preferred due to their benefits in terms of a reduced amount of computational time, ease of modeling and its simplicity, and ease of coupling with stress analysis model.

In this chapter, the authors have investigated and reported the influence of ambient welding atmosphere during fiber laser welding of austenitic stainless steel (SS 304 and SS 316) plates of 5 and 3 mm thickness. The fiber laser welding experiments are conducted in argon gas atmosphere as well in open atmospheric conditions to study the characteristic difference between them in terms of weld bead dimensions, weld bead profile, microstructures and top surface appearance.

The depth of penetration is more in argon inert gas atmosphere and weld bead profile on top surface is smoother than that in open atmospheric conditions. Moreover, finite element based conduction heat transfer model is developed for simulation of fiber laser welding in open atmospheric conditions using volumetric heat source model. Furthermore, the molten material flow has been approximated by an artificially increased thermal conductivity within weld pool cavity accounting enhanced heat transport for temperatures above the melting point.

This chapter is organized into five different sections to elaborate and refer to all the aspects of fiber laser welding process. In the '*introduction*' section, a general background and critical literature review of existing efforts on the influence of different welding atmospheres in fusion welding process are reported. Moreover, a brief literature in the direction of heat transfer and fluid flow analysis in fusion welding process are incorporated along with overview of the present work. In section '*materials and methods*', a detailed experimental procedure of fiber laser welding in a controlled atmosphere of argon and in open atmospheric conditions under similar welding conditions are presented. Furthermore, the authors also reported in this section the materials used, experimental requirements for controlled atmosphere and process variables. In the '*theoretical background*' section, a detailed mathematical overview for developing conduction heat transfer model using finite element method is presented. In section '*results and discussion*', the experimental observations in both open and controlled atmosphere are presented. The numerically computed results and validation of the same with experimentally measured data are testified. Subsequently, concluding remarks of the present work are presented at the end of the chapter followed by selective references used to construct the foundation of the chapter.

2 Materials and Methods

A 2.0 kW continuous wave fiber laser based welding system was used to perform bead-on-plate welding on commercially available type 304 and 316 stainless steel plates of 5 and 3 mm thickness respectively. The experimental setup used for controlled atmosphere welding is shown schematically in Fig. 1. The welding experimental setup consists of four major subsystems. They are computerized numerically controlled (CNC) workstation in a Glove box, 2 kW ytterbium continuous wave fiber laser source, vacuum pump and CNC controller. The experiments were conducted with similar welding process variables both in controlled and open atmospheric conditions to determine the characteristic difference between them. Table 1 depicts the welding process variables for both controlled atmosphere of argon and open atmospheric conditions. A mixed mode fiber laser beam power which is the mixture of two fundamental modes, 60 % of TEM01, and 40 % of TEM00, which gives a near flat top beam profile (Kumar et al. 2008, 2012), is considered (refer Fig. 2). The combined intensity profile for a mixed mode laser beam is given by following equation:

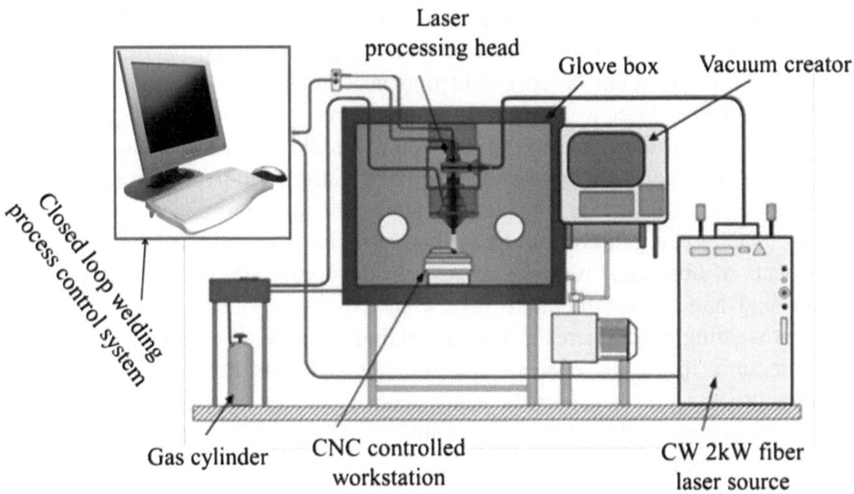

Fig. 1 Schematic representation of controlled atmosphere welding experimental setup

Table 1 Fiber laser welding process variables corresponding to **a** SS 304, 5 mm plate and **b** SS 316, 3 mm plate (Yadaiah et al. 2014)

(a)	Data set	Laser power (W)	Velocity (mm s^{-1})	Heat input per unit length (J/mm)
	1	2000	18.33	109.11
	2	2000	16.67	119.98
	3	2000	15.00	133.33
	4	2000	13.33	150.04
(b)	Data set	Laser power (W)	Velocity (mm s^{-1})	Heat input per unit length (J/mm)
	1	1000	18.33	54.55
	2	1000	13.33	75.00
	3	900	18.33	49.09
	4	900	13.33	67.50
	5	800	18.33	43.64
	6	800	13.33	60.00

$$I_{fl}(r) = \frac{2P_{fl}}{\pi r_{ofl}^2}[a_y + (1 - a_y)\frac{2r^2}{r_{ofl}^2}]\exp[-\frac{2r^2}{r_{ofl}^2}] \quad (1)$$

where P_{fl}, r_{ofl}, a_y and r depict the fiber laser power on the substrate (W), radius of the Gaussian (TEM00) beam component (m), fraction of the lower mode (TEM00) in mixed mode laser beam and the distance from the center of the laser beam (m) respectively.

The workstation is CNC controlled and positioned stationary during welding process where as the laser head is mounted in Z-direction vertically and movable.

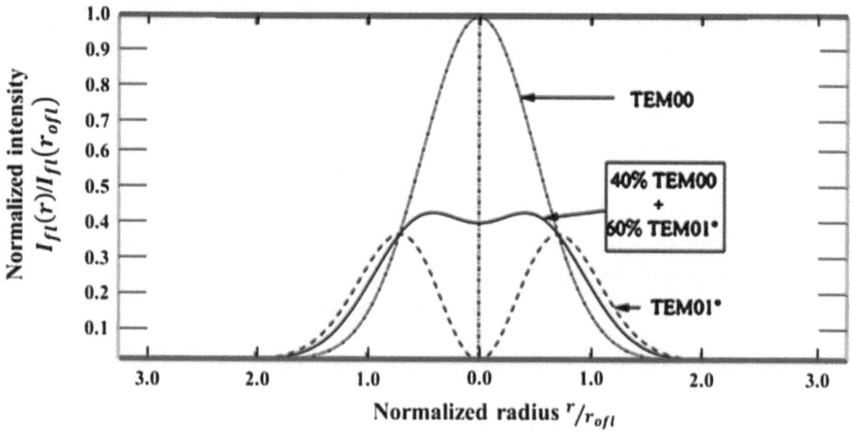

Fig. 2 Schematic representation of intensity profiles of two different fundamental modes and their mixture

Prior to the welding operation the work pieces were cleaned to remove oil, grease, and water vapour using an organic solvent such as acetone. The work piece is kept normal to heat source. The laser welding workstation is also equipped with laser power on/off mode and gas on/off mode to preserve the necessities. The fiber laser welding experimental setup characteristics during performance of the welding experiments are specified in Table 2.

During welding in controlled atmosphere of argon, the five axes CNC workstation was in a glove box. The glove box is essential for controlling atmospheric conditions during processing. To protect the weldment from the atmosphere, the welding experiments were performed inside a high purity argon gas filled glove box. Air and moisture particle are the main impurities in atmosphere which affects the properties of the weld joint. Henceforth, a controlled atmosphere of argon fiber laser welding system was also integrated with oxygen and moisture analyzers. The desired purity levels are accomplished by purging high purity grade argon gas. The purity level of the glove box is retained by keeping the differential pressure just above the atmospheric pressure.

Table 2 Experimental setup conditions

Laser spot diameter	200 μm
Mode of beam	Multi-mode and flat on top surface
Laser power mode	Continuous wave
Wavelength	1080 nm
Beam angle	90°
Focal point	1 mm below to the top surface
Fiber core diameter	50 μm
Shielding gas	Argon

After carrying out the welding experiments in both atmospheres, the metallographic analysis was done to measure the weld pool shape and dimensions, and microstructure for welded samples of both atmospheres. In the course of metallographic analysis, the welded samples were sectioned perpendicular to welding direction, polished with different grades of polishing papers (220, 400, 600, 800, 1000 and diamond polish) and etched with Villella's reagent. The etched samples were analyzed on optical microscope to take macro and microstructures. The weld bead shape and dimensions are calculated for each welded sample. Top bead profile, appearance of the weld joint and microstructures for various welding speeds under both the atmospheric conditions are also examined.

3 Theoretical Background

The conservation of energy is the fundamental principle in heat transfer analysis of fusion welding process. The general heat conduction equation in Cartesian coordinate system for linear laser welding process, if the heat source moves along Y-axis is

$$k_m\left(\frac{\partial^2 T}{\partial x^2}+\frac{\partial^2 T}{\partial y^2}+\frac{\partial^2 T}{\partial z^2}\right)+\dot{Q}=\rho C_p\left(\frac{\partial T}{\partial t}+v\frac{\partial T}{\partial y}\right) \tag{2}$$

where (x, y, z) is coordinate system attached to the heat source, k_m is the thermal conductivity of the material (W m^{-1} K^{-1}) and is given by the following equation to compensate the fluid flow of the molten material as

$$k_m=\begin{cases} k_o & T<T_m \\ k_o+k' & T\geq T_m \end{cases} \tag{3}$$

where k_o is the thermal conductivity of the material used; k' is the additional value by which the convection heat transfer capability is equally considered in the present proposed thermal modeling; T_m is the melting point of the material used. T is the temperature variable (K), $\dot{Q}$ is the rate of heat generation per unit volume (W m^{-3}), ρ is the density of the material (kg m^{-3}), C_p is the specific heat capacity of the material (J kg^{-1} K^{-1}), t is time variable (s) and v is the welding velocity (m s^{-1}). Figure 3 schematically represents the weldment transverse cross-section along with applied thermal boundary conditions. Because of symmetric nature of the welding process, the half of the weld plate was considered for simulation. The temperature gradient is zero at symmetric surface which is normal to the weld interface. The top surface of the work piece subjected to a specified heat flux due to fiber laser beam and the remaining surfaces, (except symmetric surface) are subjected to convection and radiation heat losses. The natural boundary condition can be represented mathematically as

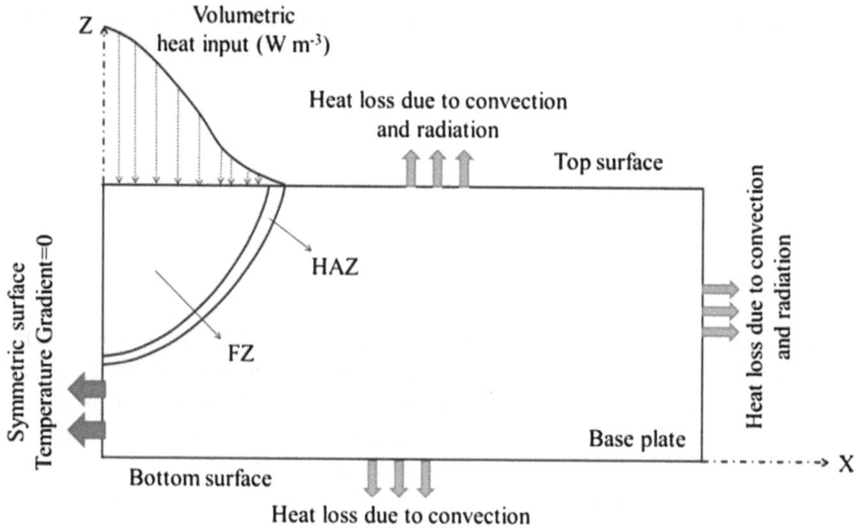

Fig. 3 Schematic representation of solution domain and applied boundary conditions

$$k_n \frac{\partial T}{\partial n} - q_s + h(T - T_o) + \sigma\varepsilon(T^4 - T_o^4) = 0 \tag{4}$$

where k_n is thermal conductivity normal to the surface, q_s is the imposed heat flux, h is the convection heat transfer coefficient (W m^{-2} K^{-1}), σ is Stefan-Boltzmann constant, ε is emissivity and T_o is ambient temperature (K). To avoid the non-linear nature of the Eq. (4), 'a lumped heat transfer coefficient' is considered in this simulation, this include the convection and radiation heat losses. A lumped heat transfer coefficient can be represented mathematically below (Goldak et al. 984)

$$h_{eff} = 2.4 \times 10^{-3} \times \varepsilon \times T^{1.61} \tag{5}$$

where h_{eff}, T, ε depict the lumped heat transfer coefficient (W m^{-2} K^{-1}), temperature (K) and emissivity respectively. Henceforth, the Eq. (4) can be modified by combining Eqs. (4) and (5) as given below

$$K_n \frac{\partial T}{\partial n} - q + h_{eff}(T - T_o) = 0 \tag{6}$$

The representation of the heat source in the present work has been considered a double ellipsoidal volumetric heat source model with Gaussian distribution (Goldak et al. 1984). The main characteristics of this heat source model are the digging action and the movement of the heat source by nonsymmetrical distribution of heat flux density in the front part and rear portions of the heat source. When the finer laser heat source move along the Y-axis, the power density distribution inside the front (f) and rear (r) portions are given by following equations

$$q_f(x, y, z) = \frac{6 \times \mathrm{sqrt}(3) \times f_f \times Q}{\pi^{3/2} \times a_f bc} \times e^{-\left[\frac{3x^2}{b^2} + \left(\frac{3y^2}{a_f^2} + \frac{3z^2}{c^2}\right)\right]} \quad (7)$$

$$q_r(x, y, z) = \frac{6 \times \mathrm{sqrt}(3) \times f_r \times Q}{\pi^{3/2} \times a_r bc} \times e^{-\left[\frac{3x^2}{b^2} + \left(\frac{3y^2}{a_r^2} + \frac{3z^2}{c^2}\right)\right]} \quad (8)$$

where v, τ, t, f_f and f_r depict the welding speed (m s^{-1}), lag factor, time (s), the fractions of heat deposited in the front and the rear portions respectively. a_f, a_r, b, c are heat source model parameters which are chosen based on experimental weld bead dimensions. Accurate results are obtained when the computed weld pool dimensions are slightly larger than the ellipsoid dimensions. Moreover, heat intensity Q can be expressed as

$$Q = \eta P \quad (9)$$

where η is the laser absorption coefficient and P is the fiber laser beam power (W). The fraction of heat deposition parameter should satisfy below equation

$$f_f + f_r = 2 \quad (10)$$

Nevertheless, the actual heat input as a result of laser beam is replaced by double-ellipsoidal volumetric heat source. Figure 4 represents the Gaussian distribution double ellipsoidal heat source model. From Fig. 4 it is obvious that peak heat intensity is at the Centre of heat source and the distribution is Gaussian in nature over the volume of ellipsoid.

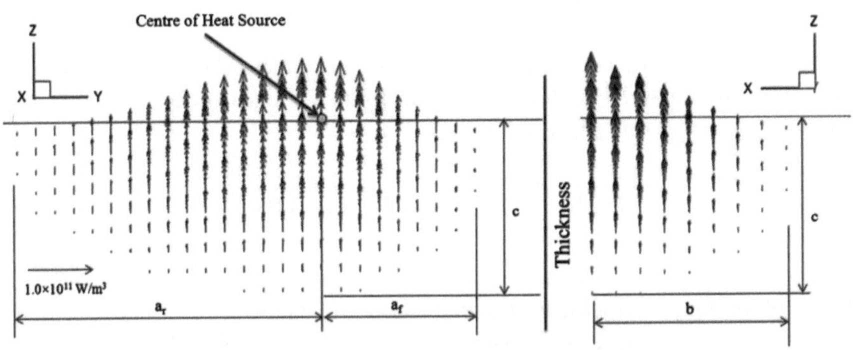

Fig. 4 Schematic representation of volumetric double ellipsoidal heat source model

4 Results and Discussion

Bead-on-plate welds were generated on commercially available SS 304, 5 mm plate and SS 316, 3 mm plate using a 2 kW ytterbium continuous wave fiber laser source. Then, metallographic analysis was performed to obtain weld bead dimensions and microstructure of the welds for both welding atmospheric conditions and various process variables.

Figure 5 describes the measured weld macrograph corresponding to laser power 2000 W and welding speed, 18.33 mm/s in open and controlled atmosphere of argon respectively. From this figure it is observed that the weld bead dimensions and aspect ratio is more in case of controlled atmosphere corresponding to similar welding conditions. Moreover, the weld bead dimensions increases with decreasing the welding speed at constant laser power. However, the weld bead shapes and dimensions of open atmosphere and controlled atmosphere of argon welds are dissimilar. Figure 6 refer to the experimental weld macrographs corresponding to SS 316 of laser power 1000 W and laser scan speed 18.33 mm/s. It is obvious that irrespective of weld material, the weld bead dimensions and aspect ratio is more in controlled atmosphere of argon as compared to open atmospheric condition.

The most notable observation in this experimental investigation is the appearance of weld bead surface quality. Figure 7 represents the top view weld bead appearance corresponding to data set 1 and data set 2 given in Table 1a, respectively. More neat and clean surface is obtained in controlled atmosphere of argon welding as compared to open atmosphere welds with similar welding process variables. In case of welding in open atmospheric condition, air particles such as hydrogen, oxygen and nitrogen react with molten weld pool at high temperature gradients. These reactions with atmosphere particles may result in high degree of porosity, rough surfaces and gas pores. However, welding in argon controlled atmosphere, the appearance of weld surface is clean with less porosity and low degree of surface roughness, when compared with the open atmosphere weld

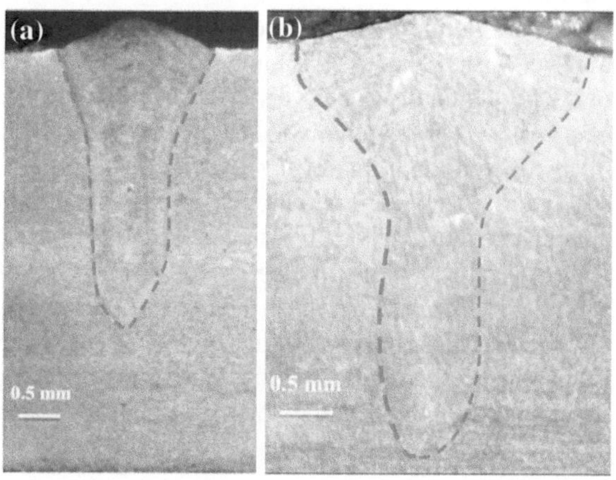

Fig. 5 Weld macrographs corresponding to data set 1 given in Table 1a; **a** open atmosphere and **b** controlled atmosphere welding

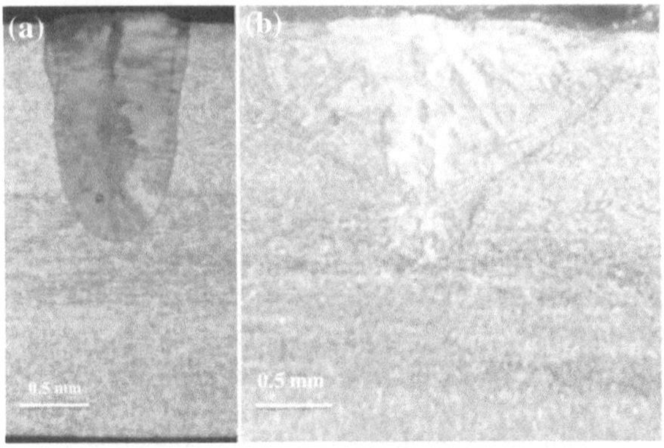

Fig. 6 Comparison of experimental weld macrographs corresponding to SS 316 of data set 1 given in Table 1b; **a** open atmosphere and **b** controlled

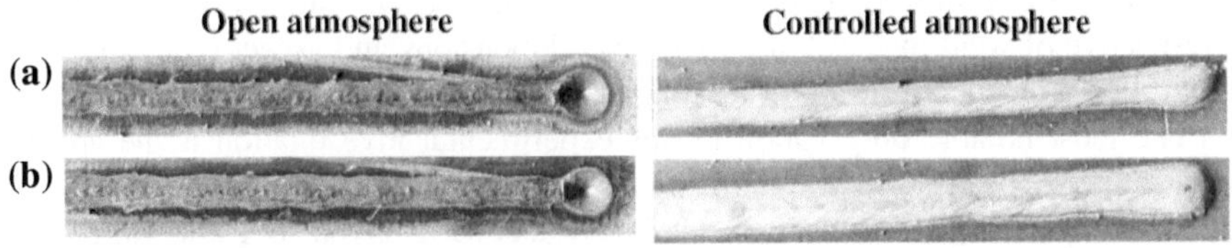

Fig. 7 Top view appearance comparisons for open and controlled atmosphere welding corresponding to **a** data set 1 and **b** data set 2 given in Table 1a

surface. In open atmosphere welding the air particles may dissolve in molten weld pool and they may try to escape from the weld pool during solidification. This may leads to cracking of the weld joint and reduces the weld joint quality. In case of controlled atmosphere, the defects due to the reaction of molten pool with oxygen, hydrogen, and nitrogen present in the ambient atmosphere can be minimized or nullified.

Microstructure of the as-received material is shown in Fig. 8a. In general the transverse section of weld sample after fiber laser welding shows that the extension of the heat affected zone is very small. However, this is the distinctive advantageous effect associated to a highly focused welding heat source (Zambon et al. 2006). The microstructure analysis reveals that there are no cracks formed in any one of the specimens examined. Even though some isolated pores were found mostly in open atmosphere weldment rather than a controlled atmosphere welds. No noticeable inclusions are observed in a controlled atmosphere of argon welds when compared to open atmosphere welds. In both atmosphere welds, the obvious feature is the highly directional nature of the microstructure around the axis of the fiber laser beam due to the solidification of weld metal at high cooling rate. Figure 8b, c describes the optical micrographs of heat affected zone (HAZ) and fusion zone (FZ) in open atmosphere weld made using laser power is 2000 W and welding speed is

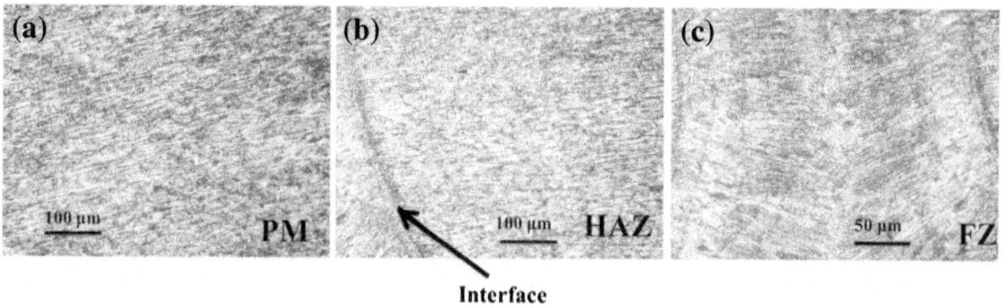

Fig. 8 Optical microstructure of weldment in open atmospheric condition corresponding to data set 1 given in Table 1 at various zones, **a** parent material (PM), **b** heat affected zone (HAZ and **c** fusion zone (FZ)

18.33 mm/s. Controlled atmosphere of argon welds microstructure (Fig. 9) show that the solidification structure is dendritic and contains austenite and a few percent of delta ferrite at the dendritic boundaries.

Figure 9a, b illustrates the HAZ and FZ micrographs of weld metal in a controlled atmosphere of argon corresponding to data set 1 given in Table 1a. The variation of the microstructure in open and a controlled atmosphere of argon from

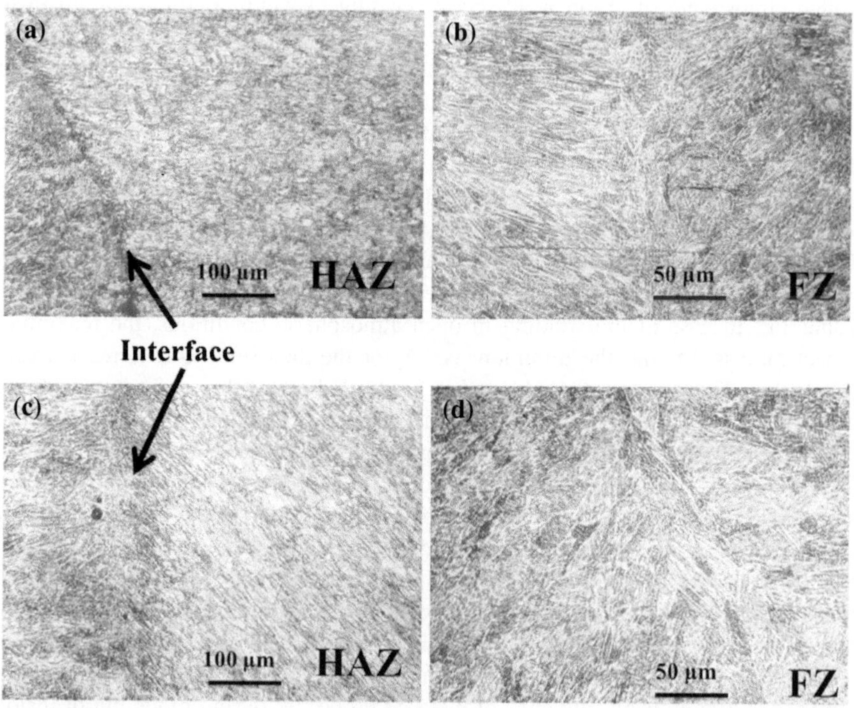

Fig. 9 Microstructure of weldment in controlled atmosphere of argon corresponding to **a** data set 1, HAZ, **b** data set 1, FZ, **c** data set 4, HAZ and **d** data 4, FZ

Figs. 8b, c and 9a, b with the similar process variables is clearly detected. This difference in the microstructure may be due to the solidification of the weld metal varies for both the atmospheres; in addition to this, the chemical reactions between the molten weld pool and surrounding atmosphere can be avoided in a controlled atmosphere of argon. This elimination of chemical reaction with atmosphere particle such as nitrogen, oxygen and hydrogen lead to minimum porosity or pin-holes formation in controlled atmosphere welds. Figure 9c, d represents the HAZ and FZ optical micrographs of weld metal made using the laser power of 2000 W and welding speed of 13.33 mm/s in a controlled atmosphere of argon. It can be found that the trend of the influence of the laser welding speed in a controlled atmosphere is following same trend of open atmosphere welds. However, the microstructures in the both welds are not same. From Fig. 9a–d, it is obvious that at the higher welding speed the dendritic structure is finer as compared to lower welding speed structure. Due to fine protection of weld pool in a controlled atmosphere; oxide, nitride, porosity and inclusion were formed at minimum level in the weld metal. This may lead to the high efficiency of the formed weld joint in controlled atmosphere. Moreover, the finer dendritic structure is observed when the welding speed is relatively higher; since the cooling rate is higher at high welding speeds.

Figure 10a–c describes the quantitative comparison of the weld bead dimensions, depth of penetration, bead width and aspect ratio between open and controlled atmosphere of argon welds corresponding to welding conditions given in Table 1a. For each data set index/number given in Table 1a, weld bead dimensions and aspect ratio are more in case of controlled atmosphere of argon as compared with open atmosphere welds. This resembles the fact that argon helps to constrict the laser beam and results in more concentrated heat flux. The maximum weld bead dimensions are achieved for the data set 4 given in Table 1a. The aspect ratio which is well-defined by the ratio of weld depth of penetration to bead width of a weld joint is a characteristic quantity to specify the effectiveness of the formed weld joint. The maximum aspect ratio, 2.998 is achieved in controlled atmosphere of argon welding (data set 4 in Table 1a) whereas the minimum is 2.02 (data set 1 in Table 1a). In case of the welding in open atmospheric conditions, the maximum aspect ratio is 2.8 and the minimum is 1.8 for the data set 4 and 1, respectively (Table 1a). The aspect ratio achieved in controlled atmosphere of argon welding process is more when compared with the process of welding under open atmospheric conditions for the similar welding conditions given in Table 1.

A Gaussian distributed double ellipsoidal heat source model is implemented using ANSYS APDL (ANSYS 14.0) for determining the transient thermal behavior and weld bead dimensions during fiber laser welding process of SS 316. Element SOLID 70 (ANSYS 14.0), which is eight node brick element with single degree of freedom, temperature, at each node has been used in thermal analysis. The thermal properties vary with temperature. In modeling, the temperature dependent materials properties provide accurate results as compared to temperature independent material properties. A significant influence of material properties during modeling of fusion welding process was studied by Zhu and Chao (2002). The temperature dependent material properties of SS 316 used in the present work are taken from an

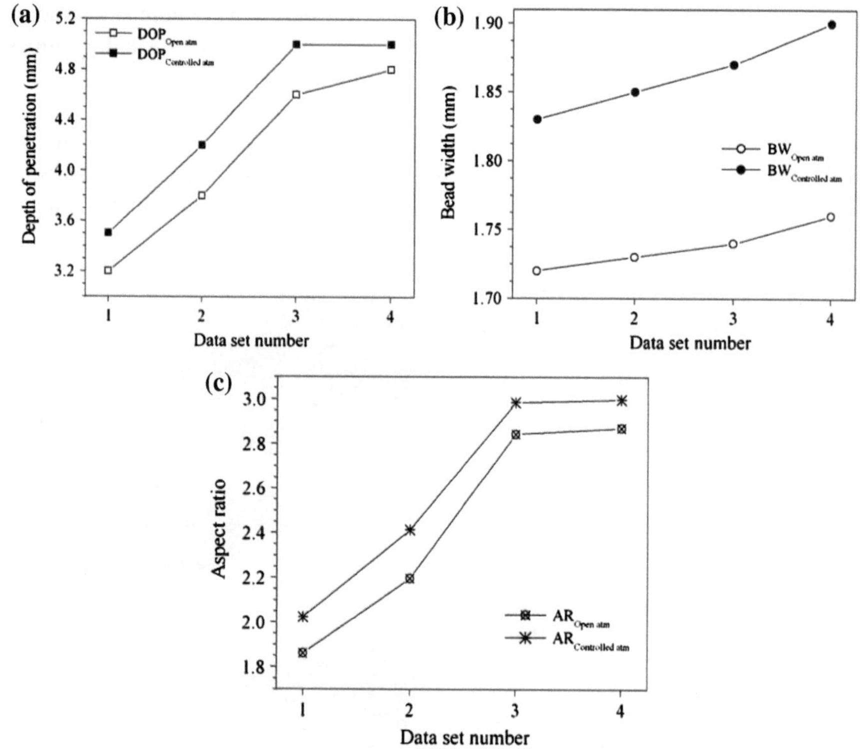

Fig. 10 Comparison of weld bead dimensions of open and controlled atmosphere welding corresponding to welding conditions given in Table 1a

independent literature (Wang and Felicelli 2007; Lindgren et al. 1999) and the same are given in Fig. 11. The density of the material SS 316 is considered as 7200 kg m^{-3}.

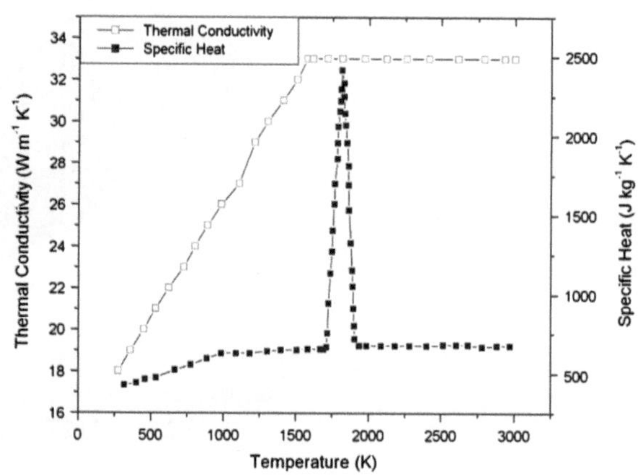

Fig. 11 Temperature dependent thermal conductivity and specific heat of SS 316 material (Yadaiah et al. 2014)

Figure 12 refers to three dimensional (3D) transient temperature distributions corresponding to data set 2 given in Table 1b at three different welding times. Figure 12a–c represents 3D temperature distribution at 0.9, 2.25, and 3.45 s respectively. The region surrounded by the liquidus temperature, 1727 K, signifies the weld pool dimensions; and its intercepts along X, Y and Z axes represents the half bead width, weld length and depth of penetration, respectively. It is also being seen from Fig. 12 that the temperature distribution is symmetrical about YZ plane;

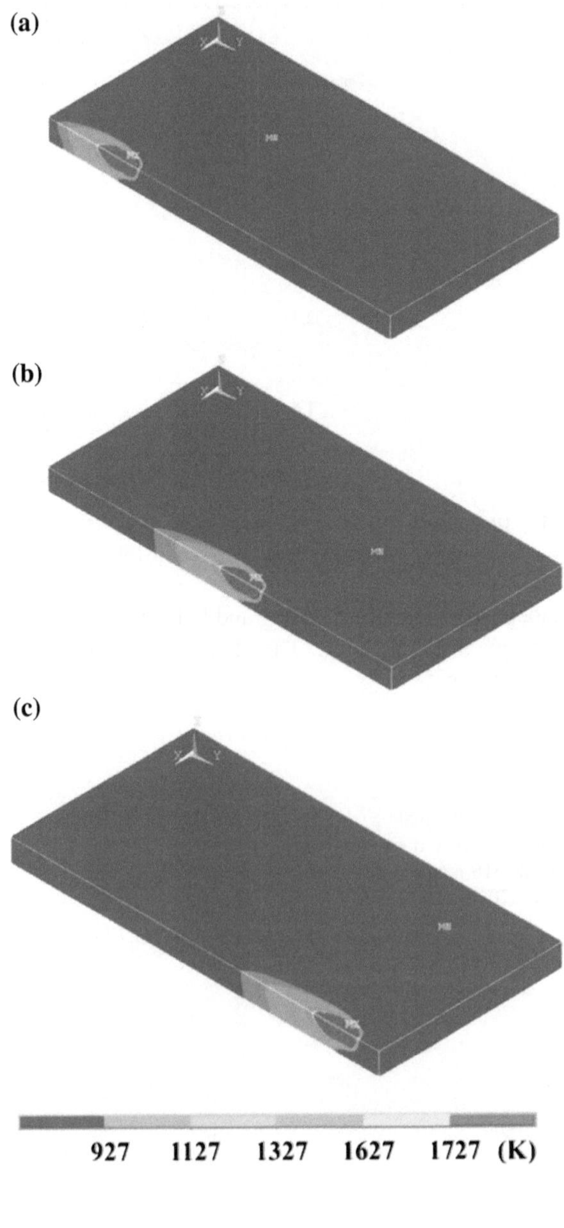

Fig. 12 3D transient temperature distribution corresponding to data set 2 given in Table 1b at three different times; **a** at 0.9 s, **b** at 2.25 s and **c** at 3.45 s (Yadaiah et al. 2014)

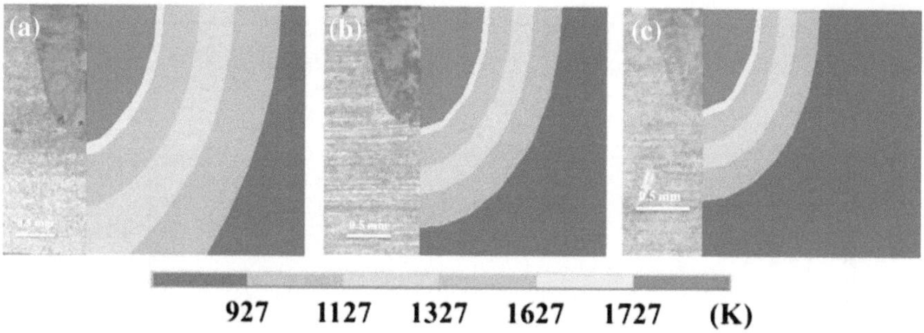

Fig. 13 Comparison of experimental and calculated transverse cross-section weld macrographs corresponding to **a** data set 2, **b** data set 4 and **c** data set 6 (Yadaiah et al. 2014)

on the other hand, it is asymmetric nature with respect to other planes as a result of the laser beam movement along the positive Y-axis.

Figure 13a–c indicates the comparison of experimentally measured weld macro with equivalent computed weld macrographs corresponding to data set 2, 4 and 6 given in Table 1b respectively. Moreover, the area enclosed by the base material liquidus temperature, 1727 K, specifies the molten pool and its intercepts along X- and Z-axes depict half bead width and depth of penetration respectively. A fair agreement between actual and computed weld macrographs is achieved at similar processing variables.

Figure 14 illustrates the quantitative comparison of calculated and experimental weld bead dimensions along with percentage error corresponding to welding process variables given in Table 1b. It can be noticed from these results (refer Fig. 14) that keeping laser power as constant and decreasing the welding velocity; the weld depth of penetration increases substantially while bead width increment is nominal. Furthermore, when welding velocity is constant and decreases the fiber laser power; weld bead width decreases substantially whereas weld depth of penetration decreases nominally.

The maximum weld depth of penetration achieved corresponding to data set 2 given in Table 1b and is 1.64 mm. as well as minimum penetration is observed for data set 5 given in Table 1b and is 0.6 mm. The percentage error between calculated and experimental weld bead dimensions are below 8 % and this can be realized from Fig. 14a, b. A fair agreement of calculated weld bead dimensions with measured results enables the robustness of the finite element model.

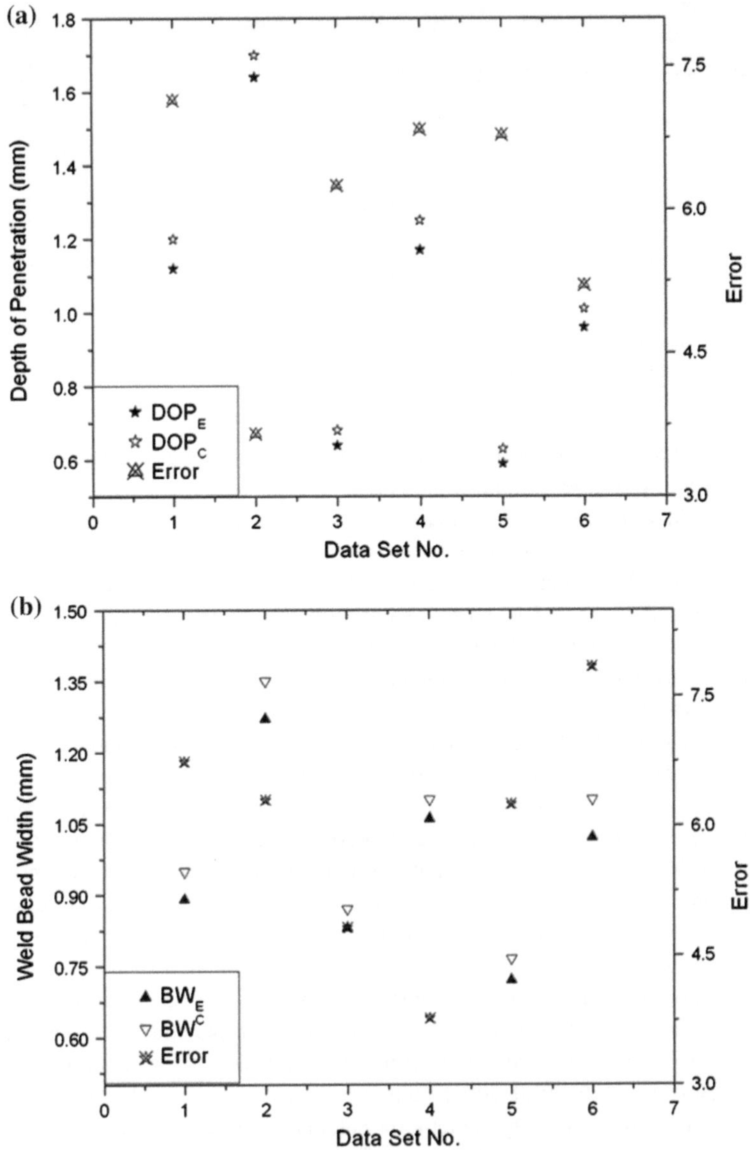

Fig. 14 Comparison of experimental and computed weld bead dimensions along with percentage error corresponding to welding conditions given in Table 1b, **a** depth of penetration and **b** bead width (Yadaiah et al. 2014)

5 Conclusions

This work demonstrates the performance of fiber laser welding in two different atmospheres, namely, argon and open atmospheres. The characteristic difference between two different atmospheres fiber laser welding on SS 304, 5 mm thickness and SS 316, 3 mm thickness plates are reported. The results showed that in controlled atmosphere of argon, the weld bead dimensions, depth of penetration and bead width are more as compared to open atmosphere welds. Moreover, the aspect ratio is also more in case of controlled atmosphere of argon. The most significant outcome of this investigation is the top surface profile and top view appearance that is clean and neat in controlled atmosphere of argon. The full depth of penetration is achieved in controlled atmosphere for welding speed, 13.33 mm/s and laser power, 2000 W for SS 314, 5 mm thickness plate. The metallographic analysis revealed that microstructure of welds in controlled atmosphere has less porosity as compared to open atmosphere welds. This is due to a fine protection of molten weld pool from oxygen and hydrogen etc. atmosphere particles. This may lead to high efficiency of formed weld joint in controlled atmosphere. The microstructure, at higher welding speed, the dendritic structure is finer as compared to relatively lower welding speed structure in both welding atmospheres. From this experimental investigation, it is recommended that the welding in controlled atmosphere of argon is far better than open atmospheric condition. In the present work, the authors also developed an efficient 3D transient heat transfer model using finite element method for open atmosphere welding of SS 316. From the numerical modeling it is concluded that: the inclusion of temperature dependent material properties along with effective thermal conductivity of liquid material enhances the scope and utilization of conduction mode heat transfer model. Moreover, laser power and welding velocity are most significant process variables in fiber laser welding that actually estimates the input energy per unit length to the process. At constant laser power, depth of penetration and bead width increase with decrease in weld velocity. However, the change of weld depth is more as compared to bead width. A fair agreement between computed weld bead dimensions with experimental results show the robustness of numerical model.

Acknowledgments The authors gratefully acknowledge the financial support provided by Science & Engineering Research Board, India (Grant no. SERB/F/0797/2013-2014 dated 20.05.2013) to carry out this research work. The authors also express thanks to Dr. Fanrong Kong affiliated with Research Center for Advanced Manufacturing, Southern Methodist University, Dallas, United States for his help at some stages of numerical work.

References

ANSYS 14.0, User manual.

Bag, S., Trivedi, A., & De, A. (2009). Development of a finite element based heat transfer model for conduction mode laser spot welding process using an adaptive volumetric heat source. *Internal Journal of Thermal Science, 48*, 1923–1931.

Bag, S., & De, A. (2010). Probing reliability of transport phenomena based heat transfer and fluid flow analysis in autogenous fusion welding process. *Metallurgical and Materials Transactions A, 41*(9), 2337–2347.

Bayram, K., Ramazan, K. K., Suleyman, G., & Fatih, H. (2008). An effect of heat input, weld atmosphere and weld cooling conditions on the resistance spot weldability of 316L austenitic stainless steel. *Journal of Materials Processing Technology, 195*, 327–335.

Christensen, N., Davies, V.L., & Gjermundsen, K. (1965). Distribution of temperatures in arc welding. *British Welding Journal, 12*(2), 54–75.

De, A., Maiti, S. K., Walsh, C., & Bhadeshia, H. D. K. H. (2003). Finite element modelling of laser spot welding. *Science and Technology of Welding and Joining, 8*(5), 377–384.

De, A., & DebRoy, T. (2006). Improving reliability of heat and fluid flow calculations during conduction model laser spot welding by multi-variable optimization. *Science and Technology of Welding and Joining, 11*(2), 143–153.

De, A., & DebRoy, T. (2005). Reliable calculations of heat and fluid flow during conduction mode laser welding through optimization of uncertain parameters. *Welding Journal, 84*(7), 101–112.

Dong, W., Kokawa, H., Tsukamoto, S., & Yutaka, S. S. (2005). Nitrogen desorption by high-nitrogen steel weld metal during CO_2 laser welding. *Metallurgical and Materials Transactions B, 36*, 677–681.

Dong, W., Kokawa, H., Yutaka, S. S., & Tsukamoto, S. (2003). Nitrogen absorption by iron and stainless steels during CO_2 laser welding. *Metallurgical and Materials Transactions B, 34*, 75–82.

Dursun, O. (2008). An effect of weld current and weld atmosphere on the resistance spot weldability of 304L austenitic stainless steel. *Materials and Design, 29*, 597–603.

Frewin, M. R., & Scott, D. A. (1999). Finite element model of pulsed laser welding. *Welding Research Supplement, 78*(1), 15–22.

Goldak, J., Chakravarti, A., & Bibby, M. (1984). A new finite element model for welding heat sources. *Metallurgical Transactions B, 15*, 299–305.

Kang, B. Y., Yarlagadda, K. D. V., Kang, M. J., Kim, H. J., & Kim, I. S. (2009). The effect of alternate supply of shielding gases in austenite stainless steel GTA welding. *Journal of Materials Processing Technology, 209*, 4722–4727.

Kim, H. J., Frost, H. R., & Olson, D. L. (1998). Electrochemical oxygen transfer during direct current arc welding. *Welding Journal, 77*(12), 488–493.

Kou, S. (2002). *Welding Metallurgy* (3rd ed.). New York: Willey Inter Science.

Kumar, A., Paul, C. P., Pathak, A. K., Bhargava, P., & Kukreja, L. M. (2012). A finer modeling approach for numerically predicting single track geometry in two dimensions during laser rapid manufacturing. *Optics and Laser Technology, 44*(3), 555–565.

Kumar, S., Roy, S., Paul, C. P., & Nath, A. K. (2008). Three-dimensional conduction heat transfer model for laser cladding process. *Numerical Heat Transfer, Part B: Fundamentals: An International Journal of Computation and Methodology, 53*, 271–287.

Lindgren, L. E., Runnemalm, H., & Nasstrom, M. O. (1999). Simulation of multipass welding of a thick plate. *International Journal for Numerical Methods in Engineering, 44*, 1301–1316.

Ostsemin, A. A. (2009). Estimating the temperature of an electrode-metal drop when welding in a carbon-dioxide atmosphere. *Russian Engineering Research, 29*(7), 668–670.

Pavelic, V., Tanbakuchi, R., Uyehara, O. A., & Myers, P. S. (1969). Experimental and computed temperature histories in gas tungsten-arc welding of thin plates. *Welding Journal, 48*(7), 295–305.

Ramazan, K., & Koray, K. (2005). Effect of controlled atmosphere on the mig-mag arc weldment properties. *Materials and Design, 26*, 508–516.

Ramirez, J. E., Han, B., & Liu, S. (1994). Effect of welding variables and solidification substructure on weld metal porosity. *Metallurgical and Materials Transactions A, 25*, 2285–2294.

Rosenthal, D. (1946). The theory of moving sources of heat and its application to metal treatments. *Transactions of ASME, 43*(11), 849–865.

Rosenthal, D. (1947). Mathematical theory of heat distribution during welding and cutting. *Welding Journal, 20*(5), 220–234.

Rykalin, R. R. (1974). Energy sources for welding. *Welding in the World, 12*, 227–248.

Sahoo, P., Collur, M. M., & DebRoy, T. (1988). Effects of oxygen and sulfur on alloying element vaporization rates during laser welding. *Metallurgical Transactions B, 19*, 967–972.

Wang, L., & Felicelli, S. (2007). Process modeling in laser deposition of multilayer SS410 steel. *Transactions of the ASME, 129*(1), 261–270.

Wu, C. S., Wang, H. G., & Zhang, Y. M. (2006). A new heat source model for keyhole plasma arc welding in FEM analysis of the temperature profile. *Welding Journal, 85*, 284–291.

Yadaiah, Y., & Bag, S. (2014). Development of egg-configuration heat source model in numerical simulation of autogenous fusion welding process. *International Journal of Thermal Sciences, 86*, 125–138.

Yadaiah, N., & Bag, S. (2012). Effect of heat source parameters in thermal and mechanical analysis of linear GTA welding process. *ISIJ International, 52*(11), 2069–2075.

Yadaiah, N., & Bag, S. (2013). Role of oxygen as surface-active element in linear GTA welding process. *Journal of Materials Engineering and Performance, 22*(11), 3199–3209.

Yadaiah, N., Bag, S., Paul, C. P., & Kukreja, L. M. (2014). Efficient finite element modeling of fiber laser welding process under conduction regime on 316 stainless steel plate. In *International Conference on Advances in Mechanical Sciences* (pp. 24–31), Hyderabad, India.

Zambon, A., Ferro, P., & Bonollo, F. (2006). Microstructural, compositional and residual stress evaluation of CO_2 laser welded superaustenitic AISI 904L stainless steel. *Materials Science and Engineering A, 424*, 117–127.

Zhu, X. K., & Chao, Y. J. (2002). Effect of temperature-dependent material properties on welding simulation. *Computers and Structures, 80*, 967–976.

A 3-D Finite Element Analysis of Transient Temperature Profile of Laser Welded Ti-6Al-4V Alloy

Chandan Kumar, Manas Das and Pankaj Biswas

Abstract In this work, a numerical investigation of transient temperature profile of Laser beam welding process is carried out. A 3-D finite element modelling is developed considering combined double-ellipsoidal heat source model for both spot and moving heat sources. The temperature dependent thermo-physical material properties of Ti-6Al-4V alloy are incorporated. The effect of latent heat of fusion and convective and radiative boundary conditions are considered. The effect of laser beam power on the transient temperature profile and the dimensions of the heat affected zone are analysed. From finite element simulation, it is observed that the peak temperature in the fusion zone increases with increased laser beam power. Also, the size of the heat affected zone strongly depends on the power of the laser beam.

Keywords Laser beam welding · Power density · Titanium alloy · Thermal analysis

1 Introduction

Titanium is the ninth most copious element on the planet and fourth most abundant structural material after aluminium, iron, and magnesium. Titanium name comes from the word "Titan" who is the powerful son of the earth according to Greek methodology. It was discovered simultaneously in England and Germany in 1790 and purified in early 1900s. Unalloyed titanium exists in two crystal forms i.e. low

C. Kumar · M. Das (✉) · P. Biswas
Department of Mechanical Engineering, IIT Guwahati, Guwahati, India
e-mail: manasdas@iitg.ernet.in

C. Kumar
e-mail: chandan.kumar@iitg.ernet.in

P. Biswas
e-mail: pankaj.biswas@iitg.ernet.in

S.N. Joshi and U.S. Dixit (eds.), *Lasers Based Manufacturing*,
Topics in Mining, Metallurgy and Materials Engineering,
DOI 10.1007/978-81-322-2352-8_21

temperature α phase having hexagonal close packed (HCP) structure, and high temperature β phase having body centered cubic structure (BCC). The allotropic transformation takes place in unalloyed titanium at above 882 °C known as transition temperature (Moiseyev 2006). Different alloying element are added in pure titanium to enhance its mechanical properties. The alloying elements for titanium alloys are classified in two categories i.e. α stabilizer and β stabilizer depending on their effects on α and β phases. The alloying elements either stabilize α phase by raising the α-β transition temperature or stabilize the β phase by suppressing the α-β transition temperature, and can act as solid solution strengtheners, without affecting the transition temperature. Carbon, nitrogen, and oxygen are more soluble in α phase and thus known as α stabilizer. These elements form interstitial solid solution with titanium. Hydrogen is more soluble in the β phase, and it behaves as a β stabilizing element (Pederson et al. 2001). Some of the most common alloying elements and their stabilizing effects are discussed in the next paragraph.

Aluminium and tin are the most common alloying elements which stabilize the α phase and their concentration (in wt%) in Ti alloy are in the range of 2–7 % and 2–6 %, respectively. Some of the most common β phase stabilizing elements in Ti alloy are vanadium (2–20 wt%), copper (2–6 wt%), chromium (2–12 wt%), and molybdenum (2–20 wt%) etc. Whereas, silicon is a neutral element having concentration in the range of 0.2–1 wt% and it is mainly used to improve the creep resistance of Ti alloy. Also, zirconium (2–8 wt%) is used for strengthening both α and β phase of Ti alloy (Matthew and Donachie 2000).

The interstitial elements like carbon, hydrogen, nitrogen, and oxygen etc. are the impurities that contaminate titanium during welding. Nitrogen is the most effective strengtheners for unalloyed titanium. Oxygen is next in effectiveness, followed by carbon. In two-phase, α + β titanium alloys, the action of interstitial elements becomes more complex. The strengthening effect of interstitials on α + β alloys is greatly dependent on the relative amount and distribution of α phase. Hydrogen is slightly soluble in titanium at room temperature. β titanium dissolves substantially more hydrogen than α titanium. α + β alloys are able to dissolve a considerable amount of hydrogen without the formation of hydrides.

Titanium and its alloys serve as a bridge between the ideal properties of aluminium and steel. Titanium alloys are divided into four categories, alpha (α) alloy, near alpha alloy, alpha and beta (α + β) alloy, and beta alloy (β). Table 1 shows the qualitative comparison of characteristics for different types of titanium alloy. α alloy shows higher ductility, but lower strength. On the other hand, β alloys exhibit higher strength but lower ductility and poor weldability. In contrast, α + β titanium alloy (Ti-6Al-4V) shows a good balance between strength and ductility.

Ti-6Al-4V (Ti64) alloy is very popular and its production is about half of all titanium alloys. Ti64 is a α + β alloy in which 6 wt% aluminium which stabilizes α phase and 4 wt% vanadium which stabilizes β phase. At room temperature, the microstructure in equilibrium consists of mainly α phase (HCP) with some retained β phase (BCC). The β transus temperature for Ti64 alloy is approximately 995 °C.

Depending on the cooling rate and the prior heat treatment condition, the micro constituents and microstructures are divided into several types, namely grain

Table 1 Comparisons of α, α + β, and β titanium alloys

Mechanical properties	α	β	α + β
Ductility	Moderate	Low	High
Strength	Moderate	Very high	High
Weldability	High	Low	Moderate

Data source Leyens and Peters (2003)

boundary allotriomorph α, globular or primary α (called bi-modal microstructure when the globular α is surrounded by Widmanstatten platelets), Widmanstatten, basket weave, and martensitic. A recently described microstructure is the bi-lamellar, in which the retained β phase, lying between α platelets in a Widmanstatten structure, itself contains thinner secondary α platelets. The length and width of these α platelets depend on the cooling rate. An increase in cooling rate, the length and width of α platelets are reduced. However, the nucleation rate is increased which decreases the growth rate. If quenched from the β phase, the β grain will fully or partially transform into a martensitic α, denoted as α′ (Wanjara et al. 2005). The minimum cooling rate for full martensitic structure of 410 °C/s is reported for Ti64 (Ahmed and Rack 1998).

Ti64 alloy is characterized by its high specific weight, better corrosion resistance, and excellent biocompatibility. Therefore, it is widely used in many areas such as defence, aerospace, nuclear, chemical, and medical industries, etc. The thermal conductivity of Ti64 alloy is very poor. However, the heat of fusion and the heat of vaporization of Ti64 alloy are almost double of iron. Due to these properties, the welding of Ti64 alloy using conventional welding technique is very difficult because of its higher tendency to react with atmospheric gases such as H_2, O_2, and N_2 above 550 °C which increases the weld defects (Yang et al. 2010).

1.1 Weldability of Titanium Alloy

The weldability of an alloy is defined as the ability to produce a good quality weld joint using specific welding process parameters. During welding of the materials, the microstructural and mechanical properties of the base materials change at the welded region. Titanium and its alloys generally require advanced joining techniques i.e. Laser beam welding, electron beam welding, and plasma welding. The study of weldability and characterization of the welded joints of titanium and its alloys is a promising challenge for research and development sector.

Destefani (1992) reported that unalloyed titanium and all α titanium alloys have good weldability, toughness, and strength. (α + β) titanium alloys show good formability and weldability, and their properties are highly influenced by the heat treatment. Ti64 is reported as the best weldable among all α + β titanium alloys. β alloys show bad weldability due to the degradation of strength after welding. Some interesting properties of welding of titanium alloys are their low thermal conductivity which prevents heat dissipation during welding. Due to the lower value of coefficient of thermal expansion of titanium alloy, less residual stress, and thermal distortion are observed during welding. Proper joining technique for Ti64 can make the joint ductile and corrosion resistant like the base metal however improper joining process may lead to the embrittlement of the welded structure (Costa et al. 2007).

Titanium alloys also possess high rate of laser beam absorption (0.4 %) and a high melting point (around 1670 °C). These properties show that high energy is required for welding these alloys. Hence, welding of titanium alloy by using laser beam is the most suitable process as it allows small size of fusion as well as heat affected zone (Quintino et al. 2007).

There are some specific requirements for welding of titanium alloys using laser heat source. The cleaning of samples before welding is very crucial due to the high reactivity of titanium alloy with gases like, oxygen and nitrogen at high temperature which results in contamination by the interstitial impurities in the weld. Titanium alloys have high reactivity with grease, dirt, refractories, and with other materials at high temperature which results in the embrittlement of the welded joint (Costa et al. 2007). The use of shielding gas is compulsory during Ti welding to prevent oxidation of weld metal pool. Insignificant oxygen contamination occurs during Laser beam welding under shielding gas environment.

2 Literature Survey on Welding of Ti Alloy

Welding is the most important materials processing technology in manufacturing industries. It is defined by the American Welding Society (AWS) as a localized coalescence of metals or non-metals produced by either heating of the materials to a suitable temperature with or without the application of pressure, or by the application of pressure alone, with or without the use of filler metal (Mackerle 2002). Various advanced welding technologies have been evolved for last few years to achieve better weld quality, higher weld efficiency, reduced welding cost, environment-friendly features, and to ensure the safety and reliability of the welded structures.

Laser beam welding (LBW) is the most widely used process in industry as it is versatile and cost effective. The theoretical and experimental study of laser beam is started in 1962. However, the use of LBW is reported in 1971 using CO_2 laser heat source. Since, then the use of LBW has grown rapidly in the manufacturing industries (Banas 1971).

LBW is one of the most important types of non-conventional and advanced welding methods for joining similar or dissimilar materials. In LBW process, high-power density heat source is utilized which is absorbed by the materials to be joined. LBW process can weld a variety of metals or plastics ranging from thin sheets of about 0.01 mm thickness to thick plates of about 50 mm in the presence of shielding gas such as helium, argon or sometimes nitrogen under atmospheric condition. The process parameters that can be controlled during LBW process are spot size, focused position of the laser beam, volumetric flow rate of the shielding gas, laser beam power, and welding speed. A hemispherical weld bead and heat affected zone (HAZ) is formed during LBW similar to conventional fusion arc welding processes. Due to the rapid change in thermal cycles during LBW, it produces fine-grained weld bead with excellent mechanical properties (Sakagawa et al. 2011).

LBW gained great popularity among all welding processes as a promising joining technology having high quality, high precision, high performance, high speed, good flexibility, and low deformation (Hongping et al. 2011). LBW is capable of joining carbon steels, HSLA steels, stainless steel, aluminium, titanium, and its alloys.

Depending on the availability of the input energy, two different modes of LBW process are mentioned in the literature (Nath et al. 2002). These are conduction mode and keyhole mode of LBW process. Conduction mode of welding is a low energy input process having laser beam power density below 10^3 W/cm^2. In this mode of welding, the laser energy is absorbed at the material surface and it is distributed to the bulk material by conduction only. Here, the bead depth to width ratio is generally less than 0.5. Laser conduction mode of welding produces low welding depth, small aspect ratio, and low coupling efficiency (Chen et al. 2009). The conduction mode of Laser beam welding process is shown in Fig. 1a.

The keyhole mode of welding is a high energy input process having power density in the range of 10^5–10^7 W/cm^2. At the centre of the focused laser beam, the vaporization of metal takes place which creates a hole into the molten pool i.e. a keyhole is generated. In this mode of welding, the bead depth to width ratio is generally more than 0.7. The small size of keyhole is responsible for producing relatively small fusion and heat affected zone. Due to the generation of the highly

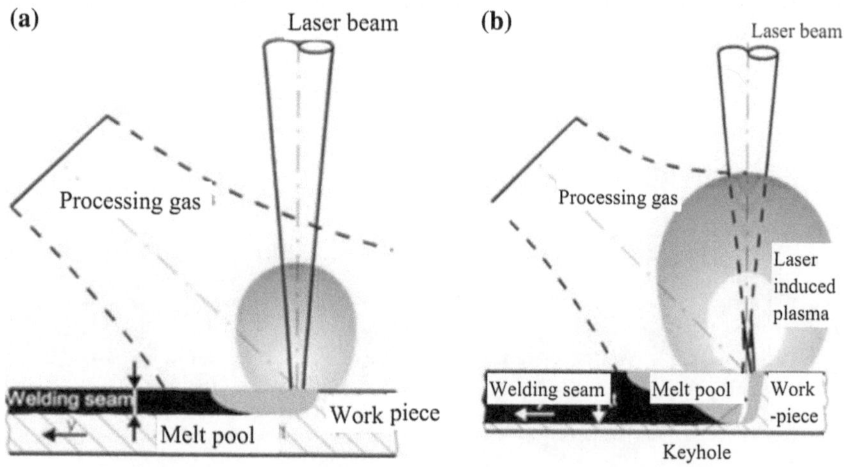

Fig. 1 **a** Conduction, and **b** Keyhole mode of Laser beam welding process

localized heat on the workpiece surface, the simultaneous heating up and the cooling down of the fusion and heat affected zones takes place very rapidly leading to the minimization of grain growth in the welded region. The keyhole mode of welding produces high welding depth, high aspect ratio, high coupling efficiency, and narrow HAZ (Manonmani et al. 2007). The keyhole mode of LBW process is shown in Fig. 1b.

The schematic diagram of LBW experimental setup is shown in Fig. 2. LBW is a complex process since large number of process parameters are involved which must be controlled for better functioning of the process. Also, due to the stochastic nature of LBW process, achieving optimum process parameters by doing experiments using trial-and-error method is difficult. During LBW, non-uniform temperature distribution takes place throughout the workpiece. The temperature distribution depends on several factors such as heat input, temperature depended material properties, latent heat of fusion, and rate of convective heat flow in the fusion zone. The heat loss to the surrounding takes place by convection and radiation. Due to these large number of variables, the experimental determination of all input parameters is very difficult for achieving higher welding performance. Hence, the demand of finite element method (FEM) based numerical simulation is progressively increasing.

Now a days, modelling and simulation have become powerful tools for welding research and development. Through combining the software based simulation approaches with less experimental validations, welding technology can be transformed from the earlier mode "theory-experiment-production" to the newer mode "theory-computer simulation production". During last two decades many of the existing welding processes have been simulated by numerical methods, especially by the finite element methods. The numerical simulation of a welding process is a complex process involving the interaction of thermal, mechanical, electrical, and metallurgical phenomena.

The use of FEM in product development is now well established and became a part of the manufacturing industries. Its use in manufacturing industries is increasing day-by-day to improve the quality of the products. It is also useful for

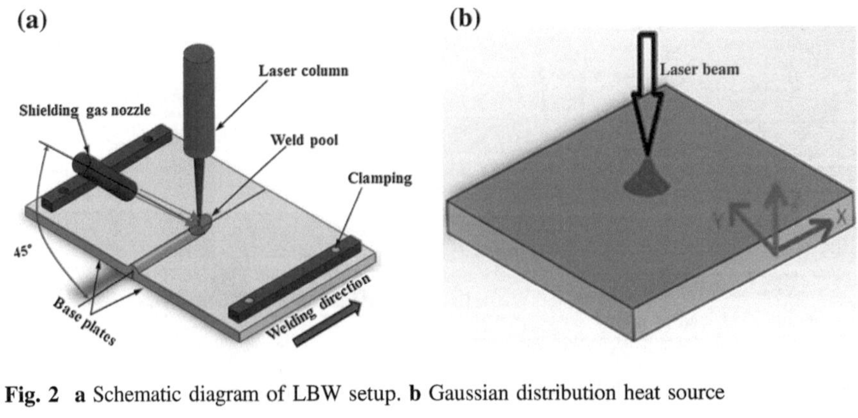

Fig. 2 **a** Schematic diagram of LBW setup. **b** Gaussian distribution heat source

better understanding of the effect of different process parameters on the quality of the products. The thermal cycle in welding is studied by analytical and numerical methods by many researchers.

The most popular analytical model for predicting transient temperature profile during welding was developed by Rosenthal in 1946 for moving heat source considering the quasi-stationary state where the heat source is concentrated at a point. The temperature dependent material properties are not included in his model. This point heat source model was incapable to explain the transient temperature profile (Rosenthal 1946).

Goldak et al. (1984) considered a 3-D non-axisymmetric double-ellipsoidal heat source model in order to simulate the welding phenomena. Na and Lee (1987) analysed the transient temperature distribution in GTAW (gas tungsten arc welding) process by using a 3-D FEM model. The combination of the convection and the boundary conditions were varied during simulation.

Frewin and Scott (1999) developed a 3-D FE model to calculate the transient temperature history at the fusion zone and HAZ considering Gaussian energy distribution of laser beam. The simulated result was in close agreement with the experimental data. Tsirkas et al. (2003) developed a finite element SYSWELD code to simulate the Laser welding process using non-linear heat transfer analysis. This model is used for coupled transient thermo-mechanical analysis considering temperature dependent material properties of AH36 shipbuilding steel plates.

Nguyen et al. (2004) derived an analytical solution for a double ellipsoidal heat source model for finite thick plate. Their solution can be used effectively to predict the thermal history of a thick welded plate and also the shape of weld pool.

Zhu and Chao (2004) developed an inverse analysis method to determine transient temperature history of friction stir welding (FSW) of 304L stainless steel based on the experimental data of transient temperature history at several specific locations. After the transient temperature field is determined, the residual stresses in the welded plate are then calculated using a three-dimensional elastic–plastic thermo-mechanical simulation. The effect of fixture release after the welding on the residual stresses is also studied. Comparison with the residual stress fields measured by the neutron diffraction technique shows that the numerical simulation results have good agreement with the test data.

Deng and Murakawa (2006) developed both 2-D and 3-D finite element codes to measure the transient temperature and residual stress in multi-pass welding of SUS304 stainless steel pipe. The simulated results shows that the 2-D model saves large amount of computational time and it can be effectively used to simulate the temperature fields and residual stresses for SUS304 stainless steel pipe. Experiments are carried out to verify simulated results and they found that both 2-D and 3-D models are in good agreement with the experimental results. Spina et al. (2007) developed a 3-D finite element model to predict the displacement of AA5083 thin sheets during LBW process at different welding speeds.

Yang et al. (2010) developed a finite element model to measure the size of heat affected zone in laser heating of Ti-6Al-4V plate. They concluded that the size of the HAZ decreases with increasing laser scan speed. Biswas et al. (2010) developed a 3-D FEM model for double-sided fillet SAW process and successfully compared with the experimental results. The temperature distribution obtained from both FEM analysis and experimental results fairly compared well with a variation of 8 % for the peak temperature.

Shanmugam et al. (2010) employed a FEM code (SYSWELD) to study the thermal field and to determine the bead shape during the LBW of T-joint. Proper fusion of base material (horizontal and vertical sheets) is achieved when the laser system is operated at 60° beam incident angle irrespective of the beam power and welding speed. A weld defect is found in the macro-graph at 30° beam angle. Comparison between experimental and simulated results reveals a very good correlation for depth of penetration and bead width with an error of 2.78 and 1.9 %, respectively.

Ranjbarnodeh (2011) developed a 3-D FEM model to predict the temperature profile during welding of dissimilar materials (CK4 and AISI 409) using automatic TIG welding process. Also, experimental measurements were performed to assess the effect of welding parameters on grain growth in HAZ. It was concluded that the grain size and its distribution were strongly dependent on the heat input during welding. The sample with the highest welding heat input (i.e. 583 J/mm) showed larger grain size (250 μm), more homogenous grain size distribution in the HAZ for AISI 409.

Double-ellipsoidal volumetric heat source with Gaussian distribution of heat intensity is most popular and it is used for the modelling of fusion welding process. It is very difficult to find out the optimum length of the front and rear part of the double-ellipsoidal volumetric heat source model. Till date, the length of these parameters are taken arbitrary.

Yadaiah and Bag (2012) suggested an optimum value of ratio of front and rear length of a double ellipsoidal heat source model over a wide range of welding current and velocity in GTAW process. They considered that this ratio is a function of weld velocity. Integrated optimization algorithm was used to optimize the ratio of front and rear length of double ellipsoidal heat source model and a good agreement between simulated and experimental results are observed for GTAW process.

Shanmugam et al. (2012) developed a 3-D FE model using Ansys™ to obtain the transient temperature profile and the shape of the molten pool. They considered a 3-D conical Gaussian heat source distribution for performing a non-linear transient thermal analysis. The thermo-physical material properties of AISI 304 stainless steel were also included in the model. Good agreement between the simulation and experimental results were observed.

Moreover, only conduction based heat transfer models are not sufficient for an accurate measurement of thermal cycles and weld bead dimensions. Transport phenomena based on heat transfer and fluid models are required for an exact estimation of transient thermal history in molten weld pool (Bag and De 2010; Yadaiah and Bag 2013).

Yadaiah and Bag (2014) proposed a new heat source model, namely, 'egg' configuration heat source model for the simulation of linear GTA and laser welding processes. The advantage of this model is having less number of heat source model parameters as compared to double ellipsoidal heat source model. This model was used to measure the transient temperature history for predicting the size of weld pool with an error of 10 % with the experimental results.

Akbari et al. (2014) developed a 3-D FEM model for the prediction of transient temperature profile of laser welded Ti64 alloy considering Gaussian energy distribution of laser beam. The average absorptivity value of 0.34 for Ti64 alloy was considered by them. The thermo physical property of Ti64 alloy was also incorporated in the model. The model prediction error was found in the range of 2–17 %. Various works on LBW process including the prediction of weld bead geometry, simulation of temperature distribution, and mechanical behaviour of laser welded sheet at different welding process parameters are carried out by many researchers. Following conclusions are observed from the above literature survey.

- Finite element analysis is the most accurate and flexible method for solving transient thermal analysis and structural analysis of LBW process.
- The use of thermo physical material properties play an important role for the estimation of the temperature distribution in different regions of the plates.
- The size of fusion zone and HAZ can be predicted by temperature distribution profile.

It is also observed that researchers used different heat source models for FE simulation of fusion welding process. Further researches are still going on to define a reliable heat source model. It is observed that double ellipsoidal heat source model yields better result than other heat source model.

From literature, it is noticed that very few literatures are available on LBW of Ti64 alloy having plate thickness between 3 and 6 mm. Also, very few experimental and simulation works on LBW process especially on transient thermal and structural analysis of Ti64 alloy are available in the literature.

Considering above mentioned literature gap, the main objectives of the present simulation work are defined as follows: Feasibility study of double ellipsoidal heat source model for the simulation of LBW of Ti64 alloy in both conduction and keyhole mode of welding; to generate transient temperature profile of Laser beam welded Ti64 alloy; and to study the effect of laser beam power and welding speed on the weld bead geometry and on the size of fusion zone and HAZ.

3 FEM Simulation of LBW Process

FEM is mostly used to simulate the transient thermal, structural as well as heat transfer phenomena in LBW process. The finite element modeling procedure consists of the following steps as shown in Fig. 3.

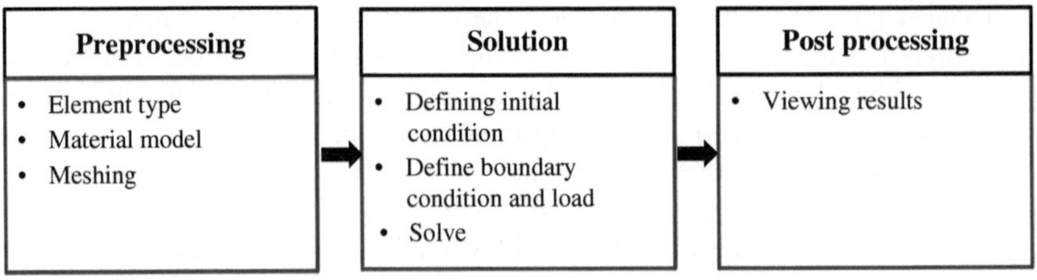

Fig. 3 Steps in finite element modelling

3.1 Preprocessing

In the preprocessing step, solid modelling or workpiece geometry is generated. For the present case, the workpiece dimension is 30 × 30 × 3 mm^3. Ansys™ provides an option of creating solid model of the workpiece on its GUI (Graphical User Interface) or it may be imported from any CAD software. After that the whole domain of the workpiece is divided into a number of small elements.

3.1.1 Element Type

Ansys™ library contains large number of different elements. The selection of proper element and its size are very crucial for obtaining higher accuracy and reducing solution time during analysis. In the present thermal analysis, 'solid 70' element (Shanmugam et al. 2012) having eight nodes and single degree of freedom is selected.

3.1.2 Material Properties

Temperature dependent thermo physical material properties yield better results. The titanium alloy of Grade 5 (Ti64) is selected for the transient thermal analysis. The mechanical properties of α–β titanium alloy depend on the distribution of α and β phases. The melting point of Ti64 ranges from 1650 to 1690 °C. The chemical compositions and mechanical properties of Ti64 alloy at room temperature and under annealed condition are listed in Tables 2 and 3, respectively. The temperature dependent material properties (i.e. thermal conductivity, enthalpy, density, specific heat, and emissivity) of Ti64 alloy are assumed to be isotropic and homogeneous (Yang et al. 2010).

Table 2 Composition of Ti64 titanium alloy *Data source* ASTM B265 (2006)

Element	Fe	Al	V	C	H	O	N	Ti
wt%	<0.30	5.5–6.75	3.5–4.5	<0.08	<0.015	<0.25	<0.03	Bal.

Table 3 Mechanical properties of Ti64 titanium alloy

Yield strength (MPa)	Ultimate tensile strength (MPa)	Young's modulus (MPa)	% elongation	Hardness (HV)
800–1100	900–1200	110–140	13–16	300–400

Data source Leyens and Peter (2003)

3.1.3 Meshing

The accuracy and the convergency of the FEM results depend on the type of element and the mesh size at different zones. In the present thermal analysis, the geometry of the workpiece is meshed with a fine meshing near the weld line and coarser meshing away from the weld line (Fig. 4). Also, convergency tests are carried out for selecting suitable number of elements particularly around the weld line and along the thickness direction. It is observed that five elements along the thickness direction give reasonable convergency with less solution time and good accuracy.

3.2 Solution

In the solution module of Ansys^{TM}, the applied load (i.e. heat source), initial condition, and boundary conditions are specified. The boundary conditions may be absolute, natural or a function of time. The absolute boundary conditions define the

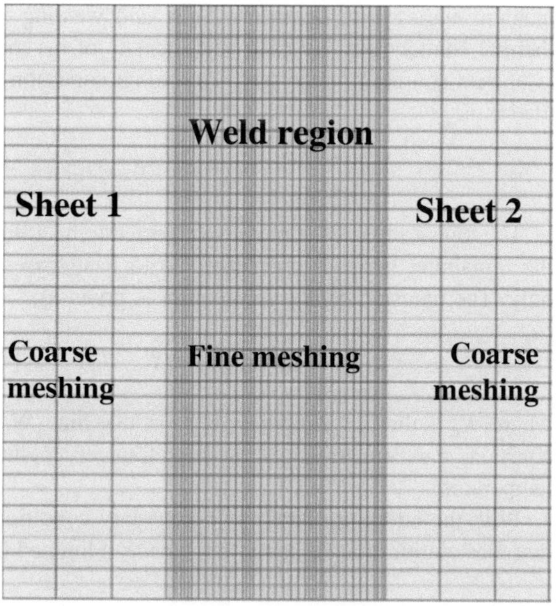

Fig. 4 Finite element meshing with fine meshing along the weld line and coarse meshing away from the weld line

values of primary variables at the boundary, whereas, the natural boundary conditions define the gradients of the primary variables (Ansys 14.5 manual).

In LBW process, the natural boundary condition is specified by the application of heat flux on the elements. The ambient temperature of 27 °C is set as initial condition. The governing equation and boundary conditions for LBW process are discussed in Sect. 4.

3.3 *Post Processing*

In post processing of LBW process, transient temperature profile, shape of the molten zone and weld bead profile are obtained. From both temperature profile and the weld bead profile, the size of the fusion zone and HAZ are measured.

4 Governing Equation and Boundary Conditions

The 3-D transient non-linear heat transfer equation for LBW process can be represented by

$$\rho c\left[\frac{\partial T}{\partial t}+(-v)\frac{\partial T}{\partial y}\right]=K\left[\frac{\partial}{\partial x}\left(\frac{\partial T}{\partial x}\right)+\frac{\partial}{\partial y}\left(\frac{\partial T}{\partial y}\right)+\frac{\partial}{\partial z}\left(\frac{\partial T}{\partial z}\right)\right]+Q(x,y,z) \quad (1)$$

where, $Q(x,y,z)$ is volumetric heat source and it varies with beam power, beam incident angle, beam exposure time, and welding speed. K, c, ρ, and v are the thermal conductivity, specific heat, density of the material, and the velocity of the laser heat source, respectively. The initial condition at time t = 0, during FE simulation is given as,

$$T(x,y,z,0)=T_0(x,y,z) \quad (2)$$

The top surface of the plate is subjected to heat flux produced by laser beam and the remaining surfaces are subjected to heat losses by convection and radiation only. The natural boundary condition is given as,

$$K_n(\partial T/\partial n)-q+h(T-T_o)+\sigma\varepsilon\left(T^4-T_o^4\right)=0 \quad (3)$$

where, K_n is thermal conductivity, q is heat flux (W/m^2 K), σ is Stefan-Boltzmann constant, ε is emissivity, h is heat transfer coefficient, and T_o is the ambient temperature.

With the addition of radiative boundary condition in the model, it is observed that the solution time is increased. Hence, a lumped heat transfer coefficient is used in the present analysis, which combines both radiative and convective heat transfer phenomena (Frewin and Scott 1999). It is given as

$$h_{lump} = 2.4 \times 10^{-3} \varepsilon T^{1.61} \tag{4}$$

In Eq. (4) h_{lump} is lumped heat transfer coefficient. Equation 3 can be modified by combining it with Eq. 4 and it is given as,

$$K_n(\partial T/\partial n) - q + h_{lump}(T - T_o) = 0 \tag{5}$$

4.1 Heat Source Model

The double-ellipsoidal Gaussian-distribution of volumetric heat source model is most suitable for the simulation of fusion welding process and it is considered in the current FE simulation (Fig. 5). The front and the rear part of this heat source model are given in Eqs. (6) and (7), respectively.

$$q_e(x, y, z) = \frac{6\sqrt{3} f_{ef} Q}{a_e b_e c_{ef} \pi^{3/2}} \exp\left\{-\left(\frac{3x^2}{a_e^2} + \frac{3y^2}{b_e^2} + \frac{3z^2}{c_{ef}^2}\right)\right\} \tag{6}$$

$$q_e(x, y, z) = \frac{6\sqrt{3} f_{ef} Q}{a_e b_e c_{er} \pi^{3/2}} \exp\left\{-\left(\frac{3x^2}{a_e^2} + \frac{3y^2}{b_e^2} + \frac{3z^2}{c_{er}^2}\right)\right\} \tag{7}$$

The parameters, f_{ef}, and f_{er} are the fractions between the front and rear quadrants in the ellipsoid. To obtain continuity between the front and rear quadrants of the functions, they must obey the following relationships.

$$f_{ef} + f_{er} = 2 \tag{8a}$$

$$f_{ef} = \frac{2c_{ef}}{c_{ef} + c_{er}} \tag{8b}$$

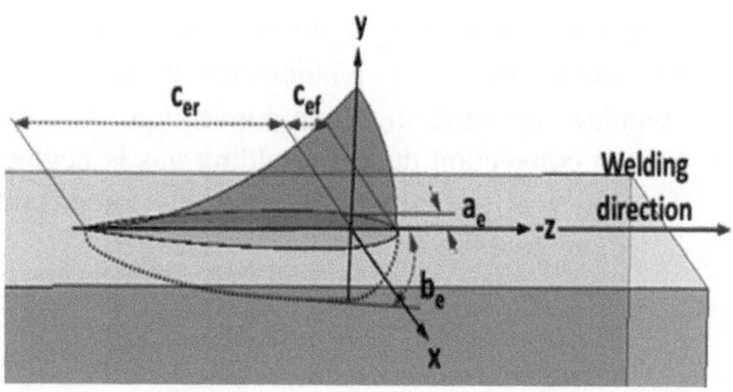

Fig. 5 Double-ellipsoidal heat source model

Table 4 Values of heat source parameters

Parameter	a_e	b_e	c_{ef}	c_{er}
Value (mm)	1.7	0.2	0.2	3

The values of the heat source parameters are given in Table 4. The total length of heat source and the length of the front ellipsoid are considered as 2×10^{-3} and 0.5×10^{-3} m, respectively.

5 Finite Element Modelling

The transient thermal analysis is performed in commercial FE package ANSYS™ 14.5. FE simulations are performed at laser beam power of 500, 1000 and 1500 W to study its effect on the workpiece temperature profile at a constant welding speed of 312 mm/min. Laser spot welding simulation is also performed on Ti64 workpiece at 500, 1000, and 1500 W laser beam power. The double ellipsoid heat source model for 25 ms duration is considered during simulation. The simulation is performed in two steps. In the first step, the heat flux is applied on the surface of the workpiece. In the second step, the heat flux is removed and the heat is dissipated from all the boundaries of the workpiece by convection and radiation. Eight-noded quadratic 3-D solid element "SOLID70" is considered for performing the transient thermal analysis.

During welding, the temperature in the fusion zone reaches beyond the boiling point of the material. Hence, a phase change phenomenon during welding is implemented in FE simulation by considering enthalpy as a function of temperature as,

$$E = \int \rho c dT \tag{9}$$

The initial temperature of the workpiece is considered as 27 °C. Following assumptions are considered for the FE simulation.

- Thermo-physical material properties are temperature dependent.
- Heat source is moving and workpiece is fixed.
- Vinokurov empirical relationship is used i.e. $h_{lump} = 2.4 \times 10^{-3} \varepsilon T^{1.61}$. It combines both radiation and convection.
- Forced convection due to shielding gas is neglected.
- The physical phenomena like viscous force, buoyancy force, convective melt flow, and Marangoni effect are neglected.

6 Results and Discussion

The transient temperature profiles for laser welded Ti64 workpieces are shown in Fig. 6a–c for 500, 1000 and 1500 W beam power, respectively at a constant welding speed of 312 mm/min, 90° beam incident angle, and 0.4 mm spot diameter. From Fig. 6a–c, it is observed that the temperature profile around the laser heat source increases rapidly from ambient temperature of 27 °C to peak temperature of around 2593, 3205 and 5946 °C at beam power of 500, 1000 and 1500 W, respectively for 0–3 s time interval. At the end of 3 s, the position of the heat source reaches approximately to the middle of the workpiece.

During this time period, the temperature gradient in the front of the heat source is comparatively lesser than the rear part of the heat source. At this stage, the workpiece is subjected to three modes of heat transfer i.e. conduction, convection, and radiation.

It is observed that the peak temperature and the temperature gradient increase while increasing laser beam power from 500 to 1500 W. It can be attributed to the fact that at higher beam power, the laser beam power density is highly concentrated over a small area (~0.4 mm spot diameter). The size of the weld bead also increases with the increase in the laser beam power. It can be attributed to the fact

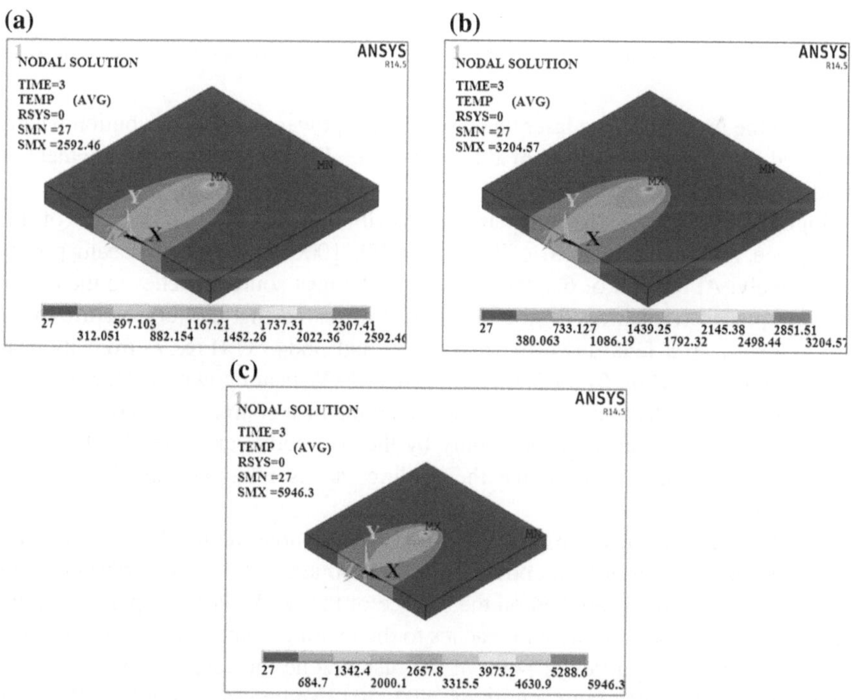

Fig. 6 Temperature distribution plot after 3 s at **a** 500 W **b** 1000 W, and **c** 1500 W beam power

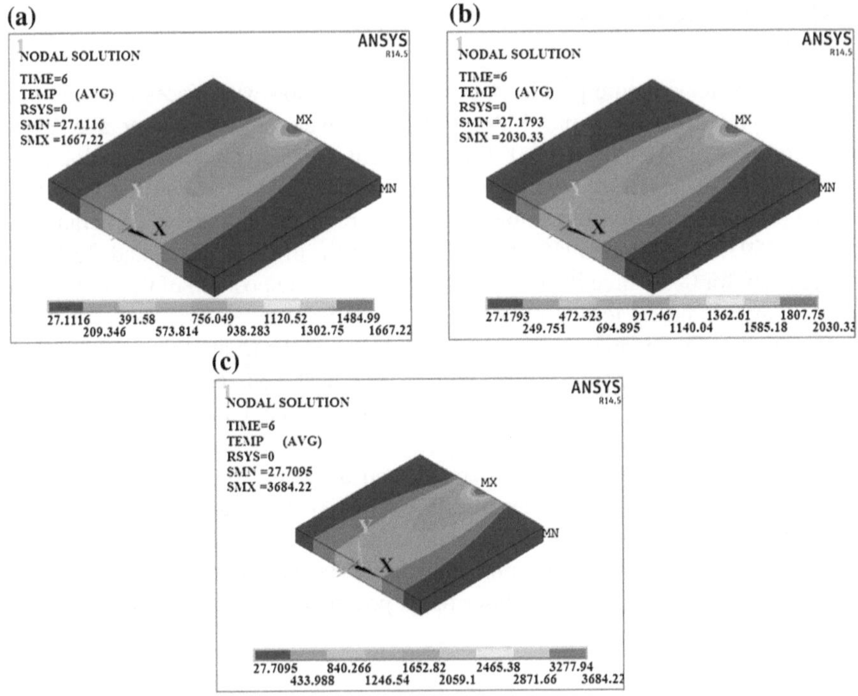

Fig. 7 Temperature distribution plot after 6 s at **a** 500 W **b** 1000 W, and **c** 1500 W beam power

that with the increase in the laser beam power, the power density distribution widely spreaded over the top surface of the workpiece. Therefore, a slight increment in bead geometry is observed.

Figure 7a–c shows the temperature distribution on the top surface of the workpiece for a time interval of 0–6 s at 500, 1000 and 1500 W beam power, respectively. At the end of 6 s, the position of the heat source reaches to the end of the workpiece. Comparing Figs. 6 and 7, it is observed that the temperature in the fusion zone at 6 s falls down to 1667, 2030 and 3684 °C (Fig. 7) from its peak temperature at 3 s (Fig. 6) for 500, 1000 and 1500 W beam power, respectively. At this final edge, the temperature decreases since the heat source is turned off and the temperature profiles are influenced only by the convective and radiative boundary conditions. It is also noticed that the cooling rate is relatively smaller than the heating rate.

The transient temperature profiles at a particular node along the weld line are shown in Fig. 8 for laser beam power of 500, 1000 and 1500 W. It is observed that when the heat source is applied on the surface of the workpiece, temperature in the fusion zone increases rapidly and reaches to the melting point at around 0.9 s. Also, the temperature at a particular point in the fusion zone increases from the ambient temperature to the peak temperature of 2492, 3047 and 5548 °C for 500, 1000 and

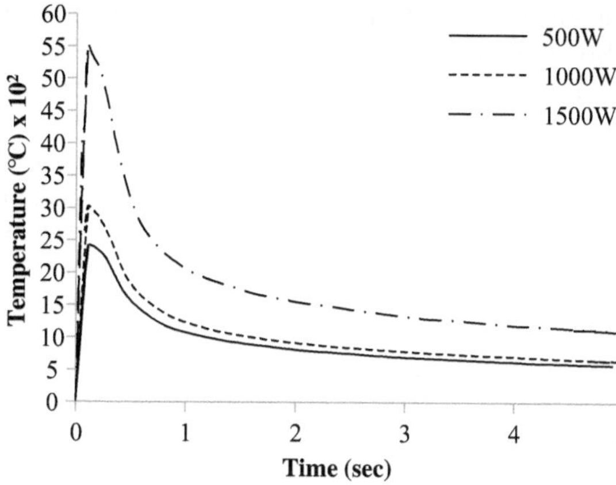

Fig. 8 Comparison between transient temperature histories at a point in the weld line at different laser beam power

1500 W beam power, respectively after that the temperature decreases continuously.

A non-linear relationship between the peak temperature and the laser beam power is noticed as shown in Fig. 8. The possible reason is due to the variation of absorption coefficient of the Ti64 alloy with temperature. The absorptivity plays an

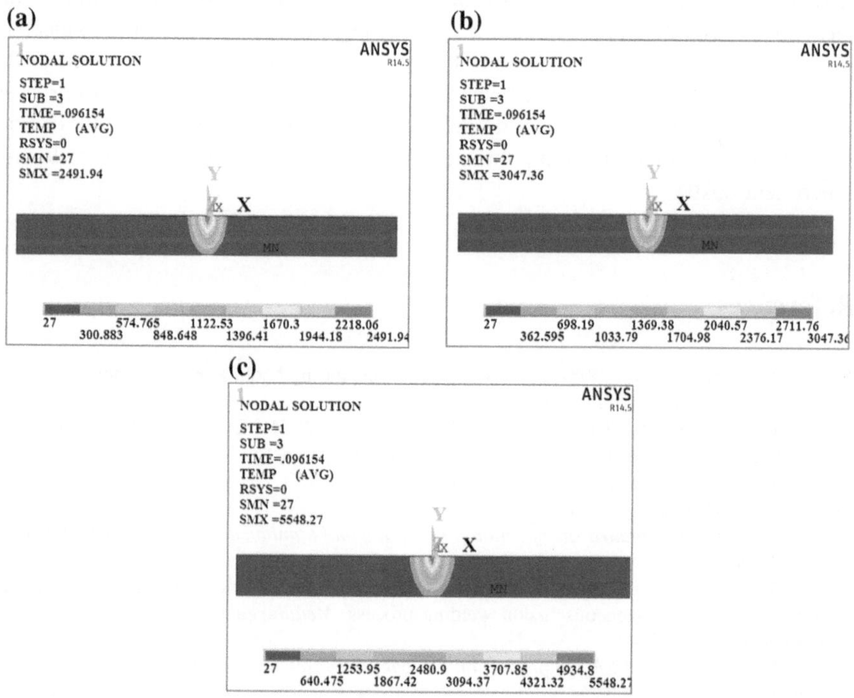

Fig. 9 Temperature distribution plot along the cross section of the workpiece for laser spot welding process after heating for 20 ms at **a** 500 W, **b** 1000 W, and **c** 1500 W beam power

important role on the thermal field since it controls the heat input to the workpiece. Also, absorptivity is a function of surface temperature.

Figure 9a–c shows the distribution of temperature along the cross section of the workpiece for laser spot welding process at 500, 1000 and 1500 W beam power, respectively after heating for 20 ms. From Fig. 9, it is observed that the temperature in the fusion zone increases from ambient temperature to the peak temperature of 2492, 3047 and 5548 °C for 500, 1000 and 1500 W beam power, respectively.

7 Conclusions

In the present work, a FEM based simulation of LBW process of Ti64 workpiece for both spot welding and moving heat source is carried out. From simulation results, it is observed that the double ellipsoidal heat source model gives reasonable result for both conduction and keyhole mode of welding. The non-uniform finite element mesh save solution time during simulation and yields better results with reasonable convergency. The temperature dependent thermo physical material properties play an important role in thermal analysis of LBW process. It is observed from the transient temperature profiles that the peak temperature increases with increasing laser beam power at constant welding speed. A nonlinear relationship between the laser beam power and the peak temperature of the welded workpiece is observed. It is also observed from the weld bead profile that the weld bead geometry is affected by the laser beam power and it increases with increasing laser beam power density. The present model can be useful to reduce the number of trial experiments before performing the actual experiment which helps in reducing experiment costs.

References

Ahmed, T., & Rack, H. J. (1998). Phase transformations during cooling in α + β titanium alloys. *Materials Science and Engineering, 243*, 206–211.

Akbari, M., Saedodin, S., Toghraie, D., & Razavi, R. K. (2014). Experimental and numerical investigation of temperature distribution and melt pool geometry during pulsed laser welding of Ti6Al4 V alloy. *Optics and Laser Technology, 59*, 52–59.

Ansys™ 14.5 Manual.

ASTM B265 (2006). *Standard specification for titanium and titanium alloys strip, sheet and plate. ASTM International.*

Bag, S., & De, A. (2010). Probing reliability of transport phenomena based heat transfer and fluid flow analysis in autogeneous fusion welding process. *Metallurgical and Materials Transactions A, 41*, 2337–2347.

Banas, M. (1971). *11th IEEE symposium on electron ion laser beam technology. UARL report,* Vol. 125.

Biswas, P., Mahapatra, M., & Mandal, N. R. (2010). Numerical and experimental study on prediction of thermal history and residual deformation of double sided fillet welding.

Proceedings of the Institution of Mechanical Engineers, Journal of Engineering Manufacture, 224, 125–134.

Chen, W., Ackerson, P., & Molian, P. (2009). CO_2 laser welding of galvanized steel sheets using vent holes. *Materials and Design, 30*, 245–251.

Costa, A., Miranda, R., Quintino, L., & Yapp, D. (2007). Analysis of beam material interaction in welding of titanium with fiber lasers. *Materials and Manufacturing Processes, 22*, 798–803.

Deng, D., & Murakawa, H. (2006). Numerical simulation of temperature field and residual stress in multi-pass welds in stainless steel pipe and comparison with experimental measurements. *Computational Materials Science, 37*, 269–277.

Destefani, J. D. (1992). *Introduction to titanium and titanium alloys: Properties and selection nonferrous alloys and special purpose materials*. Materials Park, OH: ASM International.

Donachie, M. J., Jr. (2000). *Titanium: A Technical Guide* (2nd ed.). Materials Park, OH: ASM International.

Frewin, M. R., & Scott, D. A. (1999). Finite element model of pulsed laser welding. *Welding Research Supplement, 78*, 15–22.

Goldak, J., Chakravarti, A., & Bibby, M. (1984). A new finite element model for welding heat sources. *Metallurgical Transactions, 15*, 299–305.

Hongping, G. U., Yin, G., & Shulkin, B. (2011). Laser beam welding of nitride steel components. *Physics Procedia, 12*, 40–45.

Leyens, C., & Peter, M. (2003). *Titanium and titanium alloy: Fundamentals and applications*. Weinheim: Wiley.

Mackerle, J. (2002). Finite element analysis and simulation of welding—an addendum: A bibliography (1996–2001). *Modelling and Simulation in Materials Science and Engineering, 10*, 295–318.

Manonmani, K., Murugan, N., & Buvanasekaran, G. (2007). Effects of process parameters on the bead geometry of laser beam butt welded stainless steel sheets. *International Journal of Advanced Manufacturing Technology, 32*, 1125–1133.

Moiseyev, V. N. (2006). *Titanium alloys: Russian aircraft and aerospace applications*. Boca Raton: Taylor & Francis.

Na, S. J., & Lee, S. Y. (1987). A study on the three-dimensional analysis of the transient temperature distribution in gas tungsten arc welding. *Journal of Engineering Manufacture, 201*, 149–156.

Nath, K., Sridhar, R., Ganesh, P., & Kaul, R. (2002). Laser power coupling efficiency in conduction and keyhole welding of austenitic stainless steel. *Sadhana, 27*, 383–392.

Nguyen, N. T., Mai, Y. W., Simpson, S., & Ohta, A. (2004). Analytical approximate solution for double ellipsoidal heat source in finite thick plate. *Welding Journal, 83*(3), 82s–93s.

Quintino, L., Costa, A., Miranda, R., Yapp, D., Kumar, V., & Kong, C. J. (2007). Welding with high power fiber lasers-a preliminary study. *Materials and Design, 28*, 1231–1237.

Pederson R., Babushkin, O., Skystedt, F. and Warren, R. (2001). *The use of high temperature X-ray diffractometry to study phase transitions in Ti-6Al-4V in titanium alloys at elevated temperature*. Structural Development and Service Behaviour, Institute of Materials, ISSN 1336-5510

Ranjbarnodeh, E., Serajzadeh, S., Kokabi, A. H., & Fischer, A. (2011). Prediction of temperature distribution in dissimilar arc welding of stainless steel and carbon steel. *Journal of Engineering Manufacture, 226*, 117–125.

Rosenthal, D. (1946). The theory of moving source of heat and its application to metal treatment. *Transactions ASME, 68*, 849–866.

Sakagawa, T., Nakashiba, S., & Hiejima, H. (2011). Laser micro welding system and its application to seam welding of rechargeable battery. *Physics Procedia, 12*, 6–10.

Shanmugam, N. S., Buvanashekaran, G., & Sankaranarayanasamy, K. (2012). Some studies on weld bead geometries for laser spot welding process using finite element analysis. *Materials and Design, 34*, 412–426.

Shanmugam, N. S., Shekaran, G. B., Samy, K. S. N., & Kumar, R. (2010). A transient finite element simulation of the temperature and bead profiles of T-joint laser welds. *Journal of Materials and Design, 31*, 4528–4542.

Spina, R., Tricarico, L., Basile, G., & Sibillano, T. (2007). Thermo-mechanical modelling of laser welding of AA5083 sheets. *Journal of Material Processing Technology, 191*, 215–219.

Tsirkas, S. A., Papanikos, P., & Kermanidis, T. (2003). Numerical simulation of the laser welding process in butt-joint specimens. *Journal of Material Processing Technology, 134*, 59–69.

Wanjara, P., Brochu, M., & Jahazi, M. (2005). Ti-6Al-4V electron beam weld qualification using laser scanning confocal microscopy. *Materials Characterization, 54*, 254–262.

Yadaiah, N., & Bag, S. (2012). Effect of heat source parameters in thermal and mechanical analysis of linear GTA welding process. *ISIJ International, 52*, 2069–2075.

Yadaiah, N., & Bag, S. (2013). Role of oxygen as surface-active element in linear GTA welding process. *Journal of Materials Engineering and Performance, 22*, 3199–3209.

Yadaiah, N., & Bag, S. (2014). Development of egg-configuration heat source model in numerical simulation of autogenous fusion welding process. *International Journal of Thermal Sciences, 86*, 125–138.

Yang, J., Sun, S., Brandt, M., & Yan, W. (2010). Experimental investigation and 3D finite element prediction of the heat affected zone during laser assisted machining of Ti6Al4 V alloy. *Journal of Material Processing Technology, 210*, 2215–2222.

Zhu, X. K., & Chao, Y. J. (2004). Numerical simulation of transient temperature and residual stresses in friction stir welding of 304L stainless steel. *Journal of Materials Processing Technology, 146*, 263–272.

Selective Laser Sintering: A Case Study of Tungsten Carbide and Cobalt Powder Sintering by Pulsed Nd:YAG Laser

Subrata Kumar Ghosh, Alok Kumar Das and Partha Saha

Abstract In the present study, tungsten carbide (WC) and cobalt (Co) powder mixture was sintered through selective laser sintering process using a pulsed Nd: YAG laser. Two different compositions of the powder mixture having 85 wt% WC + 15 wt%Co and 80 wt% WC + 20 wt% Co were used in the experiments. The optimum level of parameters, such as, composition of powder, layer thickness, hatching distance, pulse energy, pulse width and distance from focal plane were obtained by using the Taguchi method for achieving higher density, higher micro-hardness and minimum porosity. The Taguchi design of experiments involving an L-18 orthogonal array was followed. The effects of various sintering parameters were investigated on various responses like density, microhardness and porosity. The composition of the powder mixture and the pulse energy were found to have significant role on the microhardness. Hatching distance and cobalt percentage were the main influencing parameters on density and porosity. Surface morphology and formation of intermetallic compounds were analyzed through scanning electron microscopy (SEM) and X-ray diffraction (XRD) techniques.

Keywords Selective laser sintering · Taguchi method · Micro-hardness · Density · Porosity · SEM · XRD

S.K. Ghosh (✉)
Department of Mechanical Engineering, NIT Agartala, Agartala
Tripura 799055, India
e-mail: subratagh82@gmail.com

A.K. Das
Department of Mechanical Engineering, ISM Dhanbad,
Dhanbad, Jharkhand 826004, India
e-mail: eralok@yahoo.co.in

P. Saha
Department of Mechanical Engineering, IIT Kharagpur,
Kharagpur, West Bengal 721302, India
e-mail: psaha@mech.iitkgp.ernet.in

S.N. Joshi and U.S. Dixit (eds.), *Lasers Based Manufacturing*,
Topics in Mining, Metallurgy and Materials Engineering,
DOI 10.1007/978-81-322-2352-8_22

1 Introduction

1.1 Background

Earlier, hard metal sintering operations were generally carried out in controlled environment which was created by purging hydrogen gas. But, in certain cases, control of carbon percentage in metal and hydrogen embrittlement posed difficulties which lead to inconsistent material properties in different directions of the sintered part (Upadhyaya 1998). With the advancement in vacuum furnace technologies, the sintering of metal powder in vacuum became possible (Upadhyaya 1996). But sintering of very hard materials, such as, cemented carbide, within short period is still a challenging task. The progress in vacuum furnace technology cannot address this problem as a sintering cycle typically takes 15–20 h. Moreover, the sintering of micro parts is also a big challenge in this process. Powder metallurgy is another alternative for sintering of hard metal parts. However, it is difficult to sinter irregular or complex shaped parts with this process. Rapid prototyping (RPT) is the ultimate solution wherein components of any complex shape can be fabricated within a short interval. Rapid prototyping is the generic name of a machine which can be classified depending on the methods adopted by it to produce the desired component. The methods include solid ground curing (SGC), selective laser sintering (SLS), 3-dimensional printing (3DP), laminated object manufacturing (LOM), solid creation system (SCS), fused deposition modeling (FDM), selective adhesive and hot press (SAHP), solid object ultraviolet-laser plotter (SOUP), direct shell production casting (DSPC), multi-jet modeling system (MJM), ballistic particle manufacturing (BPM) and multiphase jet solidification (MJS) (Chua et al. 1998; Yan et al. 1996).

The SLS method can produce near-net shaped products. The method is highly applicable in batch as well as in small volume production. It is a leading commercial process to fabricate patterns for investment casting, metal moulds for injection moulding and die casting, moulds and cores for sand casting (Ghosh et al. 2010; Gu and Shen 2006; Kruth et al. 2003; Simchi 2006; Wang et al. 2002; Zhu et al. 2003). The SLS process has wide applications in bioengineering, such as, in making scaffolds (Williams et al. 2005; Tan et al. 2003, 2005; Berry et al. 1997; Wiria et al. 2007; Chua et al. 2004; Zhou et al. 2008), for dental replacements, etc. Many materials are processed by this technique and it does not require any specialized tooling as in the conventional sintering process (Wang et al. 2002). The SLS process also has applications in fabrication of components for MEMS. For example, Kathuria (1999) has fabricated 3D micro-structures on different materials with SLS. Ko et al. (2007a, b) have fabricated organic field effect transistor through this process.

In this chapter, the SLS process for producing the sintered parts from the tungsten carbide and cobalt (WC + Co) powder mixture will be explained. Tungsten carbide–cobalt composite is broadly used as a heat-resistant and wear-resistant material for cutting tools, moulds, drills for mining and geological explorations, nozzles and other applications. These applications are feasible because of the unique combination of high hardness and toughness of the WC–Co (Upadhyaya et al. 2001). The SLS technique may emerge to be the most viable one to develop the WC–Co composite. It can adopt the continuous wave and pulse mode laser. In case of pulse laser, there are a number of input parameters, such as, laser pulse energy, scan speed, pulse width, powder layer distance from focal plane, hatching distance and layer thickness. These parameters may be controlled to obtain better mechanical properties of the laser sintered specimen.

In this study, powder mixtures of tungsten carbide and cobalt in two different proportions, (i) 85 wt% WC + 15 wt% Co, and, (ii) 80 wt% WC + 20 wt% Co, were sintered. The density, micro-hardness and porosity were measured and optimized to achieve the suitable parameters. Micro-structural characterization and the XRD analysis were also carried out.

1.2 Principle of SLS Process

SLS is one of the recently developed RPT technologies by which the direct fabrication of components of arbitrary geometry can be attained in a single-step process from any materials, such as, metals, polymers, ceramics and composites. The materials should be in the form of powder mixture which may contain the binding agent or sacrificial polymer binder (Kruth et al. 2003). In the SLS process, the laser beam scans over the thin layer of powder having uniform thickness. With this, the temperature of the powder crosses the temperature of the glass transition, resulting in bonding/fusing of the particles. This leads to the consolidation of particles at the top layer and with the previous layer (Chua et al. 1998). The success of the SLS process can mainly be attributed to its ability to fabricate components of complicated shape within a short time.

1.3 Laser Material Interaction in SLS Process

SLS can be used to process almost any material, provided the material is available in powder form and that the powder particles tend to fuse or sinter when heat is applied. Powders that possess low fusion or sintering properties can be laser sintered by adding a sacrificial binder material (typically a polymer binder) to the basic powder. After sintering the full part, the sacrificial binder can be removed by debinding the 'green' part in a thermal furnace. However, the range of materials (powder) that can be laser sintered without sacrificial binder is quite large as compared to the other rapid prototyping processes.

The materials which can be used for developing laser sintered solid freeform fabricated (SFF) products are polymer coated low carbon steel, thermoplastics, bronze-nickel blend, powder, pre-alloyed bronze powder, etc. The components developed from these materials have low mechanical strength and durability.

In most cases, they are hardly used for functional prototypes of any engineering components. In a few cases, post-processing operations are necessary for achieving full density. Single metal powders have been proved to be difficult materials for SFF by laser sintering because they tend to 'ball up' in molten form. Solutions to this problem have been tried out by either pre-alloyed single phase powder system or a powder mixture of two phases with different melting temperatures instead of one so that one phase melts while the other remains solid. This is known as liquid sintering where one metal acts as a binder.

1.4 Sintering Mechanism of WC–Co Powder

The SLS process is more effective with the mixture of two metal powders comprising high and low melting points. The low melting point powder acts as a binder. The laser energy is adjusted in such a way that the temperature at the laser and metal powder mixture interaction zone crosses the melting point of one powder but does not reach the melting temperature of the other powder. The melted material flows through the void in between the powder particles, consolidating powder particles in the process. In this way, the laser scans over the powder layer, converting it to a solid layer. The density of the sintered part can be increased by reducing the porosity which can be accomplished by selecting the binding material with smaller particle size than that of the structural material. The bigger particle size of binder material has higher melting enthalpy because of its large mass which leads to partial melting. Therefore, the void is not filled up. As a result, the density of the component reduces. On the other hand, smaller particle size leads to complete melting and the melt flows through the void and forms cluster structures. In this way, the density of the sintered object can be increased in the SLS process.

1.5 Process Parameters of SLS Process

The SLS process can be performed using both continuous wave (CW) and pulse wave (Modulated) laser. Table 1 presents a few characteristics of the laser systems. The parameters of the SLS process with different lasers are given below:

1.5.1 Process Parameters for Continuous Wave Lasers (Fiber/CO_2)

- Beam energy density—it is the energy carried by the laser beam per unit area. It is denoted by J/cm^2
- Scan speed—it is the travel length of laser beam on the work piece per unit time. It is denoted by cm/s or m/min.

Table 1 Characteristics of different laser systems

Laser	Power range (W)	Wavelength (μm)	Typical industrial applications
CO_2-flowing gas (continuous wave and pulsed)	500–45,000	10.6	Cutting, welding, cladding, free forming, and hardening
CO_2-sealed (pulsed)	10–1,000	10.6	Micro-welding, cutting, scribing and drilling
Nd:YAG (continuous wave)	1000–5000	1.06	Welding, cutting, cladding and hardening
Nd:YAG (pulsed)	10–2000	1.06	Micro-welding, cutting, drilling, scribing and marking
Nd:YAG diode pumped (pulsed)	10–500	1.06	Cutting, drilling, scribing, marking and micro-machining
Excimer (pulsed)	0.001–400	0.157–351	Micro-machining, marking and photolithography

- Scan spacing (mm)—it is distance between two consecutive laser scan lines.
- Layer thickness (μm)—it is powder layer thickness deposited on substrate during the laser sintering process.
- Spot size (mm): it is spot size on work piece after interaction between laser and material.

1.5.2 Process Parameters for Modulated/Pulse Wave Lasers (Fiber/Nd:YAG)

The energy density, scan speed, scam spacing, spot size and layer thickness are the common parameters to control the sintering process while continuous or pulsed laser are used. Beside these parameters, few more parameters which are described in below are also important in case of pulsed lase used.

- Pulse energy (J)—it is laser energy per each pulse.
- Pulse width (ms)—it is duration of each pulse. It is denoted by millisecond (ms).
- Pulse off time (ms)—it is idle time between two consecutive pluses.
- Repetition rate—it is number of pulses per second.

1.6 Advantages and Disadvantages of SLS

1.6.1 Advantages of SLS

The SLS is one of a few rapid prototyping (RP) techniques which can be used to process wide range of powder materials. It is not considering the conductivity of the material. Since it is producing the product directly and process lasts for less time, overall time to obtain product will be less. The porosity and density of the parts can be controlled since the powder is deposited layer-by-layer fashion. The 3-D objects with complex shape and size can be processed easily with the use of CAD modeling. Highly accurate parts can be produced. Since the whole the process are fully automatic, it requires less skillful operator.

1.6.2 Disadvantages of SLS

Beside the advantages, few disadvantages are to be mentioned. It is costly operation because the capital cost is high. Laser produces harmful radiation which is very harmful to eyes. Temperature control in the sintering process is critical. During solidification, shrinkage can create serious residual stresses which can cause distortion or cracking. Sometimes, metallic balls are produced due to rapid cooling and high surface tension of molten material. Inert atmosphere is always required to avoid oxidation, fire or explosion. Sometimes, surface roughness of the sintered parts becomes high.

1.7 Applications of Selective Laser Sintering Process

This process is used to fabricate various parts in automotive and aerospace industry. It can produce the tool inserts of hard material. Very hard material like ceramic can also be sintered by this process. This process finds application to produce near-net-shape fabrication of prototypes and several tooling such as molds, dies etc. for injection molding and die casting. Very complicated shapes like impeller of a turbine, piston, crank shaft etc. can be produced by the process. Parts for mechanical and thermal tests and series of small plastic parts can be fabricated by the process.

2 Experimental Details

2.1 Experimental Setup for Pulsed Laser Sintering

Different components of the experimental setup are as follows:

- Pulsed laser source (Nd:YAG)
- NC machine table
- Powder spreading device
- Sintering chamber with inert gas purging facility
- Powder spreading device

Table 2 Technical specifications of Nd:YAG laser system

Manufacturer	Baasel Laser Tech., Germany
Type of laser	Solid state laser
Nd:YAG	Neodymium-doped yttrium aluminum garnet
Chemical formula	Nd^{3+}:$Y_3Al_5O_{12}$
Crystal structure	Cubic
Mass density	4.5 (g/cm^3)
Wavelength	1064 (nm)
Minimum pulse width	0.5 (ns)
Maximum pulse space	1000 (ns)
Pulse length	0.1–20 (ms)
Pulse energy, max	25 (J)
Average power, max (W)	50 (W)
Pulse power	4.8 (kW)
Beam diameter	0.6 (mm)
Beam divergence	6 (milli-rad.)
Spot size at focus point	620 (μm)

Technical specifications of the Nd:YAG laser source which has been used in the experimentation are presented in Table 2. The setup for the pulsed laser sintering process is shown in Fig. 1.

2.2 *Inert Gas Chamber with Powder Spreading Arrangement*

An inert gas chamber (IGC), manufactured by Perspex, contained the powder bed and was mounted on the NC machine table (Fig. 1). The laser entered through a quartz window mounted on the top plate of the chamber. Argon gas was fed into the

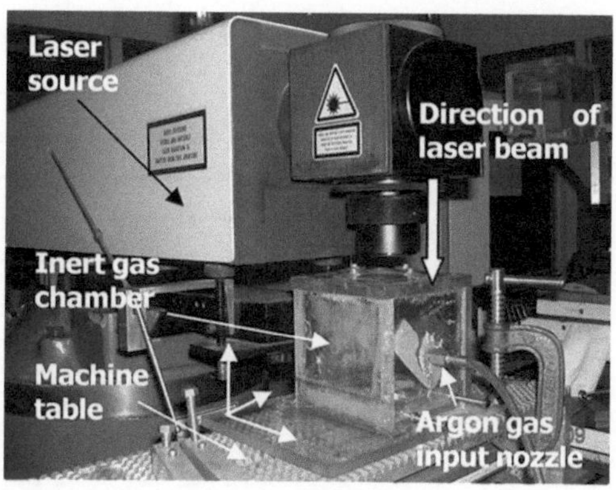

Fig. 1 Setup for pulse laser SLS process

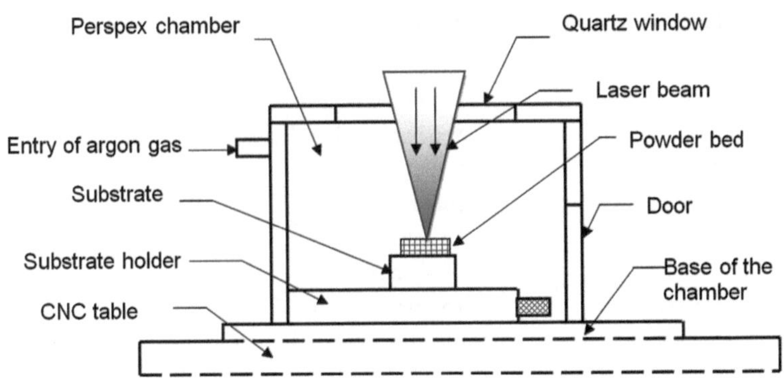

Fig. 2 Schematic of sintering chamber with powder spreading arrangement

IGC through a brass nozzle at a rate of 5 L/min in order to create inert atmosphere inside the chamber. The powder bed was in the form of a drawer that could be slid out of the box for spreading the powder layer.

The distribution of powder on the substrate surface with uniform layer thickness is very crucial for good quality sintered part. In order to accomplish this, powder spreader was fabricated. Figure 2 represents the schematic of the inert gas chamber with powder spreading arrangement. The substrate is fixed on the base and the powder was distributed layer by layer on the substrate and the laser beam fell over the surface of powder bed due to which the projected corner of the particles melted and stuck to other particles, forming a block. In this way, the first layer of the powder bed was consolidated. After consolidation of the first layer, the second layer of powder with the same thickness was put over the previously consolidated powder layer and the same process was repeated a number of times. Figure 3 shows the nomenclature of the selective laser sintering.

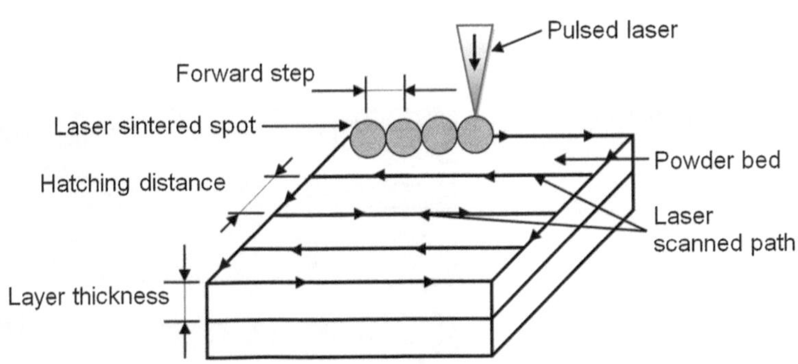

Fig. 3 Nomenclature of the pulse laser sintering process

Table 3 Controlling parameters and their levels

Parameters	Levels		
	1	2	3
Composition	85 wt% WC + 15 wt% Co	80 wt% WC + 20 wt% Co	–
Layer thickness (μm)	300	400	500
Hatching distance (μm)	400	500	600
Pulse energy (J)	15	17	19
Pulse width (ms)	14	17	20
Powder layer distance from focal plane (μm)	400	450	500

2.3 Design of Experiments

Based on the Taguchi design, L_{18} orthogonal array was chosen for carrying out the experiment. The controlling parameters, such as, composition, layer thickness, hatching distance, pulse energy, pulse width and powder layer distance from focal plane were the variable parameters. The parameters and their levels are reported in Table 3. Table 4 presents different combinations of parameters for the L_{18} orthogonal arrays (Phadke 1989; Ross 1996).

2.4 Procedure

Two homogeneous powder mixtures of tungsten carbide (with average particle size of around 50 μm) and cobalt were prepared. One contained 85 wt% WC + 15 wt% Co and the other had 80 wt% WC + 20 wt% Co powder. The loose powder mixtures were spread over a mild steel substrate of size 10 mm × 10 mm × 8 mm. The extra amount of powder was scraped off in order to maintain the layer thickness. The scraper was moved to the top of two slip-gauge-stacks of equal heights placed on two sides of the substrate. The substrate was then positioned into the inert gas chamber exactly below the quartz window on top (Fig. 1). Argon gas was continuously purged through a nozzle to eliminate air content inside the sintering chamber and to maintain an inert environment. Laser pulses were irradiated and scanned over the powder layer. After completion of one layer, the process was repeated to achieve the required height of the sintered specimen. All the eighteen experiments were conducted in a similar manner and one of the specimens is shown in Fig. 4. After carrying out all the experiments, the properties, i.e., density and micro-hardness of all specimens, were measured. The surface porosity was measured with the help of Image Pro software. At first, the sintered specimens were

Table 4 Design of experiments as per orthogonal array

Exp. no.	Composition (wt% of WC + wt% of Co)	Powder layer distance below focal plane (μm)	Layer thickness (μm)	Pulse width (ms)	Pulse energy (J)	Hatching distance (μm)
1	85 + 15	400	300	14	15	400
2	85 + 15	400	400	17	17	500
3	85 + 15	400	500	20	19	600
4	85 + 15	450	300	14	17	500
5	85 + 15	450	400	17	19	600
6	85 + 15	450	500	20	15	400
7	85 + 15	500	300	17	15	600
8	85 + 15	500	400	20	17	400
9	85 + 15	500	500	14	19	500
10	80 + 20	400	300	20	19	500
11	80 + 20	400	400	14	15	600
12	80 + 20	400	500	17	17	400
13	80 + 20	450	300	17	19	400
14	80 + 20	450	400	20	15	500
15	80 + 20	450	500	14	17	600
16	80 + 20	500	300	20	17	600
17	80 + 20	500	400	14	19	400
18	80 + 20	500	500	17	15	500

Fig. 4 One fabricated specimen

sectioned with the help of wire-cut electro discharge machine and then polished to the mirror finishing. The microscopic view of the surfaces was captured at low magnifications with the help of stereo zoom microscope. These images were processed with the Image Pro software for finding the percentage of pores (porosity) present in the sintered specimen.

3 Results and Discussion

3.1 *Taguchi Analysis for Density, Micro-hardness and Porosity*

The Taguchi method was applied to study the parametric effect and to find out the optimum experimental conditions for the best material properties. The S/N ratios were calculated based on the nature of quality characteristics. In the present study, 'the-larger-the-better' quality characteristic was desirable for both density and micro-hardness. 'The-smaller-the-better' quality characteristic was essential for porosity. Parametric optimization was achieved with the main effects plot which establishes a relationship between factor and S/N ratios. Analysis of Variance (ANOVA) was used to select significant parameters and estimate the percentage of contribution of each parameter on the responses. The significance of factors is represented in terms of F-test values or percentage of contribution.

3.1.1 ANOVA for Density

ANOVA table was analyzed according to the Taguchi method. F-test values, F-tabulated values and percentage of contribution of each factor on density are reported in the Table 5. It was observed that F-test values for hatching distance and composition were more than the F-tabulated values. Therefore, these were considered to be the most significant parameters. Hatching distance has a major contribution (58.79 %) among the all parameters. Although composition is a significant parameter based on F-test value, contribution of layer thickness (10.44 %) is slightly more than that of the composition (9.60 %).

Table 5 Analysis of variance for density

Factors[a]	DF	SS	MS	Percentage of contribution	F-test	F-tab
Composition	1	2.370117	2.370117	9.60	6.24	5.99
Distance below focal plane (μm)	2	1.370132	0.685066	5.54	1.80	
Layer thickness (μm)	2	2.581380	1.29069	10.44	3.40	
Pulse width (ms)	2	0.342555	0.171277	1.39	0.45	
Pulse energy (J)	2	1.241931	0.620965	5.02	1.63	
Hatching distance (μm)	2	14.533098	7.266549	58.79	19.13	5.14
Error	6	2.279436	0.379906	9.22		
Total	17	24.718649		100		

[a]*DF* degrees of freedom, *SS* sum of squares, *MS* mean squares, *F-test* F-test value, *F-tab* F-tabulated value

3.1.2 Optimal Parameters for Higher Density

Main effects plot between S/N ratios for density and levels of the factors is shown in Fig. 5. The highest S/N ratio in the graph was chosen to select the optimum level of factors in case of higher density. It was observed that the second level of composition (80 wt% WC + 20 wt% Co), the second level of layer distance below focal plane (450 μm), the first level of layer thickness (300 μm), the first level of pulse width (14 ms), the first level of pulse energy (15 J) and the first level of hatching distance (400 μm) are the optimum combination of parameters within the domain. Predicted density of the sintered part is 12.42 g/cc.

3.1.3 ANOVA for Microhardness

ANOVA table for micro-hardness was analyzed. F-test values, F-tabulated values and percentage of contribution of each factor on density are shown in the Table 6. It was found that F-test values for composition were more than the F-tabulated values. Hence, this was considered to be the most significant parameter. Composition had the maximum contribution of 40.40 % on microhardness among all the parameters. Pulse energy also had significant contribution of 21.90 %.

3.1.4 Optimal Parameters for Higher Microhardness

Main effects plot between S/N ratios for microhardness and levels of the factors is shown in Fig. 6. The highest S/N ratio in the graph was chosen to select optimum

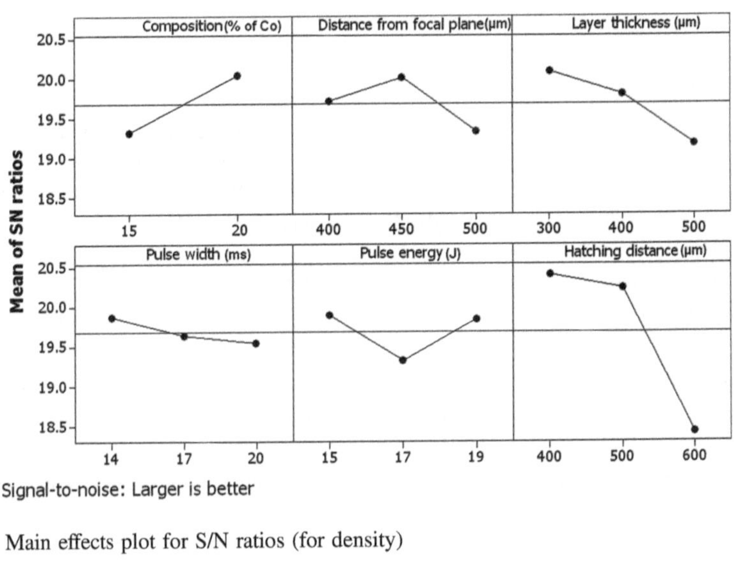

Fig. 5 Main effects plot for S/N ratios (for density)

Table 6 Analysis of variance for microhardness

Factors	DF	SS	MS	Percentage of contribution	F-test	F-tab
Composition	1	3.256473	3.256473	40.40	12.84	5.99
Layer distance below focal plane (μm)	2	0.79468	0.39734	9.86	1.57	
Layer thickness (μm)	2	0.051002	0.025501	0.63	0.10	
Pulse width (ms)	2	0.281628	0.140814	3.49	0.56	
Pulse energy (J)	2	1.76559	0.882795	21.90	3.48	
Hatching distance (μm)	2	0.390116	0.195058	4.84	0.77	
Error	6	1.522097	0.253683	18.88		
Total	17	8.061587		100		

level of factors in case of microhardness. It was observed that the first level of composition (85 wt% WC + 15 wt% Co), the third level of layer distance below focal plane (500 μm), the third level of layer thickness (500 μm), the second level of pulse width (17 ms), the third level of pulse energy (19 J) and the third level of hatching distance (600 μm) were the optimum combination of parameters within the domain. Predicted microhardness was 20.72 GPa.

3.1.5 ANOVA for Porosity

Table 7 presents the ANOVA for porosity of the sintered components. It was found that the F-test values for hatching distance and composition were more than the F-

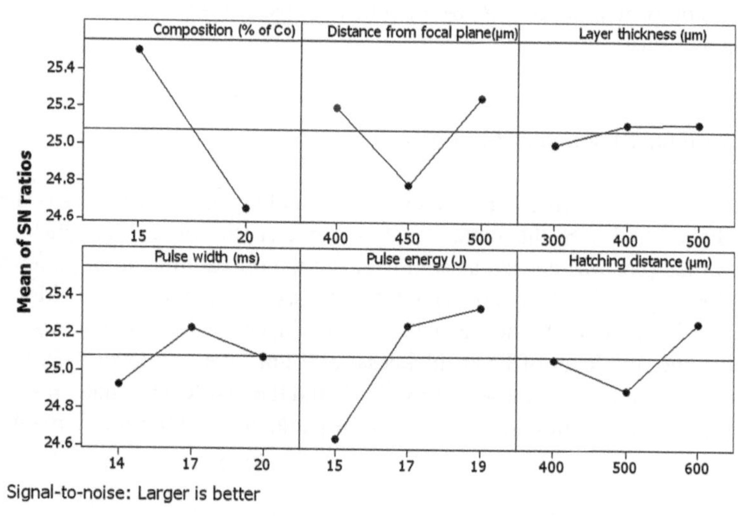

Fig. 6 Main effects plot for S/N ratios (for microhardness)

Table 7 Analysis of variance for porosity

Factors	DF	SS	MS	Percentage of contribution	F-test	F-tab
Composition	1	11.033	11.034	15.35	9.62	5.99
Distance below focal plane (μm)	2	6.600	3.300	9.18	2.88	
Layer thickness (μm)	2	5.400	2.699	7.51	2.35	
Pulse width (ms)	2	1.495	0.747	2.08	0.65	
Pulse energy (J)	2	4.531	2.265	6.30	1.97	
Hatching distance (μm)	2	35.946	17.973	50.00	15.68	5.14
Error	6	6.877	1.146	9.57		
Total	17	71.883		100		

tabulated values. Hence, these were considered to be the significant parameters to influence the porosity of the specimen. The hatching distance has major contribution of 50 % as compared to all the other parameters. The composition has a contribution of 15.35 %. It is a common knowledge that if the hatching distance is increased, then the percentage of overlapping decreases. Hence, the waviness in the specimen was increased. This kind of waviness is filled by the subsequent powder layer deposit in the previous layer. But if higher hatching distance persists, then the depth of waviness increases. Subsequently, filling of the entire depth of waviness is stopped by the rapid cooling and solidification of molten material as the interaction time between powder and laser lasts for a very short period. Although composition is also a significant parameter, its contribution is less. In this process, Co is melted due to its low melting point and spreading and holding of the WC particles. As a result, a compact product can be obtained. Subsequently, the porosity may be reduced as well.

3.1.6 Optimal Parameters for Lower Porosity

Main effect plot between S/N ratios for porosity and levels of the factors is shown in Fig. 7. Since 'the-lower–the–better' quality characteristic was chosen, the highest S/N ratio in the plot was chosen as well to select the optimum level of parameters in case of lower porosity. It was observed that the second level of composition (80 wt % WC + 20 wt% Co), the second level of layer distance below focal plane (450 μm), the first level of layer thickness (300 μm), the first level of pulse width (14 ms), the third level of pulse energy (19 J) and the first level of hatching distance (400 μm) were the optimum combination of parameters within the domain.

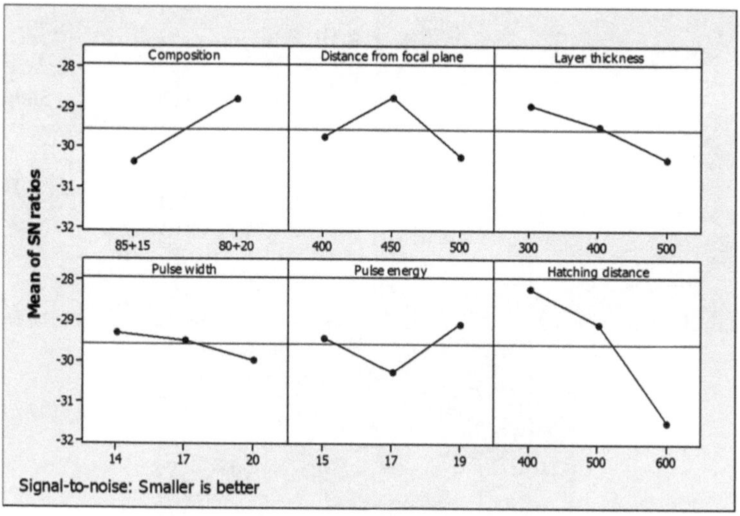

Fig. 7 Main effects plot for porosity

3.2 *Micro-structural Characterization*

The SEM micrographs of the top surfaces of three specimens (1, 2 and 3) are shown in Fig. 8a–c. The influence of the hatching distance is clearly visible in the figures. These figures exhibit that if the hatching distance is increased, then the porosity increases as well.

The WC–Co sintered specimen was initially etched with Chaporova's solution (HCl saturated with $FeCl_3$) to "darken" the cobalt-rich portion. Then Murakami's reagent was applied on the surface to outline the WC grains.

SEM micrographs of the specimen are shown in Figs. 9 and 10. It is evident from Fig. 9 that white phases (spot 1) spread in the micro-structure like particles. Based on spot EDS results, these were found to be W-rich material with elemental composition of 98.10 % W and 1.10 % Co. In comparison, although black phases (spot 2) were W-rich material as well, percentage of Co in them increased with an elemental composition of 59.41 % W and 40.59 % Co. The difference in the values of coefficient of thermal expansion of Co and WC may lead to thermal stress in the agglomerate during rapid cooling when the laser pulse is put off. The cooling rate is dependent on the laser parameters and may result in generation and propagation of micro cracks as seen in Fig. 9.

It was observed that microstructure of specimen 7 (Fig. 9) was almost the same as specimen 2 (Fig. 10). Only elemental composition of phases was slightly changed. Black phases (spot 4) were found to be containing 64.31 % W and 35.69 % Co. White phases (spot 3) were found to be comprising 98.83 % W and 1.17 % Co.

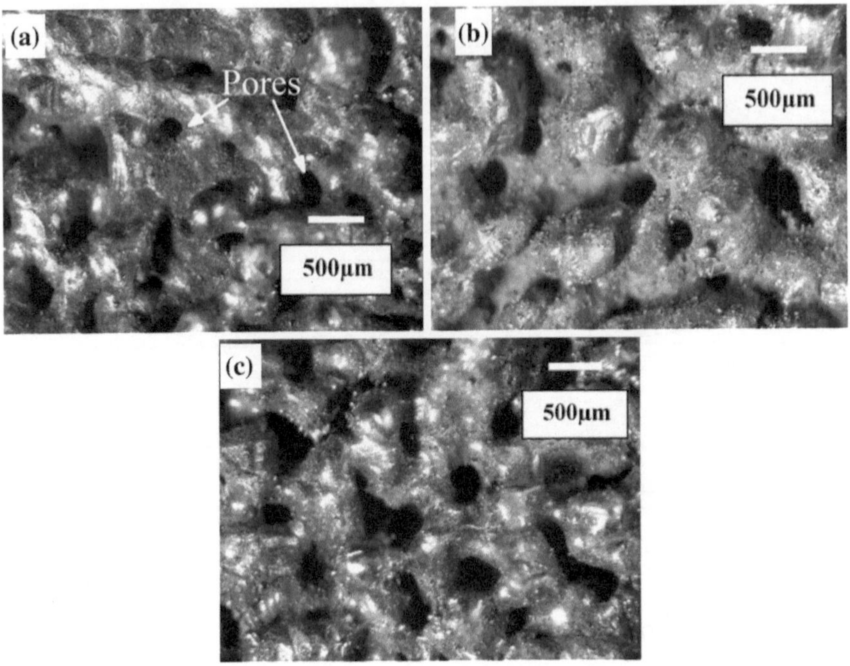

Fig. 8 SEM micrograph of **a** specimen 1, **b** specimen 2 and **c** specimen 3

Fig. 9 SEM micrograph of specimen 7

3.3 XRD Analysis

XRD analysis was done for the laser sintered specimen to investigate the phase generated by the reaction between the constituents of the material. One XRD pattern is shown in Fig. 11. It was observed that occurrence of peaks for WC was the maximum. Some inter-metallic compounds, i.e., CW_3, Co_2C, and $Co_3W_9C_4$, were detected from the pattern.

Fig. 10 SEM micrograph of specimen 2

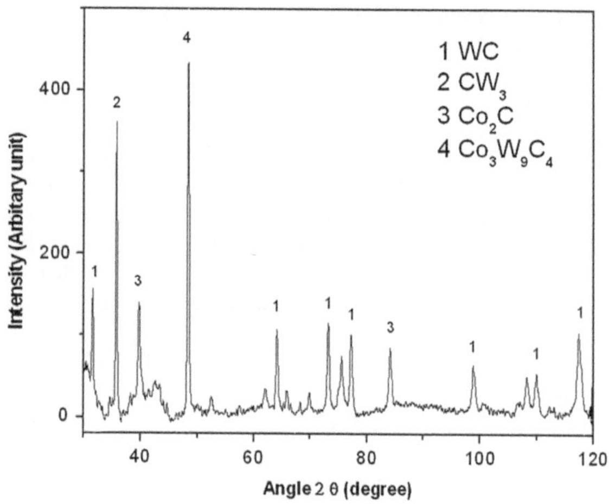

Fig. 11 XRD pattern for specimen 2

4 Conclusions

In this investigation, laser sintering of WC–Co was successfully carried out. During the laser sintering, it was observed that cobalt melts first due to its low melting point. Molten material spreads, engulfs the solid WC particles and makes a bond following resolidification. Density of the sintered part mainly depends on the hatching distance and composition. But, hatching distance has greater influence than the composition. Microhardness is mainly influenced by the percentage of WC present in the sample. Both composition and the hatching distance are influencing parameters for porosity. The composition has the highest contribution of 50 % as compared to the other parameters. The reduction in hatching distance and increase of percentage of Co content in the powder mixture may lead to a decrease of the porosity of the specimen. The SEM micrographs and XRD analysis confirm the bonding between WC and Co. Further studies are required for determining the influence of the process parameters on sintering characteristics so as to learn about the other effective properties, such as, wear resistance, fatigue strength, etc.

References

Berry, E., Brown, J. M., Connell, M., Craven, C. M., Efford, N. D., Radjenovic, A., et al. (1997). Preliminary experience with medical applications of rapid prototyping by selective laser sintering. *Medical Engineering and Physics, 19*(1), 90–96.

Chua, C. K., Chou, S. M., & Wong, T. S. (1998). A study of the state-of-the-art rapid prototyping technologies. *International Journal of Advanced Manufacturing Technology, 14*(2), 146–152.

Chua, C. K., Leong, K. F., Tan, K. H., Wiria, F. E., & Cheah, C. M. (2004). Development of tissue scaffolds using selective laser sintering of polyvinyl alcohol/hydroxyapatite biocomposite for craniofacial and joint defects. *Journal of Materials Science Materials in Medicine, 15*(10), 1113–1121.

Ghosh, S. K., Saha, P., & Kishore, S. (2010). Influence of size and volume fraction of SiC particulates on properties of ex situ reinforced Al4.5–Cu3Mg metal matrix composite prepared by direct metal laser sintering process. *Material Science Engineering A, 527*, 4694–4701.

Gu, D., & Shen, Y. (2006). Processing and microstructure of submicron WC–Co particulate reinforced Cu matrix composites prepared by direct laser sintering. *Materials Science and Engineering A, 435–436*, 54–61.

Kathuria, Y. P. (1999). Microstructuring by selective laser sintering of metallic powder. *Surface and Coatings Technology, 116*, 643–647.

Ko, S. H., Pan, H., Grigoropoulos, C. P., Luscombe, C. K., Fréchet, J. M., & Poulikakos, D. (2007a). Air stable high resolution organic transistors by selective laser sintering of ink-jet printed metal nanoparticles. *Applied Physics Letters, 90*(14), 141103.

Ko, S. H., Pan, H., Grigoropoulos, C. P., Luscombe, C. K., Fréchet, J. M., & Poulikakos, D. (2007b). All-inkjet-printed flexible electronics fabrication on a polymer substrate by low-temperature high-resolution selective laser sintering of metal nanoparticles. *Nanotechnology, 18*(34), 345202.

Kruth, J. P., Wang, X., Laoui, T., & Froyen, L. (2003). Lasers and materials in selective laser sintering. *Assembly Automation, 23*(4), 357–371.

Phadke, M. S. (1989). *Quality engineering using robust design*. Englewood Cliffs, New Jersey: Prentice Hall.

Ross, P. J. (1996). *Taguchi techniques for quality engineering*. New York: McGraw-Hill.

Simchi, A. (2006). Direct laser sintering of metal powders: Mechanism, kinetics and microstructural features. *Material Science Engineering A, 428*, 148–158.

Tan, K. H., Chua, C. K., Leong, K. F., Cheah, C. M., Cheang, P., Abu Bakar, M. S., et al. (2003). Scaffold development using selective laser sintering of polyetheretherketone–hydroxyapatite biocomposite blends. *Biomaterials, 24*(18), 3115–3123.

Tan, K. H., Chua, C. K., Leong, K. F., Cheah, C. M., Gui, W. S., Tan, W. S., et al. (2005). Selective laser sintering of biocompatible polymers for applications in tissue engineering. *Biomedical Materials and Engineering, 15*(1), 113–124.

Upadhyaya, G. S. (1996). *Nature and properties of refractory carbides*. NY, USA: Nova Science Publishers Commack.

Upadhyaya, G. S. (1998). *Cemented tungsten carbides, production, properties and testing*. Norwich, NY, USA: William Andrew Publishing.

Upadhyaya, A., Sarathy, D., & Wagner, G. (2001). Advances in sintering of hard metals. *Materials and Design, 22*(6), 499–506.

Wang, X. C., Laoui, T., Bonse, J., Kruth, J. P., Lauwers, B., & Froyen, L. (2002). Direct selective laser sintering of hard metal powders: experimental study and simulation. *International Journal of Advanced Manufacturing Technology, 19*, 351–357.

Williams, J. M., Adewunmi, A., Schek, R. M., Flanagan, C. L., Krebsbach, P. H., Feinberg, S. E., et al. (2005). Bone tissue engineering using polycaprolactone scaffolds fabricated via selective laser sintering. *Biomaterials, 26*(23), 4817–4827.

Wiria, F. E., Leong, K. F., Chua, C. K., & Liu, Y. (2007). Poly-ε-caprolactone/hydroxyapatite for tissue engineering scaffold fabrication via selective laser sintering. *Acta Biomaterialia, 3*(1), 1–12.

Yan, X., & Gu, P. E. N. G. (1996). A review of rapid prototyping technologies and systems. *Computer-Aided Design, 28*(4), 307–318.

Zhou, W. Y., Lee, S. H., Wang, M., Cheung, W. L., & Ip, W. Y. (2008). Selective laser sintering of porous tissue engineering scaffolds from poly (L-lactide)/carbonated hydroxyapatite nanocomposite microspheres. *Journal of Materials Science Materials in Medicine, 19*(7), 2535–2540.

Zhu, H. H., Lu, L., & Fuh, J. Y. H. (2003). Development and characterization of direct laser sintering Cu-based metal powder. *Journal of Material Processing Technology, 140*, 314–317.

Author Index

S.N. Joshi and U.S. Dixit (eds.), *Lasers Based Manufacturing*,
Topics in Mining, Metallurgy and Materials Engineering,
DOI 10.1007/978-81-322-2352-8

Subject Index

S.N. Joshi and U.S. Dixit (eds.), *Lasers Based Manufacturing*,
Topics in Mining, Metallurgy and Materials Engineering,
DOI 10.1007/978-81-322-2352-8